The Microbiome's Impact on the Brain

Md. Habibur Rahman • Muhasina K M
Editors

The Microbiome's Impact on the Brain

A Paradigm Shift in Neurodegenerative Research

Springer

Editors
Md. Habibur Rahman
Novel Global Community Educational Foundation
Herbersum, NSW, Australia

Muhasina K M
Department of Pharmaceutical Analysis
Prime College of Pharmacy
Erattayal, Palakkad, Kerala, India

ISBN 978-981-95-7493-3 ISBN 978-981-95-7494-0 (eBook)
https://doi.org/10.1007/978-981-95-7494-0

This Springer imprint is published by the registered company Springer Nature Singapore Pte Ltd.
The registered company address is: 152 Beach Road, #21-01/04 Gateway East, Singapore 189721, Singapore

Preface

The Microbiome's Impact on the Brain: A Paradigm Shift in Neurodegenerative Research was written to address one of the most rapidly evolving and consequential areas in modern biomedical science: the connection between the gut microbiome and neurological health. As evidence has accumulated over the past decade, it has become increasingly clear that the microorganisms residing in the gastrointestinal tract influence far more than digestion—they play a fundamental role in neuroinflammation, neurotransmission, immune regulation, and overall brain function. This book was a compendium of these emerging findings and present a comprehensive, research-driven understanding of how microbiome disruptions may contribute to neurodegenerative disorders such as Alzheimer's disease, Parkinson's disease, cerebral palsy, and amyotrophic lateral sclerosis (ALS).

This book is intended for targeting a broad interdisciplinary audience, including clinicians, neuroscientists, microbiologists, immunologists, medical researchers, and students entering the fields of neurology or microbiome science. It is designed both for readers seeking foundational knowledge about gut–brain interactions and for specialists aiming to explore the therapeutic implications of this expanding field. All the chapters are organized well to guide readers from fundamental concepts to advanced therapeutic applications. Early sections introduce the biological mechanisms linking gut microbes to brain function, followed by chapters dedicated to specific neurodegenerative diseases and the evidence connecting each condition to microbiome alterations. Later sections focus on emerging therapeutic strategies, including probiotics, dietary interventions, and fecal microbiota transplantation, and consider how precision-based, personalized approaches may shape future clinical practice. By structuring the book in this way, readers can build a layered understanding of both the science and its real-world implications.

Contributors to this volume were selected based on their expertise in microbiome research, neurology, immunology, or clinical translation. Each author brings a unique perspective grounded in active research or clinical experience, ensuring that the content reflects current knowledge and ongoing scientific progress. Ultimately, this book aims to integrate knowledge of the gut microbiome with traditional treatment modalities for neurodegenerative disorders, considering the microbiome as a dynamic and potentially transformative factor in brain health. We hope that the insights compiled here will inspire continued investigation in the dynamic field of gut microbiome interactions with brain cells and significantly contribute to the advancement of the scientific threshold in various target audiences, and that may also result in breakthrough research in the area of neurodegenerative disorders.

Herbersum, NSW, Australia
Palakkad, Kerala, India

Md. Habibur Rahman
Muhasina K M

Contents

The Microbiome-Body Connection: Unraveling the Gut-Brain Axis

1

Md Minarul Islam, Nasir Uddin Mahbub, and Mohammmad Amirul Islam

1.1 Introduction

The gut-brain axis (GBA) is now regarded as a key focus in modern biomedical science, playing a crucial role in affecting numerous health outcomes, such as mental well-being, cognitive performance, metabolic control, and immune system responses [1]. This complex and bidirectional signaling network connects the enteric and central nervous systems (CNS), representing an intricate communication network in understanding the physiological and pathological processes involved in neurological and systemic disorders [2, 3]. The gut-brain axis enables the CNS to alter gastrointestinal (GI) functions, particularly during times of stress, influencing immune and secretory activities [4]. Neurological disorders, which affect the brain and spinal cord, have multifaceted origins involving genetics, environment, and lifestyle [5]. Recent researches highlight the crucial role of the gut-brain axis in the initiation and advancement of neurological diseases such as Parkinson's,

M. M. Islam
Department of Food Science and Biotechnology, Gachon University, Seongnam, Gyeonggi-do, Republic of Korea
e-mail: minarul2026@gachon.ac.kr

N. U. Mahbub
Department of Biomedical Sciences and Institute for Medical Science, Jeonbuk National University Medical School, Jeonju, Republic of Korea
e-mail: nasir.mahbub@jbnu.ac.kr

M. A. Islam (✉)
Department of Biochemistry and Molecular Biology, University of Rajshahi, Rajshahi, Bangladesh
e-mail: maislam14@ru.ac.bd

M. H. Rahman, Muhasina K M (eds.), *The Microbiome's Impact on the Brain*,
https://doi.org/10.1007/978-981-95-7494-0_1

Alzheimer's, multiple sclerosis (MS), and psychiatric conditions [6]. Evidence indicates that disruption in gut health may lead to and contribute to the progression of these neurological conditions [7].

The gut microbiome, a heterogeneous ecosystem of trillions of microscopic organisms in the intestines, plays a substantial role in the GBA, influencing both physical and mental health [8]. Clinical, epidemiological, and immunological evidence demonstrate that the gut microbiota (GMB) affects behaviors such as emotion, pain perception, social interaction, and eating [9]. Changes in microbiota composition are strongly associated with various diseases, including those affecting distant organs like the brain. These effects go beyond neuroinflammation and psychiatric disorders to also impact brain development [10]. Ongoing research is uncovering how the microbiota communicates with and influences the emotional and cognitive functions of the brain within these communication systems. For instance, neurodegenerative conditions such as Parkinson's disease (PD), Alzheimer's disease (AD), prion diseases, and autism spectrum disorders (ASD) are now closely associated with functional GI disturbances [11–16]. In contrast, GI disorders, including irritable bowel syndrome, inflammatory bowel disease, diabetes, and obesity, often coexist with psychological issues, which are commonly linked to alterations in the gut microbiome [17, 18]. Brain injury is a frequent problem in preterm infants, causing harm to crucial brain structures and functions, leading to neurodevelopmental disorders. Furthermore, studies suggest that irregular colonization of GMB in these infants may impede proper brain development [19, 20].

Animal studies have highlighted the indispensable role of the gut microbiome in normal neurological function and behavior, and perturbations in the microbiome are involved in the progression of neurodegenerative diseases. Research using germ-free (GF) mice produces strong evidence that GMB holds a central role in gut-brain communication. GF mice display heightened stress responses and elevated glucocorticoid levels, which return to normal after microbial colonization [21]. They typically exhibit reduced anxiety-like behavior, although results may vary between studies [22, 23]. These behavioral changes can be reversed by introducing microbiota and are linked to changes in brain neurotransmitter levels and gene expression, underscoring the microbiome's impact on nervous system function [24, 25]. The GBA is recognized for its impact on host behavior, particularly mediated via the microbiota-gut-behavior axis, which plays a pivotal role in regulating social behavior [26]. Research indicates that fungi associated with the intestinal lining can influence social interactions in mice by modulating neuroimmune responses [27]. It has been demonstrated that GF mice exhibit notable social deficits, such as decreased sociability, impaired social cognition, social avoidance, and increased repetitive grooming. These problems are more severe in males. Bacterial colonization after weaning can improve sociability and decrease social avoidance, but it does not completely reverse impairments in social cognition [28].

While evidence linking the gut microbiome to neurological function through the brain-gut-microbiome (BGM) axis is growing steadily, clear causality has not yet been established. Most studies focus on associations rather than direct cause-and-effect relationships, especially in humans and clinical populations. This makes it challenging to translate these findings into effective treatments for conditions like ASD, schizophrenia, anxiety, and depression [29]. Despite interest in gut microbiome-focused therapies, like prebiotics, probiotics, postbiotics, and specialized diets, interest is increasing rapidly [30]. However, their clinical effectiveness is still uncertain, and there is a risk that these interventions may not provide significant benefits without a more comprehensive understanding of the underlying mechanisms. The multifactorial communication pathways between the gut and the brain are not well understood, highlighting the need for more rigorous, mechanistic, and longitudinal research. This chapter delves into the complex interplay between the gut microbiome and the human body, with particular attention to the microbiota-gut-brain axis. It highlights the significant influence of the gut microbiome on brain health, the management of neurodegenerative disease, and the potential for therapeutic strategies by influencing this axis.

1.2 Pathways for Microbiota-Gut-Brain Communication

The GBA represents a reciprocal signaling system that involves the central, enteric, and autonomic nervous systems, along with the hypothalamic-pituitary-adrenal (HPA) axis. The microbiota-gut-brain axis incorporates GMB (bacteria, viruses, fungi, archaea) and their by-products into this communication network [31]. Despite being physically separated, the GMB influences the brain through immune modulation, the vagus nerve, the enteric and neuroendocrine networks, and circulatory systems, utilizing neuroactive compounds such as neurotransmitters (Fig. 1.1) [32]. Together, these interrelated pathways represent what is known as the microbiota-gut-brain axis.

The vagus nerve is a crucial signaling pathway in the microbiota-gut-brain axis, linking the gut and the brain through enteroendocrine cells and vagal afferents that detect and transmit signals. Activated by microbial metabolites and neurotransmitters, it transmits the message to the brain's central autonomic system and regulates functions such as inflammation, gut motility, food intake, and energy balance [33, 34]. The vagus nerve also influences the cholinergic anti-inflammatory pathway (CAIP), which is involved in sustaining gut health and mental wellness, including conditions such as depression. Its impact extends to psychiatric disorders, obesity, and stress-related illnesses. Therapeutic approaches such as vagus nerve stimulation and meditation hold promise due to the nerve's anti-inflammatory and calming properties [35].

The immune system is a substantial determinant in regulating gut-brain communication through the microbiota-gut-brain (MGB) axis. Both central (brain-resident) and peripheral immune cells play active roles in this intricate interaction. Disruptions in immune signaling within this axis can lead to inflammatory conditions and hinder neurogenesis in the CNS [36]. Recent studies suggest that specific gut microbial species can influence neuroinflammation by triggering immune responses. A well-balanced gut microbiome is vital for the normal maturity and functionality of microglia, the brain's resident immune cells. Conversely, alterations in the

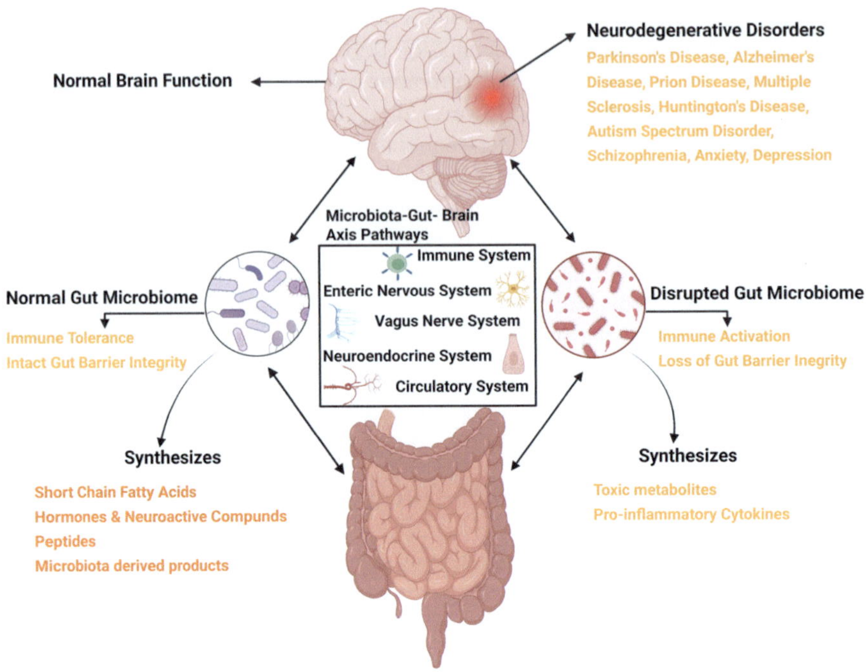

Fig. 1.1 Microbiota-gut-brain axis. The communication between the brain and the gut microbiota involves various systems such as the immune system, neuroendocrine system, enteric nervous system, circulatory system, and vagus nerve. These systems transmit neuroactive compounds like microbial metabolites, peptides, gut hormones, and signaling molecules. In a healthy gut, these compounds help maintain immune tolerance and gut barrier integrity, supporting normal brain function. However, disruptions in gut balance can lead to the production of toxic metabolites and pro-inflammatory cytokines that can reach the brain and contribute to neurodegenerative disorders like multiple sclerosis, Parkinson's disease, Alzheimer's disease, Huntington's disease, prion diseases, autism spectrum disorder, depression, anxiety, and schizophrenia. (Created with biorender.com)

microbial community (dysbiosis) linked to neurological disorders are characterized by microglial dysfunction [37, 38]. Chronic activation of microglia leads to persistent, low-grade neuroinflammation, which can harm neurons and synapses, ultimately contributing to neurodegenerative processes [39]. This ongoing inflammatory state, commonly seen in conditions like AD, often reflects underlying immune dysfunction and shifts in gut microbiome composition. The GMB serves a pivotal function in maintaining immune balance in experimental autoimmune encephalomyelitis (EAE), a model of MS. Modifying gut bacteria, either through antibiotics or germ-free conditions, can decrease CNS inflammation and severity of the disease [40, 41]. This implies that manipulating the gut microbiome has the potential to serve as an effective strategy for treating MS and similar inflammatory diseases.

The role of gut-derived hormones is increasingly acknowledged in the coordination of gut-brain signaling, with evidence linking GMB and mood disorders such as anxiety and depressive disorders, which are frequently linked to obesity [42]. The microbiota can impact neuroendocrine function and behavior by producing neuroactive metabolites and influencing inflammation, immune responses, and hormone secretion [43]. In the gut-brain axis, serotonin (5-HT) acts in the capacity of a significant signaling mediator. The GMB serves a key function in controlling 5-HT production within the gut by triggering its release from enterochromaffin (EC) cells. GF mice exhibit reduced 5-HT levels, which quickly rise when exposed to microbiota, highlighting the microbiota's impact on GBA signaling [44]. Dopamine, another important neurotransmitter in the CNS, is notably lower inside the hippocampus, striatum, as well as frontal cortex of GF mice relative to specific pathogen-free (SPF) mice, indicating that the absence of GMB impacts the central dopaminergic system [45]. Interestingly, 5-HT can have both stimulating and inhibiting effects on the growth of beneficial intestinal microbes, depending on its levels and the particular microbial species present [46]. This reveals a novel influence exerted by the gut-derived serotonin in configuring the composition of the GMB.

The enteric nervous system (ENS), often referred to as the "second brain," is a sophisticated neural network located in the gut. It functions autonomously while also interacting with the CNS through the GBA. The ENS crucially plays in regulating gut functions, maintaining barrier integrity, and ensuring overall homeostasis. Alterations in the gut microbiota, along with genetic factors, may affect the ENS, CNS, and immune system, potentially contributing to conditions such as IBS, IBD, and neurodegenerative disorders [47].

The GBA is influenced by systemic circulation, which comprises several vascular and epithelial interfaces, contributes to gut homeostasis, and mediates signaling between the brain and the gastrointestinal tract. The circulatory system facilitates the transport of nutrients, oxygen, and signaling molecules, allowing for the exchange of metabolites, neurotransmitters, hormones, and immune mediators between the gut and the brain [48]. Even microbes can penetrate the CNS by breaching the blood-brain barrier (BBB), especially when its integrity is compromised. They can pass through endothelial cells in the transcellular route to reach the CNS [49].

1.3 Gut-Brain Axis Pathway Signaling Molecules

The GBA communicates through a range of signaling molecules such as neurotransmitters, hormones, and metabolites, released by the GMB and host cells [50]. These molecules regulate digestion, immunity, mood, cognition, and brain development by transmitting signals along the gut-brain axis and contributing significantly to maintaining overall physiological and mental health. Neurotransmitters like serotonin, dopamine, norepinephrine, and GABA are produced within the CNS and the gut, involving host cells, enteroendocrine cells, and GMB. These neurotransmitters are crucial for communication along the GBA, affecting the nervous and GI functions.

Serotonin (5-hydroxytryptamine, 5-HT) is a key neurotransmitter responsible for regulating mood, sleep, behavior, and GI physiology. Interestingly, enterochromaffin cells in the gut secrete about 95% of the body's serotonin, influenced by the GMB. While serotonin itself cannot cross the BBB, its molecular precursors, including tryptophan and 5-hydroxytryptophan, traverse the BBB and affect the CNS [51]. Tryptophan, an essential amino acid required for 5-HT production, acts as a major factor in modulating microglial activity and neuroinflammation induced by aryl hydrocarbon receptor (AhR) signaling [52]. The GMB significantly influences serotonin production by regulating tryptophan metabolism, primarily via the kynurenine pathway [53]. This pathway generates neuroactive compounds like quinolinic and kynurenic acids, which affect both the CNS and the ENS [54]. Several bacterial strains, including spore-forming clostridial species, have been shown to elevate serotonin levels [55]. Additionally, certain genera—such as *Candida, Streptococcus, Escherichia, Enterococcus*, and *Pseudomonas*—are capable of producing serotonin [56]. Moreover, serotonin facilitates communication along the GBA by activating vagal afferent pathways and regulating immune cell functions, including cytokine release. The dynamic relation between the GMB and the host's serotonergic system is essential for upholding the balance in the GI and CNS [57]. Imbalances in this axis are increasingly linked to neuropsychiatric disorders and functional GI conditions, underscoring the significance of microbiota-mediated serotonergic signaling in health and disease [58].

Gamma-aminobutyric acid (GABA) functions as the brain's dominant antagonistic neurotransmitter. The GMB can produce GABA and could be involved in maintaining intestinal and systemic host health. Notably, *Lactobacillus brevis* and *Bifidobacterium dentium* are effective producers of GABA [59]. A deficiency or impaired function of GABA is linked to a variety of brain disorders, encompassing neurological and psychiatric conditions, and addiction. According to a meta-analysis, levels of GABA are lower in people with ASD and major depressive disorder (MDD) than in controls [60]. $GABAB_1b$ expression is altered differentially in different parts of the brain by prolonged exposure to *Lactobacillus rhamnosus (JB-1)*, falling in the hippocampus, amygdala, and locus coeruleus and rising in cortical areas [61]. Moreover, GMB can impact anxiety-like behavior induced by ethanol by modulating GABA function. Chronic ethanol exposure alters microbiota, reduces GABA activity in the prefrontal cortex, and increases anxiety-like symptoms [62].

Dopamine constitutes an indispensable catecholamine neurotransmitter that significantly contributes to fundamental physiological and neurological processes such as reward, motivation, cognition, emotion, memory, motor control, and appetite. It is synthesized within both the central and peripheral nervous systems and serves as a biochemical precursor to norepinephrine and epinephrine [63]. Maintaining the optimal levels of dopamine is crucial for healthy brain function and preventing neurological disorders. Imbalance in dopamine levels can result in neurodegenerative disorders, such as PD, schizophrenia, and addiction, and may also lead to abnormal movements (like chorea, dystonia, and tics) and thought disorders [64, 65]. GMB substantially controls dopamine levels via the microbiota-gut-brain axis, as around

50% of dopamine is generated in the GI tract by enteric neurons and intestinal epithelial cells [66]. A study demonstrated that a high-fat diet reshapes GMB by increasing the richness of Firmicutes (e.g., *Lachnospiraceae*), *Rothia* (Actinobacteria), and pro-inflammatory Proteobacteria (e.g., RF32), leading to a reduction in food-induced dopamine signaling and motivated behavior [67]. Certain bacteria, particularly in the genus *Staphylococcus*, can synthesize dopamine through an enzyme known as staphylococcal aromatic amino acid decarboxylase encoded by the *sadA* gene. This enzyme transforms aromatic amino acids, such as L-DOPA and 5-HTP, into trace amines like dopamine and serotonin [68]. Dopamine production has also been observed in the biomass of several bacterial species, including *Staphylococcus aureus*, *Bacillus cereus*, *Bacillus mycoides*, *Bacillus subtilis*, *Proteus vulgaris*, *Serratia marcescens*, and *Escherichia coli* [69].

Histamine is a major neurotransmitter produced from L-histidine by histidine decarboxylase. The gut microbiome is capable of producing histamine, especially via species that possess pyruvoyl-dependent histidine decarboxylase, which are present in six phyla: Proteobacteria, Actinobacteriota, Bacteroidota, Firmicutes, Fusobacteriota, and Verrucomicrobiota [70]. The central histamine system plays a critical role in managing several brain functions, including arousal, hormone release, appetite control, cognition, sleep-wake patterns, learning, memory, and reactions to stress. These effects are facilitated by G-protein-coupled histamine H1-H4 receptors [71, 72]. Research has shown that reduced levels of histamine in the brain are linked to anxiety-like behavior and impair working memory [73, 74].

Short-chain fatty acids (SCFAs), including acetate, propionate, and butyrate, serve as primary energy sources synthesized by gut microbes by fermenting dietary fiber and intricate plant polysaccharides. In addition to their metabolic function, SCFAs behave as signaling molecules by engaging G-protein-coupled receptors and suppressing histone deacetylases, which strengthens gut barrier integrity and influences immune responses [75]. Besides, SCFAs can penetrate the BBB and are important for its integrity and function. GF mice show heightened BBB permeability and lower amounts of essential tight junction proteins such as claudin-5 and occludin. Inoculation with SCFA-producing strains in these mice reestablishes tight junction protein expression and lowers BBB permeability, underscoring the importance of GMB and SCFAs in preserving BBB integrity [76]. Growing data imply that SCFAs have a crucial impact on the host's brain. SCFAs, including acetate, propionate, and butyrate, produced from the fermentation of dietary fibers by intestinal bacteria, are being recognized more for their effects on the nervous system. These SCFAs are essential for regulating microglial development and function, which are the brain's resident immune cells, by enhancing their maturation, altering their morphology, and diminishing excessive inflammatory reactions [77]. In GF mice lacking GMB and SCFA production, microglia exhibit underdevelopment and hyperreactivity, which can be reversed by administering SCFAs orally [78]. Mechanistically, SCFAs influence processes via G-protein-coupled receptors (such as GPR41 and GPR43) and by suppressing histone deacetylases (HDACs), resulting in epigenetic modifications and reduced expression of pro-inflammatory cytokines, including IL-1β, IL-6, and TNF-α [79, 80]. Apart from immune regulation, SCFAs also impact neuronal signaling by modulating neurotransmitter levels, including glutamate, GABA, serotonin, and dopamine, and neurotrophic factors. For example, acetate can alter neurotransmitter levels in the hypothalamus [81], while butyrate enhances histone acetylation, a change linked to improved synaptic plasticity, learning, and long-term memory [82]. Animal studies indicate that SCFAs, particularly butyrate, can improve long-term potentiation (LTP) and cognitive performance through these mechanisms [83]. A range of neurological disorders, including AD, PD, ASD, and affective disorders, are associated with the alteration of SCFAs [84]. Given the association of dysbiosis and altered SCFA profiles with various neuropsychiatric and neurodevelopmental disorders, targeting these microbial metabolites offers potential for treatment approaches focused on reestablishing brain balance and functionality.

1.4 Microbiome Influence on Mental Health and Neurological Disorders

A healthy GMB is crucial for sustaining host health through the production of microbial metabolites and neurotransmitters that interact with host cells, such as intestinal epithelial cells (IECs) and immune cells. Recent evidences demonstrate that the gut microbiome functions like a virtual organ, impacting the CNS and significantly affecting the regulation of neurological activities, mental health, and emotional wellness [85, 86]. The microbiome-gut-brain axis concept emphasizes the reciprocal communication between the GMB and the brain. Changes in the delicate balance of GMB and its activity are tied to numerous immune-related neurological and psychiatric conditions, such as neurodevelopmental disorders, neurodegenerative diseases, and mood and behavioral issues. Microbial imbalances can disrupt the production of neuroactive compounds and trigger neuroinflammation, contributing to conditions like AD, PD, MS, and ASD [87]. Additionally, shifts in GMB diversity and function are involved in psychiatric disorders like major depressive disorder (MDD), bipolar disorder (BD), schizophrenia, anxiety disorders, attention deficit hyperactivity disorder (ADHD), post-traumatic stress disorder (PTSD), and anorexia nervosa [88, 89]. Specific microbial species may offer protective benefits, while others could increase disease susceptibility [90]. Ongoing research aims to unveil the complex connection between the GMB and CNS function, indicating that concentrating on the microbiome may lead to groundbreaking therapies for preventing and managing various neuropsychiatric and neurodegenerative conditions.

AD is characterized by a progressive and irreversible neurodegenerative condition and is the foremost cause of dementia in the elderly. It is marked by neuroinflammation, gradual cognitive decline, and dysfunction in the CNS, resulting in memory loss, learning difficulties, and behavioral changes that eventually lead to significant disability in daily activities [91]. In AD, amyloid beta peptides build up outside neurons as plaques, while tau proteins form tangles inside cells. These changes spark inflammation in the brain, leading to damage in nerve connections, cell death, and ongoing brain degeneration [92]. Alterations in GMB composition are linked to the progression of AD. Research indicates that individuals with AD exhibit a unique GMB profile compared to those who are healthy, and altering the GMB in mouse models of AD can impact disease progression and inflammation in the brain [93, 94]. The GMB contributes to AD by generating neuroactive substances, influencing immune responses and metabolism, regulating the BBB, and affecting the formation and clearance of amyloid beta (Aβ) plaques [95, 96]. Research in both humans and animals shows a robust connection between alterations in GMB and AD progression. In individuals with brain amyloidosis and cognitive difficulties, there is a greater presence of pro-inflammatory bacteria (*Escherichia/Shigella*) and a lower presence of anti-inflammatory bacteria (*Eubacterium rectale*), along with an increased amount of pro-inflammatory cytokines [97]. Similarly, in AD mouse models, changes in GMB were observed 1–3 months before amyloid plaque formation and neuroinflammation, with an increase in inflammation-related bacteria like *Escherichia-Shigella*, *Desulfovibrio*, *Akkermansia*, and *Blautia* at later stages. These findings suggest that GMB imbalance may not only be a consequence but also a possible contributor to the mechanisms underlying AD pathology [98]. Additionally, GMB might contribute to AD progression by discharging significant quantities of amyloids and lipopolysaccharides, which can impact signaling pathways and enhance the secretion of pro-inflammatory cytokines [99–102]. Bacterial amyloids are able to interact with human amyloids, which could promote the aggregation and accumulation of amyloid fibrils associated with the disease [103]. Moreover, alterations in GMB might also influence the synthesis of gut-derived amyloid-β (Aβ), affecting the advancement of AD [100].

PD is a chronic neurodegenerative condition marked by the degeneration of dopaminergic neurons in the substantia nigra and the buildup of misfolded alpha-synuclein proteins, which form Lewy bodies. These pathological changes give rise to motor symptoms including tremors, muscular rigidity, and impaired gait [104]. Oxidative stress caused by dopamine (DA) oxidation plays a critical role in the initiation of PD, as it generates detrimental by-products such as reactive oxygen species (ROS), dopamine quinones (DAQs), and 3,4-dihydroxyphenylacetaldehyde, which contribute to neuronal damage [105]. Emerging evidences suggest a connection between changes in GMB and PD. Research using a mouse model that overexpresses α-synuclein (αSyn) indicates that gut bacteria are essential for the development of motor symptoms, brain inflammation, and αSyn-related brain pathology in PD. Treatment with antibiotics in these mice alleviates symptoms, while reintroduction of gut microbes exacerbates them, highlighting the significance of gut-brain communication in disease progression. Additionally, administering specific bacterial metabolites to GF mice induces brain inflammation and movement impairments. Notably, colonization of mice with gut bacteria from PD patients results in more intense symptoms in comparison to colonization with bacteria from healthy individuals [106]. A different study found that MPTP-induced PD in mice led to the loss of dopaminergic neurons and GI dysfunction, along with changes in GMB. Specifically, there were higher levels of *Aerococcus*, *Staphylococcus*, and Ruminococcaceae and lower levels of *Lactobacillus*, *Lachnospiraceae*, and *Adlercreutzia*. Taurine-related metabolites were decreased and linked to the severity of the disease. Supplementing with taurine improved motor function and provided neuronal protection, highlighting its promise as a treatment choice for PD [107]. In PD patients, the GMB displays a distinct profile compared with their healthy counterparts. There is a noticeable decline in anti-inflammatory and neuroprotective bacteria, particularly from the *Lachnospiraceae* family (such as *Butyrivibrio* and *Blautia*). Analysis of fecal metabolites has shown significant alterations in lipids, vitamins, amino acids, and other compounds. These metabolic changes are closely associated with the reduced abundance of *Lachnospiraceae*, indicating a potential connection between gut microbial imbalance and disrupted metabolism in PD. In addition, toll-like receptors TLR2 and TLR4 can mediate the progression of PD via gut impairment, inflammation, and neurodegeneration. In PD patients, these receptors are dysregulated and overexpressed, particularly in the colon. They are activated by bacterial components and misfolded α-synuclein, leading to increased gut permeability and inflammation [108]. TLR4 triggers microglial activation and oxidative stress when binding α-synuclein [109], while TLR2 heterodimer induces a pro-inflammatory response through the MyD88-NF-κB pathway, resulting in elevated TNF-α and IL-1β production [110]. The compromised gut barrier in PD increases exposure to microbes, and the interplay between GMB and TLR signaling further hinders α-synuclein clearance, ultimately hastening neurodegeneration [111].

MS is a long-term autoimmune disorder that targets the CNS, leading to inflammation, loss of myelin, scarring (gliosis), and nerve cell damage. It affects an estimated 2.8 million individuals worldwide [112]. The damage to myelin causes various neurological issues, such as motor, sensory, visual, autonomic, and cognitive problems [113]. Researches highlight the GMB's function in the progression of MS via the GBA, and T cells serve a vital function in MS. Microbial compounds activate innate immune reactions through TLRs, resulting in cytokine production that influences adaptive immunity. When TLRs activate dendritic cells (DCs), they drive the maturation of CD4+ T cells into pro-inflammatory Th1 and Th17 cells, contributing to the inflammation seen in MS [114]. Overactive Th1/Th17 responses and impaired regulatory T cell (Treg) function contribute to autoimmune attacks on the myelin sheath in MS. Experimental models suggest that these Treg dysfunctions are linked to an altered GMB composition [115]. Gut dysbiosis worsens MS by promoting inflammation and generating reactive oxygen species (ROS) [116]. Notably, approximately 40% of MS patients experience GI issues, underscoring the robust link between intestinal health and MS progression [117].

The extensive role of gut dysbiosis in MS progression was first implicated through studies using EAE, an established animal model of MS [118]. Studies with GF mice and preventive antibiotic therapy have demonstrated that the GMB significantly impacts both pro-inflammatory and anti-inflammatory immune reactions in the CNS. GF mice exhibit a less severe variant of EAE, probably because of heightened numbers of regulatory T cells (Tregs), especially $CD25^{+}FoxP3^{+}$ T cells, alongside a decline in pro-inflammatory Th1 and Th17 cell populations [40, 119]. Interestingly, GF mice colonized with microbiota from specific-pathogen-free (SPF) mice restore disease susceptibility, leading to CNS inflammation and the production of anti-myelin autoantibodies [120]. Not all microbes have the same impact on disease development; for instance, segmented filamentous bacteria (SFBs), which promote Th17 differentiation, worsen EAE only in specific microbial or genetic contexts [121]. Clinical studies also showed that MS patients possess modified intestinal microbial communities, with elevated amounts of pro-inflammatory groups such as *Akkermansia muciniphila* and *Acinetobacter calcoaceticus*, and a reduction in potentially beneficial species like *Parabacteroides distasonis*. Exposure of immune cells from healthy individuals to MS-associated microbiota or extracts from *A. calcoaceticus* promotes Th1 polarization and reduces the proportion of $CD25^{+}FoxP3^{+}$ Tregs, while *P. distasonis* extracts encourage a regulatory phenotype [122]. A comprehensive genome study has uncovered gut microbes linked to the risk or protection of MS, with *Ruminococcus* torques posing the highest risk and *Akkermansia muciniphila* being strongly protective. Other risk-associated taxa include Actinobacteria, Bifidobacteriaceae, and *Lactobacillus*, while *Prevotella*, *Lachnospiraceae*, and *Negativibacillus* are protective, indicating a potential causal role of GMB in MS [123]. Transplanting microbiota from MS patients into GF mice worsens EAE, further supporting the role of dysbiosis in CNS autoimmunity [122]. Probiotic interventions, such as VSL#3 (mixture of *Lactobacillus, Bifidobacterium*, and *Streptococcus* strains), can temporarily restore microbial balance, reduce peripheral inflammation, and enhance regulatory immune responses, although these effects may diminish after treatment discontinuation [124]. 16S rRNA-based sequencing analysis of the gut microbiome has identified notable differences between MS patients and healthy individuals, including altered abundances of *Methanobrevibacter*, *Akkermansia*, and *Butyricimonas* in MS patients. Furthermore, disease-modifying therapies were linked to increased abundance of *Prevotella* and *Sutterella* and a decrease in *Sarcina* [125]. Another study reported moderate GMB dysbiosis in MS patients, with significant changes in bacterial species, particularly Clostridia (clusters XIVa and IV) and Bacteroidetes [126]. These findings highlight the significance of GMB in MS pathogenesis and indicate that focusing on the GBA through microbial modulation or dietary changes might present promising therapeutic approaches.

ASD is a complex neurodevelopmental condition marked by challenges in social interaction, communication, sensory processing, and repetitive behaviors, with a broad range of cognitive and behavioral impairments [127, 128]. It is thought that genetic and environmental factors are involved in the initiation of ASD, some of which can trigger neuroinflammation [129]. Genetic factors include rare mutations (e.g., MECP2, SHANK3, TSC1/2) [130–133]. Genes, like PTEN and CX3CR1, associated with ASD are linked to immune pathways that influence microglial activation and cytokine signaling, thereby contributing to the chronic neuroinflammation seen in ASD brains [134, 135]. Increased amount of pro-inflammatory cytokines and hyperactive microglia can disrupt synaptic development and neural connectivity [136]. Therefore, a gene-immune-brain axis substantially contributes to ASD, with neuroinflammation emerging as a key pathological feature in genetically predisposed individuals. Recent research on the microbiota-gut-brain axis shows a two-way relationship between the gut and the brain, suggesting that GMB might influence brain development and behavior, possibly playing a role in ASD. Individuals with ASD frequently experience gastrointestinal symptoms like constipation, diarrhea, and irritable bowel syndrome, implying a potential association between GMB imbalance and the neurobehavioral manifestations of ASD [137]. Neuroinflammation is a hallmark of ASD. Gut microbial imbalance may initiate immune activation, resulting in the release of pro-inflammatory substances like interleukin-6 (IL-6) and tumor necrosis factor-alpha (TNF-α), contributing to systemic inflammation and subsequent neuroinflammation [138]. It is commonly acknowledged that a disruption in GMB significantly affects how the GBA contributes to the development and progression of ASD. People with ASD frequently exhibit alterations in their gut microbiome, marked by reduced diversity and a disparity between helpful and harmful bacteria [139, 140]. These fluctuations in gut bacteria may influence ASD by altering brain function via microbial by-products and neurotransmitters, affecting immune responses, and increasing intestinal permeability, potentially leading to neuroinflammation [141]. People with ASD possess a unique gut microbial profile in contrast to neurotypical individuals. Several studies have revealed significant differences in GMB and metabolism between individuals with ASD and neurotypical individuals. People with ASD exhibit decreased microbial diversity, diminished levels of *Sutterella*, *Prevotella*, and *Bacteroides*, and imbalances in neurotransmitter-related metabolites such as serotonin, dopamine, histidine, and GABA [142]. They also exhibit reduced amounts of fecal acetic acid and butyrate, increased levels of valeric acid, and fewer bacteria that produce butyrate. Constipation, a common GI symptom, has been associated with an abundance of valeric acid-related microbes [143]. Additionally, significant changes include an altered Firmicutes/Bacteroidetes ratio, likely due to decreased Bacteroidetes and increased levels of *Paraprevotella*, *Granulicatella*, *Butyricimonas*, *cc_115*, *Peptoniphilus*, and a decrease in *Eubacterium* [144]. Differences in fungal composition in the gut have also been observed between autistic patients and normal controls. Research has demonstrated that children with ASD showed significant dysbiosis in GMB and mycobiota, characterized by a high prevalence of *Candida*, decreased levels of *Lactobacillus* and *Clostridium*, elevated intestinal permeability, and chronic low-level inflammation. These factors are implicated in the severity of ASD and GI symptoms [145]. Furthermore, notable shifts in fungal communities have been observed, featuring a rise in *Saccharomyces cerevisiae* and a decline in *Aspergillus versicolor*, indicating a possible involvement of fungi in immune reactions associated with ASD [146]. In brief, the GBA is crucial in influencing ASD through intricate interactions involving GMB, immune responses, metabolite production, and neural communication. These elements can affect the development and severity of ASD symptoms. Further research should explore this axis more thoroughly, as enhancing gut health shows potential for enhancing ASD outcomes.

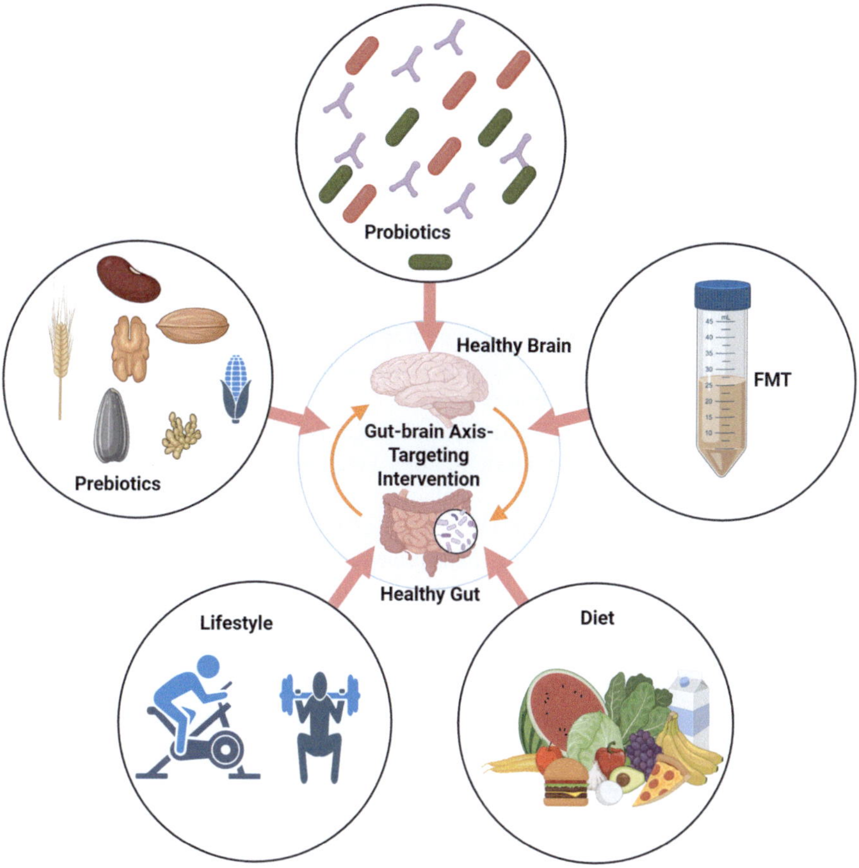

Fig. 1.2 Various therapeutic interventions can potentially alleviate neurological disorders by targeting the gut-brain axis. These interventions include prebiotics such as resistant starch, which encourage the growth of beneficial gut bacteria, and probiotics, live microbes present in fermented foods that help maintain microbial balance. A balanced diet consisting of whole foods, regular physical activity, and healthy lifestyle choices also plays a crucial role in promoting gut-brain health. Furthermore, fecal microbiota transplantation (FMT) shows promise as a method to restore healthy gut microbiota, potentially lowering the risk of neurological conditions. (Created with biorender.com)

1.5 Microbiome Targeting Intervention: Modulating the Gut-Brain Axis

Accumulated research results revealed that GMB plays a fundamental role in the GBA, having a substantial impact on brain function and the development of neurological and psychiatric illnesses like AD, depression, schizophrenia, and ASD. Imbalances in GMB have been strongly associated with these conditions, while restoring a healthy microbial balance seems to enhance brain health, reduce stress, and improve overall quality of life [6, 147]. Various therapeutic approaches, including probiotics, prebiotics, synbiotics, postbiotics, dietary supplements, exercise, and fecal microbiota transplantation (FMT), attempt to reinstate a beneficial microbial population and have demonstrated promise in modulating GMB and reducing symptoms related to brain health (Fig. 1.2) [6, 148]. Prebiotics stimulate the proliferation of beneficial bacteria, whereas probiotics introduce helpful microbes that produce postbiotics—bioactive compounds that support both gut and brain health [149]. A balanced diet can also influence GMB, aiding in preventing and managing neurological and psychiatric disorders [150]. These interventions may help decrease harmful metabolites and external substances, thereby reducing the impact of diseases [30]. Ongoing clinical and experimental research continues to confirm the efficacy of microbiota-targeted strategies, highlighting their natural origins, minimal side effects, and therapeutic benefits in addressing neurodegenerative and neuropsychiatric conditions [151].

Probiotics (psychobiotics), beneficial live organisms, can enhance the diversity and quantity of gut bacteria, potentially improving neurological disorders, offering a secure and supportive approach to balance the microbiome in these conditions. Probiotics indirectly promote brain health by influencing the GBA through several mechanisms, including regulating neurotransmitter production, modulating the immune system, producing SCFAs, and potentially affecting the development of brain circuits. Several probiotic strains have demonstrated neuroprotective effects by influencing neurotransmitters and the GBA. For instance, *Lactobacillus plantarum MTCC1325* has been found to increase acetylcholine levels, decrease acetylcholinesterase activity, enhance cognition, and repair brain tissue in AD models, indicating potential anti-Alzheimer properties [152]. *Levilactobacillus brevis* P30021 and *Lactiplantibacillus plantarum* P30025, known for their production of GABA and acetylcholine, have been shown to reduce cognitive reactivity to sad moods, thereby improving depressive symptoms [153]. Additionally, *Lactobacillus plantarum* L5 was shown to improve motor function in mice with essential tremor by modulating the GMB, increasing GABA levels in the brain, and decreasing neuroinflammation [154]. Furthermore, *Bifidobacterium adolescentis* is a significant GABA-producing bacterium identified within the human GI tract, potentially influencing crosstalk at the gut-brain communication pathway. Research has linked its presence to psychological disorders, including depression and anxiety, indicating a potential role in regulating mood and emotional health [155]. Serotonin serves as an essential neurotransmitter in the CNS, essential for regulating brain functions. Probiotic strains have been found to impact serotonin pathways and support mental health. *Akkermansia muciniphila*, a next-generation probiotic, has been reported to reduce injury to tissues and impairment in cognitive function while increasing the expression of brain-derived neurotrophic factor and serotonin [156]. *Bifidobacterium longum subsp. infantis E41* and *Bifidobacterium breve M2CF22M7* have been linked to increased Tph1 expression and 5-HTP secretion, resulting in reduced depressive behaviors in mice, indicating an antidepressant effect through serotonin regulation [157]. Similarly, *Lactobacillus plantarum strains* (RTA 8 and LRCC5314) have been found to elevate gut and brain serotonin, norepinephrine, and dopamine levels, potentially preventing neurological complications and playing a role in stress regulation [158, 159]. *Lactobacillus* and *Bifidobacterium* have been found to influence dopaminergic neurotransmission, providing potential neuroprotective and stress-relieving effects. For example, *Lactobacillus plantarum PS128* was shown to enhance dopamine metabolism, increase norepinephrine levels, and counteract neurodegeneration in a PD mouse model, indicating its potential for PD prevention [160]. Similarly, *Lactobacillus plantarum* DR7 mitigates stress by downregulating plasma cortisol and the enzymatic activity of dopamine β-hydroxylase and tyrosine hydroxylase, both essential in dopamine metabolic pathways. Imbalance in these enzymes is associated with increased stress, underscoring DR7's potential in stress management [161]. Moreover, *Bifidobacterium longum CECT 30763* has been demonstrated to improve behavioral outcomes related to depression and anxiety in a chronic social defeat stress model by lowering inflammation and supporting T regulatory cell function. This probiotic also normalizes dopamine synthesis, its metabolism, and signaling, resulting in the reversal of behavioral changes and suggesting long-term benefits for mental and cardiometabolic health [162].

Probiotic supplements are reported to have a fundamental role in regulating the immune system by interacting with various immune cells, including lymphocytes, macrophages, dendritic cells, and intestinal epithelial cells. They boost gut immunity by stimulating IgA production through IL-6-mediated B cell activation, particularly via TLR2 pathways [163, 164]. Strains like *Lactobacillus casei CRL 431* and *L. helveticus R389* IL-6 elevate production in the gut, leading to the expansion of IgA-producing B cells without impacting CD4+ T cells [164]. *Bifidobacterium breve MCC1274* reduces neuroinflammation and enhances cognitive function by reducing hippocampal Aβ levels and suppressing pro-inflammatory cytokines [165]. Similarly, *L. fermentum JDFM216* improves immunity and behavior by modulating GMB [166]. In autoimmune conditions like EAE, IRT5 probiotics—a blend of *L. casei*, *L. acidophilus*, *L. reuteri*, *B. bifidum*, and *S. thermophilus*—delay disease onset, suppress Th1/Th17 responses, and boost IL-10-producing and Foxp3+ regulatory T cells [167]. The combined use of *L. plantarum A7* and *B. animalis* further enhances these effects by promoting the development of Th2 and regulatory T cells via increased expression of Foxp3 and GATA3 while suppressing pro-inflammatory Th1/Th17 responses [168]. Additionally, *Faecalibacterium prausnitzii* with prebiotics (FOS + GOS) improved symptoms resembling treatment-resistant depression (TRD) in rats. The synbiotic treatment enhanced behavior, restored SCFA levels, increased beneficial gut bacteria, improved gut health, reduced neuroinflammation, and normalized neurotransmitter levels [169]. Overall, these results highlight the promise of probiotics in enhancing mucosal immunity, regulating systemic inflammation, and providing therapeutic benefits in immune-related and neuroinflammatory conditions.

Probiotics exert a central role in enhancing CNS functions by producing SCFAs like butyrate by fermenting prebiotic fibers and carbohydrates (prebiotics), which are indigestible by humans. These SCFAs contribute to gut health by nourishing colon cells, maintaining the gut barrier, and influencing brain function through the GBA [170]. For instance, *Bifidobacterium breve A1* and its by-product acetate have been found to optimize cognitive performance in preclinical models of AD [171].

Faecalibacterium prausnitzii, a major butyrate producer, helps regulate brain-derived neurotrophic factor levels and neurogenesis in the hippocampus, which can alleviate depressive symptoms [172]. Similarly, *Clostridium butyricum* produces butyrate that is able to penetrate the BBB to reduce neuroinflammation, modulate microglial activity, and enhance neuroplasticity, offering neuroprotective effects in AD and other brain disorders [173]. Moreover, *Lactobacillus acidophilus*, *Bifidobacterium bifidum*, *Lactobacillus reuteri*, and *Lactobacillus fermentum* probiotic strains demonstrated the promise in improving motor symptoms and providing essential antioxidant vitamins for brain health in PD [174]. These findings underscore the importance of SCFA-producing probiotics in supporting cognitive function, emotional well-being, and overall brain health through intricate gut-brain communication.

FMT is a novel therapeutic approach that entails transferring thoroughly screened fecal matter from a healthy donor into the gastrointestinal tract of a patient to reestablish microbial balance and function. Initially developed as a highly effective intervention for recurrent *Clostridioides difficile* infections, with success rates surpassing 90%, FMT has increasingly attracted interest for its potential to treat various conditions linked to gut dysbiosis [175]. Growing evidence underscores the significance of the GBA, suggesting that changes in GMB can impact neurological and neuropsychiatric disorders. A significant study by Kang et al. in 2017 showcased sustained relief from digestive symptoms and behavioral outcomes within the pediatric ASD population following an 8-week microbiota transfer therapy (MTT) regimen [176]. These benefits persisted for years posttreatment, indicating the potential long-term advantages of microbiome modulation on brain function. Other research in major depressive disorder has shown that FMT can alleviate depressive symptoms and reduce neuroinflammation [177, 178], while in PD, FMT has enhanced microbial diversity and improved motor and GI symptoms [179]. Similarly, in AD models, FMT was found to decrease neuroinflammation and cognitive decline by adjusting gut microbial composition and related inflammatory pathways [180]. Despite the promising implications of FMT in treating brain-related conditions, the evidence is still preliminary. Many studies are limited to small-scale trials or animal models, with some yielding mixed or inconclusive results. Therefore, extensive, double-blind randomized controlled trials and the creation of uniform treatment protocols are essential to thoroughly investigate the potency, safety, and lasting impacts of FMT in neurological and psychiatric conditions. Nonetheless, current studies highlight the considerable influence of GMB on neurological health and offer hopeful opportunities for upcoming microbiome-related treatments.

1.6 Future Directions and Therapeutic Potential

Advancements in microbiome research have illuminated the complex relationships among the immune system, GMB, and the GBA, offering new possibilities for treating neurological disorders. While targeting the GBA shows therapeutic potential, the field is still evolving, with many unresolved queries and challenges in translating discoveries to clinical practice in humans. Future research on the GBA ought to include both longitudinal and cross-sectional methods that combine genetic, environmental, cognitive, immune, and social elements to fully grasp the intricate interactions affecting health and disease. It is of utmost importance to delve deeper into how gut microbes produce neuroactive metabolites, modulate systemic and local inflammation, and communicate with the CNS through the vagus nerve and the ENS. The progress of multi-omics platforms, such as metagenomics, metabolomics, and transcriptomics, will enable more personalized, systems-level investigations of the GBA and its role in neurological disorders.

Therapeutically, the GBA offers promising opportunities, including the psychobiotics (probiotics that confer mental benefits), FMT, targeted dietary interventions, and microbial-derived postbiotics. Cutting-edge approaches, like genetically engineered probiotics, next-generation probiotics, are under investigation for their possible ability to deliver therapeutic molecules directly in the gut, metabolize harmful substances, and provide neuroprotective effects, such as antioxidant activity and inhibition of amyloid beta fibril formation [181–183]. These probiotics possess the strength to reshape the GMB, regulate immune function, and influence gut-brain signaling, offering new possibilities for treating neurodegenerative disorders.

Additionally, artificial intelligence (AI) is poised to revolutionize the field by analyzing vast and complex datasets to uncover disease mechanisms, predict individual responses, guide personalized probiotic and dietary interventions, and optimize clinical outcomes [184]. AI-powered tools can improve our comprehension of the gut-microbiome-immune-brain system and aid in creating novel therapies aimed at the microbiome [185]. When combined with digital health technologies and behavioral interventions that promote gut health, these advancements hold transformative potential for personalized, effective, and safe treatments aimed at enhancing mental and neurological well-being.

Declarations

Competing Interests The author(s) has no competing interests to declare that are relevant to the content of this manuscript.

Ethics Approval The author(s) declares that this manuscript does not include primary research with human participants.

The author(s) declares that this manuscript does not include primary studies involving animals.

References

1. Zhang R, Ding N, Feng X, Liao W. The gut microbiome, immune modulation, and cognitive decline: insights on the gut-brain axis. Front Immunol. 2025;16:1529958. https://doi.org/10.3389/fimmu.2025.1529958.
2. O'Riordan KJ, Moloney GM, Keane L, Clarke G, Cryan JF. The gut microbiota-immune-brain axis: therapeutic implications. Cell Rep Med. 2025;6:101982. https://doi.org/10.1016/j.xcrm.2025.101982.

3. Barbosa PM, Barbosa ER. The gut brain-axis in neurological diseases. Int J Cardiovasc Sci. 2020;33:528–36. https://doi.org/10.36660/ijcs.20200039.
4. Lu S, et al. The communication mechanism of the gut-brain axis and its effect on central nervous system diseases: a systematic review. Biomed Pharmacother. 2024;178:117207. https://doi.org/10.1016/j.biopha.2024.117207.
5. Gao W, et al. Advancements in neurodegenerative diseases: pathogenesis and novel neurorestorative interventions. J Neurorestoratol. 2025;13:100176. https://doi.org/10.1016/j.jnrt.2024.100176.
6. Ashique S, et al. Gut-brain axis: a cutting-edge approach to target neurological disorders and potential synbiotic application. Heliyon. 2024;10:e34092. https://doi.org/10.1016/j.heliyon.2024.e34092.
7. Ullah H, et al. The gut microbiota-brain axis in neurological disorder. Front Neurosci. 2023;17:1225875. https://doi.org/10.3389/fnins.2023.1225875.
8. Carabotti M, Scirocco A, Maselli MA, Severi C. The gut-brain axis: interactions between enteric microbiota, central and enteric nervous systems. Ann Gastroenterol. 2015;28:203–9.
9. Osadchiy V, Martin CR, Mayer EA. The gut-brain axis and the microbiome: mechanisms and clinical implications. Clin Gastroenterol Hepatol. 2019;17:322–32. https://doi.org/10.1016/j.cgh.2018.10.002.
10. Rutsch A, Kantsjö JB, Ronchi F. The gut-brain axis: how microbiota and host inflammasome influence brain physiology and pathology. Front Immunol. 2020;11 https://doi.org/10.3389/fimmu.2020.604179.
11. Mahbub NU, Islam MM, Hong ST, Chung HJ. Dysbiosis of the gut microbiota and its effect on α-synuclein and prion protein misfolding: consequences for neurodegeneration. Front Cell Infect Microbiol. 2024;14:1348279. https://doi.org/10.3389/fcimb.2024.1348279.
12. Islam MM, Mahbub NU, Hong S-T, Chung H-J. Gut bacteria: an etiological agent in human pathological conditions. Front Cell Infect Microbiol. 2024;14 https://doi.org/10.3389/fcimb.2024.1291148.
13. Kwon D, et al. Diet and the gut microbiome in patients with Parkinson’s disease. npj Parkinson's Dis. 2024;10:89. https://doi.org/10.1038/s41531-024-00681-7.
14. MahmoudianDehkordi S, et al. Altered bile acid profile associates with cognitive impairment in Alzheimer’s disease-An emerging role for gut microbiome. Alzheimers Dement. 2019;15:76–92. https://doi.org/10.1016/j.jalz.2018.07.217.
15. Yang D, et al. Implications of gut microbiota dysbiosis and metabolic changes in prion disease. Neurobiol Dis. 2020;135:104704. https://doi.org/10.1016/j.nbd.2019.104704.
16. Peralta-Marzal LN, et al. A robust microbiome signature for autism spectrum disorder across different studies using machine learning. Sci Rep. 2024;14:814. https://doi.org/10.1038/s41598-023-50601-7.
17. Person H, Keefer L. Psychological comorbidity in gastrointestinal diseases: update on the brain-gut-microbiome axis. Prog Neuro-Psychopharmacol Biol Psychiatry. 2021;107:110209. https://doi.org/10.1016/j.pnpbp.2020.110209.
18. Kar N, Kar S. Exploring the complex relationship between obesity and mental health: a narrative review. Odisha J Psychiatry. 2022;18:74–82. https://doi.org/10.4103/ojp.Ojp_8_23.
19. Seki D, et al. Aberrant gut-microbiota-immune-brain axis development in premature neonates with brain damage. Cell Host Microbe. 2021;29:1558–1572.e1556. https://doi.org/10.1016/j.chom.2021.08.004.
20. Lu J, et al. Early preterm infant microbiome impacts adult learning. Sci Rep. 2022;12:3310. https://doi.org/10.1038/s41598-022-07245-w.
21. Sudo N, et al. Postnatal microbial colonization programs the hypothalamic-pituitary-adrenal system for stress response in mice. J Physiol. 2004;558:263–75. https://doi.org/10.1113/jphysiol.2004.063388.
22. Neufeld KA, Kang N, Bienenstock J, Foster JA. Effects of intestinal microbiota on anxiety-like behavior. Commun Integr Biol. 2011;4:492–4. https://doi.org/10.4161/cib.4.4.15702.

23. Bercik P, et al. The intestinal microbiota affect central levels of brain-derived neurotropic factor and behavior in mice. Gastroenterology. 2011;141(599–609):609.e591–3. https://doi.org/10.1053/j.gastro.2011.04.052.
24. Heijtz RD, et al. Normal gut microbiota modulates brain development and behavior. Proc Natl Acad Sci USA. 2011;108:3047–52. https://doi.org/10.1073/pnas.1010529108.
25. Clarke G, et al. The microbiome-gut-brain axis during early life regulates the hippocampal serotonergic system in a sex-dependent manner. Mol Psychiatry. 2013;18:666–73. https://doi.org/10.1038/mp.2012.77.
26. Ntranos A, Casaccia P. The microbiome-gut-behavior axis: crosstalk between the gut microbiome and oligodendrocytes modulates behavioral responses. Neurotherapeutics. 2018;15:31–5. https://doi.org/10.1007/s13311-017-0597-9.
27. Leonardi I, et al. Mucosal fungi promote gut barrier function and social behavior via Type 17 immunity. Cell. 2022;185:831–846.e814. https://doi.org/10.1016/j.cell.2022.01.017.
28. Desbonnet L, Clarke G, Shanahan F, Dinan TG, Cryan JF. Microbiota is essential for social development in the mouse. Mol Psychiatry. 2014;19:146–8. https://doi.org/10.1038/mp.2013.65.
29. Ma Y-Y, Li X, Yu J-T, Wang Y-J. Therapeutics for neurodegenerative diseases by targeting the gut microbiome: from bench to bedside. Transl Neurodegener. 2024;13:12. https://doi.org/10.1186/s40035-024-00404-1.
30. Ji J, Jin W, Liu SJ, Jiao Z, Li X. Probiotics, prebiotics, and postbiotics in health and disease. MedComm. 2023;4:e420. https://doi.org/10.1002/mco2.420.
31. Rusch JA, Layden BT, Dugas LR. Signalling cognition: the gut microbiota and hypothalamic-pituitary-adrenal axis. Front Endocrinol. 2023;14:1130689. https://doi.org/10.3389/fendo.2023.1130689.
32. Chakrabarti A, et al. The microbiota-gut-brain axis: pathways to better brain health. Perspectives on what we know, what we need to investigate and how to put knowledge into practice. Cell Mol Life Sci. 2022;79:80. https://doi.org/10.1007/s00018-021-04060-w.
33. Han Y, et al. Vagus nerve and underlying impact on the gut microbiota-brain axis in behavior and neurodegenerative diseases. J Inflamm Res. 2022;15:6213–30. https://doi.org/10.2147/jir.S384949.
34. Berthoud H-R. The vagus nerve, food intake and obesity. Regul Pept. 2008;149:15–25. https://doi.org/10.1016/j.regpep.2007.08.024.
35. Browning KN, Verheijden S, Boeckxstaens GE. The vagus nerve in appetite regulation, mood, and intestinal inflammation. Gastroenterology. 2017;152:730–44. https://doi.org/10.1053/j.gastro.2016.10.046.
36. Fung TC, Olson CA, Hsiao EY. Interactions between the microbiota, immune and nervous systems in health and disease. Nat Neurosci. 2017;20:145–55. https://doi.org/10.1038/nn.4476.
37. Rea K, Dinan TG, Cryan JF. The microbiome: a key regulator of stress and neuroinflammation. Neurobiol Stress. 2016;4:23–33. https://doi.org/10.1016/j.ynstr.2016.03.001.
38. Hao W, et al. Gut dysbiosis induces the development of depression-like behavior through abnormal synapse pruning in microglia-mediated by complement C3. Microbiome. 2024;12:34. https://doi.org/10.1186/s40168-024-01756-6.
39. Xie J, et al. Low-grade peripheral inflammation affects brain pathology in the AppNL-G-Fmouse model of Alzheimer's disease. Acta Neuropathologica Commun. 2021;9:163. https://doi.org/10.1186/s40478-021-01253-z.
40. Ochoa-Repáraz J, et al. Role of gut commensal microflora in the development of experimental autoimmune encephalomyelitis. J Immunol (Baltimore, Md: 1950). 2009;183:6041–50. https://doi.org/10.4049/jimmunol.0900747.
41. Lee YK, Menezes JS, Umesaki Y, Mazmanian SK. Proinflammatory T-cell responses to gut microbiota promote experimental autoimmune encephalomyelitis. Proc Natl Acad Sci USA. 2011;108:4615–22. https://doi.org/10.1073/pnas.1000082107.

42. Gariepy G, Nitka D, Schmitz N. The association between obesity and anxiety disorders in the population: a systematic review and meta-analysis. Int J Obes. 2010;34:407–19. https://doi.org/10.1038/ijo.2009.252.
43. Cussotto S, Sandhu KV, Dinan TG, Cryan JF. The neuroendocrinology of the microbiota-gut-brain axis: a behavioural perspective. Front Neuroendocrinol. 2018;51:80–101. https://doi.org/10.1016/j.yfrne.2018.04.002.
44. Hata T, et al. Regulation of gut luminal serotonin by commensal microbiota in mice. PLoS One. 2017;12:e0180745. https://doi.org/10.1371/journal.pone.0180745.
45. Wang Y, Qiao H, Zhang Y. Changes of dopamine and tyrosine hydroxylase levels in the brain of germ-free mice. Iran J Biotechnol. 2023;21:e2798. https://doi.org/10.30498/ijb.2022.236732.2798.
46. Kwon YH, et al. Modulation of gut microbiota composition by serotonin signaling influences intestinal immune response and susceptibility to colitis. Cell Mol Gastroenterol Hepatol. 2019;7:709–28. https://doi.org/10.1016/j.jcmgh.2019.01.004.
47. Hajjeh O, et al. Enteric nervous system dysfunction as a driver of central nervous system disorders: the Forgotten brain in neurological disease. Neuroscience. 2025;572:232–47. https://doi.org/10.1016/j.neuroscience.2025.03.015.
48. Gwak MG, Chang SY. Gut-brain connection: microbiome, gut barrier, and environmental sensors. Immune Netw. 2021;21:e20. https://doi.org/10.4110/in.2021.21.e20.
49. Vollmuth N, Sin J, Kim BJ. Host-microbe interactions at the blood-brain barrier through the lens of induced pluripotent stem cell-derived brain-like endothelial cells. MBio. 2024;15:e0286223. https://doi.org/10.1128/mbio.02862-23.
50. Bellono NW, et al. Enterochromaffin cells are gut chemosensors that couple to sensory neural pathways. Cell. 2017;170:185–198.e116. https://doi.org/10.1016/j.cell.2017.05.034.
51. Kanova M, Kohout P. Serotonin-its synthesis and roles in the healthy and the critically ill. Int J Mol Sci. 2021;22 https://doi.org/10.3390/ijms22094837.
52. Yang Y, et al. Aryl hydrocarbon receptor dependent anti-inflammation and neuroprotective effects of tryptophan metabolites on retinal ischemia/reperfusion injury. Cell Death Dis. 2023;14:92. https://doi.org/10.1038/s41419-023-05616-3.
53. Agus A, Planchais J, Sokol H. Gut microbiota regulation of tryptophan metabolism in health and disease. Cell Host Microbe. 2018;23:716–24. https://doi.org/10.1016/j.chom.2018.05.003.
54. Gao K, Mu C-L, Farzi A, Zhu W-Y. Tryptophan Metabolism: A Link Between the Gut Microbiota and Brain. Adv Nutr. 2020;11:709–23. https://doi.org/10.1093/advances/nmz127.
55. Yano JM, et al. Indigenous bacteria from the gut microbiota regulate host serotonin biosynthesis. Cell. 2015;161:264–76. https://doi.org/10.1016/j.cell.2015.02.047.
56. Roshchina VV. In: Lyte M, Freestone PPE, editors. Microbial endocrinology: interkingdom signaling in infectious disease and health. New York: Springer; 2010. p. 17–52.
57. Hwang YK, Oh JS. Interaction of the vagus nerve and serotonin in the gut–brain axis. Int J Mol Sci. 2025;26:1160.
58. Del Colle A, Israelyan N, Gross Margolis K. Novel aspects of enteric serotonergic signaling in health and brain-gut disease. Am J Physiol Gastrointest Liver Physiol. 2020;318:G130–g143. https://doi.org/10.1152/ajpgi.00173.2019.
59. Barrett E, Ross RP, O'Toole PW, Fitzgerald GF, Stanton C. γ-Aminobutyric acid production by culturable bacteria from the human intestine. J Appl Microbiol. 2012;113:411–7. https://doi.org/10.1111/j.1365-2672.2012.05344.x.
60. Schür RR, et al. Brain GABA levels across psychiatric disorders: a systematic literature review and meta-analysis of (1) H-MRS studies. Hum Brain Mapp. 2016;37:3337–52. https://doi.org/10.1002/hbm.23244.
61. Bravo JA, et al. Ingestion of *Lactobacillus* strain regulates emotional behavior and central GABA receptor expression in a mouse via the vagus nerve. Proc Natl Acad Sci USA. 2011;108:16050–5. https://doi.org/10.1073/pnas.1102999108.
62. Yao H, et al. Chronic ethanol exposure induced anxiety-like behaviour by altering gut microbiota and GABA system. Addict Biol. 2022;27:e13203. https://doi.org/10.1111/adb.13203.

63. Klein MO, et al. Dopamine: functions, signaling, and association with neurological diseases. Cell Mol Neurobiol. 2019;39:31–59. https://doi.org/10.1007/s10571-018-0632-3.
64. Cramb KML, Beccano-Kelly D, Cragg SJ, Wade-Martins R. Impaired dopamine release in Parkinson's disease. Brain. 2023;146:3117–32. https://doi.org/10.1093/brain/awad064.
65. Marsden CD. Dopamine and basal ganglia disorders in humans. Semin Neurosci. 1992;4:171–8. https://doi.org/10.1016/1044-5765(92)90015-T.
66. Eisenhofer G, et al. Substantial production of dopamine in the human gastrointestinal tract. J Clin Endocrinol Metab. 1997;82:3864–71. https://doi.org/10.1210/jcem.82.11.4339.
67. Kim JS, et al. The gut-brain axis mediates bacterial driven modulation of reward signaling. Mol Metab. 2023;75:101764. https://doi.org/10.1016/j.molmet.2023.101764.
68. Luqman A, Nega M, Nguyen M-T, Ebner P, Götz F. SadA-expressing staphylococci in the human gut show increased cell adherence and internalization. Cell Rep. 2018;22:535–45. https://doi.org/10.1016/j.celrep.2017.12.058.
69. Tsavkelova EA, Botvinko IV, Kudrin VS, Oleskin AV. Detection of neurotransmitter amines in microorganisms with the use of high-performance liquid chromatography. Doklady Biochem Proc Acad Sci USSR Biochem Sec. 2000;372:115–7.
70. Mou Z, Yang Y, Hall AB, Jiang X. The taxonomic distribution of histamine-secreting bacteria in the human gut microbiome. BMC Genomics. 2021;22:695. https://doi.org/10.1186/s12864-021-08004-3.
71. Kárpáti A, et al. Histamine H1 receptor on astrocytes and neurons controls distinct aspects of mouse behaviour. Sci Rep. 2019;9:16451. https://doi.org/10.1038/s41598-019-52623-6.
72. Nuutinen S, Panula P. In: Thurmond RL, editor. *Histamine in inflammation*. Springer; 2010. p. 95–107.
73. Yoshikawa T, et al. Insufficient intake of L-histidine reduces brain histamine and causes anxiety-like behaviors in male mice. J Nutr. 2014;144:1637–41. https://doi.org/10.3945/jn.114.196105.
74. Nakamura T, et al. Oral histidine intake improves working memory through the activation of histaminergic nervous system in mice. Biochem Biophys Res Commun. 2022;609:141–8. https://doi.org/10.1016/j.bbrc.2022.04.016.
75. Zhang D, et al. Short-chain fatty acids in diseases. Cell Commun Signal. 2023;21:212. https://doi.org/10.1186/s12964-023-01219-9.
76. Braniste V, et al. The gut microbiota influences blood-brain barrier permeability in mice. Sci Transl Med. 2014;6:263ra158. https://doi.org/10.1126/scitranslmed.3009759.
77. Silva YP, Bernardi A, Frozza RL. The role of short-chain fatty acids from gut microbiota in gut-brain communication. Front Endocrinol (Lausanne). 2020;11 https://doi.org/10.3389/fendo.2020.00025.
78. Erny D, et al. Host microbiota constantly control maturation and function of microglia in the CNS. Nat Neurosci. 2015;18:965–77. https://doi.org/10.1038/nn.4030.
79. Li M, et al. Pro- and anti-inflammatory effects of short chain fatty acids on immune and endothelial cells. Eur J Pharmacol. 2018;831:52–9. https://doi.org/10.1016/j.ejphar.2018.05.003.
80. Macfarlane S, Cleary S, Bahrami B, Reynolds N, Macfarlane GT. Synbiotic consumption changes the metabolism and composition of the gut microbiota in older people and modifies inflammatory processes: a randomised, double-blind, placebo-controlled crossover study. Aliment Pharmacol Ther. 2013;38:804–16. https://doi.org/10.1111/apt.12453.
81. Frost G, et al. The short-chain fatty acid acetate reduces appetite via a central homeostatic mechanism. Nat Commun. 2014;5:3611. https://doi.org/10.1038/ncomms4611.
82. Intlekofer KA, et al. Exercise and sodium butyrate transform a subthreshold learning event into long-term memory via a brain-derived neurotrophic factor-dependent mechanism. Neuropsychopharmacology. 2013;38:2027–34. https://doi.org/10.1038/npp.2013.104.
83. Levenson JM, et al. Regulation of histone acetylation during memory formation in the hippocampus*. J Biol Chem. 2004;279:40545–59. https://doi.org/10.1074/jbc.M402229200.
84. Mirzaei R, et al. Role of microbiota-derived short-chain fatty acids in nervous system disorders. Biomed Pharmacother. 2021;139:111661. https://doi.org/10.1016/j.biopha.2021.111661.

85. Trevelline BK, et al. Convergent remodelling of the gut microbiome is associated with host energetic condition over long-distance migration. Funct Ecol. 2023;37:2840–54. https://doi.org/10.1111/1365-2435.14430.
86. Charitos IA, et al. The gut microbiota's role in neurological, psychiatric, and neurodevelopmental disorders. Nutrients. 2024;16:4404.
87. Kandpal M, et al. Dysbiosis of gut microbiota from the perspective of the gut-brain axis: role in the provocation of neurological disorders. Metabolites. 2022;12 https://doi.org/10.3390/metabo12111064.
88. Ma Z, Zhao H, Zhao M, Zhang J, Qu N. Gut microbiotas, inflammatory factors, and mental-behavioral disorders: a mendelian randomization study. J Affect Disord. 2025;371:113–23. https://doi.org/10.1016/j.jad.2024.11.049.
89. Ahmed GK, Ramadan HK-A, Elbeh K, Haridy NA. Bridging the gap: associations between gut microbiota and psychiatric disorders. Middle East Curr Psychiatry. 2024;31:2. https://doi.org/10.1186/s43045-024-00395-9.
90. Dash S, Syed YA, Khan MR. Understanding the role of the gut microbiome in brain development and its association with neurodevelopmental psychiatric disorders. Front Cell Dev Biol. 2022;10:880544. https://doi.org/10.3389/fcell.2022.880544.
91. Zhang J, et al. Recent advances in Alzheimer's disease: mechanisms, clinical trials and new drug development strategies. Signal Transduct Target Ther. 2024;9:211. https://doi.org/10.1038/s41392-024-01911-3.
92. Chandra S, Sisodia SS, Vassar RJ. The gut microbiome in Alzheimer's disease: what we know and what remains to be explored. Mol Neurodegener. 2023;18:9. https://doi.org/10.1186/s13024-023-00595-7.
93. Vogt NM, et al. Gut microbiome alterations in Alzheimer's disease. Sci Rep. 2017;7:13537. https://doi.org/10.1038/s41598-017-13601-y.
94. Shukla PK, et al. Alterations in the gut-microbial-inflammasome-brain axis in a mouse model of Alzheimer's disease. Cells. 2021;10 https://doi.org/10.3390/cells10040779.
95. Onisiforou A, Charalambous EG, Zanos P. Shattering the amyloid illusion: the microbial enigma of Alzheimer's disease pathogenesis – from gut microbiota and viruses to brain. Biofilms. 2025;13:90.
96. Seo D-O, Holtzman DM. Current understanding of the Alzheimer's disease-associated microbiome and therapeutic strategies. Exp Mol Med. 2024;56:86–94. https://doi.org/10.1038/s12276-023-01146-2.
97. Cattaneo A, et al. Association of brain amyloidosis with pro-inflammatory gut bacterial taxa and peripheral inflammation markers in cognitively impaired elderly. Neurobiol Aging. 2017;49:60–8. https://doi.org/10.1016/j.neurobiolaging.2016.08.019.
98. Chen Y, et al. Gut microbiome alterations precede cerebral amyloidosis and microglial pathology in a mouse model of Alzheimer's disease. Biomed Res Int. 2020;2020:8456596. https://doi.org/10.1155/2020/8456596.
99. Chapman MR, et al. Role of Escherichia coli curli operons in directing amyloid fiber formation. Science (New York, N.Y.). 2002;295:851–5. https://doi.org/10.1126/science.1067484.
100. Jin J, et al. Gut-derived β-amyloid: likely a centerpiece of the gut-brain axis contributing to Alzheimer's pathogenesis. Gut Microbes. 2023;15:2167172. https://doi.org/10.1080/19490976.2023.2167172.
101. Zhao Y, Cong L, Lukiw WJ. Lipopolysaccharide (LPS) accumulates in neocortical neurons of Alzheimer's Disease (AD) brain and impairs transcription in human neuronal-glial primary co-cultures. Front Aging Neurosci. 2017;9 https://doi.org/10.3389/fnagi.2017.00407.
102. Brown GC, Heneka MT. The endotoxin hypothesis of Alzheimer's disease. Mol Neurodegener. 2024;19:30. https://doi.org/10.1186/s13024-024-00722-y.
103. Elkins M, Jain N, Tükel Ç. The menace within: bacterial amyloids as a trigger for autoimmune and neurodegenerative diseases. Curr Opin Microbiol. 2024;79:102473. https://doi.org/10.1016/j.mib.2024.102473.

104. MacMahon Copas AN, McComish SF, Fletcher JM, Caldwell MA. The pathogenesis of Parkinson's disease: a complex interplay between astrocytes, microglia, and T lymphocytes? Front Neurol. 2021;12 https://doi.org/10.3389/fneur.2021.666737.
105. Zhang X, Tang B, Guo J. Parkinson's disease and gut microbiota: from clinical to mechanistic and therapeutic studies. Transl Neurodegener. 2023;12:59. https://doi.org/10.1186/s40035-023-00392-8.
106. Sampson TR, et al. Gut microbiota regulate motor deficits and neuroinflammation in a model of Parkinson's disease. Cell. 2016;167:1469–1480.e1412. https://doi.org/10.1016/j.cell.2016.11.018.
107. Cui C, et al. Gut microbiota-associated taurine metabolism dysregulation in a mouse model of Parkinson's disease. mSphere. 2023;8:e0043123. https://doi.org/10.1128/msphere.00431-23.
108. Gorecki AM, Anyaegbu CC, Anderton RS. TLR2 and TLR4 in Parkinson's disease pathogenesis: the environment takes a toll on the gut. Transl Neurodegener. 2021;10:47. https://doi.org/10.1186/s40035-021-00271-0.
109. Fellner L, et al. Toll-like receptor 4 is required for α-synuclein dependent activation of microglia and astroglia. Glia. 2013;61:349–60. https://doi.org/10.1002/glia.22437.
110. Daniele SG, et al. Activation of MyD88-dependent TLR1/2 signaling by misfolded α-synuclein, a protein linked to neurodegenerative disorders. Sci Signal. 2015;8:ra45. https://doi.org/10.1126/scisignal.2005965.
111. Zhang Z, Liu Z, Lv A, Fan C. How Toll-like receptors influence Parkinson's disease in the microbiome–gut–brain axis. Front Immunol. 2023;14 https://doi.org/10.3389/fimmu.2023.1154626.
112. Walton C, et al. Rising prevalence of multiple sclerosis worldwide: insights from the Atlas of MS, third edition. Multiple Sclerosis (Houndmills, Basingstoke, England). 2020;26:1816–21. https://doi.org/10.1177/1352458520970841.
113. Kamma E, Lasisi W, Libner C, Ng HS, Plemel JR. Central nervous system macrophages in progressive multiple sclerosis: relationship to neurodegeneration and therapeutics. J Neuroinflammation. 2022;19:45. https://doi.org/10.1186/s12974-022-02408-y.
114. Duan T, Du Y, Xing C, Wang HY, Wang RF. Toll-Like receptor signaling and its role in cell-mediated immunity. Front Immunol. 2022;13:812774. https://doi.org/10.3389/fimmu.2022.812774.
115. Shim JA, Ryu JH, Jo Y, Hong C. The role of gut microbiota in T cell immunity and immune mediated disorders. Int J Biol Sci. 2023;19:1178–91. https://doi.org/10.7150/ijbs.79430.
116. van Horssen J, Witte ME, Schreibelt G, de Vries HE. Radical changes in multiple sclerosis pathogenesis. Biochim Biophys Acta (BBA) - Mol Basis Dis. 2011;1812:141–50. https://doi.org/10.1016/j.bbadis.2010.06.011.
117. Nusrat S, Gulick E, Levinthal D, Bielefeldt K. Anorectal dysfunction in multiple sclerosis: a systematic review. ISRN Neurol. 2012;2012:376023. https://doi.org/10.5402/2012/376023.
118. Croxford AL, Kurschus FC, Waisman A. Mouse models for multiple sclerosis: historical facts and future implications. Biochim Biophys Acta. 2011;1812:177–83. https://doi.org/10.1016/j.bbadis.2010.06.010.
119. Lee YK, Menezes JS, Umesaki Y, Mazmanian SK. Proinflammatory T-cell responses to gut microbiota promote experimental autoimmune encephalomyelitis. Proc Natl Acad Sci USA. 2011;108(Suppl 1):4615–22. https://doi.org/10.1073/pnas.1000082107.
120. Berer K, et al. Commensal microbiota and myelin autoantigen cooperate to trigger autoimmune demyelination. Nature. 2011;479:538–41. https://doi.org/10.1038/nature10554.
121. Ivanov II, et al. Induction of intestinal Th17 cells by segmented filamentous bacteria. Cell. 2009;139:485–98. https://doi.org/10.1016/j.cell.2009.09.033.
122. Cekanaviciute E, et al. Gut bacteria from multiple sclerosis patients modulate human T cells and exacerbate symptoms in mouse models. Proc Natl Acad Sci USA. 2017;114:10713–8. https://doi.org/10.1073/pnas.1711235114.
123. Zancan V, et al. Gut microbiota composition is causally linked to multiple sclerosis: a mendelian randomization analysis. Microorganisms. 2024;12:1476.

124. TankouSK,etal.Investigationofprobioticsinmultiplesclerosis.MultipleSclerosis(Houndmills, Basingstoke, England). 2018;24:58–63. https://doi.org/10.1177/1352458517737390.
125. Jangi S, et al. Alterations of the human gut microbiome in multiple sclerosis. Nat Commun. 2016;7:12015. https://doi.org/10.1038/ncomms12015.
126. Miyake S, et al. Dysbiosis in the gut microbiota of patients with multiple sclerosis, with a striking depletion of species belonging to Clostridia XIVa and IV clusters. PLoS One. 2015;10:e0137429. https://doi.org/10.1371/journal.pone.0137429.
127. Taniya MA, et al. Role of gut microbiome in autism spectrum disorder and its therapeutic regulation. Front Cell Infect Microbiol. 2022;12:915701. https://doi.org/10.3389/fcimb.2022.915701.
128. Gonçalves CL, et al. The many faces of microbiota-gut-brain axis in autism spectrum disorder. Life Sci. 2024;337:122357. https://doi.org/10.1016/j.lfs.2023.122357.
129. Than UTT, et al. Inflammatory mediators drive neuroinflammation in autism spectrum disorder and cerebral palsy. Sci Rep. 2023;13:22587. https://doi.org/10.1038/s41598-023-49902-8.
130. Alonso-Gonzalez A, Rodriguez-Fontenla C, De Carracedo A. Novo Mutations (DNMs) in Autism Spectrum Disorder (ASD): pathway and network analysis. Front Genet. 2018;9:406. https://doi.org/10.3389/fgene.2018.00406.
131. Bey AL, et al. Subacute neuropsychiatric syndrome in girls with SHANK3 mutations responds to immunomodulation. Pediatrics. 2020;145 https://doi.org/10.1542/peds.2019-1490.
132. Nagarajan RP, Hogart AR, Gwye Y, Martin MR, LaSalle JM. Reduced MeCP2 expression is frequent in autism frontal cortex and correlates with aberrant MECP2 promoter methylation. Epigenetics. 2006;1:e1–11. https://doi.org/10.4161/epi.1.4.3514.
133. Uysal SP, Şahin M. Tuberous sclerosis: a review of the past, present, and future. Turk J Med Sci. 2020;50:1665–76. https://doi.org/10.3906/sag-2002-133.
134. Skelton PD, Stan RV, Luikart BW. The role of PTEN in neurodevelopment. Mol Neuropsychiatry. 2020;5:60–71. https://doi.org/10.1159/000504782.
135. Ishizuka K, et al. Rare genetic variants in CX3CR1 and their contribution to the increased risk of schizophrenia and autism spectrum disorders. Transl Psychiatry. 2017;7:e1184. https://doi.org/10.1038/tp.2017.173.
136. Cornell J, Salinas S, Huang HY, Zhou M. Microglia regulation of synaptic plasticity and learning and memory. Neural Regen Res. 2022;17:705–16. https://doi.org/10.4103/1673-5374.322423.
137. McElhanon BO, McCracken C, Karpen S, Sharp WG. Gastrointestinal symptoms in autism spectrum disorder: a meta-analysis. Pediatrics. 2014;133:872–83. https://doi.org/10.1542/peds.2013-3995.
138. Zarimeidani F, et al. Gut microbiota and autism spectrum disorder: a neuroinflammatory mediated mechanism of pathogenesis? Inflammation. 2025;48:501–19. https://doi.org/10.1007/s10753-024-02061-y.
139. Młynarska E, et al. The gut–brain–microbiota connection and its role in autism spectrum disorders. Nutrients. 2025;17:1135.
140. Zuffa S, et al. Early-life differences in the gut microbiota composition and functionality of infants at elevated likelihood of developing autism spectrum disorder. Transl Psychiatry. 2023;13:257. https://doi.org/10.1038/s41398-023-02556-6.
141. Zhou M, et al. Intervention and research progress of gut microbiota-immune-nervous system in autism spectrum disorders among students. Front Microbiol. 2025;16:1535455. https://doi.org/10.3389/fmicb.2025.1535455.
142. Dan Z, et al. Altered gut microbial profile is associated with abnormal metabolism activity of Autism Spectrum Disorder. Gut Microbes. 2020;11:1246–67. https://doi.org/10.1080/19490976.2020.1747329.
143. Liu S, et al. Altered gut microbiota and short chain fatty acids in Chinese children with autism spectrum disorder. Sci Rep. 2019;9:287. https://doi.org/10.1038/s41598-018-36430-z.

144. Kong X, et al. New and preliminary evidence on altered oral and gut microbiota in individuals with Autism Spectrum Disorder (ASD): implications for ASD diagnosis and subtyping based on microbial biomarkers. Nutrients. 2019;11 https://doi.org/10.3390/nu11092128.
145. Iovene MR, et al. Intestinal dysbiosis and yeast isolation in stool of subjects with autism spectrum disorders. Mycopathologia. 2017;182:349–63. https://doi.org/10.1007/s11046-016-0068-6.
146. Zou R, et al. Dysbiosis of gut fungal microbiota in children with autism spectrum disorders. J Autism Dev Disord. 2021;51:267–75. https://doi.org/10.1007/s10803-020-04543-y.
147. Che Mohd Nassir CMN, et al. The microbiota–gut–brain axis: key mechanisms driving glymphopathy and cerebral small vessel disease. Life (Basel). 2025;15:3.
148. Wadan A-HS, El-Aziz MKA, Ellakwa DE-S. The microbiota-gut-brain-axis theory: role of gut microbiota modulators (GMMs) in gastrointestinal, neurological, and mental health disorders. Naunyn Schmiedeberg's Arch Pharmacol. 2025; https://doi.org/10.1007/s00210-025-04155-2.
149. Mitra S, Dash R, Nishan AA, Habiba SU, Moon IS. Brain modulation by the gut microbiota: from disease to therapy. J Adv Res. 2023;53:153–73. https://doi.org/10.1016/j.jare.2022.12.001.
150. Berding K, et al. Diet and the microbiota-gut-brain axis: sowing the seeds of good mental health. Adv Nutr (Bethesda, Md). 2021;12:1239–85. https://doi.org/10.1093/advances/nmaa181.
151. Lee S-H, Han C, Shin C. IUPHAR review: Microbiota-gut-brain axis and its role in neuropsychiatric disorders. Pharmacol Res. 2025;216:107749. https://doi.org/10.1016/j.phrs.2025.107749.
152. Nimgampalle M, Kuna Y. Anti-Alzheimer properties of probiotic, Lactobacillus plantarum MTCC 1325 in Alzheimer's disease induced albino rats. J Clin Diagn Res JCDR. 2017;11:Kc01–kc05. https://doi.org/10.7860/jcdr/2017/26106.10428.
153. Casertano M, et al. Gaba-producing lactobacilli boost cognitive reactivity to negative mood without improving cognitive performance: a human Double-Blind Placebo-Controlled Cross-Over study. Brain Behav Immun. 2024;122:256–65. https://doi.org/10.1016/j.bbi.2024.08.029.
154. Zhong H-J, et al. Supplementation with high-GABA-producing Lactobacillus plantarum L5 ameliorates essential tremor triggered by decreased gut bacteria-derived GABA. Transl Neurodegener. 2023;12:58. https://doi.org/10.1186/s40035-023-00391-9.
155. Duranti S, et al. Bifidobacterium adolescentis as a key member of the human gut microbiota in the production of GABA. Sci Rep. 2020;10:14112. https://doi.org/10.1038/s41598-020-70986-z.
156. Kang EJ, et al. Akkermansia muciniphila improve cognitive dysfunction by regulating BDNF and serotonin pathway in gut-liver-brain axis. Microbiome. 2024;12:181. https://doi.org/10.1186/s40168-024-01924-8.
157. Tian P, Wang G, Zhao J, Zhang H, Chen W. Bifidobacterium with the role of 5-hydroxytryptophan synthesis regulation alleviates the symptom of depression and related microbiota dysbiosis. J Nutr Biochem. 2019;66:43–51. https://doi.org/10.1016/j.jnutbio.2019.01.007.
158. Kaur A, Kaur IP, Chopra K, Rishi P. Bi-directional elucidation of Lactiplantibacillus plantarum (RTA 8) intervention on the pathophysiology of gut-brain axis during Salmonella brain infection. Gut Pathog. 2022;14:11. https://doi.org/10.1186/s13099-022-00484-2.
159. Jeong J, Lee Y, Yoon S, Kim J-H, Kim W. Lactiplantibacillus plantarum LRCC5314 includes a gene for serotonin biosynthesis via the tryptophan metabolic pathway. J Microbiol. 2021;59:1092–103. https://doi.org/10.1007/s12275-021-1472-2.
160. Liao JF, et al. Lactobacillus plantarum PS128 alleviates neurodegenerative progression in 1-methyl-4-phenyl-1,2,3,6-tetrahydropyridine-induced mouse models of Parkinson's disease. Brain Behav Immun. 2020;90:26–46. https://doi.org/10.1016/j.bbi.2020.07.036.

161. Liu G, Chong HX, Chung FY, Li Y, Liong MT. Lactobacillus plantarum DR7 modulated bowel movement and gut microbiota associated with dopamine and serotonin pathways in stressed adults. Int J Mol Sci. 2020;21 https://doi.org/10.3390/ijms21134608.
162. Tamayo M, et al. Bifidobacterium longum CECT 30763 improves depressive- and anxiety-like behavior in a social defeat mouse model through the immune and dopaminergic systems. Brain Behav Immun. 2025;125:35–57. https://doi.org/10.1016/j.bbi.2024.12.028.
163. Mazziotta C, Tognon M, Martini F, Torreggiani E, Rotondo JC. Probiotics mechanism of action on immune cells and beneficial effects on human health. Cells. 2023;12 https://doi.org/10.3390/cells12010184.
164. Vinderola G, Matar C, Perdigon G. Role of intestinal epithelial cells in immune effects mediated by gram-positive probiotic bacteria: involvement of toll-like receptors. Clin Diagn Lab Immunol. 2005;12:1075–84. https://doi.org/10.1128/CDLI.12.9.1075-1084.2005.
165. Abdelhamid M, Zhou C, Jung C-G, Michikawa M. Probiotic Bifidobacterium breve MCC1274 mitigates Alzheimer's disease-related pathologies in wild-type mice. Nutrients. 2022;14:2543.
166. Park MR, et al. Probiotic Lactobacillus fermentum strain JDFM216 improves cognitive behavior and modulates immune response with gut microbiota. Sci Rep. 2020;10:21701. https://doi.org/10.1038/s41598-020-77587-w.
167. Kim J, et al. Clinical effect of IRT-5 probiotics on immune modulation of autoimmunity or alloimmunity in the eye. Nutrients. 2017;9 https://doi.org/10.3390/nu9111166.
168. Salehipour Z, et al. Bifidobacterium animalis in combination with human origin of Lactobacillus plantarum ameliorate neuroinflammation in experimental model of multiple sclerosis by altering CD4+ T cell subset balance. Biomed Pharmacother. 2017;95:1535–48. https://doi.org/10.1016/j.biopha.2017.08.117.
169. Palepu MSK, Bhalerao HA, Sonti R, Dandekar MP. Faecalibacterium prausnitzii, FOS and GOS loaded synbiotic reverses treatment-resistant depression in rats: restoration of gut-brain crosstalk. Eur J Pharmacol. 2024;983:176960. https://doi.org/10.1016/j.ejphar.2024.176960.
170. Munir MU, et al. Reverse engineering the Gut-Brain Axis and microbiome-metabolomics for symbiotic/pathogenic balance in neurodegenerative diseases. Gut Microbes. 2024;16:2422468. https://doi.org/10.1080/19490976.2024.2422468.
171. Kobayashi Y, et al. Therapeutic potential of Bifidobacterium breve strain A1 for preventing cognitive impairment in Alzheimer's disease. Sci Rep. 2017;7:13510. https://doi.org/10.1038/s41598-017-13368-2.
172. Suda K, Matsuda K. How microbes affect depression: underlying mechanisms via the gut-brain axis and the modulating role of probiotics. Int J Mol Sci. 2022;23 https://doi.org/10.3390/ijms23031172.
173. Sun J, et al. Effect of Clostridium butyricum against microglia-mediated neuroinflammation in Alzheimer's disease via regulating gut microbiota and metabolites butyrate. Mol Nutr Food Res. 2020;64:1900636. https://doi.org/10.1002/mnfr.201900636.
174. Tamtaji OR, et al. Clinical and metabolic response to probiotic administration in people with Parkinson's disease: a randomized, double-blind, placebo-controlled trial. Clin Nutr. 2019;38:1031–5. https://doi.org/10.1016/j.clnu.2018.05.018.
175. Massaro M, Vansia J, McGill S. Ulcerative proctitis in a patient with a history of fecal microbiota transplant for clostridioides difficile infection. ACG Case Rep J. 2020;7
176. Kang D-W, et al. Microbiota transfer therapy alters gut ecosystem and improves gastrointestinal and autism symptoms: an open-label study. Microbiome. 2017;5:10. https://doi.org/10.1186/s40168-016-0225-7.
177. Rao J, et al. Fecal microbiota transplantation ameliorates stress-induced depression-like behaviors associated with the inhibition of glial and NLRP3 inflammasome in rat brain. J Psychiatr Res. 2021;137:147–57. https://doi.org/10.1016/j.jpsychires.2021.02.057.
178. Doll JPK, et al. Fecal Microbiota Transplantation (FMT) as an adjunctive therapy for depression – case report. Front Psychiatry. 2022;13 https://doi.org/10.3389/fpsyt.2022.815422.

179. DuPont HL, et al. Fecal microbiota transplantation in Parkinson's disease – A randomized repeat-dose, placebo-controlled clinical pilot study. Front Neurol. 2023;14:1104759. https://doi.org/10.3389/fneur.2023.1104759.
180. Li X, et al. Fecal microbiota transplantation attenuates Alzheimer's disease symptoms in APP/PS1 transgenic mice via inhibition of the TLR4-MyD88-NF-κB signaling pathway-mediated inflammation. Behav Brain Funct. 2025;21:2. https://doi.org/10.1186/s12993-024-00265-8.
181. Abouelela ME, Helmy YA. Next-generation probiotics as novel therapeutics for improving human health: current trends and future perspectives. Microorganisms. 2024;12 https://doi.org/10.3390/microorganisms12030430.
182. Kurhaluk N, et al. Role of antioxidants in modulating the microbiota–gut–brain axis and their impact on neurodegenerative diseases. Int J Mol Sci. 2025;26:3658.
183. Pan I, Umapathy S. Probiotics an emerging therapeutic approach towards gut-brain-axis oriented chronic health issues induced by microplastics: a comprehensive review. Heliyon. 2024;10:e32004. https://doi.org/10.1016/j.heliyon.2024.e32004.
184. Han X, et al. Synergizing artificial intelligence and probiotics: a comprehensive review of emerging applications in health promotion and industrial innovation. Trends Food Sci Technol. 2025;159:104938. https://doi.org/10.1016/j.tifs.2025.104938.
185. D'Urso F, Broccolo F. Applications of artificial intelligence in microbiome analysis and probiotic interventions – an overview and perspective based on the current state of the art. Appl Sci. 2024;14:8627.

Gut Microbiome Imbalance: The Root Causes of Neurodegeneration?

2

Taskia Azad Konika, Md Shamsuzzaman, Md Ekhtiar Uddin, and Jungmin Kim

2.1 Introduction

The human gastrointestinal tract harbors a highly complex and dynamic microbial ecosystem called the gut microbiota. This community, comprising trillions of microorganisms including bacteria, archaea, viruses, fungi, and protozoa, plays a fundamental role in maintaining host homeostasis by modulating digestion, immune function, and energy metabolism [1]. Beyond these local effects, compelling evidence over the past two decades highlights the gut microbiota's influence on distant organs, especially the CNS. This interaction is mediated via the gut-brain axis, a sophisticated bidirectional communication network integrating neural, endocrine, immune, and metabolic pathways [2, 3].

The GBA is now increasingly recognized as a critical regulator of neurological health, influencing behavior, cognition, emotion, and neurodevelopment. Mechanistically, it comprises pathways such as the vagus nerve, the enteric nervous system, the hypothalamic-pituitary-adrenal axis, immune cell signaling, and microbial metabolite production [4]. Disruptions in this network can profoundly impact CNS integrity, underscoring the importance of gut microbiota balance for brain health.

Dysbiosis, defined as reduced microbial diversity and compositional alterations, has been implicated in various chronic diseases, including metabolic disorders,

T. A. Konika (✉)
Department of Pharmacy, State University of Bangladesh, Dhaka, Bangladesh

M. Shamsuzzaman · J. Kim
Department of Microbiology, School of Medicine, Kyungpook National University, Daegu, Republic of Korea
e-mail: minkim@knu.ac.kr

M. E. Uddin
The School of Medicine, Medical Sciences and Nutrition, University of Aberdeen, Aberdeen, UK

M. H. Rahman, Muhasina K M (eds.), *The Microbiome's Impact on the Brain*,
https://doi.org/10.1007/978-981-95-7494-0_2

inflammatory bowel disease (IBD), and, more recently, neurodegenerative diseases [5]. Neurodegenerative disorders, such as AD, PD, ALS, and HD, are characterized by progressive neuronal loss, abnormal protein aggregation, synaptic dysfunction, and chronic neuroinflammation [6, 7]. These conditions lead to severe cognitive and motor impairments, impose substantial socioeconomic burdens worldwide, and currently lack effective disease-modifying treatments.

Historically, research on neurodegeneration focused primarily on intrinsic CNS pathologies, including amyloid plaques, tau tangles, and alpha-synuclein aggregates. However, recent advances in systems biology and multi-omics approaches have shifted this paradigm. Emerging evidence suggests that peripheral factors, particularly the gut microbiome, may act as modulators and potentially as initiators of neurodegenerative processes [8].

Experimental studies in germ-free mice have demonstrated the critical role of microbiomes in neurodevelopment and neuroinflammation. Germ-free animals exhibit aberrant stress responses, increased blood-brain barrier permeability, impaired microglial maturation, and altered synaptic protein expression, emphasizing the essential role of microbial signals in shaping CNS function [9]. Moreover, fecal microbiota transplantation (FMT) studies show that transferring microbiota from PD patients into healthy mice induces motor deficits and promotes alpha-synuclein pathology in the brain, directly linking gut microbes to disease phenotypes [10].

Human studies further support these findings, revealing disease-specific microbial signatures. For instance, patients with AD show reduced anti-inflammatory commensals such as *Bifidobacterium* and increased pro-inflammatory taxa like *Escherichia/Shigella* [11]. In PD, decreased *Prevotella* and increased *Enterobacteriaceae* have been consistently observed, contributing to gut barrier disruption and systemic inflammation [12]. Alterations in microbial metabolites, including SCFAs, LPS, and tryptophan derivatives, are implicated in neuroinflammation, microglial activation, and neurotransmitter modulation [13].

The hypothesis that gut dysbiosis may precede neurodegenerative pathology has significant early detection and intervention implications. Unlike the relatively inaccessible and slowly regenerating brain, the gut microbiome is highly plastic and modifiable through dietary changes, prebiotics, probiotics, antibiotics, and FMT. This opens new avenues for noninvasive strategies to mitigate or delay neurodegenerative progression [14].

Recent technological advancements have been instrumental in elucidating the complexity of the gut-brain axis. High-throughput sequencing and metagenomics allow precise taxonomic and functional profiling of gut microbes. Metabolomics and lipidomics provide insights into microbe-derived biochemical mediators, while integrative systems biology approaches map host-microbiota networks at unprecedented resolution [15]. Combining microbiome data with neuroimaging biomarkers, cerebrospinal fluid analyses, and behavioral assessments is beginning to reveal specific mechanistic links between gut health and brain pathology [16].

Despite these advances, significant challenges remain. Establishing causality in human studies is inherently difficult due to the multifactorial nature of

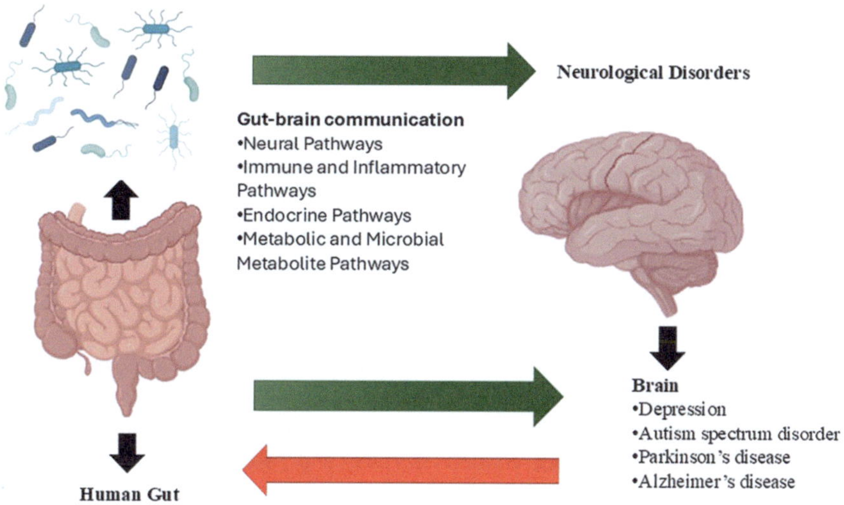

Fig. 2.1 Understanding the correlation between gut microbiota and neurological disorders [NDs]

neurodegenerative diseases. Inter-individual variability, confounding factors such as diet and medication, and predominantly cross-sectional study designs limit definitive conclusions. Large-scale longitudinal studies with standardized methodologies and multi-omics integration are essential to move from association to causation [17].

Nevertheless, emerging evidence positions the gut microbiome not merely as a bystander but potentially as a central driver of neurodegeneration and possibly a root cause. Recognizing and harnessing this connection could transform our approach to neurological diseases, shifting from symptomatic management to preventive and precision-targeted interventions.

This chapter aims to provide a comprehensive and integrative overview of gut microbiome imbalance and its role in neurodegenerative diseases. We explore mechanistic pathways, examine preclinical and clinical evidence, highlight disease-specific microbial alterations, and discuss therapeutic strategies to restore microbial homeostasis. By synthesizing these multidisciplinary insights, we hope to advance a more holistic understanding of neurodegenerative disease etiology and foster the development of innovative microbiome-targeted therapies (Fig. 2.1).

2.2 Understanding Healthy Gut Microbiome

The human gut microbiome, primarily in the large intestine, constitutes one of the body's most complex and metabolically active microbial ecosystems. In a healthy state, it maintains a dynamic equilibrium characterized by high taxonomic diversity, functional redundancy, and immune tolerance. Coevolved with its human host, the gut microbiota performs essential functions related to digestion, energy harvest, immune system development, and neurological modulation functions crucial for systemic and brain health [18].

2.2.1 Structural Organization and Microbial Taxa

Gut colonization begins at birth and is influenced by delivery mode (vaginal vs. cesarean), infant feeding practices (breastfeeding vs. formula), and early-life environmental exposures. By approximately 3 years of age, the gut microbiota reaches adultlike complexity and relative stability [19].

In healthy adults, over 90% of gut microbes belong to the phyla *Firmicutes* and *Bacteroidetes*, with smaller contributions from *Actinobacteria, Proteobacteria, Verrucomicrobia*, and *Fusobacteria* [20]. *Bacteroides, Prevotella, Faecalibacterium, Ruminococcus, Lactobacillus, Bifidobacterium*, and *Akkermansia* are commonly predominant at the genus level. These abundances vary among individuals and are shaped by diet, geography, age, genetics, and host physiology [21].

Is *Faecalibacterium prausnitzii* particularly interesting? It is a key *Firmicute* species with potent anti-inflammatory properties that produces butyrate, an SCFA essential for gut and brain health [22]. The gut ecosystem also includes nonbacterial organisms such as fungi (e.g., *Candida* spp.), protozoa, and a diverse virome, mainly composed of bacteriophages. While less is known about their physiological roles, emerging research suggests they influence immune regulation and host-microbe metabolic interactions [23, 24]. A summary of key microbial taxa and their neuroactive functions is provided in Table 2.1.

2.2.2 Metabolic Functions of the Gut Microbiota

The gut microbiota's metabolic capacity exceeds that of the human liver in enzymatic diversity [25]. A key activity is the fermentation of dietary fibers and resistant starches into SCFAs, primarily acetate, propionate, and butyrate. These metabolites fuel colonocytes, maintain epithelial barrier integrity, regulate immune responses, and modulate gene expression through epigenetic mechanisms such as histone deacetylase (HDAC) inhibition [26].

Beyond SCFAs, gut microbes metabolize amino acids, bile acids, and polyphenols into bioactive molecules that exert local and systemic effects, including the CNS. These compounds activate G-protein-coupled receptors (GPR41, GPR43) and nuclear receptors such as PPARs and FXR, influencing immune balance, lipid metabolism, and neuroendocrine signaling [27]. Gut bacteria also synthesize essential vitamins (e.g., K, B12, folate) required for neurotransmitter synthesis and neural development [28].

Significantly, microbial regulation of amino acid metabolism, including choline, tryptophan, and tyrosine, affects the biosynthesis of neurotransmitters, such as

Table 2.1 Key gut microbes and their roles in metabolism, immunity, and brain function

Microbial taxa	Phylum	Key functions	Neuroactive role	References
Faecalibacterium prausnitzii	*Firmicutes*	Butyrate production, anti-inflammatory	Enhances BDNF, reduces neuroinflammation	[5, 22, 49, 104]
Bifidobacterium longum	*Actinobacteria*	SCFA production, vitamin synthesis	Modulates serotonin and GABA signaling	[6, 33, 34, 147]
Lactobacillus rhamnosus	*Firmicutes*	Immune modulation, GABA production	Improves mood and anxiety-like behavior (animal models)	[6, 33, 34, 144]
Akkermansia muciniphila	*Verrucomicrobia*	Mucin degradation, barrier integrity	Preserves BBB and gut barrier function	[45, 103, 105]
Bacteroides fragilis	*Bacteroidetes*	Polysaccharide A production, immune tolerance	Reduces systemic inflammation	[19, 35, 48]
Prevotella copri	*Bacteroidetes*	Fiber degradation, propionate production	Associated with behavioral modulation (context-dependent)	[13, 25, 58, 157]
Clostridium spp. (cluster IV and XIVa)	*Firmicutes*	Treg induction, SCFA production	Maintains microglial homeostasis	[19, 29, 49]

acetylcholine, serotonin, dopamine, and norepinephrine, highlighting the microbiota's far-reaching influence on brain function and behavior [29].

2.2.3 Neuroactive Metabolites and Their Influence on Brain Function

Gut microbes influence brain physiology through neuroactive metabolites, some crossing the blood-brain barrier (BBB) or signaling via the vagus nerve. SCFAs, particularly butyrate, stimulates brain-derived neurotrophic factor (BDNF), maintain BBB integrity, and regulate microglial function, offering neuroprotective and anti-inflammatory effects [30].

Tryptophan metabolism via the kynurenine and indole pathways also links the gut to CNS health. Imbalanced kynurenine activity can lead to neurotoxic byproducts like quinolinic acid, contributing to neuroinflammation and excitotoxicity [31]. In contrast, indole derivatives from bacterial tryptophanase activate aryl hydrocarbon receptors (AhR), modulating mucosal and CNS immune responses [32].

Certain microbes, including *Lactobacillus, Bifidobacterium*, and *Enterococcus*, can synthesize or modulate levels of GABA, serotonin, dopamine, and acetylcholine [33]. Although peripheral serotonin does not cross the BBB, it affects central serotonergic circuits indirectly via vagal signaling, reflecting the complexity of gut-brain communication [34].

2.2.4 Immune and Barrier-Modulatory Roles

The gut microbiota is essential for developing and regulating mucosal and systemic immunity. Commensals influence T cell differentiation, Toll-like receptor (TLR) signaling, and overall immune homeostasis [35]. Through anti-inflammatory signaling, these microbes suppress pathogen-associated molecular pattern (PAMP) recognition and inhibit NF-κB-driven inflammatory cascades—mechanisms implicated in neurodegenerative disease pathology [36].

Moreover, a healthy microbiota supports intestinal epithelial barrier integrity, preventing endotoxin (e.g., LPS) translocation into systemic circulation. Elevated systemic LPS levels can activate TLR4, trigger widespread inflammation, compromise the BBB, and stimulate microglial activation, all contributing to neurodegeneration [37].

2.3 Mechanisms Linking Gut Dysbiosis to Neurodegeneration

The gut microbiome is a pivotal regulator of neuroimmune and neurochemical signaling, profoundly influencing the CNS via the gut-brain axis. In a healthy state, this bidirectional system promotes neuronal integrity, neurogenesis, and cognitive stability. However, disturbances in microbial equilibrium caused by infections, poor diet, aging, chronic stress, antibiotic overuse, or genetic predisposition lead to gut dysbiosis. This dysregulated state is characterized by reduced microbial diversity, loss of beneficial commensals, and overgrowth of pathobionts [38]. Gut dysbiosis triggers pathological cascades that collectively contribute to neurodegenerative disease onset and progression, including AD, PD, and ALS [39]. This section examines five principal mechanistic pathways: neuroinflammation, blood-brain barrier disruption, microglial activation, oxidative stress and mitochondrial dysfunction, and vagal nerve signaling.

2.3.1 Neuroinflammation

Chronic neuroinflammation is a hallmark of many neurodegenerative diseases and is increasingly linked to gut microbiota alterations. In dysbiosis, the intestinal barrier becomes more permeable allowing the translocation of PAMPs, such as LPS, peptidoglycans, and bacterial amyloids, into the systemic circulation [40].

These endotoxins activate peripheral immune cells and can penetrate a compromised BBB, triggering microglial activation in the CNS [41]. LPS, for instance, binds to TLR4 on microglia and astrocytes, initiating NF-κB signaling and upregulating pro-inflammatory cytokines including TNF-α, IL-1β, and IL-6 [42]. These cytokines amplify neuroinflammation, disrupt synaptic plasticity, and promote neuronal apoptosis, exacerbating neurodegenerative progression [43]. Animal studies support this causative link, showing that oral LPS administration or FMT from PD patients induces gliosis, motor deficits, and α-synuclein aggregation in mice [44].

2.3.2 Blood-Brain Barrier Disruption

The BBB is a highly selective barrier that protects the brain and maintains CNS homeostasis. Gut dysbiosis compromises BBB integrity through multiple mechanisms.

Commensal bacteria produce SCFAs, especially butyrate, which help preserve tight junction proteins (occludin and claudin-5) [45]. In dysbiosis, reduced SCFA production weakens tight junctions. Concurrent systemic inflammation induces matrix metalloproteinases (MMPs), degrading extracellular matrix components and further disrupting BBB structure [46]. A compromised BBB permits infiltration of peripheral immune cells, microbial products, and pro-inflammatory cytokines, creating an environment conducive to neuronal injury [47]. Germ-free mice studies reveal increased BBB permeability and impaired CNS development, partially restored upon colonization with SCFA-producing bacteria, highlighting the gut's protective role [48].

2.3.3 Microglial Activation: Neuroimmune Cross Talk

Microglia are CNS-resident immune cells essential for synaptic pruning, neurogenesis, and surveillance. Under normal conditions, they maintain a homeostatic, anti-inflammatory phenotype.

Dysbiosis disrupts this balance via:

- Reduced SCFA levels, which are essential for microglial quiescence and anti-inflammatory signaling [49]
- Increased systemic LPS, promoting a pro-inflammatory M1 microglial phenotype characterized by reactive oxygen species (ROS), nitric oxide (NO), TNF-α, and IL-1β production [50]
- Enhanced gut permeability, leading to chronic microglial priming and a hyper-responsive, neurotoxic state [51]

This chronic activation contributes to synaptic dysfunction, myelin degradation, and neuronal loss, features common in AD, PD, and multiple sclerosis (MS) [52]. Notably, germ-free mice exhibit underdeveloped microglia, which can be partially corrected by SCFA supplementation [53].

2.3.4 Oxidative Stress and Mitochondrial Dysfunction

Neurons are particularly vulnerable to oxidative damage due to high metabolic demand and low regenerative capacity. Gut dysbiosis exacerbates oxidative stress via:

1. Systemic inflammation, driving ROS and reactive nitrogen species production, damaging neuronal lipids, proteins, and DNA [54]
2. Depletion of antioxidant-producing microbes (e.g., *Lactobacillus* and *Bifidobacterium*), lowering glutathione and polyphenol metabolites [55]
3. Mitochondrial dysfunction, impairing ATP synthesis, promoting cytochrome c release, and activating apoptotic pathways involving caspase-3 [56]

Oxidative stress also enhances amyloid β aggregation and tau hyperphosphorylation in AD pathology. Additionally, microbial siderophores may influence iron metabolism, triggering ferroptosis—an iron-dependent cell death implicated in neurodegeneration [57].

2.3.5 Vagal Nerve Signaling and Neuromodulation

The vagus nerve provides a direct, bidirectional link between the gut and brain, conveying afferent signals to CNS nuclei and modulating autonomic and neuroendocrine responses [58].

Gut-derived metabolites, including SCFAs and neurotransmitter analogs (e.g., GABA, serotonin, dopamine), influence vagal tone and central signaling. In dysbiosis:

- Pathogenic bacteria (e.g., *Clostridium, Escherichia coli*) produce toxins or degrade tryptophan, reducing serotonin bioavailability [59].
- Reduced SCFA levels impair vagal afferent stimulation, diminishing CNS anti-inflammatory signals [60].
- Elevated peripheral cytokines (e.g., IL-1β, TNF-α) modulate vagal signaling, inducing sickness behavior and depressive symptoms [61].

The Braak's hypothesis suggests that PD pathology may originate in the enteric nervous system and ascend to the CNS via the vagus nerve. Epidemiological studies show truncal vagotomy significantly reduces PD risk, reinforcing the gut-origin concept of neurodegeneration [58]. These intertwined mechanisms highlight how gut dysbiosis can act as an initiator rather than merely a correlation of neurodegenerative disease processes. A conceptual summary of these pathways is presented in Fig. 2.2.

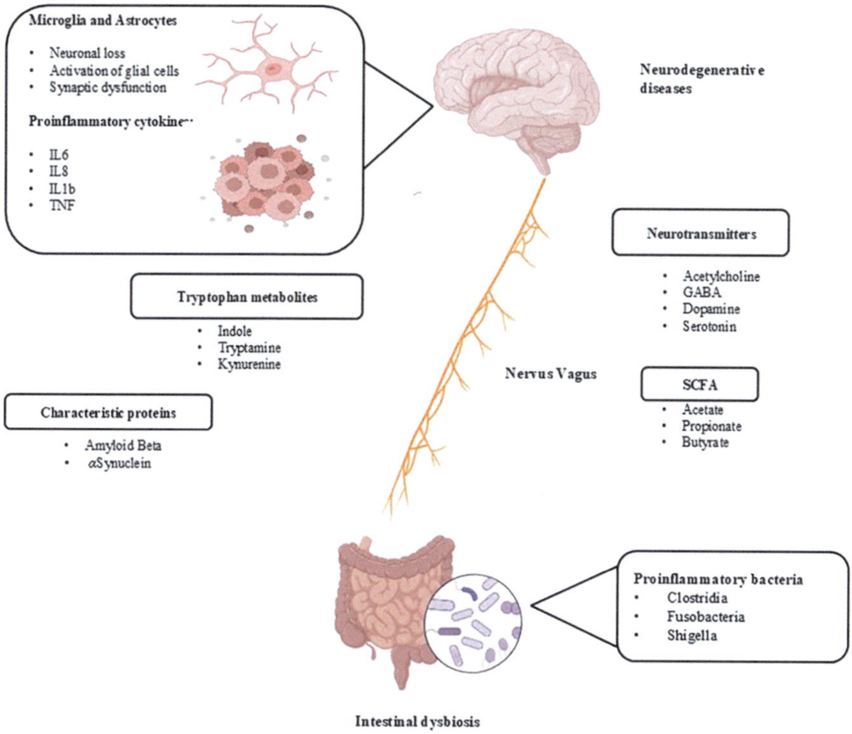

Fig. 2.2 Pathways connecting gut dysbiosis to neurodegeneration. Altered gut microbiota can increase intestinal permeability, promote systemic inflammation, disrupt the blood-brain barrier, activate microglia, and impair vagal signaling. These changes contribute to neuronal damage and neurodegenerative disease progression

2.4 Evidence from Preclinical Models

Preclinical animal studies have been pivotal in demonstrating that gut microbiome dysbiosis is not merely associated with neurodegeneration but may actively initiate and exacerbate it. By using germ-free mice, fecal microbiota transplantation, antibiotic-induced dysbiosis, and gnotobiotic colonization, researchers have shown that altered microbial composition can profoundly affect CNS function, behavior, and neuroinflammatory responses. This section summarizes key findings in AD, PD, and MS, highlighting the causative roles of gut dysbiosis.

2.4.1 Alzheimer's Disease (AD)

Animal models of AD have consistently demonstrated that gut dysbiosis aggravates amyloid pathology and neuroinflammation. For example, germ-free APP/PS1 transgenic mice exhibit significantly reduced cerebral amyloid-β (Aβ) deposition compared to conventionally raised counterparts, indicating that the presence of microbiota promotes Aβ accumulation [62].

FMT from aged wild-type mice or AD patients into young transgenic mice leads to worsened cognitive impairment and increased Aβ burden, implicating microbial metabolites and systemic inflammation in driving disease progression [63]. In contrast, transplantation from healthy donors or treatment with probiotics (e.g., *Lactobacillus plantarum*, *Bifidobacterium breve*) improves spatial memory and reduces plaque deposition [64].

Neurochemical analyses further reveal that gut dysbiosis elevates pro-inflammatory cytokines (IL-1β, TNF-α) and reduces BDNF in the hippocampus, contributing to synaptic degeneration [65]. Additionally, dysbiotic mice show disrupted tryptophan metabolism and impaired cholinergic signaling, further linking microbiota changes to neuronal dysfunction [66].

2.4.2 Parkinson's Disease (PD)

The α-synuclein-overexpressing mouse model has provided valuable insights into gut-brain interactions in PD. These mice develop progressive motor deficits and α-synuclein aggregation in both the enteric nervous system (ENS) and CNS, mimicking human PD pathology. Remarkably, germ-free α-synuclein transgenic mice exhibit attenuated motor symptoms and reduced α-synuclein pathology, which are restored upon colonization with gut microbiota from PD patients, suggesting a causal role of the microbiota in disease progression [67].

PD-associated dysbiosis has been shown to enhance microglial activation and striatum-related neuroinflammation, promoting dopaminergic neuronal loss. SCFAs, particularly butyrate, acetate, and propionate, are altered in PD models and may modulate microglial activation via GPR41/43 signaling, although their exact role (beneficial vs. detrimental) appears to depend on context and disease stage [68].

Importantly, α-synuclein pathology is hypothesized to begin in the gut, with misfolded aggregates propagating to the brain via the vagus nerve, in line with Braak's hypothesis. Experimental studies show that vagotomy reduces α-synuclein spread and brain pathology, strengthening the gut-origin theory of PD [69]. Moreover, antibiotic-induced depletion of the gut microbiota has been shown to reduce α-synuclein aggregation and improve motor function in PD mouse models, further underscoring the influence of gut microbes on neurodegenerative processes [70].

2.4.3 Multiple Sclerosis (MS)

Experimental autoimmune encephalomyelitis (EAE), a widely used MS model, highlights the immunomodulatory role of gut microbiota in demyelinating diseases. Germ-free mice are resistant to EAE induction, whereas colonization with segmented filamentous bacteria (SFB) restores susceptibility by promoting Th17 cell differentiation [71].

In EAE, dysbiosis is characterized by reduced *Lactobacillus* and *Bifidobacterium* and increased *Akkermansia* and *Clostridium*, which are associated with elevated intestinal permeability and systemic inflammation [72]. Probiotic supplementation with *Lactobacillus reuteri* or *Bifidobacterium infantis* ameliorates clinical scores and expands regulatory T cells (Tregs) in the CNS [73].

Behaviorally, dysbiotic EAE mice exhibit increased anxiety-like behaviors and cognitive deficits, correlating with elevated TNF-α and IL-6 expressions in the prefrontal cortex and hippocampus [74]. Moreover, FMT from MS patients into EAE mice exacerbates clinical symptoms and demyelination, while FMT from healthy donors provides protective effects, further implicating gut microbiota in disease modulation [75].

2.4.4 Mechanistic Themes Across Models

Several conserved mechanisms emerge across these neurodegenerative disease models:

- Microglial priming and activation triggered by systemic LPS and microbial metabolites
- Increased gut permeability, allowing pathogen-associated molecular patterns (PAMPs) to enter circulation and stimulate neuroinflammation
- Altered neurotransmitter biosynthesis (e.g., serotonin, dopamine, GABA) due to dysregulated microbial metabolism of tryptophan and other precursors
- Impaired neurogenesis and synaptic plasticity linked to reduced beneficial SCFAs and elevated pro-inflammatory cytokines
- Vagus nerve-mediated propagation of gut-originating pathology to the CNS, particularly evident in PD models

A summary of preclinical studies highlighting microbiota-based interventions and neurodegenerative outcomes is provided in Table 2.2.

Table 2.2 Key preclinical findings linking gut dysbiosis to neurodegenerative disorders

Disease model	Microbiome intervention	Key findings	References
AD	APP/PS1, 3xTg-AD	GF mice show ↓ Aβ plaque burden; FMT from AD ↑ neuroinflammation; probiotics improve cognition	[62, 63, 93, 101, 145]
PD	α-synuclein transgenic	PD microbiota ↑ motor deficits and α-syn aggregation; vagotomy blocks CNS propagation	[10, 67–70, 102]
MS	EAE	GF mice resistant to EAE; SFB ↑ Th17 cells; MS FMT ↑ demyelination; probiotics ↓ inflammation	[72–75, 87, 88]

2.4.5 Limitations and Future Directions

While these animal models provide compelling evidence of a causal role for gut dysbiosis in neurodegeneration, there are significant limitations. Differences in microbial composition and immune responses between mice and humans limit direct translation. Moreover, many studies lack standardization in microbiota manipulation protocols and often neglect sex-specific effects.

Future research should incorporate multi-omics approaches (e.g., metagenomics, metabolomics, transcriptomics) to unravel complex microbe-host interactions. Development of humanized mouse models colonized with patient-derived microbiota and longitudinal behavioral assessments can enhance translational relevance.

Additionally, personalized microbiome-targeted interventions, including customized probiotics, dietary modifications, and precision microbiome editing, represent promising strategies to prevent or mitigate neurodegenerative diseases in humans.

2.5 Clinical Evidence in Human Studies

Understanding the contribution of gut microbiome alterations to neurodegenerative diseases in humans requires robust clinical investigation. Advances in high-throughput sequencing, metabolomics, and integrative multi-omics approaches have revealed distinct microbial signatures and metabolite profiles in patients with AD, PD, MS, and ALS. These discoveries stem from cross-sectional case-control studies, longitudinal cohort analyses, and interventional trials, including probiotic administration and fecal microbiota transplantation.

2.5.1 Alzheimer's Disease (AD)

Multiple case-control studies using 16S rRNA gene sequencing and shotgun metagenomics have shown that patients with AD exhibit reduced abundances of *Firmicutes* and *Bifidobacterium*, along with increased levels of *Bacteroides* and *Escherichia/Shigella* compared to healthy controls [76]. These changes are accompanied by lower levels of SCFAs, especially butyrate, which supports blood-brain barrier integrity and modulates neuroinflammation [45].

A metagenomic analysis from a Chinese AD cohort demonstrated significant enrichment of pro-inflammatory *Proteobacteria* and depletion of anti-inflammatory *Eubacterium rectale* [77]. Dysbiosis in these patients has been associated with elevated plasma levels of LPS, TNF-α, and IL-6, as well as with hippocampal atrophy and cognitive impairment on neuroimaging [78].

Longitudinal evidence further suggests that gut microbial signatures may precede cognitive decline. The Age-Well study reported that individuals harboring higher abundances of SCFA-producing bacteria experienced better cognitive preservation over 5 years [79].

Preliminary interventional data are promising. In a double-blind, randomized trial, 12-week administration of *Lactobacillus* and *Bifidobacterium* strains improved Mini-Mental State Examination (MMSE) scores and reduced oxidative stress biomarkers in AD patients [80].

2.5.2 Parkinson's Disease (PD)

Among neurodegenerative disorders, PD has the most extensive clinical microbiome research. Patients with PD consistently display a decreased abundance of anti-inflammatory bacteria such as *Prevotella* and increased levels of mucin-degrading and endotoxin-producing genera including *Akkermansia, Proteus*, and *Enterobacteriaceae* [81]. These shifts are associated with greater intestinal permeability, systemic inflammation, and more severe motor symptoms [82]. A landmark multi-cohort study involving over 500 PD patients from the USA, Finland, and Germany found that higher abundances of *Lactobacillus* and *Bifidobacterium* were inversely correlated with disease duration and rigidity scores [83]. Metabolomic analyses indicate decreased fecal SCFA levels and reductions in tryptophan-derived indoles in PD patients, impairing intestinal barrier integrity and suppressing anti-inflammatory signaling [84]. Elevated calprotectin and zonulin levels further support gut inflammation and barrier dysfunction. Clinical trials have begun to explore microbial modulation strategies. For example, a 12-week intervention with *Lactobacillus plantarum* PS128 improved motor function (Unified Parkinson's Disease Rating Scale scores) and sleep quality [85]. Additionally, a pilot FMT study in PD patients demonstrated improvements in constipation, non-motor symptoms, and reductions in fecal α-synuclein levels [86].

2.5.3 Multiple Sclerosis (MS)

In MS patients, gut dysbiosis is characterized by decreased microbial diversity and altered abundances of key taxa. Specifically, reductions in *Faecalibacterium prausnitzii* and *Bifidobacterium* species, alongside increases in *Methanobrevibacter* and *Akkermansia*, have been observed across several cohorts [87].

Miyake et al. identified enrichment of *Clostridium* perfringens, a producer of epsilon toxin, which may contribute to demyelination in MS patients [88]. Longitudinal data from the MS Microbiome Consortium show that baseline gut dysbiosis correlates with higher relapse rates and increased MRI lesion load over 2 years [89].

Interventional studies using *Lactobacillus reuteri*, *Bifidobacterium longum*, and prebiotics (e.g., resistant starch) have reported modest reductions in serum IL-17 and C-reactive protein (CRP), alongside improvements in fatigue and quality-of-life metrics [90].

2.5.4 Amyotrophic Lateral Sclerosis (ALS)

Although less extensively studied, emerging evidence suggests that gut microbiome alterations may also play a role in ALS progression. ALS patients show reduced abundances of *Butyrivibrio* and *Eubacterium*, with increases in *Ruminococcus* and *Dorea* [91]. These shifts correspond to reduced SCFA levels and elevated gut-derived endotoxins, which may exacerbate blood-brain barrier permeability and contribute to motor neuron damage [41].

A small-scale human FMT trial in ALS patients demonstrated mild symptomatic improvements and reduced fecal LPS levels, encouraging further investigation in larger studies [92].

2.5.5 Integrative Multi-omics Evidence

Recent integrative multi-omics studies have elucidated systemic links between gut microbiota alterations and host neuroimmune signaling. Combining metagenomic, transcriptomic, proteomic, and metabolomic data, these approaches provide a holistic view of gut-brain interactions in neurodegeneration. For example, a 2023 study in AD patients linked *Bacteroides fragilis* abundance with higher plasma TNF-α levels, altered kynurenine/tryptophan ratios, and reduced cognitive scores [93]. In PD, metabolomic profiling reveals downregulation of SCFA biosynthesis and bile acid metabolism pathways, correlating with worsening rigidity and depression symptoms [94]. A comprehensive summary of key clinical findings and microbiota-based associations with neurodegenerative features is provided in Table 2.3.

2.5.6 Limitations and Considerations

Although clinical evidence strongly supports a gut-brain connection in neurodegenerative diseases, several challenges limit its translational impact. Many studies are cross-sectional in nature, which restricts causal inference. Variability in microbiota composition driven by geography, diet, age, and medication further complicates standardization. Additionally, interventional approaches such as FMT and probiotics lack uniform protocols, and data on strain-specific efficacy remain limited.

Future clinical research should prioritize:

- Large, multicenter longitudinal studies with diverse cohorts
- Standardized probiotic trials using well-defined microbial consortia
- Incorporation of multi-omics biomarkers and advanced machine learning models to predict disease progression and therapeutic response
- Personalized intervention strategies tailored to individual microbiome profiles

Table 2.3 Summary of clinical trials and cohort studies linking gut microbiome alterations to neurodegeneration

Disease	Study type	Sample size	Key findings	References
AD	Case control (China)	70 AD vs. 70 controls	↑ *Bacteroides*, ↓ *Bifidobacterium*, ↑ LPS, ↓ SCFAs	[78, 93, 95]
AD	RCT	$n = 52$	Probiotics improved MMSE, ↓ oxidative stress	[80, 93, 147]
PD	Multi-center cohort	>500	↓ *Prevotella*, ↑ *Akkermansia* correlated with symptoms	[67, 68, 83, 84]
PD	RCT (*L. plantarum*)	$n = 60$	Improved UPDRS, sleep, reduced IL-6	[85, 86]
MS	Longitudinal cohort	$n = 150$	Dysbiosis linked to relapses and MRI lesions	[87–89]
MS	Probiotic trial	$n = 40$	↓ IL-17, CRP; improved fatigue and QoL	[74, 85, 87]
ALS	FMT pilot	$n = 12$	Mild improvement in GI symptoms and LPS	[92, 113, 114]

2.6 Microbiome Changes in Specific Neurodegenerative Diseases

The gut microbiota composition and its metabolic outputs differ distinctly across neurodegenerative diseases, indicating disease-specific dysbiosis that contributes to diverse pathophysiological mechanisms. Characterizing these unique microbial alterations can help elucidate disease pathways, provide diagnostic biomarkers, and support the development of targeted microbiome-based therapies. This section discusses alterations in AD, PD, ALS, and HD, emphasizing links to neuroinflammation, neurotransmitter imbalance, and disease progression.

2.6.1 Alzheimer's Disease (AD)

AD is characterized by progressive cognitive decline, extracellular amyloid-beta (Aβ) plaques, and intracellular tau tangles. Recent evidence implicates gut dysbiosis as a significant contributor to systemic inflammation and metabolic disturbances that exacerbate AD pathology.

Key Microbial Alterations in AD

- Reduced overall microbial diversity (alpha diversity) [95]
- Increased abundance of pro-inflammatory taxa such as *Bacteroides*, *Escherichia/Shigella*, *Proteobacteria*, and *Desulfovibrio* [96]
- Decreased abundance of anti-inflammatory genera including *Firmicutes, Bifidobacterium, Eubacterium*, and *Ruminococcaceae* [97]

These compositional shifts increase intestinal permeability, enabling endotoxins like lipopolysaccharide (LPS) to enter circulation and promote chronic neuroinflammation [98]

Mechanistic Insights
LPS from Gram-negative bacteria triggers peripheral cytokine production (e.g., IL-1β, TNF-α), which can cross a compromised blood-brain barrier and activate microglia, leading to synaptic dysfunction and neuronal death [99]. Butyrate deficiency impairs regulatory T cell (Treg) differentiation, reduces tight junction protein expression at the BBB, and diminishes neuroprotective mechanisms [45]. Additionally, gut-derived amyloids (e.g., Curli proteins) from *E. coli* may cross-seed with Aβ in the brain, accelerating plaque formation [100].

FMT from AD patients into germ-free mice induces memory deficits, hippocampal inflammation, and increased amyloid deposition, supporting a causal role of gut microbiota in AD [101].

2.6.2 Parkinson's Disease (PD)

PD involves progressive dopaminergic neuron loss in the substantia nigra and widespread α-synuclein aggregation. Braak's hypothesis suggests that PD pathology may originate in the enteric nervous system and ascend to the brain via the vagus nerve [102].

Microbiome Signatures in PD

- Increased abundance of *Akkermansia muciniphila, Proteobacteria, Enterobacteriaceae*, and *Lactobacillus* [103]
- Decreased abundance of *Prevotella, Blautia, Faecalibacterium,* and *Roseburia*—all-important SCFA producers [104]

Prevotella depletion impairs mucin synthesis, increasing gut barrier permeability and systemic inflammation [51]. Elevated *Enterobacteriaceae* levels correlate with motor severity and systemic endotoxemia.

Mechanistic Insights

Reduced SCFA levels diminish anti-inflammatory signaling and Treg-mediated immune regulation [12]. Elevated fecal calprotectin and zonulin suggest persistent gut barrier dysfunction and local inflammation [105].

Certain bacterial strains, such as *Desulfovibrio* and *Enterococcus faecalis*, may promote α-synuclein aggregation through metabolites and extracellular amyloid-like proteins [106]. Longitudinal studies suggest that these microbial shifts may precede motor symptoms, providing a potential diagnostic window [107].

Clinical trials using probiotics (e.g., *Lactobacillus plantarum* PS128) have demonstrated improvements in motor and non-motor symptoms, suggesting gut microbiota as a therapeutic target [108].

2.6.3 Amyotrophic Lateral Sclerosis (ALS)

ALS involves degeneration of upper and lower motor neurons, resulting in progressive paralysis. Though its exact etiology remains unclear, gut dysbiosis is emerging as a potential modifier of disease severity and progression.

Microbiome Alterations in ALS

- Increased abundance of pro-inflammatory taxa such as *Dorea*, *Ruminococcus*, *Clostridium*, and *Desulfovibrio* [109]
- Decreased abundance of butyrate-producing genera including *Butyrivibrio*, *Lachnospira*, and *Faecalibacterium prausnitzii* [110]

These shifts foster a pro-inflammatory environment, systemic endotoxemia, and compromised SCFA-mediated gut-brain signaling.

Mechanistic Insights

SCFA and bile acid deficiencies impair gut and BBB integrity, facilitating LPS translocation and NF-κB activation, which drives IL-6 and TNF-α production [111]. Disturbances in gut-derived NAD+ biosynthesis can further impair neuronal energy metabolism, potentially accelerating neurodegeneration [112].

Notably, *Akkermansia muciniphila* supplementation in ALS mouse models delayed disease onset and extended survival, supporting microbiota modulation as a therapeutic strategy [113].

Although human data are limited, early studies suggest that FMT and probiotics may reduce gastrointestinal symptoms and systemic inflammation in ALS patients [114].

2.6.4 Huntington's Disease (HD)

HD is a monogenic disorder caused by CAG repeat expansions in the huntingtin gene, leading to striatal neurodegeneration and progressive motor and cognitive impairment.

Microbiome Alterations in HD

- Decreased overall microbial diversity [115]
- Increased abundance of *Clostridium* cluster XIVa and *Erysipelotrichaceae* [116]
- Decreased abundance of *Lactobacillus, Bifidobacterium*, and *Butyrivibrio* [117]

HD patients exhibit abnormal fecal metabolic signatures, including reduced SCFA and altered bile acid profiles, affecting mitochondrial energetics and neuroinflammation.

Mechanistic Insights
Neurotoxic metabolites from *Clostridium* species can exacerbate mitochondrial dysfunction and neuronal apoptosis [118]. Altered tryptophan metabolism, with an increased kynurenine/tryptophan ratio, may exacerbate excitotoxicity via NMDA receptor hyperactivation [119].

Animal studies demonstrate that HD mice with dysbiotic microbiota have worse motor outcomes and increased neuroinflammatory markers compared to germ-free or antibiotic-treated mice [120]. These findings suggest that, despite its monogenic nature, HD progression is modulated epigenetically via gut microbiota interactions.

2.7 The Role of Microbial Metabolites and Neurotoxins in Neuronal and Glial Cell Function

The gut-brain axis is a bidirectional communication network between the gastrointestinal tract and the CNS, mediated partly by microbial metabolites and neurotoxins. These molecules, including SCFAs, LPS, ammonia, D-lactate, tryptophan derivatives, and GABA, profoundly affect neuronal and glial cell function, influencing brain health and disease progression.

This section provides an in-depth analysis of how these microbial-derived compounds modulate CNS activity, highlighting their dual roles as neuroprotective or neurotoxic agents.

2.7.1 Short-Chain Fatty Acids (SCFAs)

SCFAs, such as acetate, propionate, and butyrate, are produced through the anaerobic fermentation of dietary fibers by gut bacteria. These metabolites play critical roles in CNS homeostasis.

- Neurons: SCFAs enhance synaptic plasticity by regulating BDNF, a key regulator of neuronal survival and connectivity [30]. Butyrate also modulates histone deacetylase (HDAC) activity, leading to epigenetic changes that promote neurogenesis and protect against neurodegeneration [121].
- Glial Cells: SCFAs suppress microglial activation and reduce pro-inflammatory cytokines, such as IL-6 and TNF-α, thereby mitigating neuroinflammation and supporting glial cell homeostasis [122].

2.7.2 Lipopolysaccharides (LPS)

LPS, derived from Gram-negative bacterial outer membranes, is a potent inducer of systemic and neuroinflammation.

- Neurons: LPS disrupts blood-brain barrier integrity, allowing infiltration of peripheral immune cells and inflammatory mediators, ultimately causing neuronal apoptosis and cognitive impairment [99].
- Glial Cells: LPS activates microglia via TLR4, leading to the release of IL-1β, IL-6, and TNF-α. Chronic activation promotes neurodegenerative pathology, particularly in Alzheimer's disease [123].

2.7.3 Ammonia

Ammonia, a byproduct of protein metabolism by gut bacteria, can accumulate in the brain under conditions of liver dysfunction or gut dysbiosis.

- Neurons: Elevated ammonia disrupts glutamate metabolism, induces excitotoxicity, impairs mitochondrial function, and increases oxidative stress, contributing to neuronal injury [124].
- Glial Cells: Astrocytes detoxify ammonia by converting it to glutamine; however, excessive ammonia overwhelms this pathway, resulting in astrocyte swelling and dysfunction—hallmarks of hepatic encephalopathy [125].

2.7.4 D-Lactate

Additional gut-derived metabolites, such as tryptophan derivatives and GABA, also affect CNS function.

- Neurons: High D-lactate levels impair mitochondrial function, increase oxidative stress, and disrupt neurotransmission, leading to cognitive dysfunction and neuronal loss [126].
- Glial Cells: D-lactate promotes microglial activation and the release of reactive oxygen species (ROS), exacerbating neuroinflammation and neuronal injury [127].

Table 2.4 Microbial metabolites and their neuroactive effects on neurons and glial cells

Metabolite	Source	Effect on neurons	Effect on glial cells	References
SCFAs	Fermentation of dietary fibers	Enhances synaptic plasticity, increases BDNF	Reduces microglial activation, anti-inflammatory	[22, 30, 45, 122]
LPS	Gram-negative bacteria	Induces neuronal apoptosis, disrupts BBB	Activates microglia, pro-inflammatory	[35, 42, 54, 99]
Ammonia	Protein metabolism	Causes excitotoxicity, impairs mitochondria	Overwhelms astrocyte detoxification	[124, 125]
D-lactate	Bacterial overgrowth	Impairs mitochondrial function, oxidative stress	Activates microglia, increases ROS	[126, 127]
Tryptophan derivatives	Tryptophan metabolism	Modulates mood (serotonin), neurotoxicity (kynurenine)	Influences neuroinflammation	[32, 119, 128]
GABA	Gut bacteria	Inhibitory neurotransmitter, reduces excitability	Modulates glial signaling	[33, 34, 129]

2.7.5 Other Microbial Metabolites

Additional gut-derived metabolites, such as tryptophan derivatives and GABA, also affect CNS function.

- Tryptophan Derivatives: Gut bacteria metabolize tryptophan into serotonin and kynurenine. While serotonin modulates mood and cognition, kynurenine crosses the BBB and induces neurotoxicity via quinolinic acid, an NMDA receptor agonist [128].
- GABA: Produced by certain gut bacteria, GABA acts as an inhibitory neurotransmitter, reducing neuronal excitability and modulating anxiety and stress responses [129].

A summary of key microbial metabolites and their neuroactive effects on neurons and glial cells is presented in Table 2.4, which clearly illustrates their dual protective and detrimental roles.

Microbial metabolites and neurotoxins play a complex, context-dependent role in brain health. While SCFAs generally support neuroprotection and anti-inflammatory responses, compounds such as LPS, ammonia, and D-lactate can exacerbate neuroinflammation and degeneration. Understanding these mechanisms is critical for designing therapeutic strategies that mitigate harmful effects while enhancing beneficial pathways. Future research should explore targeted microbiome modulation, precision metabolite therapies, and personalized interventions for neurological disorders, including Alzheimer's disease, Parkinson's disease, and depression.

2.8 The Influence of Diet, Antibiotics, and Lifestyle on the Microbiome and Neurodegeneration Risk

The gut microbiome is a dynamic ecosystem profoundly shaped by external factors, including diet, antibiotic use, and lifestyle. Modern dietary patterns, widespread antibiotic exposure, and chronic stress disrupt the composition and function of gut microbiota, contributing to dysbiosis and increasing neurodegenerative disease risk. This section explores how these factors modulate the microbiome and influence brain health.

2.8.1 Modern Diets and the Microbiome

The Western diet, rich in processed foods, refined sugars, and saturated fats, is strongly linked to gut microbial imbalances.

- High-Fat and High-Sugar Diets: These diets reduce microbial diversity and promote pro-inflammatory taxa, such as *Firmicutes*, while suppressing beneficial *Bacteroidetes* [130]. This imbalance increases intestinal permeability, allowing translocation of microbial metabolites and endotoxins (LPS) into the bloodstream, which triggers systemic and neuroinflammation [40].
- Low-Fiber Diets: Dietary fiber is essential to produce SCFAs, which maintain gut barrier integrity and exert neuroprotective effects. A fiber-deficient diet reduces SCFA levels, compromising gut and brain health [45].

2.8.2 Antibiotics and the Microbiome

Antibiotics, while essential for treating bacterial infections, have broad-spectrum effects that can disrupt the gut microbiota.

- Short-Term Effects: Antibiotic use reduces microbial diversity and alters key bacterial taxa, leading to temporary or persistent dysbiosis. These changes can last weeks to months depending on antibiotic type and duration [131].
- Long-Term Consequences: Repeated or prolonged antibiotic exposure results in sustained loss of beneficial bacteria such as *Lactobacillus* and *Bifidobacterium*, which maintain barrier function and modulate immune responses [132]. The resulting dysbiosis elevates systemic and neuroinflammatory risk, contributing to neurodegenerative processes.

2.8.3 Lifestyle Factors and the Microbiome

Lifestyle factors, including stress, sleep deprivation, and physical inactivity, also influence the gut microbiota and brain health.

- Chronic Stress: Stress activates the hypothalamic-pituitary-adrenal (HPA) axis, increasing cortisol and other stress hormones that disrupt microbial composition. This dysbiosis can impair gut barrier function, promote systemic inflammation, and directly initiate neuroinflammation and neurodegeneration [133, 134].
- Sleep Deprivation: Poor sleep quality and duration reduce microbial diversity and disrupt gut circadian rhythms, impairing metabolic and immunomodulatory functions [135].
- Physical Inactivity: Sedentary behavior decreases microbial diversity and compromises gut barrier integrity. In contrast, regular exercise enhances beneficial bacteria and supports SCFA production, promoting gut and brain health [136].

2.8.4 Mechanisms Linking Microbiome Dysbiosis to Neurodegeneration

Dysbiosis, induced by diet, antibiotics, and lifestyle factors, promotes neurodegeneration through multiple mechanisms:

1. Neuroinflammation: Increased gut permeability enables microbial metabolites and endotoxins, such as LPS, to enter circulation and activate neuroinflammatory pathways [99].
2. Oxidative Stress: Dysbiosis promotes reactive oxygen species (ROS) production, impairing mitochondrial function and causing neuronal damage [55].
3. Impaired Neurogenesis: Reduced SCFA levels and increased inflammation decrease BDNF, disrupting neurogenesis and synaptic plasticity [137].

2.9 Emerging Therapeutic Approaches Targeting the Microbiome

The gut microbiome is a promising therapeutic target for neurodegenerative, metabolic, and inflammatory diseases. Recent advances have enabled novel interventions to restore microbial balance and support brain health. Table 2.5 summarizes the key microbiota-targeted therapeutic strategies—such as probiotics, prebiotics, dietary interventions, FMT, postbiotics, and precision microbiome editing—along with their mechanisms, applications, and clinical trial status. This section outlines these strategies in more detail, including their mechanisms of action, therapeutic potential, and associated challenges.

2.9.1 Probiotics

Probiotics are live microorganisms that provide health benefits when administered in adequate amounts.

Table 2.5 Therapeutic interventions and trial status

Intervention	Mechanism	Applications	Trial status	References
Probiotics	Enhance gut barrier, modulate immunity	IBS, IBD, neurodegenerative disorders	Phase II/III trials	[65, 73, 80, 91, 147, 158]
Prebiotics	Stimulate SCFA production	Metabolic health, cognitive function	Phase II trials	[65, 140, 147]
FMT	Restore microbial diversity	*C. difficile* infections, IBD, neurological disorders	Phase III trials	[86, 92, 143]
Dietary interventions	Promote beneficial bacteria, reduce inflammation	Metabolic syndrome, neurodegenerative diseases	Observational and interventional studies	[97, 130, 136, 145]
Postbiotics	Anti-inflammatory, immunomodulatory effects	Metabolic disorders, neurodegenerative diseases	Preclinical and early-phase trials	[146, 147]
Precision microbiome editing	Targeted gene editing in bacteria	Antibiotic-resistant infections, metabolic disorders	Preclinical studies	[148, 155]

- Mechanisms: Enhance gut barrier integrity, produce antimicrobial compounds, and modulate immune responses. *Lactobacillus* and *Bifidobacterium* strains reduce inflammation and support gut health [138].
- Applications: Investigated for irritable bowel syndrome, inflammatory bowel disease, and neurodegenerative conditions such as AD [139].
- Challenges: Strain-specific effects, individual variability, and limited long-term data.

2.9.2 Prebiotics

Prebiotics are nondigestible food components that selectively stimulate beneficial gut bacteria.

- Mechanisms: Serve as substrates for beneficial bacteria, supporting SCFA production and anti-inflammatory effects [140]
- Applications: Explored for metabolic health, cognitive enhancement, and neurodegenerative disease prevention [141]
- Challenges: Individual variability in responses and optimal dosing

2.9.3 Fecal Microbiota Transplantation (FMT)

FMT involves transferring fecal material from a healthy donor to restore microbial balance in the recipient.

- Mechanisms: Replaces dysbiotic microbiota with a diverse, healthy community, correcting gut-related pathologies [142]
- Applications: Effective in recurrent *Clostridioides difficile* infections; investigation for IBD, metabolic syndromes, and neurological disorders [143]
- Challenges: Safety concerns (e.g., pathogen transmission), long-term effects, and regulatory hurdles

2.9.4 Dietary Interventions

Dietary modifications offer a noninvasive strategy to modulate the microbiome.

- Mechanisms: Diets rich in fiber, polyphenols, and omega-3 fatty acids enhance beneficial bacteria and support SCFA production, reducing inflammation and oxidative stress [144].
- Applications: Mediterranean and ketogenic diets are being studied for gut and brain health benefits and for mitigating neurodegenerative risk [145].
- Challenges: Adherence and individualized responses.

2.9.5 Postbiotics

Postbiotics are bioactive compounds produced during probiotic fermentation, including SCFAs, enzymes, and peptides.

- Mechanisms: Exert anti-inflammatory, immunomodulatory, and antimicrobial effects [146]
- Applications: Explored for metabolic, inflammatory, and neurodegenerative disorders [147]
- Challenges: Standardization and understanding complex mechanisms

2.9.6 Precision Microbiome Editing

Precision microbiome editing uses advanced tools such as CRISPR/Cas systems to modify gut microbial genes.

- Mechanisms: Enables targeted deletion or modification of bacterial genes, allowing precise modulation of microbial functions [148]
- Applications: Investigated for antibiotic-resistant infections, metabolic syndromes, and neurodegenerative diseases [149]
- Challenges: Ethical issues, off-target effects, and ecological complexity

2.10 Limitations, Challenges, and Future Directions in Microbiome Research

Although the gut microbiome has emerged as a critical player in health and disease, substantial limitations and challenges remain in fully understanding its complex roles. Overcoming these challenges is essential to advance microbiome-based therapies and ensure successful clinical translation. This section discusses key obstacles—including causality, inter-individual variability, small sample sizes, and reproducibility—and explores promising future directions such as omics integration and AI-based models.

2.10.1 Causality

Establishing causal relationships between the gut microbiome and specific diseases remains a fundamental challenge.

- Correlation vs. Causation: Many human studies rely on observational designs, which can only identify associations without proving direct causation [150].
- Animal Models: While valuable for mechanistic insights, animal microbiomes differ significantly from humans, limiting translational relevance [151].
- Future Directions: Randomized controlled trials (RCTs), interventional studies, and Mendelian randomization approaches are needed to validate causality and identify specific microbial strains or metabolites that drive disease. Personalized interventional trials, designed with individual microbial profiles in mind, will further strengthen causal inferences.

2.10.2 Inter-individual Variability

The gut microbiome varies markedly among individuals due to genetics, diet, lifestyle, and environmental exposures.

- Personalized Approaches: One-size-fits-all interventions, such as probiotics or fecal microbiota FMT, may not be effective for all individuals due to this variability [152].
- Future Directions: Precision microbiome therapies, tailored to an individual's unique microbial profile, hold promise for improving efficacy and reducing adverse effects.

2.10.3 Sample Size Issues

Many microbiome studies suffer from small sample sizes, limiting their statistical power and generalizability.

- Power and Reproducibility: Small studies are prone to false positives and may fail to replicate in larger cohorts [153].
- Future Directions: Large-scale, multicenter studies with standardized protocols are needed to enhance reproducibility and validate findings across diverse populations.

2.10.4 Reproducibility

Reproducibility remains a significant challenge in microbiome research due to variations in experimental design, sample collection, and data analysis.

- Standardization: Lack of standardized protocols for sample collection, storage, and sequencing can lead to inconsistent results [154].
- Future Directions: Developing universal guidelines and open-access databases for microbiome research can improve transparency and reproducibility.

2.10.5 Omics Integration

Integrating multi-omics data (e.g., metagenomics, metabolomics, proteomics) is essential for a comprehensive understanding of microbiome-host interactions.

- Holistic Insights: Combining omics data can reveal complex interactions between microbial communities, host metabolism, and immune responses [155].
- Future Directions: Advanced computational tools and pipelines for omics integration will enable the identification of novel biomarkers and therapeutic targets.

2.10.6 AI-Based Models

Artificial intelligence (AI) and machine learning offer powerful tools for analyzing complex microbiome data and predicting disease outcomes.

- Predictive Modeling: AI algorithms can identify patterns in large datasets, enabling the prediction of disease risk and treatment responses based on microbial profiles [156].
- Future Directions: Developing AI-driven platforms for personalized microbiome analysis and intervention design will enhance precision medicine approaches.

Addressing current limitations in microbiome research is crucial for advancing our understanding of the gut-brain axis and developing effective, individualized therapies. Priorities include establishing causality through robust trials, embracing inter-individual variability with precision approaches, increasing sample sizes for stronger evidence, and improving reproducibility through standardized methods.

Looking forward, integrating multi-omics data and harnessing AI-based predictive models hold tremendous promises for overcoming these challenges and unlocking the full therapeutic potential of the microbiome. Collaborative, interdisciplinary efforts will be essential to translate microbiome science from bench to bedside, ultimately shaping the future of personalized medicine and neurodegenerative disease prevention.

2.11 Conclusion

The gut microbiome has emerged as a central player in the pathogenesis of neurodegenerative diseases such as AD, PD, ALS, and HD. Microbial metabolites and neurotoxins, including SCFAs, LPS, ammonia, and D-lactate, through the gut-brain axis, significantly affect neuronal and glial function. These compounds influence neuroinflammation, oxidative stress, and synaptic plasticity. Dysbiosis, characterized by reduced microbial diversity and compositional shifts, disrupts these processes and accelerates neurodegeneration. Modern diets, antibiotic use, and lifestyle factors further exacerbate dysbiosis and increase neurodegenerative disease risk. High-fat, high-sugar diets and low-fiber intake reduce SCFA production, impair gut barrier integrity, and promote systemic inflammation. Chronic stress and sleep deprivation alter microbial composition and compromise mucosal defenses. Although antibiotics are essential for treating infections, they can induce long-term dysbiosis, ultimately affecting brain health. Emerging microbiome-targeted therapies, including probiotics, prebiotics, FMT, dietary interventions, postbiotics, and precision microbiome editing, offer promising avenues to restore microbial balance and mitigate neurodegeneration. However, inter-individual variability, lack of standardization, and safety concerns must be addressed to optimize these interventions and improve their translational success. Future research should prioritize establishing causal relationships between microbiome changes and neurodegenerative disease progression. Integrating multi-omics data, including metagenomics, metabolomics, and transcriptomics, will provide comprehensive insights into host-microbe interactions. Furthermore, leveraging artificial intelligence (AI) and machine learning can enable personalized disease risk predictions and therapeutic responses. Large-scale longitudinal studies and multi-center clinical trials are essential to validate and translate findings into clinical practice. In conclusion, the gut microbiome is not merely a passive bystander but an active, dynamic, and modifiable factor in neurodegeneration. Deciphering its role in disease pathogenesis opens new avenues for early diagnosis, prevention, and treatment. By strategically targeting the gut-brain axis, we may unlock novel therapeutic strategies to delay or even prevent neurodegenerative diseases, ultimately improving the quality of life for millions worldwide.

Declarations

Competing Interests The author(s) has no competing interests to declare that are relevant to the content of this manuscript.

Ethics Approval The author(s) declares that this manuscript does not include primary research with human participants.

The author(s) declares that this manuscript does not include primary studies involving animals.

References

1. Mishra A, Handa AS, Handa AK. Intricate interplay of bacteria, virus & our immune system. EPH-Int J Med Health Sci. 2025;11(1):9–19.
2. Bakshi I, et al. Exploring the gut-brain axis: a comprehensive review of interactions between the gut microbiota and the central nervous system. Int J Multidiscip Res. 2024;6(3):1–15.
3. Lu S, et al. The communication mechanism of the gut-brain axis and its effect on central nervous system diseases: a systematic review. Biomed Pharmacother. 2024;178:117207.
4. Maiuolo J, et al. The contribution of gut microbiota–brain axis in the development of brain disorders. Front Neurosci. 2021;15:616883.
5. Padhi P, et al. Mechanistic insights into gut microbiome dysbiosis-mediated neuroimmune dysregulation and protein misfolding and clearance in the pathogenesis of chronic neurodegenerative disorders. Front Neurosci. 2022;16:836605.
6. Pathak N, et al. Neurodegenerative disorders of Alzheimer, Parkinsonism, amyotrophic lateral sclerosis and multiple sclerosis: an early diagnostic approach for precision treatment. Metab Brain Dis. 2022;37:67–104.
7. Tag SH, et al. Neuropathological and behavioral features of an APP/PS1/MAPT (6xTg) transgenic model of Alzheimer's disease. Mol Brain. 2022;15(1):51.
8. Ferguson D. AI-driven integrative design of ADAMTS partial agonists within a multi hallmark therapeutic framework for neurodegenerative disorders. 2025. https://doi.org/10.26434/chemrxiv-2025-4n2nb-v4.
9. Tabatabaee M. Microbiome-glia crosstalk: bridging the communication divide in the central nervous system. Neuroglia. 2024;5(2):89–104.
10. Fitzgerald E, Murphy S, Martinson HA. Alpha-synuclein pathology and the role of the microbiota in Parkinson's disease. Front Neurosci. 2019;13:369.
11. Onisiforou A, Charalambous EG, Zanos P. Shattering the amyloid illusion: the microbial enigma of Alzheimer's disease pathogenesis—from gut microbiota and viruses to brain biofilms. Microorganisms. 2025;13(1):90.
12. Aho VT, et al. Relationships of gut microbiota, short-chain fatty acids, inflammation, and the gut barrier in Parkinson's disease. Mol Neurodegener. 2021;16:1–14.
13. Missiego-Beltrán J, Beltrán-Velasco AI. The role of microbial metabolites in the progression of neurodegenerative diseases—therapeutic approaches: a comprehensive review. Int J Mol Sci. 2024;25(18):10041.
14. Kołodziej J, Krzemińska P. The wheel of microbiome rises, and it falls: the role and results of microbiome transplantation and the recent reports. Qual Sport. 2024;15:53020.
15. Dessì A, et al. Integrative multiomics approach to skin: the sinergy between individualised medicine and futuristic precision skin care? Metabolites. 2024;14(3):157.
16. Liu P, et al. Crosstalk between the gut microbiota and the brain: an update on neuroimaging findings. Front Neurol. 2019;10:883.
17. Mohammadi-Shemirani P, Sood T, Paré G. From 'omics to multi-omics technologies: the discovery of novel causal mediators. Curr Atheroscler Rep. 2023;25(2):55–65.
18. Colella M, et al. Microbiota revolution: how gut microbes regulate our lives. World J Gastroenterol. 2023;29(28):4368.
19. Rodríguez JM, et al. The composition of the gut microbiota throughout life, with an emphasis on early life. Microb Ecol Health Dis. 2015;26(1):26050.
20. Jandhyala SM, et al. Role of the normal gut microbiota. World J Gastroenterol: WJG. 2015;21(29):8787.

21. Syromyatnikov M, et al. Characteristics of the gut bacterial composition in people of different nationalities and religions. Microorganisms. 2022;10(9):1866.
22. Miquel S, et al. Identification of metabolic signatures linked to anti-inflammatory effects of Faecalibacterium prausnitzii. MBio. 2015;6(2):e00300–15. https://doi.org/10.1128/mBio.00300-15.
23. Young GR, et al. Bacteriophage communities are a reservoir of unexplored microbial diversity in neonatal health and disease. Curr Opin Microbiol. 2023;75:102379.
24. Pronovost GN, Hsiao EY. Perinatal interactions between the microbiome, immunity, and neurodevelopment. Immunity. 2019;50(1):18–36.
25. Koppel N, Balskus EP. Exploring and understanding the biochemical diversity of the human microbiota. Cell Chem Biol. 2016;23(1):18–30.
26. Topping DL, Clifton PM. Short-chain fatty acids and human colonic function: roles of resistant starch and nonstarch polysaccharides. Physiol Rev. 2001;81:1031.
27. Yan M, et al. Gut liver brain axis in diseases: the implications for therapeutic interventions. Signal Transduct Target Ther. 2023;8(1):443.
28. Uebanso T, et al. Functional roles of B-vitamins in the gut and gut microbiome. Mol Nutr Food Res. 2020;64(18):2000426.
29. Oleskin AV, Shenderov BA, Rogovsky VS. Role of neurochemicals in the interaction between the microbiota and the immune and the nervous system of the host organism. Probiotics Antimicrob Proteins. 2017;9(3):215–34.
30. Guo B, et al. Gut microbiota-derived short chain fatty acids act as mediators of the gut–brain axis targeting age-related neurodegenerative disorders: a narrative review. Crit Rev Food Sci Nutr. 2025;65(2):265–86.
31. Kearns R. Gut–brain axis and neuroinflammation: the role of gut permeability and the kynurenine pathway in neurological disorders. Cell Mol Neurobiol. 2024;44(1):64.
32. Tan Y-Q, et al. Host/microbiota interactions-derived tryptophan metabolites modulate oxidative stress and inflammation via aryl hydrocarbon receptor signaling. Free Radic Biol Med. 2022;184:30–41.
33. Dicks LM. Gut bacteria and neurotransmitters. Microorganisms. 2022;10(9):1838.
34. Hwang YK, Oh JS. Interaction of the vagus nerve and serotonin in the gut–brain axis. Int J Mol Sci. 2025;26(3):1160.
35. Burgueño JF, Abreu MT. Epithelial Toll-like receptors and their role in gut homeostasis and disease. Nat Rev Gastroenterol Hepatol. 2020;17(5):263–78.
36. Lee E, Chang Y. Modulating neuroinflammation as a prospective therapeutic target in Alzheimer's disease. Cells. 2025;14(3):168.
37. Vidya Bernhardt G, et al. Probiotics—role in alleviating the impact of alcohol liver disease and alcohol deaddiction: a systematic review. Front Nutr. 2024;11:1372755.
38. He Y, et al. Microbiota–gut–brain axis in health and neurological disease: interactions between gut microbiota and the nervous system. J Cell Mol Med. 2024;28(18):e70099.
39. Toader C, et al. Decoding neurodegeneration: a review of molecular mechanisms and therapeutic advances in Alzheimer's, Parkinson's, and ALS. Int J Mol Sci. 2024;25(23):12613.
40. Chmielarz M, Sobieszczańska B, Środa-Pomianek K. Metabolic endotoxemia: from the gut to neurodegeneration. Int J Mol Sci. 2024;25(13):7006.
41. Beltran-Velasco AI, Clemente-Suárez VJ. Impact of peripheral inflammation on blood–brain barrier dysfunction and its role in neurodegenerative diseases. Int J Mol Sci. 2025;26(6):2440.
42. Kumar S, Sharma V, Yadav S. TLR4 targeting: a promising therapeutic approach across multiple human diseases. Curr Protein Pept Sci. 2025;26(4):241–58.
43. Rauf A, et al. Neuroinflammatory markers: key indicators in the pathology of neurodegenerative diseases. Molecules. 2022;27(10):3194.
44. Kulcsarova K, et al. Pesticides and the microbiome-gut-brain axis: convergent pathways in the pathogenesis of Parkinson's disease. J Parkinsons Dis. 2023;13(7):1079–106.
45. Fock E, Parnova R. Mechanisms of blood–brain barrier protection by microbiota-derived short-chain fatty acids. Cells. 2023;12(4):657.

46. Yang R, et al. Blood–brain barrier integrity damage in bacterial meningitis: the underlying link, mechanisms, and therapeutic targets. Int J Mol Sci. 2023;24(3):2852.
47. Yu S, et al. Revealing the mechanisms of blood–brain barrier in chronic neurodegenerative disease: an opportunity for therapeutic intervention. Rev Neurosci. 2024;35(8):895–916.
48. Hasan A, et al. An updated and comprehensive review exploring the gut–brain axis in neurodegenerative disorders and neurotraumas: implications for therapeutic strategies. Brain Sci. 2025;15(6):654.
49. Caetano-Silva ME, et al. Inhibition of inflammatory microglia by dietary fiber and short-chain fatty acids. Sci Rep. 2023;13(1):2819.
50. Kumar M, Arora P, Sandhir R. Hydrogen sulfide reverses LPS-induced behavioral deficits by suppressing microglial activation and promoting M2 polarization. J Neuroimmune Pharmacol. 2021;16:483–99.
51. Di Vincenzo F, et al. Gut microbiota, intestinal permeability, and systemic inflammation: a narrative review. Intern Emerg Med. 2024;19(2):275–93.
52. Geloso MC, D'Ambrosi N. Microglial pruning: relevance for synaptic dysfunction in multiple sclerosis and related experimental models. Cells. 2021;10(3):686.
53. Ben-Azu B, et al. Emerging epigenetic dynamics in gut-microglia brain axis: experimental and clinical implications for accelerated brain aging in schizophrenia. Front Cell Neurosci. 2023;17:1139357.
54. Gusev E, Zhuravleva Y. Inflammation: a new look at an old problem. Int J Mol Sci. 2022;23(9):4596.
55. Kurhaluk N, et al. Role of antioxidants in modulating the microbiota–gut–brain axis and their impact on neurodegenerative diseases. Int J Mol Sci. 2025;26(8):3658.
56. Moradi Vastegani S, et al. Mitochondrial dysfunction and Parkinson's disease: pathogenesis and therapeutic strategies. Neurochem Res. 2023;48(8):2285–308.
57. Kuziak A, et al. Iron homeostasis dysregulation, oro-gastrointestinal microbial inflammatory factors, and Alzheimer's disease: a narrative review. Microorganisms. 2025;13(1):122.
58. Longo S, Rizza S, Federici M. Microbiota-gut-brain axis: relationships among the vagus nerve, gut microbiota, obesity, and diabetes. Acta Diabetol. 2023;60(8):1007–17.
59. Gonçalves S, et al. Enzyme promiscuity in serotonin biosynthesis, from bacteria to plants and humans. Front Microbiol. 2022;13:873555.
60. Moțățăianu A, Șerban G, Andone S. The role of short-chain fatty acids in microbiota–gut–brain cross-talk with a focus on amyotrophic lateral sclerosis: a systematic review. Int J Mol Sci. 2023;24(20):15094.
61. Bonaz B. Anti-inflammatory effects of vagal nerve stimulation with a special attention to intestinal barrier dysfunction. Neurogastroenterol Motil. 2022;34(10):e14456.
62. Chen C, et al. Gut microbiota regulate Alzheimer's disease pathologies and cognitive disorders via PUFA-associated neuroinflammation. Gut. 2022;71(11):2233–52.
63. Zhang Y, et al. Transmission of Alzheimer's disease-associated microbiota dysbiosis and its impact on cognitive function: evidence from mice and patients. Mol Psychiatry. 2023;28(10):4421–37.
64. Baldi S, et al. Microbiota shaping—the effects of probiotics, prebiotics, and fecal microbiota transplant on cognitive functions: a systematic review. World J Gastroenterol. 2021;27(39):6715.
65. Suganya K, Koo B-S. Gut–brain axis: role of gut microbiota on neurological disorders and how probiotics/prebiotics beneficially modulate microbial and immune pathways to improve brain functions. Int J Mol Sci. 2020;21(20):7551.
66. Zhao L-P, et al. DSS-induced acute colitis causes dysregulated tryptophan metabolism in brain: an involvement of gut microbiota. J Nutr Biochem. 2023;115:109282.
67. Menozzi E, Schapira AH, Borghammer P. The gut-brain axis in Parkinson disease: emerging concepts and therapeutic implications. Mov Disord Clin Pract. 2025;12:904.
68. He Y, et al. Citrobacter rodentium infection impairs dopamine metabolism and exacerbates the pathology of Parkinson's disease in mice. J Neuroinflammation. 2024;21(1):153.

69. Kumari S, Taliyan R, Dubey SK. Comprehensive review on potential signaling pathways involving the transfer of α-synuclein from the gut to the brain that leads to Parkinson's disease. ACS Chem Neurosci. 2023;14(4):590–602.
70. Mahbub NU, et al. Dysbiosis of the gut microbiota and its effect on α-synuclein and prion protein misfolding: consequences for neurodegeneration. Front Cell Infect Microbiol. 2024;14:1348279.
71. Janakiraman M. Regulation of spontaneous CNS autoimmunity by dietary salt. 2022. https://doi.org/10.5282/edoc.29918.
72. Bianchimano P, et al. Mining the microbiota to identify gut commensals modulating neuroinflammation in a mouse model of multiple sclerosis. Microbiome. 2022;10(1):174.
73. Abbaszadeh SH, Yousefi M, Arefhosseini SR, Mahmoodpoor A, Mameghani ME. Effect of a seven-strain probiotic on dietary intake, inflammatory markers, and T-cells in severe traumatic brain injury patients: A randomized, double-blind, placebocontrolled trial. Sci Prog. 2024;107(3):368504241259299. https://doi.org/10.1177/00368504241259299.
74. Yousof SM, et al. Modulation of gut microbiome community mitigates multiple sclerosis in a mouse model: the promising role of Palmaria palmata alga as a prebiotic. Pharmaceuticals. 2023;16(10):1355.
75. Xu D, et al. Therapeutic effects and mechanisms of fecal microbiota transplantation on EAE partly through HPA axis-mediated neuroendocrine regulation. Heliyon. 2024;10(12):e33214.
76. Gyriki D, et al. The gut microbiota and aging: interactions, implications, and interventions. Front Aging. 2025;6:1452917.
77. Huang H, et al. Gut microbiota regulation of inflammatory cytokines and microRNAs in diabetes-associated cognitive dysfunction. Appl Microbiol Biotechnol. 2023;107(23):7251–67.
78. Wang S-s, et al. The relationship between Alzheimer's disease and intestinal microflora structure and inflammatory factors. Front Aging Neurosci. 2022;14:972982.
79. Seo D-o, et al. ApoE isoform–and microbiota-dependent progression of neurodegeneration in a mouse model of tauopathy. Science. 2023;379(6628):eadd1236.
80. Den H, et al. Efficacy of probiotics on cognition, and biomarkers of inflammation and oxidative stress in adults with Alzheimer's disease or mild cognitive impairment—a meta-analysis of randomized controlled trials. Aging (Albany NY). 2020;12(4):4010.
81. Chauhan NS, Kumar S. Microbiome therapeutics: personalized therapy beyond conventional approaches. Elsevier; 2023.
82. Huang B. Gut microbiota across early stages of synucleinopathy: from high-risk relatives, REM sleep behavior disorder to early Parkinson's disease. Hong Kong: The Chinese University of Hong Kong; 2020.
83. Heinzel S, et al. Gut microbiome signatures of risk and prodromal markers of Parkinson disease. Ann Neurol. 2021;90(3):E1–E12.
84. Chen S-J, et al. Alteration of gut microbial metabolites in the systemic circulation of patients with Parkinson's disease. J Parkinsons Dis. 2022;12(4):1219–30.
85. Mincic AM, et al. Modulation of gut microbiome in the treatment of neurodegenerative diseases: a systematic review. Clin Nutr. 2024;43:1832.
86. Bruggeman A, et al. Safety and efficacy of faecal microbiota transplantation in patients with mild to moderate Parkinson's disease (GUT-PARFECT): a double-blind, placebo-controlled, randomised, phase 2 trial. EClinicalMedicine. 2024;71:102563.
87. Ferri C, et al. Gut microbiota changes during dimethyl fumarate treatment in patients with multiple sclerosis. Int J Mol Sci. 2023;24(3):2720.
88. Ma Y, et al. Epsilon toxin–producing Clostridium perfringens colonize the multiple sclerosis gut microbiome overcoming CNS immune privilege. J Clin Invest. 2023;133(9):e163239.
89. Zhu F, et al. A cross-sectional study of MRI features and the gut microbiome in pediatric-onset multiple sclerosis. Ann Clin Transl Neurol. 2024;11(2):486–96.
90. Stec A, et al. Bacterial metabolites: a link between gut microbiota and dermatological diseases. Int J Mol Sci. 2023;24(4):3494.
91. Roy S, Dhaneshwar S. Role of prebiotics, probiotics, and synbiotics in management of inflammatory bowel disease: current perspectives. World J Gastroenterol. 2023;29(14):2078.

92. Feng R, et al. Effect of fecal microbiota transplantation on patients with sporadic amyotrophic lateral sclerosis: a randomized, double-blind, placebo-controlled trial. BMC Med. 2024;22(1):566.
93. Zhang S, et al. Gut microbiota metabolites: potential therapeutic targets for Alzheimer's disease? Front Pharmacol. 2024;15:1459655.
94. Liu J, Chen Q, Su R. Interplay of human gastrointestinal microbiota metabolites: short-chain fatty acids and their correlation with Parkinson's disease. Medicine (Baltimore). 2024;103(17):e37960.
95. Sheng C, et al. Combination of gut microbiota and plasma amyloid-β as a potential index for identifying preclinical Alzheimer's disease: a cross-sectional analysis from the SILCODE study. Alzheimers Res Ther. 2022;14(1):35.
96. Kushkevych I, et al. Comparison of microbial communities and the profile of sulfate-reducing bacteria in patients with ulcerative colitis and their association with bowel diseases: a pilot study. Microb Cell. 2024;11:79.
97. Bolte LA, et al. Long-term dietary patterns are associated with pro-inflammatory and anti-inflammatory features of the gut microbiome. Gut. 2021;70(7):1287–98.
98. Mavrogeni ME, et al. Direct action of non-digestible oligosaccharides against a leaky gut. Nutrients. 2022;14(21):4699.
99. Kalyan M, et al. Role of endogenous lipopolysaccharides in neurological disorders. Cells. 2022;11(24):4038.
100. Singh AA, Khan F, Song M. Biofilm-associated amyloid proteins linked with the progression of neurodegenerative diseases. Int J Mol Sci. 2025;26(6):2695.
101. Kim M-S, et al. Transfer of a healthy microbiota reduces amyloid and tau pathology in an Alzheimer's disease animal model. Gut. 2020;69(2):283–94.
102. Casini A, et al. Tracking the importance of enteric a-syn pathology in Parkinson's disease. Asian J Tradit Complement Altern Med. 2022;2(10):50–7.
103. Pepoyan A. Gut Akkermansia muciniphila, Prevotellaceae, and Enterobacteriaceae spp. as possible markers in women-related nutritional and clinical trials: familial Mediterranean fever disease. Women's Health Rep. 2024;5(1):785–93.
104. Fusco W, et al. Short-chain fatty-acid-producing bacteria: key components of the human gut microbiota. Nutrients. 2023;15(9):2211.
105. Neurath MF, Artis D, Becker C. The intestinal barrier: a pivotal role in health, inflammation, and cancer. Lancet Gastroenterol Hepatol. 2025;10:573.
106. Muronetz VI, et al. Are gastrointestinal microorganisms involved in the onset and development of amyloid neurodegenerative diseases? Microbiol Res. 2023;14(4):1942–55.
107. Postuma RB, Berg D. Prodromal Parkinson's disease: the decade past, the decade to come. Mov Disord. 2019;34(5):665–75.
108. Lorente-Picón M, Laguna A. New avenues for Parkinson's disease therapeutics: disease-modifying strategies based on the gut microbiota. Biomolecules. 2021;11(3):433.
109. Sultan S, et al. Metabolic influences of gut microbiota dysbiosis on inflammatory bowel disease. Front Physiol. 2021;12:715506.
110. James MM, et al. Role of butyrogenic Firmicutes in type-2 diabetes. J Diabetes Metab Disord. 2022;21(2):1873–82.
111. Calzadilla N, et al. Bile acids as inflammatory mediators and modulators of intestinal permeability. Front Immunol. 2022;13:1021924.
112. Jung YH, Chae CW, Han HJ. The potential role of gut microbiota-derived metabolites as regulators of metabolic syndrome-associated mitochondrial and endolysosomal dysfunction in Alzheimer's disease. Exp Mol Med. 2024;56(8):1691–702.
113. Noor Eddin A, et al. Gut-modulating agents and amyotrophic lateral sclerosis: current evidence and future perspectives. Nutrients. 2024;16(5):590.
114. Sun J, et al. Gut microbiome and amyotrophic lateral sclerosis: a systematic review of current evidence. J Intern Med. 2021;290(4):758–88.
115. Yang Y, et al. Negative effects of multiple global change factors on soil microbial diversity. Soil Biol Biochem. 2021;156:108229.

116. Hooper MJ, et al. Gut dysbiosis in cutaneous T-cell lymphoma is characterized by shifts in relative abundances of specific bacterial taxa and decreased diversity in more advanced disease. J Eur Acad Dermatol Venereol. 2022;36(9):1552–63.
117. Devaux CA, Million M, Raoult D. The butyrogenic and lactic bacteria of the gut microbiota determine the outcome of allogenic hematopoietic cell transplant. Front Microbiol. 2020;11:1642.
118. Borbolis F, Mytilinaiou E, Palikaras K. The crosstalk between microbiome and mitochondrial homeostasis in neurodegeneration. Cells. 2023;12(3):429.
119. Alberts C, Owe-Larsson M, Urbanska EM. New perspective on anorexia nervosa: tryptophan-kynurenine pathway hypothesis. Nutrients. 2023;15(4):1030.
120. Radisavljevic N, et al. Effects of gut microbiota alterations on motor, gastrointestinal, and behavioral phenotype in a mouse model of Parkinson's disease. J Parkinsons Dis. 2022;12(5):1479–95.
121. Zhao A, et al. Role of histone modifications in neurogenesis and neurodegenerative disease development. Ageing Res Rev. 2024;98:102324.
122. Dong Y, Cui C. The role of short-chain fatty acids in central nervous system diseases. Mol Cell Biochem. 2022;477(11):2595–607.
123. Ciesielska A, Matyjek M, Kwiatkowska K. TLR4 and CD14 trafficking and its influence on LPS-induced pro-inflammatory signaling. Cell Mol Life Sci. 2021;78:1233–61.
124. Limón ID, et al. Disturbance of the glutamate-glutamine cycle, secondary to hepatic damage, compromises memory function. Front Neurosci. 2021;15:578922.
125. Simicic D, Cudalbu C, Pierzchala K. Overview of oxidative stress findings in hepatic encephalopathy: from cellular and ammonium-based animal models to human data. Anal Biochem. 2022;654:114795.
126. Meng X, et al. Lactate/Hydroxycarboxylic acid receptor 1 in Alzheimer's disease: mechanisms and therapeutic implications-exercise perspective. Mol Neurobiol. 2024;61(10):7717–31.
127. Teleanu DM, et al. An overview of oxidative stress, neuroinflammation, and neurodegenerative diseases. Int J Mol Sci. 2022;23(11):5938.
128. Huang Y, et al. Tryptophan metabolism in central nervous system diseases: pathophysiology and potential therapeutic strategies. Aging Dis. 2023;14(3):858.
129. Barakat H, Aljutaily T. Role of γ-Aminobutyric acid (GABA) as an inhibitory neurotransmitter in diabetes management: mechanisms and therapeutic implications. Biomolecules. 2025;15(3):399.
130. Bagheri S, Zolghadri S, Stanek A. Beneficial effects of anti-inflammatory diet in modulating gut microbiota and controlling obesity. Nutrients. 2022;14(19):3985.
131. Dahiya D, Nigam PS. Antibiotic-therapy-induced gut dysbiosis affecting gut microbiota—brain axis and cognition: restoration by intake of probiotics and synbiotics. Int J Mol Sci. 2023;24(4):3074.
132. Duan H, et al. Antibiotic-induced gut dysbiosis and barrier disruption and the potential protective strategies. Crit Rev Food Sci Nutr. 2022;62(6):1427–52.
133. Frankiensztajn LM, Elliott E, Koren O. The microbiota and the hypothalamus-pituitary-adrenocortical (HPA) axis, implications for anxiety and stress disorders. Curr Opin Neurobiol. 2020;62:76–82.
134. Sharma VK, et al. Dysbiosis and Alzheimer's disease: a role for chronic stress? Biomolecules. 2021;11(05):678.
135. Yang D-F, et al. Acute sleep deprivation exacerbates systemic inflammation and psychiatry disorders through gut microbiota dysbiosis and disruption of circadian rhythms. Microbiol Res. 2023;268:127292.
136. Varghese S, et al. Physical exercise and the gut microbiome: a bidirectional relationship influencing health and performance. Nutrients. 2024;16(21):3663.
137. Ribeiro MF, et al. Diet-dependent gut microbiota impacts on adult neurogenesis through mitochondrial stress modulation. Brain Commun. 2020;2(2):fcaa165.
138. Cristofori F, et al. Anti-inflammatory and immunomodulatory effects of probiotics in gut inflammation: a door to the body. Front Immunol. 2021;12:578386.

139. Aziz MNM, et al. Irritable bowel syndrome, depression, and neurodegeneration: a bidirectional communication from gut to brain. Nutrients. 2021;13(9):3061.
140. Costa GT, Vasconcelos QD, Aragão GF. Fructooligosaccharides on inflammation, immunomodulation, oxidative stress, and gut immune response: a systematic review. Nutr Rev. 2022;80(4):709–22.
141. Arora K, Green M, Prakash S. The microbiome and Alzheimer's disease: potential and limitations of prebiotic, synbiotic, and probiotic formulations. Front Bioeng Biotechnol. 2020;8:537847.
142. Michael OS, et al. Microbiota transplantation, health implications, and the way forward. In: Microbiomes and emerging applications; 2022. p. 79–97. https://doi.org/10.1201/9781003180241-5.
143. Khoruts A, Staley C, Sadowsky MJ. Faecal microbiota transplantation for Clostridioides difficile: mechanisms and pharmacology. Nat Rev Gastroenterol Hepatol. 2021;18(1):67–80.
144. Ephraim E, Brockman JA, Jewell DE. A diet supplemented with polyphenols, prebiotics and omega-3 fatty acids modulates the intestinal microbiota and improves the profile of metabolites linked with anxiety in dogs. Biology. 2022;11(7):976.
145. Nagpal R, et al. Modified Mediterranean-ketogenic diet modulates gut microbiome and short-chain fatty acids in association with Alzheimer's disease markers in subjects with mild cognitive impairment. EBioMedicine. 2019;47:529–42.
146. Russo E, et al. Immunomodulating activity and therapeutic effects of short chain fatty acids and tryptophan post-biotics in inflammatory bowel disease. Front Immunol. 2019;10:2754.
147. Li H-Y, et al. Effects and mechanisms of probiotics, prebiotics, synbiotics, and postbiotics on metabolic diseases targeting gut microbiota: a narrative review. Nutrients. 2021;13(9):3211.
148. Ahmed MM, et al. CRISPR-Cas systems in the fight against antimicrobial resistance: current status, potentials, and future directions. Infect Drug Resist. 2024;17:5229–45.
149. Elmaleh DR, et al. New approaches to profile the microbiome for treatment of neurodegenerative disease. J Alzheimer's Dis. 2021;82(4):1373–401.
150. Williams TC, et al. Directed acyclic graphs: a tool for causal studies in paediatrics. Pediatr Res. 2018;84(4):487–93.
151. Ferreira GS, et al. Levelling the translational gap for animal to human efficacy data. Animals. 2020;10(7):1199.
152. Wang J, et al. Fecal microbiota transplants in pediatric autism: opportunities and challenges. World J Pediatr. 2024;20(12):1201–4.
153. Stanley TD, Doucouliagos H, Ioannidis JP. Retrospective median power, false positive meta-analysis and large-scale replication. Res Synth Methods. 2022;13(1):88–108.
154. Bharti R, Grimm DG. Current challenges and best-practice protocols for microbiome analysis. Brief Bioinform. 2021;22(1):178–93.
155. Chetty A, Blekhman R. Multi-omic approaches for host-microbiome data integration. Gut Microbes. 2024;16(1):2297860.
156. Marcos-Zambrano LJ, et al. Applications of machine learning in human microbiome studies: a review on feature selection, biomarker identification, disease prediction and treatment. Front Microbiol. 2021;12:634511.
157. Shamsuzzaman M, et al. *Segatella intestinalis* sp. nov., and *Parabacteroides caeci* sp. nov., novel potential probiotics from the human gut microbiome. Antonie Van Leeuwenhoek. 2025;118(9):136.
158. Shamsuzzaman M, et al. Genome insight and probiotic potential of three novel species of the genus Corynebacterium. Front Microbiol. 2023;14:1225282.

Vagally Mediated Pathways: How Gut Signals Affect the Nervous System

3

T. Sengottuvel, M. A. Aysha, and V. K. Sneha

3.1 Introduction to the Microbiome-Gut-Brain Axis

The long-standing relationship between intestinal flora and central nervous system function is revealed by the microbiome-gut-brain axis (MGBA), a paradigm shift in neuroscientific and gastrointestinal research. The axis creates a bidirectional communication channel that connects the gut microbiota to the brain's emotional, cognitive, and physiological processes by integrating the neurological, hormonal, immunological, and metabolic signalling networks [1, 2]. The GI tract's "extra organ" is the trillions of bacteria that comprise the gut microbiome. These microbial communities have the genetic and metabolic ability to produce a range of neuroactive products, including neurotransmitters like dopamine, serotonin, and gamma-aminobutyric acid (GABA), as well as gut hormones and short-chain fatty acids (SCFAs). For example, SCFAs cross the blood-brain barrier to change neuroinflammation and gene expression, while serotonin, which is primarily produced in the gut, controls both local gut motility and central neurotransmission, influencing mood and cognition [3, 4]. The MGBA is dynamic: gut microbes influence brain function, and the central nervous system regulates gastrointestinal physiology via autonomic, neuroendocrine, and immunological pathways. The fact that changes in microbiome composition can result in not only digestive symptoms but also changes in mood, cognition, and even behaviour highlights the axis' significance in the pathophysiology of conditions like anxiety, depression, autism spectrum disorders, Parkinson's disease, and Alzheimer's disease. [6] Modern research using techniques such as molecular profiling, neuroimaging, and germ-free animal models continues to shed light on the MGBA's workings. These ideas pave the way for novel therapeutic approaches, such as customised probiotics, prebiotics, and faecal microbiota transplantation, that show great promise for neurological and gastrointestinal health.

T. Sengottuvel (✉) · M. A. Aysha · V. K. Sneha
Department of Pharmacology, Prime College of Pharmacy, Palakkad, Kerala, India

M. H. Rahman, Muhasina K M (eds.), *The Microbiome's Impact on the Brain*,
https://doi.org/10.1007/978-981-95-7494-0_3

3.2 Composition and Diversity of the Gut Microbiota

The gut microbiota, a multifaceted and constantly evolving microbial ecology made up of bacteria, viruses, archaea, and eukaryotes, is found in the human gut. This community is estimated to contain over 1000 distinct bacterial species-level phylotypes, although only a few phyla—primarily *Firmicutes* and *Bacteroidetes*—account for the majority of organisms found in healthy adults. Other notable constituents include Actinobacteria and Proteobacteria, with minor but functionally relevant groups such as methanogenic archaea (e.g. *Methanobrevibacter smithii*), yeasts, and viruses also present [7, 8].

3.2.1 Core Microbial Constituents and Population Differences

The core gut microbiota is characterised by the consistent presence of certain genera across individuals, such as *Faecalibacterium*, *Bacteroides*, and *Roseburia*, but there is substantial interindividual variation at the species and strain levels. Host genetics, dietary habits, age, ethnicity, and environmental exposures influence these differences. For instance, populations with diets rich in fibre (e.g. hunter-gatherers) tend to harbour more *Prevotella* and *Spirochaetes*, while urban and industrialised populations tend to be enriched in *Bacteroides* and *Firmicutes*. Even within genera, the proportions and dominant members can shift geographically and culturally; for example, Japanese populations exhibit high abundances of *Bifidobacterium*, while Indian populations frequently show *Prevotella* dominance [9–11].

3.2.2 Age, Development, and Life Stage Diversity

Gut microbiota composition and diversity evolve throughout the human lifespan. Infants are typically dominated by *Bifidobacterium* and *Akkermansia* species, with diversity increasing dramatically following early exposures such as diet (breastfeeding vs. formula), environmental microbes, and antibiotic usage. During childhood, diversity remains lower than in adults, with *Bacteroides* often more prevalent. Individuals have a peak in gut microbiota diversity, which reflects a wide range of dietary and external influences. However, diversity tends to decline again in older individuals, which may be associated with a higher risk of illness [7, 12].

3.2.3 Functional Implications of Diversity

High microbial variety in the gut is typically associated with health and resilience against disease, whereas reduced diversity (dysbiosis) is linked to conditions such as metabolic disorders, inflammatory diseases, and neuropsychiatric syndromes. Each microbial member plays a unique role in metabolism, immune modulation, and even neuroactive compound production, reinforcing the importance of both compositional and functional diversity [13, 14].

3.3 Anatomical and Functional Overview of the Vagus Nerve

The vagus nerve, the tenth cranial nerve (CN X), is the longest and most functionally diverse cranial nerve, serving as an important component of the autonomic nervous system. It originates from the medulla oblongata in the brainstem and extends through the neck, thorax, and abdomen to innervate structures throughout these regions. This mixed nerve contains both afferent (sensory) and efferent (motor) fibres, integrating somatic, visceral, and special sensory inputs and modulating parasympathetic output to multiple organ systems [15, 16].

3.3.1 Anatomical Structure and Course

Exiting the brainstem between the olive and the inferior cerebellar peduncle, the vagus nerve passes through the skull via the jugular foramen. It then descends within the carotid sheath, lying between the internal carotid artery and the internal jugular vein and coursing down the neck to the thorax and abdomen. The nerve exists as paired right and left branches that differ slightly in their anatomical pathways:

- The right vagus nerve passes anterior to the right subclavian artery, gives rise to the right recurrent laryngeal nerve looping under the artery, and then descends posterior to the right main bronchus. It contributes to the cardiac, pulmonary, and oesophageal plexuses before forming the posterior vagal trunk as it passes through the oesophageal hiatus into the abdomen.
- The left vagus nerve enters the thorax between the left common carotid and subclavian arteries, gives off the left recurrent laryngeal nerve, which loops under the arch of the aorta, and then travels posteriorly to the left lung and into the oesophageal plexus. It travels through the oesophageal hiatus and creates the anterior vagal trunk.

The throat, throat cavity, heart, lungs, gastrointestinal system, liver, pancreas, kidneys, and even the colon are all innervated by the vagus nerve.

3.3.2 Functional Divisions

About 80–90% of the fibres of the vagus nerve are afferent, which transmit sensory data from the body's sensory systems towards the central nervous system. The remaining fibres are mostly efferent, which exert parasympathetic control on visceral processes.

Important brainstem nuclei connected to the vagus nerve consist of:

- The dorsal motor nucleus, which produces parasympathetic efferent to the abdomen and thoracic organs
- The nucleus ambiguus, which sends parasympathetic signals to the heart and motor impulses to the pharyngeal and laryngeal muscles
- The single nucleus that incorporates sensory information from the viscera
- The trigeminal nucleus of the spine, which processes sensorimotor information from areas like the ear

3.3.3 Sensory and Motor Functions

- Afferent Functions: The vagus nerve conveys somatic feeling from the ear and external auditory canal, as well as visceral sensation from the heart, lungs, gastrointestinal system, and other thoracoabdominal organs. It also expresses flavour and feeling from the pharynx and epiglottis.
- Efferent Functions: Glandular secretions, heart rate, respiration rate, gastrointestinal motility, and smooth muscle function are all influenced by parasympathetic outputs. Motor fibres control the muscles involved in swallowing, phonation, and coughing.

3.3.4 Clinical Relevance and Imaging

The vagus nerve's extensive role in autonomic control highlights its significance in a variety of diseases, including gastrointestinal disorders, neuropsychiatric disorders, and cardiac arrhythmias. Imaging research and nerve stimulation therapies that target the vagus nerve have expanded treatment options for depression, epilepsy, and inflammatory disorders.

3.3.5 Bidirectional Communication: Neural, Endocrine, and Immune Pathways

The gut-brain axis is a highly dynamic and reciprocal network of communication that combines immunological, neurological, and endocrine pathways. Neural connections are mediated, in large part, by the enteric nervous system and the vagus nerve. They provide quick two-way transmission that enables the brain to receive metabolic and microbiological cues from the stomach and the central nervous system (CNS) to control gut function. For example, the vagus nerve directly transfers sensory information from gut-derived neurotransmitters and metabolites, including

GABA, serotonin, and short-chain fatty acids (SCFAs), to central brain circuits, influencing emotional and cognitive functions [17].

Neuroendocrine mediators, such as the hypothalamic-pituitary-adrenal (HPA) axis, are involved in endocrine pathways. The HPA axis releases cortisol and other hormones in response to gut and systemic signals, which alter brain and gut function. The hormonal integration of this axis is highlighted by the fact that stress-induced release of norepinephrine and adrenocorticotropic hormone (ACTH) can change microbial composition, gut permeability, and immunological status [17].

Both systemic immune cells and gut-associated lymphoid tissue produce cytokines and other immune substances to facilitate immunological communication. Proinflammatory cytokines are released when dysbiosis or gut inflammation triggers an immunological response. These cytokines can eventually affect inflammation and central nervous system function by passing through the blood-brain barrier or communicating via the vagus nerve. The development of neuropsychiatric and neurodegenerative diseases is known to be significantly influenced by changes in gut microbiota, which often lead to altered immune responses and brain homeostasis. [18] The intricate, two-way system is known as the microbiota-gut-brain axis. When combined, these pathways are crucial for maintaining physiological homeostasis and are associated with multiple disease states when disrupted [19].

3.4 Vagal Sensory Mechanisms: From the Gut to the Brain

The vagus nerve, which is essential for mediating gut-brain communication, is the main neurological pathway that carries sensory information from the gastrointestinal tract to the brain. About 80% of vagal fibres are afferent, which transmit impulses from the stomach to the central nervous system. The nodose and jugular ganglia produce these sensory fibres, which can recognise a variety of inputs from the gut, such as mechanical strain, food content, metabolic by-products, microbial metabolites, and gut hormones [20]. Vagal afferents terminate at the brainstem's nucleus tractus solitarius (NTS), where visceral information is processed and transmitted to higher brain regions. There, it interacts with loci like the hippocampus, hypothalamus, and locus coeruleus to integrate homeostatic, emotional, and cognitive processes. The direct interaction between the gut's sensory ends and enteroendocrine cells is often facilitated by specialised synaptic connections known as "neuropods", which allow for the rapid detection and transmission of chemical signals produced by food consumption and microbial metabolism [21]. Recent research suggests that vagal sensory processes are genetically varied, with particular fibre subtypes tuned to distinguish between different inputs. This makes it possible for the vagus nerve to encode and send intricate information obtained from the gut to the brain, affecting behaviour, immunological regulation, hunger, and energy balance. Because of this, the vagus nerve has therapeutic promise for treating neuroinflammatory, psychiatric, and digestive problems in addition to facilitating rapid physiological responses and helping to regulate mood, memory, and reward systems [6, 22].

3.4.1 Modulation of Neurotransmitters and Neuroactive Compounds by Gut Microbes

By regulating the synthesis, metabolism, and availability of different neurotransmitters and neuroactive metabolites, gut microbiota plays a critical role in affecting the central and enteric nervous systems. Neurotransmitters including serotonin, gamma-aminobutyric acid (GABA), dopamine, acetylcholine, and histamine can be produced by some bacterial species or stimulated to be secreted by host cells in the gut. Up to 90% of peripheral serotonin is thought to come from the gut, where levels are tightly controlled by microbial composition and metabolism. Serotonin biosynthesis disruptions have been linked to mood and cognitive changes [17]. Microbial fermentation produces short-chain fatty acids (SCFAs), such as acetate, propionate, and butyrate, which not only feed colonocytes but also pass across the blood-brain barrier and affect neuroinflammatory reactions, neural gene expression, and the integrity of the blood-brain barrier. Furthermore, gut microorganisms affect the synthesis and breakdown of tryptophan and its derivatives, which affects the kynurenine and serotonin pathways that are essential for neuroimmune and cognitive function [23]. By directly interacting with neuronal receptors or transmitting signals via the vagus nerve, neuroactive substances produced from microbiota, such as GABA and dopamine analogues, can alter brain excitability, plasticity, and behaviour. Furthermore, the microbiome generates antioxidants and neuropeptides that shield the nervous system from inflammation and oxidative stress by buffering neural tissue [24]. The pathophysiology and progression of neurodevelopmental, psychiatric, and neurodegenerative illnesses can be attributed to imbalances in these neuroactive chemicals caused by disruptions in the gut microbiome, such as those caused by antibiotics, poor food, or illness. Microbiome-based therapies have the potential to enhance neurological outcomes by restoring appropriate neurotransmitter control, according to recent research [25].

3.4.2 Vagal Signal-Based Microbial Control of the Blood-Brain Barrier

The blood-brain barrier's (BBB) integrity and functionality are greatly impacted by the gut microbiota, and the vagus nerve is an essential mediator of this control. The vagus nerve detects signals and metabolites produced from microbes that permeate the intestinal barrier, even though it is not in direct contact with the luminal contents of the gut. Central control of BBB permeability and neuroinflammatory processes is made possible via the transmission of these sensory signals to the brainstem, namely the nucleus tractus solitarius [26]. It has been demonstrated that important microbial metabolites called short-chain fatty acids (SCFAs) improve tight junction protein expression and preserve BBB integrity both directly and indirectly through vagal pathways. Vagus nerve stimulation can improve the BBB's selective permeability, lowering neuroinflammation and shielding brain tissue from harmful assaults, according to experimental research. Conversely, vagotomy impairs BBB tight junctions and increases permeability, which exacerbates neuroinflammatory conditions, underscoring the vagus nerve's indispensable role in neurovascular homeostasis linked to gut microbial signals [27].

Additionally, circulating cytokine levels that impact BBB function are influenced by vagal regulation of systemic immune responses, including the cholinergic anti-inflammatory pathway. In addition to forming an integrated network essential for preserving the homeostasis of the central nervous system, this gut microbiome-vagus nerve-brain axis may provide new therapeutic targets for neurodegenerative and neuroinflammatory disorders [6].

3.4.3 Immune System as an Intermediary: Gut-Immune-Brain Crosstalk

The gut microbiota and the brain communicate in both directions, and the immune system plays a crucial role as a mediator. When gut-associated immune cells are activated by microbial components and metabolites, cytokines and chemokines are released, which can affect neuroinflammatory and systemic reactions. The gut microbiota influences behaviour and brain function through this immunological signalling [5]. Innate immunity acts as the first line of defence by recognising microbial patterns and lowering local inflammation, while adaptive immunity provides specificity and memory, influencing long-term immune responses that affect brain functions. Activated immune mediators can cross or signal the blood-brain barrier, which can affect neurotransmitter balance, neuroinflammation, and microglial activation—all significant components of neuropsychiatric and neurodegenerative disorders [28]. Furthermore, immune system development and function are influenced by the gut microbiome; germ-free rats exhibit immature microglial states and altered immunological profiles that may be reversed by microbial colonisation. The relevance of gut-immune-brain crosstalk in preserving neurological health is highlighted by the fact that immunological dysregulation caused by dysbiosis relates to diseases including depression, anxiety, autism spectrum disorders, and multiple sclerosis [29]. As a result, the immune system provides prospective targets for therapeutic intervention in brain illnesses by combining microbial signals from the gut with reactions from the central neurological system [2].

3.4.4 Hormonal and Neuroendocrine Influence in Gut-Brain Signalling

The gut microbiota regulates the synthesis and systemic effects of hormones and neuroendocrine factors, which are essential elements of gut-brain communication. Enteroendocrine cells (EECs) interspersed within the intestinal epithelium sense microbial metabolites and nutrients, releasing hormones such as glucagon-like peptide-1 (GLP-1), peptide YY (PYY), and serotonin, which influence satiety, gut motility, and mood through interaction with vagal afferents and systemic circulation

[2]. The hypothalamic pituitary-adrenal (HPA) axis, a central neuroendocrine relay that regulates cortisol release under stress, is significantly influenced by the microbiota. Microbiota restoration can correct the high corticosterone levels and exaggerated HPA axis responses seen in germ-free and dysbiotic animal models, suggesting microbial control of stress hormone pathways. Gut microorganisms are linked to brain health through hormonal pathways, and long-term stress and HPA dysregulation have a detrimental impact on neuroplasticity and cognitive function [30]. By affecting the release of EEC hormones and the permeability of the blood-brain barrier, microbial metabolites like short-chain fatty acids also regulate neuroendocrine function. Further integrating hormonal and neurochemical signals in the gut-brain axis, the gut microbiota can also influence the production and bioavailability of neurotransmitter precursors such as tryptophan [31]. In general, important elements of hunger control, stress response, mood regulation, and cognitive function are underpinned by gut microbial modulation of hormonal and neuroendocrine pathways, providing intriguing therapeutic targets within the microbiota-gut-brain axis [32].

3.4.5 Experimental Models for Microbiota-Vagus-Brain Axis Research

To clarify the intricate relationships between gut microbiota, vagal signalling, and brain function, research on the microbiota-vagus-brain axis uses a number of experimental models. The study of the microbiota's basic functions in nervous system development and behaviour is made possible by germ-free animal models, in which microorganisms are missing from birth. The controlled study of microbial effects on vagal pathways and central brain circuits is made possible by colonising these animals with certain microbial communities or pathogenic strains [33].

Another well-known model that transfers gut microbial populations from donor to recipient animals is faecal microbiota transplantation (FMT), which aids in establishing causal links between particular microbiomes and changes in behaviour, neuroinflammation, and neurogenesis mediated via the vagus nerve. Surgical interventions like subdiaphragmatic vagotomy selectively disrupt vagal afferent signalling, demonstrating its necessity in transferring microbial signals and regulating stress-related behaviours, neurotransmitter balance, and hippocampal plasticity [6].

Furthermore, vagal sensory neurons can be selectively modulated to precisely analyse their function in gut-brain communication using neural circuit-mapping techniques like optogenetics and chemogenetics. These models offer an integrative framework to understand the molecular and cellular underpinnings of the microbiota-vagus-brain axis when combined with multi-omics techniques, such as metabolomics, transcriptomics, and proteomics. This advances the development of targeted therapies for neuropsychiatric disorders [34]. In order to coordinate gut and brain function, the gut-brain axis is a sophisticated, bidirectional communication system that integrates immune, endocrine, and neural pathways. Neural communication is significantly influenced by the enteric nervous system and the vagus nerve. They swiftly send sensory information from gut microbes and their metabolites to the brain, affecting mood, hunger, and thought processes. Microbial metabolites that influence vagal afferent signalling, which in turn influences central processing and systemic reactions, include short-chain fatty acids and neurotransmitters. Endocrine pathways are mediated by hormones derived from the gut and neuroendocrine axes, especially the hypothalamic-pituitary-adrenal (HPA) axis. Gut microbiota influence the release of hormones such as cortisol through the HPA axis, which in turn influences gut permeability, microbial structure, and the brain's responses to stress. This hormonal feedback loop combines metabolic and environmental cues to alter behaviour and homeostasis [3].

Because microbial metabolites and antigens stimulate mucosal immune cells, which then produce cytokines that can impact neuroinflammation and brain function, immune communication is crucial. Immune mediators link gut microbial balance to central nervous system health and disease aetiology through vagal pathways or the blood-brain barrier. This immune crosstalk is important in neuropsychiatric and neurodegenerative disorders influenced by gut microbiota [18]. Together, these immune, neurological, and endocrine pathways form an integrated microbiota-gut-brain axis that is essential for maintaining physiological balance and has potential for treating central nervous system disorders.

3.4.6 The Role of the Vagus Nerve in Cognitive Functions and Behaviour

Through its numerous neural connections to various brain regions involved in memory, attention, and emotional regulation, the vagus nerve plays a crucial role in regulating cognitive processes and behaviour. The nucleus tractus solitarius (NTS) of the brainstem receives signals from the gut through sensory afferents, which make up about 80% of the vagus nerve fibres. Higher-order brain regions that are essential for learning, memory consolidation, and decision-making, such as the locus coeruleus, hippocampus, amygdala, and prefrontal cortex, receive information from the NTS [35]. Preclinical and clinical research have demonstrated that vagus nerve stimulation (VNS) increases synaptic plasticity, long-term potentiation (LTP), and the expression of neurotrophic factors in the hippocampus, such as brain-derived neurotrophic factor (BDNF). These changes enable enhanced memory, executive function, and overall cognitive performance. Likewise, VNS regulates the release of key neurotransmitters such as dopamine, serotonin, and norepinephrine that impact mood, arousal and cognitive flexibility [36]. Notably, VNS has therapeutic potential for neurological disorders characterised by cognitive deficits, such as depression, epilepsy, and Alzheimer's disease. By reducing neuroinflammation, increasing cerebral blood flow, and upregulating genes linked to neuroplasticity, it has neuroprotective effects that establish the vagus nerve as a crucial modulator that connects gut signals and brain function [37].

Microbiome Disturbances and Vagal Pathways in Neuropsychiatric Disorders. The development and progression of a number of neuropsychiatric disorders, such as depression, anxiety, autism spectrum disorders, and neurodegenerative diseases, are increasingly linked to disruptions in the gut microbiome, which are commonly referred to as dysbiosis. Changes in gut microbial signals have a significant impact on behaviour and brain function through the vagus nerve. Short-chain fatty acids, catecholamines, and neurotransmitters are among the metabolites and neuroactive substances produced by gut microbiota that interact with vagal afferent fibres to modulate central nervous system (CNS) circuits involved in mood regulation, stress response, and cognitive processing. Aberrant vagal signalling brought on by dysbiosis may result in neuroinflammation, altered neurotransmitter balance, and abnormal behavior [38].

Vagus nerve stimulation (VNS) has demonstrated promise in restoring microbial balance and alleviating neuropsychiatric symptoms, whereas vagotomy or impaired vagal signalling disrupts normal microbiota-brain communication, aggravating depressive-like and anxiety-like behaviours, according to animal studies. These results highlight the potential for treating neuropsychiatric disorders by focusing on gut microbiota and vagal pathways [39].

3.5 Neurodegenerative Disorders: Parkinson's, Alzheimer's, and ALS

Parkinson's disease (PD), Alzheimer's disease (AD), and amyotrophic lateral sclerosis (ALS) are examples of neurodegenerative illnesses that cause progressive neuronal loss along with cognitive and movement impairments. There is growing evidence that the pathophysiology and course of these conditions are significantly influenced by the gut microbiota and vagally mediated pathways.

The buildup of misfolded alpha-synuclein in the enteric nervous system, which can go to the brain by the vagus nerve, is correlated with gut dysbiosis in Parkinson's disease (PD), which is characterised by decreased microbial diversity and occurs before motor symptoms. The loss of dopaminergic neurons in the substantia nigra, microglial activation, and neuroinflammation are all associated with this gut-to-brain communication [40]. Similar to this, in AD, amyloid-beta and tau pathology may be influenced by microbial metabolites and amyloidogenic proteins generated in the stomach when they are conveyed through vagal and circulatory pathways, leading to cognitive loss. Additionally, dysbiosis exacerbates neurodegeneration by promoting systemic inflammation and disruption of the blood-brain barrier [41]. Changes in gut microbial patterns have been linked to motor neuron degeneration and neuroinflammation in ALS, although vagal processes have not been as well studied as in PD and AD [42]. Probiotics, prebiotics, faecal microbiota transplantation (FMT), and vagus nerve stimulation (VNS) therapies are examples of microbiome-targeting interventions that have the potential to modulate disease progression and symptom severity by reducing neuroinflammation and restoring gut-brain communication [19].

3.5.1 Bidirectional Interactions: Vagal Effector Signals Shaping Gut Function

The vagus nerve transmits efferent signals that actively control gastrointestinal physiology in addition to carrying sensory data from the gut to the brain. These vagal efferent fibres regulate mucosal blood flow, intestinal immune response regulation, motility, and secretions from digestive glands. In order to preserve gut homeostasis, this bidirectional connection creates a carefully controlled vago-vagal reflex loop. One such mechanism is the cholinergic anti-inflammatory pathway, in which pro-inflammatory cytokines like tumour necrosis factor-alpha (TNF-α) are suppressed when acetylcholine produced from vagal efferent binds to α7-nicotinic acetylcholine receptors on macrophages. In addition to lowering intestinal inflammation, this anti-inflammatory action may maintain the integrity of the barrier and the makeup of the gut microbiota. Additionally, the vagus indirectly affects systemic immune activity by stimulating the splenic sympathetic nerve, which encourages the release of acetylcholine from splenic cells and further reduces inflammation. By influencing the enteric neurons, vagal activity also controls GI motility, which has a significant impact on nutrient absorption and digestion. All things considered, vagal efferent signalling regulates complex physiological processes in the gut, impacting the microbiological and immunological environments. In addition to providing treatment options for disorders involving neuroinflammations and gut dysfunctions, the reciprocal interactions are essential for preserving the integrity of the gut-brain axis [22].

3.5.2 Probiotics, Prebiotics, and Synbiotic Impact on Vagal Pathways

The agents such as probiotics, prebiotics, and synbiotics have a major effect on the microbiota-gut-brain axis, particularly through altering vagal nerve function. Probiotics like *Lactobacillus* and *Bifidobacterium* species improve vagal tone and autonomic regulation by stimulating afferent vagal pathways that affect brain regions responsible for mood and cognition, thereby reducing the symptoms of depression, anxiety, and stress-related disorders [43, 44].

Prebiotics serve as feeding substrates for beneficial microorganisms, indirectly increasing the vagal signalling by promoting microbial synthesis of neuroactive metabolites like short-chain fatty acids (SCFAs), which activate the vagal afferents. This results in a reduction of systemic inflammation and modification of hypothalamic, pituitary, adrenal (HPA) axis activity, which are important aspects of neuropsychiatric health.

By combining probiotics and prebiotics, synbiotics have a synergistic impact that increases vagal pathway activation and improves microbial diversity and stability. Clinical trials have shown that these treatments lessen anxiety and depression symptoms, which are partially mediated by enhanced vagal function as shown by heart rate variability measurements [43, 45].

3.5.3 Technological Advances: Imaging and Measurement of Vagal Tone

The capacity to see and quantify vagal tone, a crucial indicator of parasympathetic nervous system function, has significantly improved because of technological advancements. Heart rate variability (HRV), which indicates autonomic heart regulation but only offers an approximate assessment of total vagal nerve activity, is the main indirect approach used in traditional ways to estimate vagal tone. Recent advancements have made it possible to monitor vagal nerve activity in awake and acting animal models over an extended period of time with high resolution, thanks to direct neural recording techniques such as carbon nanotube yarn (CNTY) electrodes. These incredibly small and flexible electrodes provide a strong interface to autonomic nerve fibres, allowing for detailed analysis of tonic and phasic vagal firing patterns and revealing a more intricate relationship between HRV, respiratory cycles, and vagal nerve firing. Non-invasive methods like auricular vagus nerve stimulation (aVNS) combined with real-time monitoring of HRV and respiratory sinus arrhythmia (RSA) are promising therapeutic approaches for assessing and controlling the vagal tone. The researchers can gain insight into the neural circuits underlying vagal control by observing how the brainstem nuclei respond, and the central nervous system changes following vagal activation when combined with state-of-the-art imaging techniques like functional MRI and PET scanning [45–47]. When taken as a whole, these technological developments offer new ways to accurately describe vagal nerve functions, clarify its functions in health and illness, and customise bioelectronic treatments for disorders involving dysregulated ANS activity.

3.5.4 Translational and Therapeutic Perspectives: Vagus Nerve Stimulation

Vagus nerve stimulation (VNS), a potential treatment approach that targets the microbiome-gut-brain axis, has significant implications for neuropsychiatric and neurodegenerative diseases. Vagal afferents, which are stimulated by VNS, transmit signals from the gut microbiota to brain regions involved in mood regulation, stress response, and cognition and modulate neurotransmitter systems such as serotonin and dopamine [48]. Preclinical studies demonstrate that VNS can reverse depressive-like behaviours and neuroinflammation caused by microbiota disruptions while also restoring adult hippocampal neurogenesis and synaptic plasticity. Clinically, VNS is FDA-approved for depression and treatment-resistant epilepsy, and by using its capacity to restore gut-brain signalling, current research is examining its efficacy in a variety of brain disorders [13, 33, 49]. Mechanistically, VNS stimulates the cholinergic anti-inflammatory pathway, which reduces systemic and cerebral inflammation, a significant contributor to mood disorders and neurodegeneration. Furthermore, VNS affects neuroendocrine and autonomic balance, providing a multimodal approach to brain health based on manipulation of vagally mediated gut signalling.

A translational foundation for innovative therapies that utilise the gut-vagus-brain axis to enhance neurological and mental health may be provided by combining VNS with microbiome-targeted therapy outcomes [49].

3.6 Challenges and Limitations in Existing Research

Research on how the microbiome influences the brain through vagally mediated pathways is hampered by a number of significant issues. The complexity and diversity of the human gut microbiome and its interactions with the immune and neurological systems cannot be adequately represented by animal models, particularly germ-free or pathogen-free rats, which is a major disadvantage. This restricts the translational application of many results. It is also difficult to compare results and draw broad conclusions about specific bacterial species or metabolites linked to gut-brain transmission due to the lack of standardised procedures and comparable microbiome profiling techniques across studies. Since many studies focus on correlational data and lack a mechanistic understanding of how bacteria affect neuroplasticity and vagal responses, more comprehensive and long-term research is needed. Finding consistent microbiota-brain signals is made more difficult by interindividual variation in microbiome composition brought on by genetics, nutrition, environment, and lifestyles. The need for thorough human research to validate therapeutic potential is further highlighted by the lack of clinical evidence for microbiome-based treatments in neuropsychiatric and neurodegenerative diseases. Last but not least, new research on the brain microbiota and its direct function is still controversial and unclear, requiring the application of advanced techniques and interdisciplinary approaches to unravel this complex axis. To overcome these challenges, new experimental models, standardised protocols, and comprehensive multi-omics techniques are required to elucidate particular pathways in vagally mediated microbiome-brain communication [30, 50].

3.7 Future Directions in Microbiome-Vagus-Brain Studies

Future research in microbiome-vagus-brain studies will improve our understanding of the direct connections between specific gut microorganisms and vagal neurons by revealing the precise molecular mechanisms underlying gut-brain communication. Advances in neurogenetics and multi-omics technologies will enable the integration of microbial, brain, immunological, and metabolic data. This will enable the creation of customised methods for improving neurological and psychiatric outcomes by tailoring microbiome-targeted therapy to each patient. Comprehensive integrative frameworks that incorporate the neuroendocrine, neuroimmune, and microbiome systems must also be used in future research. Precision nutrition, microbiome engineering with genetically modified organisms, and faecal microbiota transplantation are examples of emerging treatment approaches that have enormous promise to restore normal vagal signalling and neuroplasticity. To confirm causal relationships and evaluate the long-term effectiveness of microbiome-based therapies on mood, cognition, and the progression of neurodegenerative diseases, longitudinal and large-cohort human studies are required. Future studies must also employ thorough integrative frameworks that integrate neuroimmune, neuroendocrine, and microbiome data in order to fully capture the complexity of this axis. In order to achieve the translational breakthroughs, the current experimental model limitations must be addressed.

The prevention and treatment of neurodevelopmental, psychiatric, and neurodegenerative disorders can be completely transformed by utilising the microbiota-vagus-brain axis effectively.

3.8 Conclusion

The important and complex network that illustrates how the gut microbial communities' impact behaviour, brain function, and overall neurological health is the microbiome-vagus-brain axis. The vagus nerve serves as a central conduit for bidirectional communication between microbial signals, the central autonomic nervous system, and neuroimmune pathways. This axis plays a key role in regulating mood, cognition, neurodegeneration, and neuroinflammation [22]. Recent studies emphasise the vagus nerve's dual afferent and efferent functions in detecting microbial metabolites and controlling the gut physiology and systemic inflammation, thereby maintaining homeostasis. Neuropsychiatric conditions like anxiety and depression, as well as neurodegenerative diseases like Parkinson's and Alzheimer's, are connected to dysregulation of this axis. Therapeutic approaches that target this axis, like vagus, may be beneficial for future clinical interventions. Future clinical interventions may benefit from therapeutic strategies that target the axis, such as vagus nerve stimulation and microbiome modulations.

To fully realise the potential of this axis, integrative and translational research is required because, despite advancements, there are still gaps in fully elucidating the mechanisms. At the end, comprehending how gut microbes influence brain activity through vagally mediated pathways may transform the therapeutic approaches for a range of neurological and psychiatric conditions.

The interdependence of the brain, vagus nerve, and microbiome highlights the importance of these pathways in maintaining neurological health. Continued research is needed to translate these findings into useful treatments [6, 38, 51].

Declarations

Competing Interests The author(s) has no competing interests to declare that are relevant to the content of this manuscript.

Ethics Approval The author(s) declares that this manuscript does not include primary research with human participants.

The author(s) declares that this manuscript does not include primary studies involving animals.

References

1. Ramadan YN, Alqifari SF, Alshehri K, Alhowiti A, Mirghani H, Alrasheed T, et al. Microbiome gut-brain-axis: impact on brain development and mental health. Mol Neurobiol. 2025;62(8):10813–33.
2. Loh JS, Mak WQ, Tan LKS, Ng CX, Chan HH, Yeow SH, et al. Microbiota–gut–brain axis and its therapeutic applications in neurodegenerative diseases. Signal Transduct Target Ther. 2024;9(1):37.
3. Xu J, Lu Y. The microbiota-gut-brain axis and central nervous system diseases: from mechanisms of pathogenesis to therapeutic strategies. Front Microbiol. 2025;16:1583562.
4. Doenyas C, Clarke G, Cserjési R. Gut–brain axis and neuropsychiatric health: recent advances. Sci Rep. 2025;15(1):3415. https://doi.org/10.1038/s41598-025-86858-3.
5. Carabotti M, Scirocco A, Maselli MA, Severi C. The gut brain axis: interactions between enteric microbiota, central and enteric nervous systems. Ann Gastroenterol. 2015;28(2):203–9.
6. Han Y, Wang B, Gao H, He C, Hua R, Liang C, et al. Vagus nerve and underlying impact on the gut microbiota-brain axis in behavior and neurodegenerative diseases. J Inflamm Res. 2022;15:6213–30.
7. Hou K, Wu ZX, Chen XY, Wang JQ, Zhang D, Xiao C, et al. Microbiota in health and diseases. Signal Transduct Target Ther. 2022;7(1):135.
8. Lozupone CA, Stombaugh JI, Gordon JI, Jansson JK, Knight R. Diversity, stability and resilience of the human gut microbiota. Nature. 2012;489(7415):220–30.
9. Lee Nen That LFM, Pandohee J. Composition and diversity of gut microbiota. In: Srivastava N, Ibrahim S, Chattopadhyay J, Arbab M, editors. The gut microbiota in health and disease. 1st ed. Wiley; 2023. p. 7–16. https://doi.org/10.1002/9781119904786.
10. Gupta VK, Paul S, Dutta C. Geography, ethnicity or subsistence-specific variations in human microbiome composition and diversity. Front Microbiol. 2017;8:1162.
11. Dicks LM. Our mental health is determined by an intrinsic interplay between the central nervous system, enteric nerves, and gut microbiota. Int J Mol Sci. 2023;25(1):38.
12. Radjabzadeh D, Boer CG, Beth SA, Van Der Wal P, Kiefte-De Jong JC, Jansen MAE, et al. Diversity, compositional and functional differences between gut microbiota of children and adults. Sci Rep. 2020;10(1):1040.
13. Mosca A, Leclerc M, Hugot JP. Gut microbiota diversity and human diseases: should we reintroduce key predators in our ecosystem? Front Microbiol. 2016;7:455.
14. Sharon G, Sampson TR, Geschwind DH, Mazmanian SK. The central nervous system and the gut microbiome. Cell. 2016;167(4):915–32.
15. Jackson RG. Anatomy of the vagus nerves in the region of the lower esophagus and the stomach. Anat Rec. 1949;103(1):1–18.
16. Berthoud HR, Neuhuber WL. Functional and chemical anatomy of the afferent vagal system. Auton Neurosci. 2000;85(1–3):1–17.
17. Kasarello K, Cudnoch-Jedrzejewska A, Czarzasta K. Communication of gut microbiota and brain via immune and neuroendocrine signaling. Front Microbiol. 2023;14:1118529.
18. O'Riordan KJ, Moloney GM, Keane L, Clarke G, Cryan JF. The gut microbiota-immune-brain axis: therapeutic implications. Cell Rep Med. 2025;6(3):101982.
19. Montagnani M, Bottalico L, Potenza MA, Charitos IA, Topi S, Colella M, Santacroce L. The crosstalk between gut microbiota and nervous system: a bidirectional interaction between microorganisms and metabolome. Int J Mol Sci. 2023;24(12):10322.
20. Yu CD, Xu QJ, Chang RB. Vagal sensory neurons and gut-brain signaling. Curr Opin Neurobiol. 2020;62:133–40.
21. Mörkl S, Butler MI, Wagner-Skacel J. Gut-brain-crosstalk- the vagus nerve and the microbiota-gut-brain axis in depression. A narrative review. J Affect Disord Rep. 2023;13:100607.
22. Bonaz B, Bazin T, Pellissier S. The vagus nerve at the interface of the microbiota-gut-brain axis. Front Neurosci. 2018;12:49.

23. Strandwitz P. Neurotransmitter modulation by the gut microbiota. Brain Res. 2018;1693:128–33.
24. Dicks LMT. Gut Bacteria and Neurotransmitters. Microorganisms. 2022;10(9):1838.
25. Yarandi SS, Peterson DA, Treisman GJ, Moran TH, Pasricha PJ. Modulatory effects of gut microbiota on the central nervous system: how gut could play a role in neuropsychiatric health and diseases. J Neurogastroenterol Motil. 2016;22(2):201.
26. Ohara TE, Hsiao EY. Microbiota–neuroepithelial signalling across the gut–brain axis. Nat Rev Microbiol. 2025;23(6):371–84.
27. Xie J, Bruggeman A, De Nolf C, Vandendriessche C, Van Imschoot G, Van Wonterghem E, et al. Gut microbiota regulates blood-cerebrospinal fluid barrier function and Aβ pathology. EMBO J. 2023;42(17):e111515.
28. Lyte M. Microbial endocrinology in the microbiome-gut-brain axis: how bacterial production and utilization of neurochemicals influence behavior. PLoS Pathog. 2013;9(11):e1003726.
29. Zhou SY, Guo ZN, Yang Y, Qu Y, Jin H. Gut-brain axis: mechanisms and potential therapeutic strategies for ischemic stroke through immune functions. Front Neurosci. 2023;17:1081347.
30. Al Noman A, Alhudhaibi AM, Afroza M, Tonni SD, Shehab HM, Jahan Iba N, et al. Neuroplasticity and the microbiome: how microorganisms influence brain change. Front Microbiol. 2025;16:1629349.
31. Cani PD, Knauf C. How gut microbes talk to organs: the role of endocrine and nervous routes. Mol Metab. 2016;5(9):743–52.
32. Chao J, Coleman RA, Keating DJ, Martin AM. Gut microbiome regulation of gut hormone secretion. Endocrinology. 2025;166(4):bqaf004.
33. Siopi E, Galerne M, Rivagorda M, Saha S, Moigneu C, Moriceau S, et al. Gut microbiota changes require vagus nerve integrity to promote depressive-like behaviors in mice. Mol Psychiatry. 2023;28(7):3002–12.
34. Longo S, Rizza S, Federici M. Microbiota-gut-brain axis: relationships among the vagus nerve, gut microbiota, obesity, and diabetes. Acta Diabetol. 2023;60(8):1007–17.
35. Olsen LK, Solis E, McIntire LK, Hatcher-Solis CN. Vagus nerve stimulation: mechanisms and factors involved in memory enhancement. Front Hum Neurosci. 2023;17:1152064.
36. Zhang H, Li C-L, Qu Y, Yang Y-X, Du J, Zhao Y. Effects and neuroprotective mechanisms of vagus nerve stimulation on cognitive impairment with traumatic brain injury in animal studies: a systematic review and meta-analysis. Front Neurol. 2022;13:963334.
37. Wang W, Li R, Li C, Liang Q, Gao X. Advances in VNS efficiency and mechanisms of action on cognitive functions. Front Physiol. 2024;15:1452490.
38. Lee SH, Han C, Shin C. IUPHAR review: microbiota-gut-brain axis and its role in neuropsychiatric disorders. Pharmacol Res. 2025;216:107749.
39. Ullah H, Arbab S, Tian Y, Liu C-Q, Chen Y, Qijie L, Khan MIU, Hassan IU, Li K. The gut microbiota–brain axis in neurological disorder. Front Neurosci. 2023;17:1225875.
40. Luo S, Pan M, Fu G, Meng X, Zhao Z, Yu H, et al. Vagus nerve and brain–gut communication in neurodegenerative diseases: mechanism and therapeutic perspectives. Pharmacol Res. 2025;221:107974.
41. Wei W, Wang S, Xu C, Zhou X, Lian X, He L, et al. Gut microbiota, pathogenic proteins and neurodegenerative diseases. Front Microbiol. 2022;13:959856.
42. Dandamudi BJ, Dimaano KAM, Shah N, AlQassab O, Al-Sulaitti Z, Nelakuditi B, et al. Neurodegenerative disorders and the gut-microbiome-brain axis: a literature review. Cureus. 2024;16(10):e72427.
43. Ansari F, Neshat M, Pourjafar H, Jafari SM, Samakkhah SA, Mirzakhani E. The role of probiotics and prebiotics in modulating of the gut-brain axis. Front Nutr. 2023;10:1173660.
44. Mörkl S, Narrath M, Schlotmann D, Sallmutter MT, Putz J, Lang J, et al. Multi-species probiotic supplement enhances vagal nerve function – results of a randomized controlled trial in patients with depression and healthy controls. Gut Microbes. 2025;17(1):2492377.
45. Moshfeghinia R, Nemati H, Ebrahimi A, Shekouh D, Karami S, Eraghi MM, et al. The impact of probiotics, prebiotics, and synbiotics on depression and anxiety symptoms of patients with depression: a systematic review and meta-analysis. J Psychiatr Res. 2025;188:104–16.

46. Martinez P, Grinand M, Cheggour S, Taieb J, Gourjon G. How to properly evaluate cardiac vagal tone in oncology studies: a state-of-the-art review. J Natl Cancer Cent. 2024;4(1):36–46.
47. University of California, Chopra D. A prospective on vagal tone via auricular stimulation and deep breathing. J Altern Complement Integr Med. 2024;10(1):1–12.
48. Forsythe P, Bienenstock J, Kunze WA. Vagal pathways for microbiome-brain-gut axis communication. In: Lyte M, Cryan JF, editors. Microbial endocrinology: the microbiota-gut-brain axis in health and disease, Advances in experimental medicine and biology, vol. 817. New York: Springer New York; 2014. p. 115–33.
49. Faraji N, Payami B, Ebadpour N, Gorji A. Vagus nerve stimulation and gut microbiota interactions: a novel therapeutic avenue for neuropsychiatric disorders. Neurosci Biobehav Rev. 2025;169:105990.
50. Arneth B. Gut–brain axis and brain microbiome interactions from a medical perspective. Brain Sci. 2025;15(2):167.
51. Cryan JF, O'Riordan KJ, Cowan CSM, Sandhu KV, Bastiaanssen TFS, Boehme M, et al. The microbiota-gut-brain axis. Physiol Rev. 2019;99(4):1877–2013.

Parkinson's Disease and the Microbiome: A New Frontier for Early Diagnosis

4

Amritha Chakkittukandiyil and Deepak Vasudevan Sajini

4.1 Introduction to Parkinson's Disease and Microbiome

Parkinson's disease (PD) is the second most common and incurable neurodegenerative disorder after Alzheimer's disease (AD). PD affects approximately 1.5–2.0% of the world population who are aged above 60. The disease is mainly characterized by rigidity, tremor, bradykinesia, and postural instability [1]. Apart from these motor symptoms, people with PD also have non-motor symptoms, such as gastrointestinal dysfunction, fatigue, constipation, cognitive impairment, sleep disturbances, sensory abnormalities, and depression [2]. Constipation is a common prodromal condition in PD, affecting around 60% of individuals. It can begin up to 20 years before diagnosis, indicating the close relation between the brain and the gut [3]. The symptoms of PD typically do not appear until approximately 70–80% of the neurons undergo degeneration, which makes the treatment more challenging [4]. The current treatments for PD provide only symptomatic relief, and there is no available treatment to cure or prevent the disease progression before the onset of symptoms [2]. Among the available drugs, Levodopa is the first-line choice; it gets converted to dopamine in the brain, thereby providing symptomatic relief. A few other treatments are dopamine agonists, MAO-B inhibitors, COMT inhibitors, dopamine facilitators,

A. Chakkittukandiyil · D. V. Sajini (✉)
Department of Pharmacology, National College of Pharmacy, Kerala University of Health Sciences, Manassery, Kozhikode, Kerala, India
e-mail: ndeepakvs@gmail.com

M. H. Rahman, Muhasina K M (eds.), *The Microbiome's Impact on the Brain*,
https://doi.org/10.1007/978-981-95-7494-0_4

anticholinergics, etc. [5]. Surgical options, such as deep brain stimulation (DBS), are also available, which involve the implantation of electrodes into specific areas of the brain to alleviate the symptoms. PD has a complex etiology, including both genetic and environmental factors, leading to the pathogenesis. PD is characterized by the progressive degeneration of dopaminergic neurons present in the substantia nigra pars compacta (SNpc) of the midbrain, resulting in dopamine deficiency and impaired motor control [6]. The most common factors leading to neuronal degeneration in PD are oxidative stress, mitochondrial dysfunction, and neuroinflammation. The major hallmark of PD is the aggregation of α-synuclein protein in the SNpc and the brain stem region. α-synuclein is often seen as a soluble, unfolded monomeric protein. Environmental pollutants, oxidative stress, and genetic abnormalities can all increase the process of α-synuclein misfolding. α-synuclein gets converted into β-sheet structures, which can subsequently self-assemble to form oligomers, protofibrils, and, finally, irreversible and insoluble mature fibrils. In the dopaminergic neurons of the SNpc, mature fibrils combine to produce Lewy bodies and Lewy neurites, which are believed to be the cause of anomalies in the nervous system that lead to neurodegeneration [7].

The Braak's theory, articulated by the German scientists Heiko and Eva Braak in 2003, completely changes the view of PD. The theory proposed that PD may start in the peripheral nervous system, particularly the olfactory or enteric nervous systems (ENS), before moving on to the brain. It does not always start in the motor regions of the brain. According to Braak et al., an unknown pathogen that enters the body and reaches the gut is the cause of sporadic PD and leads to local accumulation of pathological α-synuclein [8]. Later, an extension of the hypothesis arose, which suggests that gut microbiota dysregulation, which might be caused by infection, inflammation, toxins, or nutrition, affects the gut environment. This causes oxidative stress and inflammation, which can lead to α-synuclein aggregation in the ENS [7, 9].

The biggest microecosystem in the human body is the digestive tract. More than 10^{14} bacteria from 2000 known species were found in the human digestive tract [10]. Humans are thought to be superorganisms because of the intricate relationship between the host and its microbiota [11]. The gut microbiota has a crucial role in trophic, metabolic, and defensive functions, and it affects the central nervous system (CNS) via the gut-brain axis (GBA) [12]. The gut microbiome is an essential component of the GBA and is known to influence CNS homeostasis via immune system modulation. The microbiome is comprised of bacteria, viruses, fungi, protozoa, archaea, and bacteriophages. The composition and diversity of these microbes vary according to their location [13, 14]. Every person has a different microbiome, and there is more variation between people than there is in a single person's physiological changes throughout time [15]. The diversity of an individual's gut microbiome stays mostly constant throughout maturity unless there is a discernible shift in dietary practices [16]. Abnormal gut microbiota have been related to a range of neurological disorders, such as autism spectrum disorder (ASD), PD, AD, and epilepsy [17].

4.2 Diversity and Composition of the Gut Microbiota

The human gastrointestinal tract (GIT) contains around 100 trillion bacteria and is home to a diverse microbial community [10]. These microorganisms are collectively known as the gut flora or gut microbiota. Although bacteria make up the majority of the human stomach, viruses (bacteriophages), fungi (*Candida* and *Saccharomyces* species), archaea (*Methanobrevibacter smithii*), and protozoa are also present. Since the combined metabolic activity of these bacteria is equivalent to that of a virtual organ, they behave as an additional organ in the human body, the so-called forgotten organ [18].

The intestinal colon is one of the most widely distributed microbial ecosystems on Earth, with an estimated density of bacterial cells of 10^{11}–10^{12} per milliliter. With hundreds of genes in their genomes, each bacterial strain offers far greater genetic variation and, consequently, adaptability than the human genome. The two most dominant bacteria among these are the *Bacteroidetes* and the *Bacillota* (formerly known as *Firmicutes*) [19]. They make up more than 90% of the total gut microbiome population. Among the genera included in *Bacillota* are *Ruminococcus, Lactobacillus, Clostridium,* and *Faecalibacterium.* Similarly, the *Bacteroidetes* group is mainly made up of species like *Bacteroides* and *Prevotella.* All of these species have a significant function in the breakdown of complex dietary polysaccharides [20]. *Actinobacteria*, *Proteobacteria*, *Verrucomicrobia*, and *Cyanobacteria* are the other phyla present in the gut microbiota but in smaller numbers [21].

Early childhood is a time of dynamic microbiome composition evolution, influenced by a variety of prenatal and postnatal events. Individual differences in gut microbiome composition are caused by several factors, like geography, age, nutrition, lifestyle, genetics, and health conditions. Microbiome diversity declines in older populations, making people more vulnerable to infections, long-term illnesses, and frailty [15]. It is generally accepted that a high degree of microbial diversity is advantageous because it promotes immunological function, metabolic flexibility, and disease resistance. On the other hand, a variety of illnesses and impaired gut function have been connected to decreased diversity, which is commonly referred to as dysbiosis [22].

4.3 Factors Shaping the Gut Microbiome

The human gut microbiome is unique in each individual. The difference in the gut microbiome composition is influenced by genetics, diet, lifestyle, age, mode of birth, and environmental factors. Early in life, the microbiota changes dynamically and is impacted by both prenatal and postnatal factors [23].

4.3.1 Age and Mode of Delivery

Significant gut colonization starts as soon as the baby is delivered. According to studies, meconium, amniotic fluid, and the placenta all contain bacteria and their byproducts, including DNA [24]. The type of delivery, breastfeeding, and antibiotic consumption in the early stages all influence the composition of the microbiota. Infants delivered vaginally are colonized by vaginal bacteria like *Lactobacillus* and *Prevotella*, whereas those born via caesarean section are colonized by microbes in the skin microbiome [25]. Infants who are solely breastfed and born vaginally are known to have the healthiest gut microbiome. These newborns had high amounts of *Bifidobacteria* but low levels of opportunistic pathogens like *Clostridium difficile* and *Escherichia coli* [26]. The gut is expected to be colonized by 10^{13}–10^{14} microbes/ml, consisting of 500–1000 species, within the first year of life [27]. The microbiota develops significantly during childhood and adolescence. By the age of three, their gut flora resembles that of an adult [28]. It will generally become stable and diversified, while it continues to adapt to environmental factors. Microbial diversity declines in elderly persons, increasing susceptibility to infections, chronic illnesses, and frailty [15].

4.3.2 Diet

Dietary composition directly determines the nutritional availability for the growth of microbes. The baby's nutrition has the most immediate influence on the gut flora after delivery. The composition of breast milk or formula milk plays a major role. Breastfed babies' guts are mostly colonized by *Lactobacillus* and *Bifidobacterium*. Breast milk includes oligosaccharides that are easily digested by these bacteria, resulting in the synthesis of short-chain fatty acids (SCFAs), which activate the immune system and increase immunoglobulin G expression. In contrast, formula-fed infants have a gut microbiota dominated by *Enterococcus, Enterobacteria, Bacteroides, Clostridia,* and *Streptococcus* [24, 29]. There are specific diets that benefit the growth of specific species based on their metabolic capacities. A fiber-rich diet encourages the development of beneficial microorganisms, such as those generating SCFA. A diet high in fat and sugar, on the other hand, causes dysbiosis and a decrease in microbial diversity [30].

4.3.3 Lifestyle

The impact of lifestyle on the human microbiome, especially the gut microbiome, is a growing concern for human health. Lifestyle practices, including food, physical exercise, sedentary practices, and sleep, can alter the human microbiome, thereby increasing the risk of chronic illnesses. According to studies, various dietary approaches, such as omnivorous, vegetarian, vegan, low-carbohydrate, and high-fat diets, have a substantial effect on gut microbiome composition. The use of red meat, processed foods, smoking, alcohol, etc. can alter the microbial mucosa, thereby leading to dysbiosis [31]. Smoking increases *Bacteroides* and *Prevotella* in healthy individuals and has a significant effect on the makeup of the gut microbiota [32]. Stress is another lifestyle factor that affects colonic motor activity through the GBA and alters the gut microbiota. This has led to the reduction of potentially helpful *Lactobacillus* [33].

4.3.4 Environmental Factors

The geographical location also influences the gut microbiome. People living in a rural, rich, biodiverse environment have a notably different microbiome when compared to those in urban areas. Urbanization disrupts the local ecosystem and biodiversity, leading to the degradation of environmental microbial reservoirs, including bacteria, viruses, and fungi, which are involved in the immunity mediated by the gut microbiota [34]. According to the "biodiversity hypothesis," clinical illnesses brought on by immunological dysfunction, inflammation, and an unhealthy microbiome are connected to biodiversity loss [35].

Environmental pollutants change the metabolism or survival of microorganisms, disrupting the microbiome. The respiratory system, digestive tract, and mouth are important entry paths for toxic chemicals from the environment. Unlike human enzymatic systems, bacteria employ unique routes to convert these compounds. The human body typically converts these substances by oxidation and conjugation to aid in their removal, whereas bacteria mostly employ reduction, hydrolysis, and demethylation to transform them into nutrients that they may utilize for growth and multiplication [36].

4.3.5 Antibiotics

Antibiotics eliminate both harmful and helpful microorganisms randomly, enabling dysbiosis and the proliferation of undesirable microbes. Antibiotics like Clindamycin [37], ciprofloxacin [38], clarithromycin, and metronidazole [39] show long-lasting consequences on microbiota structure. Use of antibiotics during infancy has long-term impacts, including altered microbiota composition. Early empirical antibiotic treatment has been associated to reduced *Bifidobacterium* levels and increased *Enterobacteriaceae*, perhaps leading to dysbiosis [39].

4.4 Gut-Brain Axis

The GBA is the complex bidirectional communication network that connects the gastrointestinal system (GIS) with the CNS. This network comprises hormones, the immune system, neuronal pathways, and the microbial metabolites from the gut. It permits the stomach to connect with the brain and vice versa, thereby influencing brain function [40]. Grasping the GBA is crucial for understanding the physiology and pathology of various diseases, including neurodegenerative diseases.

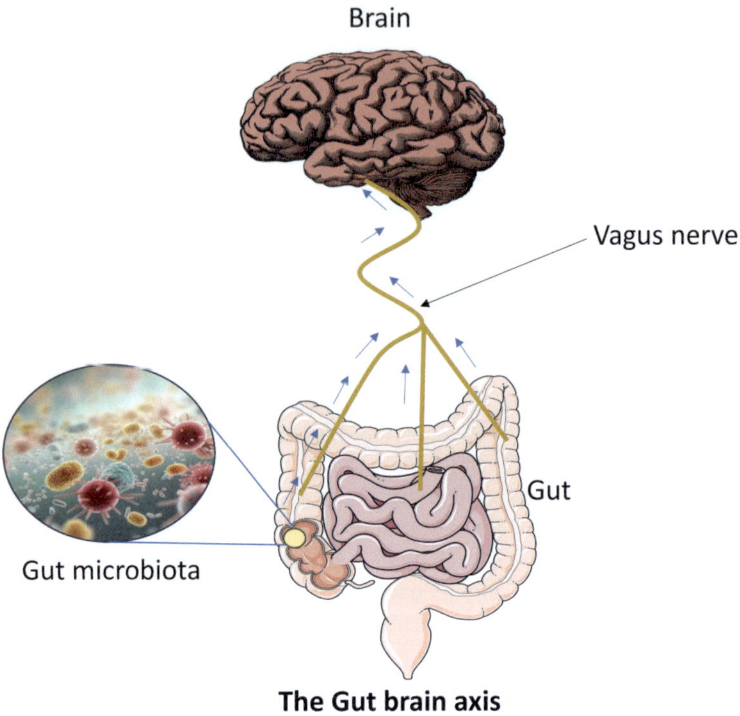

Fig. 4.1 Microbiota-gut-brain axis

The vagus nerve is the principal neurological conduit between the stomach and the brain. It transmits muscular impulses to regulate stomach function while also delivering sensory information from the gut to the brain (Fig. 4.1). The vagus nerve transmits signals that regulate gastrointestinal motility, secretion, and permeability. For example, the brain may change gut function by sending signals through the vagus nerve when it senses stress. Another important channel is the hypothalamic-pituitary-adrenal (HPA) axis. The hypothalamus releases corticotropin-releasing hormone (CRH) in response to stress, which in turn causes the pituitary gland to release adrenocorticotropic hormone (ACTH). This hormone subsequently stimulates the adrenal glands to generate corticosteroids, including cortisol. An increase in the cortisol can change the makeup of the gut bacteria by influencing the environment in the gut and immunological reactions [40, 41]. Ghrelin and leptin, hormones generated from the gut, affect the hypothalamus to control appetite and energy levels [42].

4.5 Gut Microbiome and PD

Dysbiosis, or changes in the gut microbiome, is linked to a range of illnesses, including neuronal degeneration, neuroinflammation, and gastrointestinal dysfunction [43]. Dysbiosis occurs when specific bacterial strains overgrow or the helpful bacteria decreases. Dysbiosis predominantly impacts the CNS by compromising the gut barrier's integrity. When this barrier weakens, intestinal permeability increases, resulting in "leaky gut." This further permits the toxins and other chemicals from the gut to enter the circulation, where they travel to the brain and result in neuroinflammation, which is a major consequence contributing to PD. Feces from PD patients showed high levels of leaky gut indicators like zonulin and α1-antitrypsin, as well as increased concentration of colonic inflammatory markers like calprotectin and lactoferrin [44]. This gut dysbiosis-induced neuroinflammation exacerbates both motor and non-motor symptoms [45].

As a result of a leaky gut, Gram-negative bacteria and their endotoxins, particularly lipopolysaccharide (LPS), enter the circulation, resulting in metabolic endotoxemia. Elevated LPS levels can cause systemic inflammation, leading to blood-brain barrier (BBB) disruption and allowing inflammatory chemicals from the stomach into the brain. LPS interacts with α-synuclein and promotes its clumping, leading to the creation of toxic protein aggregates [46]. The misfolding and aggregation of α-synuclein protein are the major pathological hallmarks of PD. The α-synuclein pathology was discovered in both the brain and the ENS. There is evidence that α-synuclein may originate in gut neurons and travel through the vagus nerve to the brain, leading to the progression of PD [47]. α-syn misfolding might be initiated by intestinal irritation brought on by gut bacterial pathobiont [48]. Furthermore, dysbiosis might influence the synthesis and availability of neurotransmitters and neuroactive metabolites [49]. According to molecular and cellular studies, gut dysbiosis affects the course of PD through a number of pathways, including GBA, Toll-like receptors (TLR), the humoral immune response (HIR), α-synuclein buildup, and the HPA [50].

Multiple research studies have shown that PD patients often have altered gut flora. It is thought that this change, which results in inflammation and abnormal protein aggregation, is a key factor in the pathophysiology [51]. The hypothesis that the microbiome GBA may be implicated in the underlying pathogenic processes of PD is supported by accumulating research. Non-motor symptoms, including gastrointestinal signs, usually arise before motor symptoms and the diagnosis of the illness [47]. This discovery confirms Braak's theory, which proposes that PD may begin in the ENS and spread to the CNS via the vagus nerve [52].

By reacting to microbial signals, the immune system, which includes cytokines and chemokines, enables communication. SCFAs, neurotransmitters, and neuromodulators are the gut microbiota metabolites that affect behavior and brain function [53]. SCFAs and LPS are found to be directly linked to the α-synuclein aggregation process [54].

Meta-analyses and systematic reviews have shown that opportunistic pathogens like *Porphyromonas* and *Corynebacterium*, as well as specific bacteria like

Bifidobacterium, *Lactobacillus*, and *Akkermansia*, are more common in people with PD. They also have lower concentrations of bacteria producing SCFA, such as *Faecalibacterium*, *Lachnospiraceae,* and *Prevotellaceae* [55]. A study of 39 PD patients and their healthy spouses in central China discovered an inverse association between *Prevotella* quantity and disease severity, suggesting the essential role inflammation plays in the development of the illness [56].

A family of bacteria called *Lachnospiraceae* produces butyrate, which aids in preserving intestinal health and lowering inflammation. PD patients frequently have reduced levels of these bacteria, particularly *Ruminococcus*. According to the Mini-Mental State Examination, this decline has been connected to issues with PD patient's thinking, memory, and postural stability [57, 58]. Patients with PD frequently have greater amounts of the *Verrucomicrobiaceae* family bacteria *Akkermansia muciniphila* than do healthy people. Its elevated presence may accelerate the frequency of the disease by encouraging α-synuclein to aggregate in the enteroendocrine cells of the gut [59].

A few of the dangerous molecules, like sphingolipids, trimethylamine N-oxide (TMAO), and the branched-chain fatty acid succinate, that are produced because of dysbiosis are found to be linked with neurodegeneration. TMAO has been related with increased oxidative stress, BBB rupture, mitochondrial malfunction, brain aging, and neurodegeneration [60, 61].

4.6 Microbiome Alterations as Early Biomarkers in PD

The symptoms of PD usually occur when more than 70% of the dopaminergic neurons undergo degeneration. This limited the efficiency of early disease detection [62]. Since GI symptoms arise before the beginning of disease, GI biomarkers are viewed as a possible source of early indicators for PD.

Several studies have shown that persons with PD have considerably different gut microbiomes than healthy individuals. PD patients had a substantial reduction in the number of gut flora metabolic products, which can contribute to constipation, the most prevalent GI complaint. Constipation is consequently seen as a clinical indicator for prodromal PD diagnosis [63]. In a large, well-planned case-control study, researchers observed that prodromal PD participant's gut microbiomes were like those of healthy persons and those with newly diagnosed PD. Both prodromal and recently onset PD displayed differential abundances of a variety of microbial species and microbial pathways, suggesting that these microbial changes may function as early disease markers [64].

Patients with PD frequently have low amounts of helpful bacteria such as *Prevotella* and microorganisms that produce SCFAs, as well as lower microbial diversity and distinct taxonomic shifts, including a reduction in *Firmicutes* and a rise in harmful bacteria like *Verrucomicrobiaceae* and *Akkermansia*. A decrease in *Prevotellaceae* along with an increase in *Akkermansia* has been postulated as a potential PD diagnostic feature [65]. Additionally, higher numbers of pro-inflammatory microorganisms, such as *Enterobacteriaceae*, are frequently found.

These microbiome changes could be identified noninvasively using stool samples, making microbiome profiling a viable approach for early PD identification [66, 67].

Apart from these microbial compositions, the microbiome-related molecules or their metabolites are also excellent biomarkers for PD. SCFAs, TMAO, LPS, etc. are a few such biomarkers. PD patient's plasma and CSF have been found to contain higher amounts of TMAO, which may indicate a link between disease onset and altered gut microbiota metabolism [68]. A study by He et al. on SCFAs found that lower plasma levels of acetic acid and propionic acid could distinguish MSA from PD with 80% sensitivity and 91% specificity, aiding in the differential diagnosis of synucleinopathies [69]. Succinic acid is another metabolite of intestinal microbiota. A shift in succinate levels may be a sign of disturbances in the metabolic interactions between the host and microbiota, which lead to inflammation and mitochondrial damage—two significant elements of PD pathogenesis [70].

4.7 Microbial Metabolites in Neurodegenerative Disorders

Numerous microbial metabolites with systemic effects are produced by the gut microbiota in addition to the digestive and absorptive processes. The SCFAs, which comprise butyrate, propionate, and acetate, are the most important and efficient metabolites. SCFAs are the byproducts of bacterial fermentation of indigestible dietary fibers and have a number of beneficial qualities [71]. SCFAs primarily act through three different mechanisms. The first is by binding to G-protein-coupled receptors (GPCRs), especially GPR41, GPR43, and GPR109a, which exist on the surface of the cell. The second mechanism is by regulating gene transcription via mechanisms that are independent of surface GPCR activation. Third, they immediately penetrate cells to provide energy, especially for colonocytes [72]. SCFAs that reach the circulation may enhance CNS homeostasis by strengthening the BBB, avoiding neuroinflammation, changing levels of neurotrophic factors, boosting serotonin biosynthesis, or directly stimulating the vagal afferent [73]. SCFAs are proven to maintain the quality of the mitochondrial function, thereby reducing the progression of neurodegenerative diseases. SCFAs essentially operate as cofactors for the body's mitochondrial enzymes, and they are being studied for their ability to reverse the alterations in the gut microbiota linked to a number of neurodegenerative conditions [74].

The alterations in SCFAs are a part of several neurological disorders. These changes are mostly due to the propagation of peripheral immune cells into the CNS, which results in neuroinflammation [75]. For example, multiple sclerosis (MS) is an autoimmune disease of the CNS, driven by T cells, that causes demyelination, nerve fiber damage, and long-term inflammation in the brain and spinal cord. Research indicates that people with MS have considerably decreased amounts of SCFAs in their blood and feces [76]. Interestingly, in MS mice model, the symptoms were improved upon feeding with high-fiber diet and treating with SCFAs [77].

The role of SCFAs in PD has been found to be complex. In a transgenic mouse model overexpressing human α-synuclein, it was shown that a germ-free environment prevented PD development, and oral administration of SCFAs to these mice alleviated PD symptoms, including neuroinflammation [78]. These data demonstrate the significance of SCFA in PD development. On the other hand, new research indicates that PD patients have greater plasma amounts of SCFA but lower fecal concentrations than healthy controls [79]. Interestingly, individuals with PD tend to have lower amounts of SCFAs in their feces but greater quantities in their blood than healthy people. A potential reason is that PD patients frequently have increased intestinal permeability (a "leaky gut"), which allows more SCFAs to enter the circulation. These SCFAs may subsequently get into the brain and affects neurons and support cells, adding to inflammation and disease development. Another study discovered that when animals with α-synuclein overproduction were fed gut bacteria from PD patients, their feces contained less acetate but more propionate and butyrate—different forms of SCFAs. This shows that not only the volume but also the balance of SCFAs may have a significant role in PD progression [78]. Overall, the findings suggest that SCFAs are critical mediators generated by gut bacteria that impact brain inflammation and can play a major role in the development of PD and perhaps other brain illnesses.

Another major class of compounds generated by gut bacteria is amino acid derivatives, such as those derived from tryptophan. When the gut microbiota degrades tryptophan, it produces chemicals such as indoles, indole-3-carboxylic acid, and quinolinic acid. One of the key ways by which the stomach regulates neuroinflammation is the generation of tryptophan-derived metabolites. These metabolites can help regulate the immune system and safeguard the neurological system. Quinolinic acid is remarkable for its ability to pass the BBB and affect the brain. It works by activating NMDA receptors, a kind of glutamate receptor that, if overactivated, can cause neuronal damage and increase neuroinflammation. Thus, high amounts of quinolinic acid will cause neuroinflammation and neurodegeneration [80].

Gamma-aminobutyric acid (GABA) is the brain's primary inhibitory neurotransmitter, generated both in the CNS and in the stomach. Some gut bacteria, particularly those from the *Bacteroides* and *Lactobacillus* genera, may produce or degrade GABA, altering its availability in the body. Gut microorganisms produce GABA, which supports gut health and influences brain activity [81]. Research indicates that the GABA system, which in turn affects brain function and neurological health, might be impacted by gut flora. Alterations in gut flora can interfere with GABA metabolism in diseases like PD and AD [82]. For instance, there is frequently a rise in bacteria that break down GABA in PD, which lowers its levels. Overactive neurons, mitochondrial issues, and, ultimately, brain cell death can result from this. Overall, GABA generated by microbes aids in immune response regulation, gut lining protection, and brain activity regulation, underscoring its increasing significance in the onset and progression of neurodegenerative illnesses [21].

The gut microbiota has been shown to be important in the digestion of polyphenols, which are natural bioactive chemicals found in a range of foods such as fruits, vegetables, tea, and coffee. Because polyphenols have complicated chemical structures and huge molecular sizes, they are not readily absorbed in the small intestine. The gut bacteria present in the large intestine are capable of breaking down these compounds into simpler, more absorbable molecules. Polyphenols are converted into smaller compounds like phenolic acids and simple flavonoids by bacterial activities such as hydrolysis, reduction, and demethylation [83]. These metabolites can then be absorbed by the body and even pass across the BBB to reach the CNS. Polyphenol-derived substances in the brain diminish inflammation by suppressing pro-inflammatory cytokine production and relaxing microglia, the immune cells that cause neuroinflammatory reactions. By doing so, they serve to protect neurons from the damage produced by persistent inflammation. Furthermore, these metabolites have high antioxidant activities, negating toxic ROS and safeguarding brain cells from oxidative stress, which is a major contributor to neurodegenerative illnesses [84].

4.8 Challenges and Opportunities in Microbiome-Based Diagnostics in PD

Microbiome-based diagnostics in PD are a fast-evolving field that offers promise for early identification and individualized treatment. This opens up new opportunities, but it also raises serious scientific, technological, and regulatory challenges.

4.8.1 Challenges in Microbiome-Based Diagnostics for PD

- Heterogeneity and variability: The individual gut microbiota varies significantly due to several factors such as geography, diet, lifestyle, mode of birth, etc. This heterogeneity creates complications in identifying microbial biomarkers, and it also hinders the development of robust, standardized diagnostic tools [85].
- Sampling procedures: The way how the samples are collected, as well as where and how frequently they are taken, can make a big difference. To preserve the quality and reliability of these biological samples, they must be isolated quickly and stored properly. Since the microbes can change over time, it is difficult to capture a clear picture [86].
- Limitations in technology: While breakthroughs in sequencing technologies have enabled considerable investigation into the gut microbiota, there are still obstacles in standardizing sample collection and processing and data analysis procedures. These variations can introduce biases, making it challenging to compare data from different research and laboratories.
- Lack of standardization: There is a lack of standard methodologies for sample analysis and data interpretation. This makes it more difficult to develop accurate diagnostic tools that are effective for everyone. Different research studies employ a variety of technologies, sample handling procedures, and data analysis techniques. These variances may lead to errors, making it difficult to reproduce results. As a result, it gets harder to compare the conclusions of various research projects [85].

- PD progression: The microbiome changes in PD tend to progress gradually, and not all individuals exhibit unique patterns in the early disease stages. Microbiological indicators for early detection of PD are challenging due to its temporal complexity.
- Validation of biomarkers: The majority of biomarkers discovered thus far are derived from small or minimally distributed cohorts. To verify putative biomarkers and evaluate their efficacy and predictive ability over time, extensive, multidisciplinary, and longitudinal investigations are required [87].

4.8.2 Opportunities in Microbiome-Based Diagnostics for PD

- Early detection of the disease: Some genera of gut microbes (such as *Faecalibacterium* and *Blautia*) are possibly able to differentiate PD patients from healthy controls before motor symptoms appear. This shows potential for early intervention [88].
- Noninvasive nature of biomarkers: In contrast to existing invasive or expensive neurological evaluations, fecal sample for microbial profiling provides a noninvasive diagnostic alternative [89].
- Accessibility: Sample collection may be done in a variety of situations and is rather simple, which may make it more accessible for large-scale screening or contexts with limited resources [89].
- Disease monitoring: Microbiome profiles may be used as dynamic indicators to track the course of an illness and the effectiveness of treatment, providing clinicians with insightful real-time feedback [90].
- Merging with other biomarkers: Linking metabolic, immunological, or neuroimaging biomarkers with microbiome data may enhance diagnostic precision and offer a comprehensive picture of illness causes [91].

4.9 Microbiome-Targeted Therapeutics for Parkinson's Disease

4.9.1 Dietary Interventions

The makeup of gut microbial communities is significantly influenced by nutrition and dietary intake. There are several ways that dietary changes might affect how PD develops. Changing the gut microbiota's makeup is one important strategy to influence the GBA and perhaps immune cell activity. Diet is considered to be one of the most powerful factors affecting both the composition and metabolic activity of the gut microbiota. Furthermore, certain dietary elements may help control the chronic inflammation often associated with aging. Curiously, food can also improve the absorption of L-dopa, the main therapy for PD, and relieve constipation [92]. Coffee use raises the intestinal anti-inflammatory bifidobacteria levels, which enhance intestinal mucosal function and decrease gut inflammation, hence lowering the risk of PD [93]. In individuals with PD, a Mediterranean diet that emphasizes eating fruits, vegetables, fish, whole grains, nuts, and olives has been proven to reduce the abundance of *Clostridium bolteae, Ruminococcus, Blautia, Dorea*, and *Lachnospiraceae* and alleviate symptoms of constipation [94]. The most advised diet for managing PD is the Mediterranean diet since it is rich in omega-3 fatty acids, polyphenols, and antioxidants. By encouraging the synthesis of SCFAs, a high-fiber diet can help prevent PD [95]. The ketogenic diet has become more well-known and has demonstrated potential in the management of PD. It has low carbs, high fat, and moderate protein. It promotes improved mitochondrial activity and reduces oxidative stress, two factors that are frequently compromised in PD patients. Furthermore, this diet encourages the production of ketone bodies like beta-hydroxybutyrate, which give brain cells an alternative energy source while reducing inflammation and safeguarding neurons [50]. According to research, individuals with PD who ate more omega-3 fatty acids had improved cognitive performance, reduced oxidative stress, and generated less α-synuclein. Supplementing with omega-3 also assisted in re-establishing beneficial gut flora [96]. In a similar vein, vitamin D supplements have been demonstrated to enhance PD patient's behavior, cognitive function, and gastrointestinal health [97]. Although dietary modifications cannot directly halt the course of PD, they are anticipated to have some positive effects by affecting the gut-brain-microbiota axis.

4.9.2 Probiotics

Probiotics are the living organisms that, when taken, provide positive health effects. Probiotics may help restore the GM; improve immunological homeostasis; reduce intestinal permeability, inflammation, and oxidative stress; alleviate GI symptoms including bloating, diarrhea, and abdominal pain; and prevent neurological disorders [55]. In PD, probiotics protect against neurodegeneration via regulating gut dysbiosis through the GBA. Probiotics are used to lower the primary hazards of oxidative stress and α-synuclein aggregation. *Lactobacillus, Blautia, Roseburia, Lachnospiraceae, Prevotellaceae,* and *Akkermansia* are among the helpful bacteria found in probiotics that are capable of reducing PD symptoms and halting the disease progression [98]. Ghyselinck et al. discovered that using this multi-strain probiotic, which includes *Lactobacillus acidophilus, Lactobacillus plantarum, Lactobacillus rhamnosus, and Enterococcus faecium*, can regulate the mucosal inflammation (by increasing anti-inflammatory cytokines like IL-6 and IL-10 and decreasing the pro-inflammatory chemokine IL-8), improve intestinal permeability, and alter the bacterial composition (by boosting *Firmicutes* and decreasing *Bacteroidetes*) [99].

4.9.3 Prebiotics

Prebiotics are substrates that host bacteria specially use to preserve the health benefits mediated by the microflora while also offering benefits to the host [100]. The dietary fibers found in raw oats, soybeans, unprocessed wheat, barley, and other foods constitute the primary source of prebiotics. They encourage the proliferation of helpful bacteria. The beneficial bacteria in the gut break down the prebiotics, creating SCFAs (acetate, propionate, and butyrate), which play roles in neuromodulation, anti-inflammatory processes, and the management of intestinal and blood-brain barrier function [101]. In a mouse model of PD, Perez-Pardo et al. reported that prebiotic fibers can alleviate motor symptoms, reduce synuclein levels, and recover function [102]. Polymannuronic acid, a prebiotic, has been proven to reduce dopaminergic neuronal loss through its anti-inflammatory and anti-apoptotic properties [103]. Another study discovered that taking oral supplements with resistant starch, which is digested by anaerobic bacteria to create SCFAs, resulted in higher butyrate concentrations and decreased non-motor symptoms [104].

4.9.4 Fecal Microbiota Transplantation (FMT)

FMT entails transferring fecal material from individuals who are healthy into the patient's stomach to counteract dysbiosis. This has been increasingly used for the management of neurodegenerative diseases. FMT was proven in several in vivo studies to successfully decrease inflammation, improve gut microbiome dysbiosis, and raise striatal dopamine and 5-hydroxytryptamine levels [105, 106]. FMT oral capsules have been shown to restore the gut microbiota in PD patients, alleviating GI symptoms as well as quality of life [107]. Bruggeman et al. conducted the first randomized, double-blind, placebo-controlled study on 46 PD patients treated with nasojejunal FMT. His findings suggest that in initial-stage patients, a single FMT treatment relieves motor symptoms [108]. There is a scarcity of evidence from large, long-term randomized controlled clinical studies and FMT for PD is still in its infancy. To confirm its durability, effectiveness, and safety, more clinical trials are required.

4.10 Conclusion

PD and the gut microbiome are intricately intertwined, with current research presenting the gut microbiome as a possible area for early diagnosis of PD. The GBA forms the major communication network. Gut microbiota alterations contribute to PD onset and progression through various mechanisms such as increased permeability, inflammation, α-synuclein fibril aggregation, unbalanced oxidative stress, and diminished neurotransmitter production. Microbiome-targeted therapies, including dietary interventions, probiotics, prebiotics, and fecal microbiota transplantation, provide new opportunities for controlling PD symptoms and perhaps slowing disease progression. Overall, additional study into the gut microbiota may pave the path for more tailored and successful PD diagnosis and treatment.

Declarations

Competing Interests The author(s) has no competing interests to declare that are relevant to the content of this manuscript.

Ethics Approval The author(s) declares that this manuscript does not include primary research with human participants.

The author(s) declares that this manuscript does not include primary studies involving animals.

References

1. Chakkittukandiyil A, Sajini DV, Rymbai E, Sugumar D, Mathew J, Arumugam S, Ramachandran V, Selvaraj D. Synthesis and evaluation of novel ethyl ferulate derivatives as potent Keap1 inhibitors to activate the Nrf2/ARE pathway in Parkinson's disease. Toxicol Appl Pharmacol. 2025;494:117172.
2. Kwon D, Zhang K, Paul KC, Folle AD, Del Rosario I, Jacobs JP, Keener AM, Bronstein JM, Ritz B. Diet and the gut microbiome in patients with Parkinson's disease. NPJ Parkinsons Dis. 2024;10:89.
3. Baldini F, Hertel J, Sandt E, et al. Parkinson's disease-associated alterations of the gut microbiome predict disease-relevant changes in metabolic functions. BMC Biol. 2020;18:62.
4. Kuter K, Kratochwil M, Berghauzen-Maciejewska K, Głowacka U, Sugawa MD, Ossowska K, Dencher NA. Adaptation within mitochondrial oxidative phosphorylation supercomplexes and membrane viscosity during degeneration of dopaminergic neurons in an animal model of early Parkinson's disease. Biochim Biophys Acta (BBA) – Mol Basis Dis. 2016;1862:741–53.
5. Bloem BR, Okun MS, Klein C. Parkinson's disease. Lancet. 2021;397:2284–303.
6. Chakkittukandiyil A, Sajini DV, Karuppaiah A, Selvaraj D. The principal molecular mechanisms behind the activation of Keap1/Nrf2/ARE pathway leading to neuroprotective action in Parkinson's disease. Neurochem Int. 2022;156:105325.
7. Salim S, Ahmad F, Banu A, Mohammad F. Gut microbiome and Parkinson's disease: perspective on pathogenesis and treatment. J Adv Res. 2023;50:83–105.
8. Braak H, Rüb U, Gai WP, Del Tredici K. Idiopathic Parkinson's disease: possible routes by which vulnerable neuronal types may be subject to neuroinvasion by an unknown pathogen. J Neural Transm (Vienna). 2003;110:517–36.
9. Tan AH, Hor JW, Chong CW, Lim S. Probiotics for Parkinson's disease: current evidence and future directions. JGH Open. 2020;5:414–9.
10. Thursby E, Juge N. Introduction to the human gut microbiota. Biochem J. 2017;474:1823–36.
11. Sleator RD. The human superorganism – of microbes and men. Med Hypotheses. 2010;74:214–5.
12. Carabotti M, Scirocco A, Maselli MA, Severi C. The gut-brain axis: interactions between enteric microbiota, central and enteric nervous systems. Ann Gastroenterol. 2015;28:203–9.
13. Hashish S, Salama M. The role of an altered gut microbiome in Parkinson's disease: a narrative review. Appl Microbiol. 2023;3:429–47.
14. K M, Ghosh P, Nagappan K, Palaniswamy DS, Begum R, Islam MR, Tagde P, Shaikh NK, Farahim F, Mondal TK. From gut microbiomes to infectious pathogens: neurological disease game changers. Mol Neurobiol. 2025;62:1184–204.
15. Gilbert J, Blaser MJ, Caporaso JG, Jansson J, Lynch SV, Knight R. Current understanding of the human microbiome. Nat Med. 2018;24:392–400.

16. Ley RE, Peterson DA, Gordon JI. Ecological and evolutionary forces shaping microbial diversity in the human intestine. Cell. 2006;124:837–48.
17. Chen Y, Xu J, Chen Y. Regulation of neurotransmitters by the gut microbiota and effects on cognition in neurological disorders. Nutrients. 2021;13:2099.
18. O'Hara AM, Shanahan F. The gut flora as a forgotten organ. EMBO Rep. 2006;7:688–93.
19. Ley RE, Turnbaugh PJ, Klein S, Gordon JI. Microbial ecology: human gut microbes associated with obesity. Nature. 2006;444:1022–3.
20. Moran CP, Shanahan F. Gut microbiota and obesity: role in aetiology and potential therapeutic target. Best Pract Res Clin Gastroenterol. 2014;28:585–97.
21. Missiego-Beltrán J, Beltrán-Velasco AI. The role of microbial metabolites in the progression of neurodegenerative diseases—therapeutic approaches: a comprehensive review. Int J Mol Sci. 2024;25:10041.
22. Mosca A, Leclerc M, Hugot JP. Gut microbiota diversity and human diseases: should we reintroduce key predators in our ecosystem? Front Microbiol. 2016. https://doi.org/10.3389/fmicb.2016.00455.
23. Singh Y, El-Hadidi M, Admard J, Wassouf Z, Schulze-Hentrich JM, Kohlhofer U, Quintanilla-Martinez L, Huson D, Riess O, Casadei N. Enriched environmental conditions modify the gut microbiome composition and fecal markers of inflammation in Parkinson's disease. Front Neurosci. 2019. https://doi.org/10.3389/fnins.2019.01032.
24. Hasan N, Yang H. Factors affecting the composition of the gut microbiota, and its modulation. PeerJ. 2019;7:e7502.
25. Dominguez-Bello MG, Costello EK, Contreras M, Magris M, Hidalgo G, Fierer N, Knight R. Delivery mode shapes the acquisition and structure of the initial microbiota across multiple body habitats in newborns. Proc Natl Acad Sci USA. 2010;107:11971–5.
26. Cani PD. Human gut microbiome: hopes, threats and promises. Gut. 2018;67:1716–25.
27. Weng M, Walker WA. The role of gut microbiota in programming the immune phenotype. J Dev Orig Health Dis. 2013;4:203–14.
28. Yatsunenko T, Rey FE, Manary MJ, et al. Human gut microbiome viewed across age and geography. Nature. 2012;486:222–7.
29. Yoshioka H, Iseki K, Fujita K. Development and differences of intestinal flora in the neonatal period in breast-fed and bottle-fed infants. Pediatrics. 1983;72:317–21.
30. Zhang F, Fan D, Huang J, Zuo T. The gut microbiome: linking dietary fiber to inflammatory diseases. Med Microecol. 2022;14:100070.
31. Conlon MA, Bird AR. The impact of diet and lifestyle on gut microbiota and human health. Nutrients. 2014;7:17–44.
32. Benjamin JL, Hedin CRH, Koutsoumpas A, et al. Smokers with active Crohn's disease have a clinically relevant dysbiosis of the gastrointestinal microbiota. Inflamm Bowel Dis. 2012;18:1092–100.
33. Lutgendorff F, Akkermans LMA, Söderholm JD. The role of microbiota and probiotics in stress-induced gastro-intestinal damage. Curr Mol Med. 2008;8:282–98.
34. Tasnim N, Abulizi N, Pither J, Hart MM, Gibson DL. Linking the gut microbial ecosystem with the environment: does gut health depend on where we live? Front Microbiol. 2017;8:1935.
35. (PDF) Urbanization, biodiversity and ecosystem services: challenges and opportunities: a global assessment. https://www.researchgate.net/publication/297563873_Urbanization_biodiversity_and_ecosystem_services_challenges_and_opportunities_a_global_assessment. Accessed 24 July 2025.
36. Ahn J, Hayes RB. Environmental influences on the human microbiome and implications for noncommunicable disease. Annu Rev Public Health. 2021;42:277–92.
37. Jernberg C, Löfmark S, Edlund C, Jansson JK. Long-term ecological impacts of antibiotic administration on the human intestinal microbiota. ISME J. 2007;1:56–66.
38. Dethlefsen L, Relman DA. Incomplete recovery and individualized responses of the human distal gut microbiota to repeated antibiotic perturbation. Proc Natl Acad Sci USA. 2011;108(Suppl 1):4554–61.

39. Jakobsson HE, Jernberg C, Andersson AF, Sjölund-Karlsson M, Jansson JK, Engstrand L. Short-term antibiotic treatment has differing long-term impacts on the human throat and gut microbiome. PLoS One. 2010;5:e9836.
40. Ashique S, Mohanto S, Ahmed MG, Mishra N, Garg A, Chellappan DK, Omara T, Iqbal S, Kahwa I. Gut-brain axis: a cutting-edge approach to target neurological disorders and potential synbiotic application. Heliyon. 2024;10:e34092.
41. Santos JCCD, Lima MPP, de Castro Brito GA, de Barros Viana GS. Role of enteric glia and microbiota-gut-brain axis in Parkinson disease pathogenesis. Ageing Res Rev. 2023;84:101812.
42. Yeung AY, Tadi P. Physiology, obesity neurohormonal appetite and satiety control. In: StatPearls. Treasure Island (FL): StatPearls Publishing; 2025.
43. Dos Santos JCC, Lima MPP, de Castro Brito GA, de Barros Viana GS. Role of enteric glia and microbiota-gut-brain axis in Parkinson disease pathogenesis. Ageing Res Rev. 2023;84:101812.
44. Schwiertz A, Spiegel J, Dillmann U, Grundmann D, Bürmann J, Faßbender K, Schäfer K-H, Unger MM. Fecal markers of intestinal inflammation and intestinal permeability are elevated in Parkinson's disease. Parkinsonism Relat Disord. 2018;50:104–7.
45. Shannon KM. Gut-derived sterile inflammation and Parkinson's disease. Front Neurol. 2022;13:831090.
46. Bhattacharyya D, Mohite GM, Krishnamoorthy J, et al. Lipopolysaccharide from gut microbiota modulates α-synuclein aggregation and alters its biological function. ACS Chem Neurosci. 2019;10:2229–36.
47. Klann EM, Dissanayake U, Gurrala A, Farrer M, Shukla AW, Ramirez-Zamora A, Mai V, Vedam-Mai V. The gut-brain axis and its relation to Parkinson's disease: a review. Front Aging Neurosci. 2021;13:782082.
48. Willyard C. How gut microbes could drive brain disorders. Nature. 2021;590:22–5.
49. Qian Y, Yang X, Xu S, Wu C, Song Y, Qin N, Chen S-D, Xiao Q. Alteration of the fecal microbiota in Chinese patients with Parkinson's disease. Brain Behav Immun. 2018;70:194–202.
50. Alam M, Abbas K, Mustafa M, Usmani N, Habib S. Microbiome-based therapies for Parkinson's disease. Front Nutr. 2024. https://doi.org/10.3389/fnut.2024.1496616.
51. Keshavarzian A, Green SJ, Engen PA, Voigt RM, Naqib A, Forsyth CB, Mutlu E, Shannon KM. Colonic bacterial composition in Parkinson's disease. Mov Disord. 2015;30:1351–60.
52. Rietdijk CD, Perez-Pardo P, Garssen J, van Wezel RJA, Kraneveld AD. Exploring Braak's hypothesis of Parkinson's disease. Front Neurol. 2017;8:37.
53. Cryan JF, O'Riordan KJ, Cowan CSM, et al. The microbiota-gut-brain axis. Physiol Rev. 2019;99:1877–2013.
54. Fitzgerald E, Murphy S, Martinson HA. Alpha-synuclein pathology and the role of the microbiota in Parkinson's disease. Front Neurosci. 2019;13:369.
55. Varesi A, Campagnoli LIM, Fahmideh F, Pierella E, Romeo M, Ricevuti G, Nicoletta M, Chirumbolo S, Pascale A. The interplay between gut microbiota and Parkinson's disease: implications on diagnosis and treatment. Int J Mol Sci. 2022;23:12289.
56. Mao L, Zhang Y, Tian J, Sang M, Zhang G, Zhou Y, Wang P. Cross-sectional study on the gut microbiome of Parkinson's disease patients in Central China. Front Microbiol. 2021;12:728479.
57. Lin A, Zheng W, He Y, et al. Gut microbiota in patients with Parkinson's disease in southern China. Parkinsonism Relat Disord. 2018;53:82–8.
58. Barichella M, Severgnini M, Cilia R, et al. Unraveling gut microbiota in Parkinson's disease and atypical Parkinsonism. Mov Disord. 2019;34:396–405.
59. Amorim Neto DP, Bosque BP, Pereira de Godoy JV, Rodrigues PV, Meneses DD, Tostes K, Costa Tonoli CC, Faustino de Carvalho H, González-Billault C, de Castro FM. Akkermansia muciniphila induces mitochondrial calcium overload and α-synuclein aggregation in an enteroendocrine cell line. iScience. 2022;25:103908.

60. Li D, Ke Y, Zhan R, et al. Trimethylamine-N-oxide promotes brain aging and cognitive impairment in mice. Aging Cell. 2018;17:e12768.
61. Voigt RM, Wang Z, Brown JM, et al. Gut microbial metabolites in Parkinson's disease: association with lifestyle, disease characteristics, and treatment status. Neurobiol Dis. 2022;170:105780.
62. Vasudevan Sajini D, Thaggikuppe Krishnamurthy P, Chakkittukandiyil A, Mudavath RN. Orientin modulates Nrf2-ARE, PI3K/Akt, JNK-ERK1/2, and TLR4/NF-kB pathways to produce neuroprotective benefits in Parkinson's disease. Neurochem Res. 2024;49:1577–87.
63. Shen T, Yue Y, He T, Huang C, Qu B, Lv W, Lai H-Y. The association between the gut microbiota and Parkinson's disease, a meta-analysis. Front Aging Neurosci. 2021;13:636545.
64. Palacios N, Wilkinson J, Bjornevik K, Schwarzschild MA, McIver L, Ascherio A, Huttenhower C. Metagenomics of the gut microbiome in Parkinson's disease: prodromal changes. Ann Neurol. 2023;94:486–501.
65. Vidal-Martinez G, Chin B, Camarillo C, Herrera GV, Yang B, Sarosiek I, Perez RG. A pilot microbiota study in Parkinson's disease patients versus control subjects, and effects of FTY720 and FTY720-mitoxy therapies in parkinsonian and multiple system atrophy mouse models. J Parkinsons Dis. 2020;10:185–92.
66. Suresh SB, Malireddi A, Abera M, Noor K, Ansar M, Boddeti S, Nath TS. Gut microbiome and its role in Parkinson's disease. Cureus. 2024;16:e73150.
67. Silva YP, Bernardi A, Frozza RL. The role of short-chain fatty acids from gut microbiota in gut-brain communication. Front Endocrinol. 2020;11:25.
68. Gątarek P, Pawełczyk M, Bobrowska-Korczaka B, Giebułtowicz J, Głąbiński A, Kałużna-Czaplińska J. The determination of trimethylamine N-oxide and organic acids connected with gut microbiota in the urine of Parkinson's disease patients: a pilot study. Int J Mol Sci. 2025;26:4575.
69. He X, Qian Y, Xu S, Zhang Y, Mo C, Guo W, Yang X, Xiao Q. Plasma short-chain fatty acids differences in multiple system atrophy from Parkinson's disease. J Parkinsons Dis. 2021;11:1167–76.
70. Fernández-Veledo S, Vendrell J. Gut microbiota-derived succinate: friend or foe in human metabolic diseases? Rev Endocr Metab Disord. 2019;20:439–47.
71. Jin K, Chen B, Han S, Dong J, Cheng S, Qin B, Lu J. Repetitive transcranial magnetic stimulation (rTMS) improves cognitive impairment and intestinal microecological dysfunction induced by high-fat diet in rats. Research (Wash D C). 2024;7:0384.
72. Prado C, Pacheco R. Targeting short-chain fatty acids receptors signalling for neurological disorders treatment. Explor Neuroprot Ther. 2024;4:100–7.
73. Jaworska K, Kopacz W, Koper M, Szudzik M, Gawryś-Kopczyńska M, Konop M, Hutsch T, Chabowski D, Ufnal M. Enalapril diminishes the diabetes-induced changes in intestinal morphology, intestinal RAS and blood SCFA concentration in rats. Int J Mol Sci. 2022;23:6060.
74. Yadav S, Dwivedi A, Tripathi A, Tripathi AK. Therapeutic potential of short-chain fatty acid production by gut microbiota in neurodegenerative disorders. Nutr Res. 2022;106:72–84.
75. Manjarres Z, Calvo M, Pacheco R. Regulation of pain perception by microbiota in Parkinson disease. Pharmacol Rev. 2023;76:7–36.
76. Saresella M, Marventano I, Barone M, et al. Alterations in circulating fatty acid are associated with gut microbiota dysbiosis and inflammation in multiple sclerosis. Front Immunol. 2020;11:1390.
77. Mizuno M, Noto D, Kaga N, Chiba A, Miyake S. The dual role of short fatty acid chains in the pathogenesis of autoimmune disease models. PLoS One. 2017;12:e0173032.
78. Sampson TR, Debelius JW, Thron T, et al. Gut microbiota regulate motor deficits and neuroinflammation in a model of Parkinson's disease. Cell. 2016;167:1469–1480.e12.
79. Chen S-J, Chen C-C, Liao H-Y, Lin Y-T, Wu Y-W, Liou J-M, Wu M-S, Kuo C-H, Lin C-H. Association of fecal and plasma levels of short-chain fatty acids with gut microbiota and clinical severity in patients with Parkinson disease. Neurology. 2022;98:e848–58.

80. Wang X, Sun G, Feng T, et al. Sodium oligomannate therapeutically remodels gut microbiota and suppresses gut bacterial amino acids-shaped neuroinflammation to inhibit Alzheimer's disease progression. Cell Res. 2019;29:787–803.
81. Marzouk NH, Rashwan HH, El-Hadidi M, Ramadan R, Mysara M. Proinflammatory and GABA eating bacteria in Parkinson's disease gut microbiome from a meta-analysis perspective. NPJ Parkinsons Dis. 2025;11:145.
82. Conn KA, Borsom EM, Cope EK. Implications of microbe-derived ɣ-aminobutyric acid (GABA) in gut and brain barrier integrity and GABAergic signaling in Alzheimer's disease. Gut Microbes. 2024;16:2371950.
83. Vita AA, Roberts KM, Gundersen A, Farris Y, Zwickey H, Bradley R, Weir TL. Relationships between habitual polyphenol consumption and gut microbiota in the INCLD health cohort. Nutrients. 2024;16:773.
84. Tayab MA, Islam MN, Chowdhury KAA, Tasnim FM. Targeting neuroinflammation by polyphenols: a promising therapeutic approach against inflammation-associated depression. Biomed Pharmacother. 2022;147:112668.
85. Allaband C, McDonald D, Vázquez-Baeza Y, et al. Microbiome 101: studying, analyzing, and interpreting gut microbiome data for clinicians. Clin Gastroenterol Hepatol. 2018;17:218.
86. Rudi K, Zhao L. Grand challenges in understanding gut microbes. Front Microbiol. 2021;12:752829.
87. Bonassi S, Neri M, Puntoni R. Validation of biomarkers as early predictors of disease. Mutat Res. 2001;480–481:349–58.
88. Rob M, Yousef M, Lakshmanan AP, Mahboob A, Terranegra A, Chaari A. Microbial signatures and therapeutic strategies in neurodegenerative diseases. Biomed Pharmacother. 2025;184:117905.
89. Karu N, Deng L, Slae M, Guo AC, Sajed T, Huynh H, Wine E, Wishart DS. A review on human fecal metabolomics: methods, applications and the human fecal metabolome database. Anal Chim Acta. 2018;1030:1–24.
90. Zhu M, Liu X, Ye Y, Yan X, Cheng Y, Zhao L, Chen F, Ling Z. Gut microbiota: a novel therapeutic target for Parkinson's disease. Front Immunol. 2022;13:937555.
91. Nair AT, Ramachandran V, Joghee NM, Antony S, Ramalingam G. Gut microbiota dysfunction as reliable non-invasive early diagnostic biomarkers in the pathophysiology of Parkinson's disease: a critical review. J Neurogastroenterol Motil. 2018;24:30–42.
92. Jackson A, Forsyth CB, Shaikh M, Voigt RM, Engen PA, Ramirez V, Keshavarzian A. Diet in Parkinson's disease: critical role for the microbiome. Front Neurol. 2019;10:1245.
93. Jaquet M, Rochat I, Moulin J, Cavin C, Bibiloni R. Impact of coffee consumption on the gut microbiota: a human volunteer study. Int J Food Microbiol. 2009;130:117–21.
94. Rusch C, Beke M, Tucciarone L, et al. Mediterranean diet adherence in people with Parkinson's disease reduces constipation symptoms and changes fecal microbiota after a 5-week single-arm pilot study. Front Neurol. 2021;12:794640.
95. Makki K, Deehan EC, Walter J, Bäckhed F. The impact of dietary fiber on gut microbiota in host health and disease. Cell Host Microbe. 2018;23:705–15.
96. Kumar M, Pal N, Sharma P, et al. Omega-3 fatty acids and their interaction with the gut microbiome in the prevention and amelioration of type-2 diabetes. Nutrients. 2022;14:1723.
97. Barichella M, Garrì F, Caronni S, Bolliri C, Zocchi L, Macchione MC, Ferri V, Calandrella D, Pezzoli G. Vitamin D status and Parkinson's disease. Brain Sci. 2022;12:790.
98. Rout M, Prusty SK, Kar DM. Probiotic: a gut microbiota-based therapeutic approaches for the treatment of Parkinson's disease. Curr Rev Clin Exp Pharmacol. 2024. https://doi.org/10.2174/0127724328332572241219102122.
99. Ghyselinck J, Verstrepen L, Moens F, et al. Influence of probiotic bacteria on gut microbiota composition and gut wall function in an in-vitro model in patients with Parkinson's disease. Int J Pharm X. 2021;3:100087.
100. Gibson GR, Hutkins R, Sanders ME, et al. Expert consensus document: the International Scientific Association for Probiotics and Prebiotics (ISAPP) consensus statement on the definition and scope of prebiotics. Nat Rev Gastroenterol Hepatol. 2017;14:491–502.

101. Wang L, Cui Y, Han B, et al. Gut microbiota and Parkinson's disease. Chin Med J. 2025;138:289–97.
102. Perez-Pardo P, de Jong EM, Broersen LM, van Wijk N, Attali A, Garssen J, Kraneveld AD. Promising effects of neurorestorative diets on motor, cognitive, and gastrointestinal dysfunction after symptom development in a mouse model of Parkinson's disease. Front Aging Neurosci. 2017;9:57.
103. Liu X, Du ZR, Wang X, Sun XR, Zhao Q, Zhao F, Wong WT, Wong KH, Dong X-L. Polymannuronic acid prebiotic plus Lacticaseibacillus rhamnosus GG probiotic as a novel synbiotic promoted their separate neuroprotection against Parkinson's disease. Food Res Int. 2022;155:111067.
104. Becker A, Schmartz GP, Gröger L, et al. Effects of resistant starch on symptoms, fecal markers, and gut microbiota in Parkinson's disease — the RESISTA-PD trial. Genomics Proteomics Bioinformatics. 2022;20:274–87.
105. Zhao Z, Ning J, Bao X-Q, Shang M, Ma J, Li G, Zhang D. Fecal microbiota transplantation protects rotenone-induced Parkinson's disease mice via suppressing inflammation mediated by the lipopolysaccharide-TLR4 signaling pathway through the microbiota-gut-brain axis. Microbiome. 2021;9:226.
106. Sun M-F, Zhu Y-L, Zhou Z-L, Jia X-B, Xu Y-D, Yang Q, Cui C, Shen Y-Q. Neuroprotective effects of fecal microbiota transplantation on MPTP-induced Parkinson's disease mice: gut microbiota, glial reaction and TLR4/TNF-α signaling pathway. Brain Behav Immun. 2018;70:48–60.
107. Cheng Y, Tan G, Zhu Q, et al. Efficacy of fecal microbiota transplantation in patients with Parkinson's disease: clinical trial results from a randomized, placebo-controlled design. Gut Microbes. 2023;15:2284247.
108. Bruggeman A, Vandendriessche C, Hamerlinck H, et al. Safety and efficacy of faecal microbiota transplantation in patients with mild to moderate Parkinson's disease (GUT-PARFECT): a double-blind, placebo-controlled, randomised, phase 2 trial. EClinicalMedicine. 2024;71:102563.

ALS and the Microbiome: Understanding the Gut's Impact on Motor Neuron Degeneration

5

Deepika N P and Muhasina K M

5.1 Human Motor System and Motor Neuron Diseases (MNDs)

The motor system is a very complex system that governs voluntary movement. Upper motor neurons (UMNs), lower motor neurons (LMNs), and interneurons make up its pyramidal and extrapyramidal systems. Large pyramidal cells in layer 5 of the motor cortex called upper motor neurons (UMNs) send signals to interneurons or lower motor neurons (LMNs) via the corticospinal tract to start voluntary movement. Direct motor control is provided by pyramidal tracts as opposed to indirect extrapyramidal tracts. Because of fiber decussation, UMNs regulate muscles on the other side of the body in a somatotopic arrangement that mirrors the cortical homunculus: legs medially, face laterally [20, 21].

Skeletal muscles are directly controlled by lower motor neurons (LMNs), which are found in the brainstem and spinal cord. They are categorized into motor pools, each of which innervates a single muscle: αMNs are in charge of the contraction of extrafusal fibers, are essential for motor function, and are most impacted by neurodegeneration; γMNs adjust muscle spindle sensitivity by modifying intrafusal fibers; and rare βMNs innervate both types of fibers [22]. By the type of fiber, αMNs are further separated as SFR (slow-twitch, small, for posture), FF (big, fast-twitch, for powerful, quick movements), and FFR (intermediate traits) [6, 12].

Deepika N P (✉)
Department of Pharmacognosy, MGM Silver Jubilee College of Pharmaceutical Sciences and Research, Pampakuda, Ernakulum, Kerala, India

Muhasina K M
Department of Pharmaceutical Analysis, Prime College of Pharmacy, Erattayal, Palakkad, Kerala, India
e-mail: km.muhasina@gmail.com

M. H. Rahman, Muhasina K M (eds.), *The Microbiome's Impact on the Brain*,
https://doi.org/10.1007/978-981-95-7494-0_5

The cells that regulate skeletal muscle function, including walking, breathing, speaking, and swallowing, are called motor neurons, and they are destroyed in motor neuron diseases (MNDs), a class of degenerative neurological disorders. Amyotrophic lateral sclerosis, progressive bulbar palsy, primary lateral sclerosis, progressive muscular atrophy, spinal muscular atrophy, Kennedy's disease, and post-polio syndrome are among the conditions that fall under this category [41]. MNDs can be categorized as either sporadic or inherited, and they can impact either the lower or upper motor neurons or both. Despite the fact that MNDs come in a variety of forms, they all result in muscle weakness that progressively gets worse over time and can cause physical handicap. Breathlessness, shortness of breath while lying down, frequent chest infections, sleep disturbances, poor memory and/or attention, disorientation, headaches in the morning, and exhaustion are some of the symptoms.

MNDs will occur in both adults and children. In the majority of cases, MNDs cannot be diagnosed using particular tests. Individual differences in symptoms and their early resemblance to other illnesses might make diagnosis challenging. However, SMA, Kennedy's disease, and certain hereditary causes of ALS can be tested genetically. Most MNDs have no known cure, while new therapies are being developed. Some symptoms can be treated, and supportive care can help people keep their independence and standard of living. In the treatment of people with MNDs, multidisciplinary clinics comprising experts in neurology, physical and occupational therapy, respiratory therapy, speech therapy, and social work are very crucial [28].

5.2 Amyotrophic Lateral Sclerosis

The main hallmark of amyotrophic lateral sclerosis (ALS) is progressive muscular paralysis resulting from the loss of motor neurons in the primary motor cortex, corticospinal tracts, brainstem, and spinal cord. It affects the human motor system and is idiopathic [24, 37]. The degeneration of both upper motor neurons (which extend from the cortex to the spinal cord and brainstem) and lower motor neurons (which project from the brainstem or spinal cord to the muscles) causes motor and extra-motor symptoms in ALS [33].

Considering the epidemiology of ALS, the data differs greatly between geographic regions and ancestral backgrounds; ALS is regarded as a rare disease with a global incidence rate of 1–2 per 100,000 people annually because of the disease's rapid development and death [25]. Approximately 18,000 Americans are afflicted with ALS at any given moment. About 5 out of every 100,000 persons have this condition each year, and it typically lasts 3 years before dying. Males are more likely than females to have this illness [3]. In Europe, there are two to three cases per 100,000 people. Defined geographic regions are best suited to estimating incidence and prevalence as well as supporting more thorough research on ALS risk, clinical trajectory, outcome, and service consumption [33]. In India, the mean age

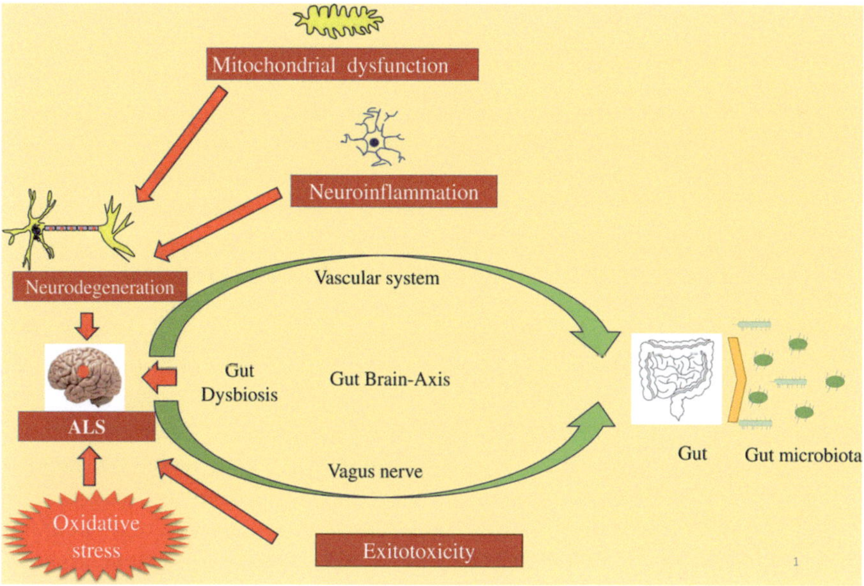

Fig. 5.1 Different risk factors in ALS

of the overall cohort was 55 years, and the male-to-female ratio was 2.5:1. The mean duration of illness for the cohort was 16 months [35].

The main etiological factors of the disease are complicated interaction of excitotoxicity, mitochondrial failure, oxidative stress, protein aggregation, and neuroinflammation (Fig. 5.1), which results in motor neuron loss and paralysis [16]. Because of its rarity, a variety of causes, and lack of accurate biomarkers, ALS treatment is still difficult. Nevertheless, the therapeutic pipeline is growing, focusing on inflammation, protein aggregation, retroviruses, and genetic abnormalities. Tofersen has the potential to slow the advancement of SOD1 mutations. TDP-43 pathology may be broadly impacted by BIIB105 (Ataxin-2 ASO), while frequent intrathecal administration is a drawback. AROSOD1 and other RNAi treatments may provide longer-lasting effects at lower dosages. The involvement of retroviruses may be clarified by a phase III antiviral therapy experiment [25].

5.3 Motor Neuron Degeneration and Gut Microbiome

The ecological community of pathogenic, commensal, and symbiotic bacteria that coexist in human body space is known as the microbiome. The greatest concentration of microorganisms is found in the human gut. If homeostasis is disrupted and the balance between beneficial and opportunistic microorganisms is not maintained, it may lead to Crohn's disease, ulcerative colitis, inflammatory bowel disease, irritable bowel disease, colorectal cancer, metabolic disorders, and neurodegenerative diseases, including diseases of the motor neurons [9]. There is increasing evidence that the gut microbiota has a contribution to the emergence of neurodegeneration through the gut-brain-microbiota axis [26]. By means of the gut-brain-microbiota (MGB) axis, changes in the gut microbiome might affect psychological and cognitive processes. Its primary roles include monitoring gut activity and establishing connections between peripheral gut mechanisms, including immune activation, gut permeability, and enteric reflexes. It is well known that the gut microbiota is extremely sensitive to a variety of environmental stimuli, and that factors like age and food have a significant impact on the gut microbiome's makeup [8].

Over the past 20 years, a number of studies have emphasized the critical function that the microbiota plays in preserving health. With an estimated human to bacterial cell ratio of about 1:1, the gut flora makes up 90% of the entire microbiota and is mostly made up of six clusters: *Firmicutes*, *Bacteroidetes*, *Actinobacteria*, *Proteobacteria*, *Fusobacteria*, and *Verrucomicrobia*. Neurotransmitters that affect mood and cognitive function, including gamma-aminobutyric acid and serotonin, are produced by gut microorganisms. Systemic inflammation brought on by the microbial imbalance may exacerbate neuroinflammation and further degenerate the brain. It also causes autoimmune illnesses by altering immunological responses [32, 36].

5.4 ALS and the Gut Microbiome

ALS is linked to altered gut microbiota, which impacts immunological, metabolic, and neurological processes (Fig. 5.1). Gut microbiota alterations are linked to ALS, with changes in the *Firmicutes/Bacteroidetes* ratio varying by disease subtype and sample type. Dysbiosis of the gut, lowered melatonin levels, and glyphosate exposure could all contribute to impairing this pathway [1]. Further, the changes in the gut microbiome may play a role in neuroinflammation associated with ALS. In ALS patients, shifts in gut microbial composition may affect immune responses and drive neuroinflammatory processes that contribute to disease progression. The gut microbiome could be crucial in the development of ALS and offers a potential target for therapy aimed at altering neuroinflammation and reducing the speed of disease progression [15].

5.5 Gut Microbiota Alterations in ALS Patients

Bacteroidetes (*Bacteroidota*) and *Firmicutes* are the major phyla in the gut microbiota, and numerous studies have observed a changed *Firmicutes/Bacteroidetes* ratio (F/B) in the majority of ALS patients. Three separate studies demonstrated a reduction in the F/B ratio among ALS patients [13, 34, 39]. ALS patients often show reduced levels of beneficial butyrate-producing bacteria like *Eubacterium rectale* and *Roseburia intestinalis*, along with increased *Prevotella copri*. Certain gut microbes have been associated with increased ALS risk, and greater microbial diversity with reduced *Prevotella* and other beneficial taxa is common. Integrated microbiome-metabolome studies reveal correlations with lipid metabolites implicated in ALS pathogenesis. Additionally, the cyanobacterial toxin BMAA can induce ALS-like neurodegeneration in animal models, which is mitigated by L-serine, suggesting a role for microbial toxins in disease progression [23].

Fecal samples from 20 ALS patients and 20 healthy controls were examined using 16S rDNA sequencing. The gut microbiota of ALS patients was altered, characterized by an increase in *Bacteroidetes* and a decrease in *Firmicutes* and *Megamonas* [39]. Another study, using 16S rRNA sequencing, compared the fecal microbiota of 25 ALS patients with that of 32 matched healthy controls. The study did not identify significant differences in microbiota diversity or composition between the groups, apart from a decrease in microbial species and changes in the abundance of uncultured *Ruminococcaceae* among ALS patients [4].

Patients with ALS show specific alterations in their gut microbiome that are linked to changes in plasma metabolites, particularly lipids. Specific bacterial genera were associated with lipid-related metabolites, and network analysis revealed strong microbe-metabolite connections. Mendelian randomization indicated that lipids may play causal roles in ALS, and the results emphasize the interactions between the gut microbiome and metabolome in ALS, providing potential biomarkers and understanding of disease mechanisms [18]. In another study, gut microbiomes of ALS patients differ from those of their spouses, characterized by increased microbial diversity and a decrease in *Prevotella* spp. Although the inflammatory markers were alike, ALS patients demonstrated a reduction in microbial activity within crucial metabolic pathways, such as butyrate and carbon metabolism [19]. Compared with healthy controls, individuals with ALS exhibit significantly lower levels of essential butyrate-producing gut bacteria like *Eubacterium rectale* and *Roseburia intestinalis*. These bacteria help maintain gut integrity and control inflammation [30].

5.6 Gut Dysbiosis and Disease Mechanisms in ALS Animal Models

In Sod1 transgenic (Sod1-Tg) mice that are prone to ALS, gut dysbiosis and changes in metabolites were noted before the onset of symptoms, and the disease worsened in germ-free or antibiotic-treated conditions. In relation to ALS severity, 11 gut bacteria were identified, with *Akkermansia muciniphila* (AM) alleviating symptoms and *Ruminococcus torques* and *Parabacteroides distasonis* exacerbating them. Supplementing with AM raised levels of nicotinamide in the central nervous system, while systemic nicotinamide enhanced motor function and gene expression in Sod1-Tg mice [2]. Gut microbiome imbalance and intestinal dysfunction contribute to ALS progression. Butyrate treatment restored microbial balance, improved gut integrity, and reduced pathological changes in ALS mice, while extending their lifespan [40].

In a mouse model, early changes to the gut microbiome, immune activation, and epigenetic alterations all contribute to the progression of ALS. The results highlight the interrelated functions of the microbiome, immune system, and epigenetics in ALS and may inform the creation of new biomarkers and treatment approaches [14]. By allowing the accumulation of toxic hemoglobin and iron, which causes oxidative stress, the breakdown of the blood-spinal cord barrier (BSCB) contributes to the early degeneration of motor neurons in ALS mouse models (SOD1G93A mice). Motor neuron defect and onset of disease were delayed by restoring BSCB integrity with an activated protein C analog or reducing iron toxicity through chelation and antioxidants. The results indicate that the disruption of the BSCB is crucial for the progression of ALS and could serve as a therapeutic target, possibly relating to human ALS [38].

Mutations in C9ORF72 lead to ALS and frontotemporal dementia via toxic gain and loss of function mechanisms. Diminishing the presence of immune-stimulating gut bacteria in mutant mice leads to decreased inflammation and enhanced survival. This implies that the gut microbiota may affect the severity of disease and could serve as a target for therapy [5]. In SOD1G93A ALS mice, alterations in the gut microbiome were observed prior to the onset of motor symptoms, with subsequent immune activation and changes in brain epigenetics occurring at the disease's end-stage [14]. Chronic exposure to LPS (a component of bacteria) exacerbates the progression of disease and degeneration of motor neurons in SOD1G37R ALS mice. This effect is associated with increased activation of the innate immune system, specifically the upregulation of Toll-like receptor 2 and proinflammatory cytokines in degenerating regions of the CNS. Environmental influences and innate immunity can work together to hasten neurodegeneration in ALS [29]. There was early gut dysbiosis in ALS mice, with disease exacerbation occurring under germ-free or antibiotic-treated conditions. Outcomes were improved by supplementing with *Akkermansia muciniphila*, probably due to the production of nicotinamide. ALS patients exhibited similar microbiome alterations and decreased nicotinamide levels, indicating a possible connection to human disease and prospective therapeutic significance [17].

5.7 Nutritional Approaches to Regulate the Gut Microbiota

Through immune, circulatory, and neuronal pathways, the gut microbiota engages in two-way communication with the central nervous system, establishing the gut-brain axis. This interaction could affect the progression of ALS, given that gut dysbiosis can impact immune responses and neuronal survival. If changes in the gut microbiota play a role in ALS, then nutritional strategies involving probiotics, prebiotics, or postbiotics could represent promising treatment options [7]. Greater consumption of antioxidants and carotenes primarily sourced from fruits and vegetables improves functional and respiratory status in ALS patients [31]. High intake of fat and glutamate, as well as β-methylamino-L-alanine, was linked to a higher risk of ALS. In fact, a diet rich in antioxidant and anti-inflammatory substances like curcumin, creatine, coenzyme Q10, vitamins E, A, and C, and phytochemicals could lower the risk of ALS [11].

ALS progression is significantly influenced by metabolic disturbances and gut dysbiosis, as they exacerbate energy deficits, oxidative stress, and neuroinflammation. Promising approaches to alleviating disease burden include dietary strategies that incorporate antioxidants and reduce inflammation such as the Mediterranean diet as well as microbiome-targeted treatments (e.g., prebiotics, probiotics, postbiotics). It is important to conduct more studies on dietary and microbiome-based treatments, as tailored strategies that integrate immunonutrition and microbiota modulation could potentially decelerate ALS progression and enhance patient outcomes [10]. Promising results are also seen with nutritional strategies like the ketogenic diet and microbiome modulation. To enhance patient outcomes and quality of life, a multidisciplinary approach that integrates early diagnosis, targeted treatment, and supportive care is crucial [27].

5.8 Conclusion and Future Perspectives

The gut microbiota is essential for overall health maintenance via the gut-brain axis and could have a major impact on ALS development. Recent discoveries in ALS mouse models and clinical studies demonstrate that intestinal dysfunction manifests before the onset of the disease, indicating that a breakdown in gut neuromuscular communication might play a role in the progression of ALS. Dysbiosis interrupts the signaling between the brain and gut, leading to problems with the intestinal barrier, increased systemic inflammation, and neurodegeneration. Grasping this aspect of gut involvement could present new options for therapy by modulating the gut microbiota with nutritional approaches, which may reduce inflammation and neuronal damage, thereby slowing disease progression.

Declarations

Competing Interests The author(s) has no competing interests to declare that are relevant to the content of this manuscript.

Ethics Approval The author(s) declares that this manuscript does not include primary research with human participants.

The author(s) declares that this manuscript does not include primary studies involving animals.

References

1. Anderson G. Amyotrophic lateral sclerosis pathoetiology and pathophysiology: roles of astrocytes, gut microbiome, and muscle interactions via the mitochondrial melatonergic pathway, with disruption by glyphosate-based herbicides. Int J Mol Sci. 2022;24(1):587.
2. Blacher E, Bashiardes S, Shapiro H, Rothschild D, Mor U, Dori-Bachash M, et al. Potential roles of gut microbiome and metabolites in modulating ALS in mice. Nature. 2019;572(7770):474–80.
3. Blumenau H. Basic epidemiology of ALS. Eukaryon. 2024;20:1–6.
4. Brenner D, Hiergeist A, Adis C, Mayer B, Gessner A, Ludolph AC, Weishaupt JH. The fecal microbiome of ALS patients. Neurobiol Aging. 2018;61:132–7.

5. Burberry A, Wells MF, Limone F, Couto A, Smith KS, Keaney J, et al. C9orf72 suppresses systemic and neural inflammation induced by gut bacteria. Nature. 2020;582(7810):89–94.
6. Burke RE, Levine DN, Tsairis P, Zajac FE 3rd. Physiological types and histochemical profiles in motor units of the cat gastrocnemius. J Physiol. 1973;234(3):723–48.
7. Calvo AC, Valledor-Martín I, Moreno-Martínez L, Toivonen JM, Osta R. Lessons to learn from the gut microbiota: a focus on amyotrophic lateral sclerosis. Genes. 2022;13(5):865.
8. Carabotti M, Scirocco A, Maselli MA, Severi C. The gut-brain axis: interactions between enteric microbiota, central and enteric nervous systems. Ann Gastroenterol Q Publ Hellenic Soc Gastroenterol. 2015;28(2):203.
9. Chatterjee K, Banerjee S. Microbiome and motor neuron diseases. Prog Mol Biol Transl Sci. 2020;176:111–22.
10. Cuffaro F, Lamminpää I, Niccolai E, Amedei A. Nutritional and microbiota-based approaches in amyotrophic lateral sclerosis: from prevention to treatment. Nutrients. 2024;17(1):102.
11. D'Antona S, Caramenti M, Porro D, Castiglioni I, Cava C. Amyotrophic lateral sclerosis: a diet review. Foods. 2021;10(12):3128.
12. Duchateau J, Enoka RM. Human motor unit recordings: origins and insight into the integrated motor system. Brain Res. 2011;1409:42–61.
13. Fang X, Wang X, Yang S, Meng F, Wang X, Wei H, Chen T. Evaluation of the microbial diversity in amyotrophic lateral sclerosis using high-throughput sequencing. Front Microbiol. 2016;7:1479.
14. Figueroa-Romero C, Guo K, Murdock BJ, Paez-Colasante X, Bassis CM, Mikhail KA, et al. Temporal evolution of the microbiome, immune system and epigenome with disease progression in ALS mice. Dis Model Mech. 2020;13(2):dmm041947.
15. Fournier CN, Houser M, Tansey MG, Glass JD, Hertzberg VS. The gut microbiome and neuroinflammation in amyotrophic lateral sclerosis? Emerging clinical evidence. Neurobiol Dis. 2020;135:104300.
16. Fymat AL. Amyotrophic lateral sclerosis: II. Etiology and action mechanisms. J Neurol Psychol Res. 2025;6(1):1–23.
17. Gotkine M, Kviatcovsky D, Elinav E. Amyotrophic lateral sclerosis and intestinal microbiota—toward establishing cause and effect. Gut Microbes. 2020;11(6):1833–41.
18. Guo K, Figueroa-Romero C, Noureldein MH, Murdock BJ, Savelieff MG, Hur J, et al. Gut microbiome correlates with plasma lipids in amyotrophic lateral sclerosis. Brain. 2024;147(2):665–79.
19. Hertzberg VS, Singh H, Fournier CN, Moustafa A, Polak M, Kuelbs CA, et al. Gut microbiome differences between amyotrophic lateral sclerosis patients and spouse controls. Amyotroph Lateral Scler Frontotemporal Degener. 2022;23(1–2):91–9.
20. Hooks BM, Papale AE, Paletzki RF, Feroze MW, Eastwood BS, Couey JJ, et al. Topographic precision in sensory and motor corticostriatal projections varies across cell type and cortical area. Nat Commun. 2018;9(1):3549.
21. Jang SH. The corticospinal tract from the viewpoint of brain rehabilitation. J Rehabil Med. 2014;46(3):193–9.
22. Kanning KC, Kaplan A, Henderson CE. Motor neuron diversity in development and disease. Annu Rev Neurosci. 2010;33(1):409–40.
23. Kaul M, Mukherjee D, Weiner HL, Cox LM. Gut microbiota immune cross-talk in amyotrophic lateral sclerosis. Neurotherapeutics. 2024;21(6):e00469.
24. Kiernan MC, Vucic S, Cheah BC, Turner MR, Eisen A, Hardiman O, et al. Amyotrophic lateral sclerosis. Lancet. 2011;377(9769):942–55.
25. Li X, Bedlack R. Evaluating emerging drugs in phase II & III for the treatment of amyotrophic lateral sclerosis. Expert Opin Emerg Drugs. 2024;29(2):93–102.
26. Liu X, Liu Y, Liu J, Zhang H, Shan C, Guo Y, et al. Correlation between the gut microbiome and neurodegenerative diseases: a review of metagenomics evidence. Neural Regen Res. 2024;19(4):833–45.

27. Nabakhteh S, Lotfi A, Afsartaha A, Khodadadi ES, Abdolghaderi S, Mohammadpour M, et al. Nutritional interventions in amyotrophic lateral sclerosis: from ketogenic diet and neuroprotective nutrients to the microbiota-gut-brain axis regulation. Mol Neurobiol. 2025;62(7):9216–39.
28. National Institute of Neurological Disorders and Stroke. Motor neuron diseases. U.S. Department of Health and Human Services, National Institutes of Health; 2025. Retrieved 4 Aug 2025, from https://www.ninds.nih.gov/health-information/disorders/motor-neuron-diseases
29. Nguyen MD, D'Aigle T, Gowing G, Julien JP, Rivest S. Exacerbation of motor neuron disease by chronic stimulation of innate immunity in a mouse model of amyotrophic lateral sclerosis. J Neurosci. 2004;24(6):1340–9.
30. Nicholson K, Bjornevik K, Abu-Ali G, Chan J, Cortese M, Dedi B, et al. The human gut microbiota in people with amyotrophic lateral sclerosis. Amyotroph Lateral Scler Frontotemporal Degener. 2021;22(3–4):186–94.
31. Nieves JW, Gennings C, Factor-Litvak P, Hupf J, Singleton J, Sharf V, et al. Association between dietary intake and function in amyotrophic lateral sclerosis. JAMA Neurol. 2016;73(12):1425–32.
32. Park KJ, Gao Y. Gut-brain axis and neurodegeneration: mechanisms and therapeutic potentials. Front Neurosci. 2024;18:1481390.
33. Robberecht W, de Jong JMBV. Amyotrophic lateral sclerosis. Amyotroph Lateral Scler. 2000;256:3.
34. Rowin J, Xia Y, Jung B, Sun J. Gut inflammation and dysbiosis in human motor neuron disease. Physiol Rep. 2017;5(18):e13443.
35. Srivastava K, Arshad F, Mujawar WJ, Cranberg L, Rajeshwaran J, Afsar M, et al. Cognitive and behavioral profile of patients with amyotrophic lateral sclerosis spectrum in the Indian context. Dement Geriatr Cogn Disord. 2024;53(6):310–20.
36. Ullah H, Arbab S, Tian Y, Liu CQ, Chen Y, Qijie L, et al. The gut microbiota–brain axis in neurological disorder. Front Neurosci. 2023;17:1225875.
37. Wijesekera LC, Nigel Leigh P. Amyotrophic lateral sclerosis. Orphanet J Rare Dis. 2009;4(1):3.
38. Winkler EA, Sengillo JD, Sagare AP, Zhao Z, Ma Q, Zuniga E, et al. Blood–spinal cord barrier disruption contributes to early motor-neuron degeneration in ALS-model mice. Proc Natl Acad Sci. 2014;111(11):E1035–42.
39. Zeng Q, Shen J, Chen K, Zhou J, Liao Q, Lu K, et al. The alteration of gut microbiome and metabolism in amyotrophic lateral sclerosis patients. Sci Rep. 2020;10(1):12998.
40. Zhang YG, Wu S, Yi J, Xia Y, Jin D, Zhou J, Sun J. Target intestinal microbiota to alleviate disease progression in amyotrophic lateral sclerosis. Clin Ther. 2017;39(2):322–36.
41. Zuccaro E, Piol D, Basso M, Pennuto M. Motor neuron diseases and neuroprotective peptides: a closer look to neurons. Front Aging Neurosci. 2021;13:723871.

The Gut Microbiome Alterations in Children with Cerebral Palsy and Epilepsy

6

Faheem Q. AL-Mojahid, Muhasina K M, Priya Prasannakumar, and Teena Merlin

6.1 Introduction

Known as the gut microbiota, the billions of bacteria that live in the human gut create a physiologically active environment that is essential to host health. Through the gut–brain axis, a sophisticated, two-way communication system involving neural, hormonal, metabolic, and immunological pathways, these microbes have a major impact on the central nervous system in addition to their well-established functions in digestion, metabolism, and immune modulation. Through the vagus nerve, microbial metabolites, and systemic cytokine signaling, this axis promotes constant communication between the brain and the gastrointestinal tract, impacting neurological function, behavior, and brain development [4, 26].

The gut microbiome's early colonization and maturation are essential for normal neurodevelopment, according to recent studies. Fundamental processes that are important in forming the growing brain include synapse pruning, myelination,

F. Q. AL-Mojahid
School of Biosciences, Mar Athanasios College for Advanced Studies, Tiruvalla, Mahatma Gandhi University, Kottayam, Kerala, India

Department of Medical Laboratory, Institute for Continuous Education - Albaydha University, Rada'a, Yemen
e-mail: faheem@macfast.ac.in

Muhasina K M
Department of Pharmaceutical Analysis, Prime College of Pharmacy, Erattayal, Palakkad, Kerala, India

P. Prasannakumar
Government Arts and Science College, Nattukkal, Palakkad, Kerala, India

T. Merlin (✉)
School of Biosciences, Mar Athanasios College for Advanced Studies, Tiruvalla, Mahatma Gandhi University, Kottayam, Kerala, India
e-mail: teena@macfast.org

M. H. Rahman, Muhasina K M (eds.), *The Microbiome's Impact on the Brain*,
https://doi.org/10.1007/978-981-95-7494-0_6

neurotransmitter synthesis, and microglial activation, all of which are influenced by the microbiota. These processes may be disrupted by dysbiosis, which is characterized as an imbalance in the structure or function of the microbial population. This may lead to altered neuronal signaling, increased neuroinflammation, and worse cognitive and motor outcomes. Therefore, it seems that preserving a diverse and balanced microbiota in early childhood is crucial for the best possible brain development and neurological resiliency [17, 27].

Among the most prevalent neurological illnesses affecting children that have long-term effects are neurodevelopmental disorders, including CP and epilepsy. In contrast to epilepsy, which is characterized by repeated seizures brought on by abnormal electrical activity in the brain, CP is characterized by nonprogressive abnormalities in motor function that result from brain injury or malformation that occurs during early development. Constipation, difficulty eating, and gastroesophageal reflux are common gastrointestinal symptoms that accompany these illnesses and can further alter the composition of gut microbes [15, 30].

Reduced microbial diversity and an overrepresentation of pro-inflammatory taxa are two examples of the unique patterns of gut dysbiosis seen in children with CP and epilepsy, according to recent research. Increased intestinal permeability, or "leaky gut," systemic inflammation, and changed synthesis of neuroactive metabolites, like serotonin and gamma-aminobutyric acid (GABA), are all linked to these imbalances, which may exacerbate neurological symptoms [12, 22].

Understanding the role of the gut microbiota in these conditions is crucial not only for elucidating disease mechanisms but also for developing novel microbiota-targeted interventions. This chapter aims to explore the current evidence on gut microbiome alterations in children with CP and epilepsy and to examine their implications for clinical practice and therapeutic innovation.

6.2 Cerebral Palsy and Epilepsy: An Overview

Cerebral palsy (CP) is a collection of nonprogressive neurodevelopmental problems brought on by early brain damage, frequently as a result of prenatal or perinatal insults, including intracranial hemorrhage, hypoxia, or infection [21]. CP is divided into four different motor subtypes according to the distribution and type of motor impairment: mixed, ataxic, dyskinetic, and spastic. Along with physical dysfunction, cognitive impairments, language delays, behavioral problems, and sensory difficulties are common in children with cerebral palsy. Children with cerebral palsy frequently have epilepsy, a comorbid condition marked by frequent spontaneous seizures brought on by aberrant neural discharges. From focal to generalized seizures, children's seizure types can vary greatly and are frequently controlled with long-term antiepileptic medication [25]. Epilepsy is especially common in children with dyskinetic and quadriplegic forms of cerebral palsy. More severe neurological outcomes and more complex clinical presentations are linked to the co-occurrence of epilepsy and cerebral palsy. Both disorders affect immunological dysregulation, gastrointestinal function, and other aspects of systemic physiology in addition to the

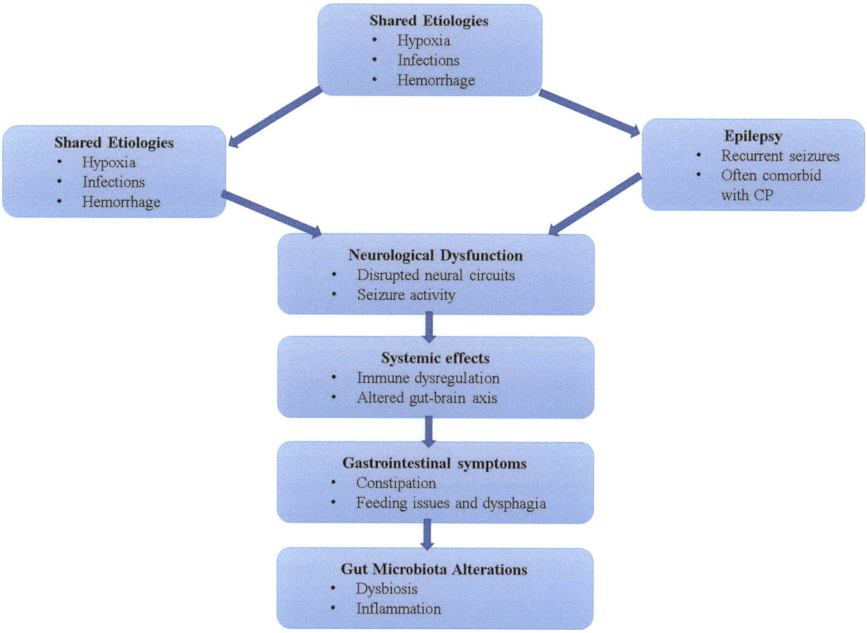

Fig. 6.1 Interplay between CP, epilepsy, and gut microbiota

central nervous system, as illustrated in Fig. 6.1. Gastrointestinal symptoms, like dysphagia, eating issues, and constipation, are common in children with CP and epilepsy. These symptoms might affect the gut microbiota's composition and jeopardize nutritional status. Neurological dysfunction can be made worse by disruption of the gut–brain axis through these routes, which emphasizes the significance of a comprehensive therapy strategy [2].

A schematic representation of the interrelationships between CP, epilepsy, and gut physiology. Shared prenatal and perinatal etiologies, such as hypoxia, infection, and hemorrhage, lead to neurological dysfunction characterized by disrupted neural circuitry and seizure activity. This dysfunction contributes to systemic immune dysregulation and an altered gut–brain axis. The resulting gastrointestinal symptoms, including constipation and feeding difficulties, can drive changes in gut microbiota composition and function, potentially feeding back into the neurological disease process.

6.3 Gut Microbiota

The billions of microorganisms that make up the human gut microbiota, which includes bacteria, viruses, fungi, and archaea, all work together to maintain host homeostasis. *Bacteroides*, *Firmicutes*, *Lactobacillus*, and *Bifidobacterium* make up the majority of the gut microbiota in healthy children; the composition is impacted by early dietary habits, environmental exposures, and the mode of birth. The gut

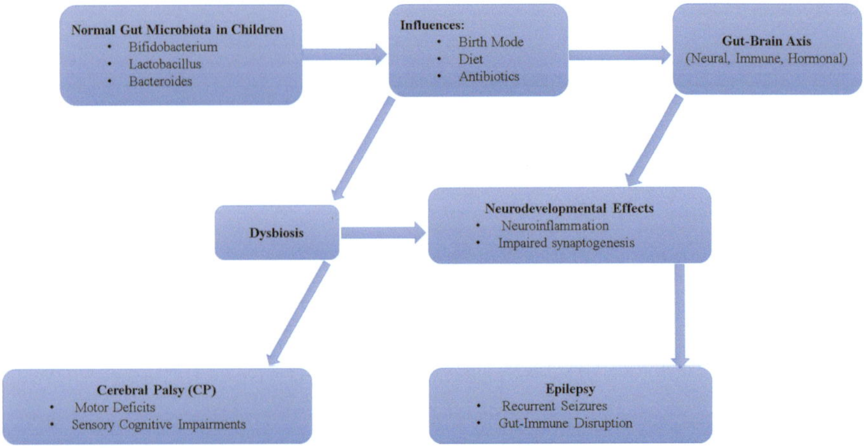

Fig. 6.2 Conceptual diagram illustrating the role of gut microbiota in neurodevelopment. Normal microbiota in children is influenced by factors such as diet, birth mode, and antibiotics. Disruptions in microbial balance (dysbiosis) can impair gut–brain axis signaling, leading to neuroinflammation, impaired synaptogenesis, and increased risk of neurological disorders such as cerebral palsy and epilepsy

microbiota experiences dynamic growth between infancy and childhood, marking an increase in diversity and complexity. By the time the child is 3 years old, the microbiome resembles that of an adult [16, 31].

Throughout development, diet has a significant impact on the composition of microbes. As an example, the transition to solid foods introduces microorganisms involved in complex glucose metabolism, while breast milk encourages colonization by beneficial *Bifidobacterium* species. High-fat, high-sugar Western diets have been linked to a rise in pro-inflammatory taxa and a decrease in microbial diversity [5].

Critically, the gut microbiota influences brain development and function via the gut–brain axis, a complex, bidirectional communication network involving neural (vagus nerve), endocrine (HPA axis), immune, and metabolic pathways, as represented in Fig. 6.2. Microbial metabolites, such as short-chain fatty acids (SCFAs), neurotransmitter precursors, and inflammatory mediators, are central to this cross talk. These molecules can modulate blood–brain barrier integrity, synaptic plasticity, microglial activity, and neurogenesis, highlighting the gut microbiota's integral role in neurodevelopmental processes [4].

6.4 The Gut–Brain Axis in Neurological Disorders

6.4.1 Mechanisms of Microbiota–Neuroimmune Interaction

The gut–brain axis is a complicated, two-way signaling channel that connects the central nervous system (CNS) and the gastrointestinal tract via metabolic, immunological, endocrine, and neurological pathways, as displayed in Fig. 6.3. Neuroimmune

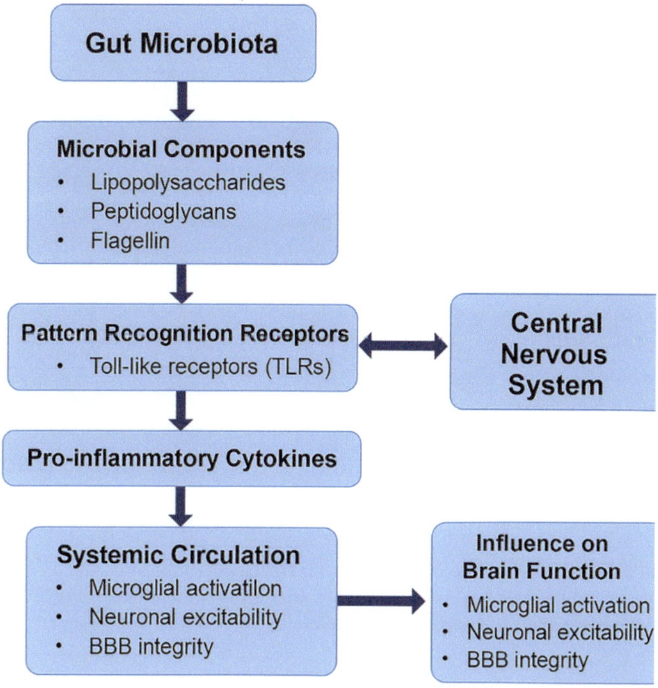

Fig. 6.3 Mechanisms of gut microbiota–CNS interaction via the gut–brain axis. The figure shows how gut microbiota influence brain function through immune signaling. Microbial components trigger immune receptors, leading to inflammation that affects the brain by altering microglial activity, neuronal excitability, and blood–brain barrier integrity

regulation mediated by the gut bacteria is a major mechanism behind this connection. On intestinal epithelium and immunological cells, microbial components, such as lipopolysaccharides (LPS), peptidoglycans, and flagellin, interact with pattern recognition receptors like toll-like receptors (TLRs) to cause the release of pro-inflammatory cytokines. These cytokines can enter the bloodstream and change the blood–brain barrier (BBB) integrity, neuronal excitability, and microglial activation to affect brain function [9, 20]. In this way, the gut microbiome can exert systemic and central effects, modulating inflammation and neural signaling pathways.

6.4.2 Role of Gut Dysbiosis in Neurodevelopmental Conditions

An increasing number of neurodevelopmental illnesses, including CP, epilepsy, and autism spectrum disorder (ASD), have been linked to gut dysbiosis, which is characterized as a disturbance in the variety, stability, or function of gut microbial populations, as demonstrated in Fig. 6.4. Microbial compositions are often changed in children with these diseases, with an overrepresentation of potentially pathogenic or pro-inflammatory species and a decrease in helpful commensals (e.g.,

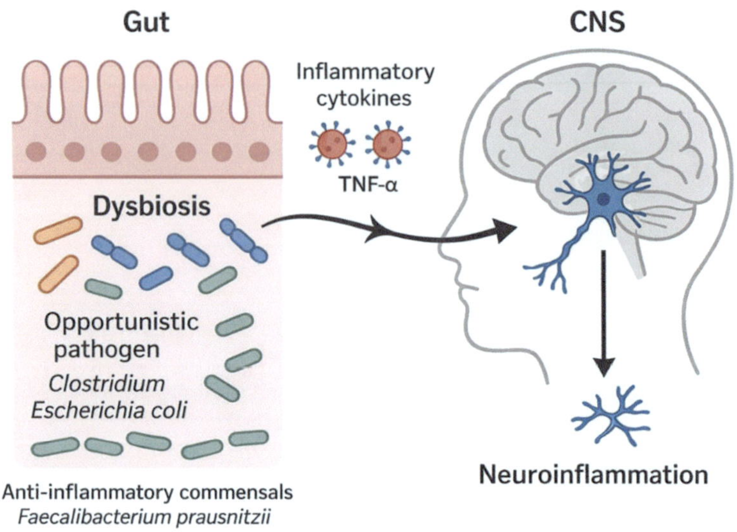

Fig. 6.4 Gut dysbiosis-induced neuroinflammation via the gut–brain axis. Gut dysbiosis increases harmful bacteria and reduces beneficial microbes, triggering inflammatory cytokines like TNF-α that travel to the brain and cause neuroinflammation

Bifidobacterium and *Lactobacillus*). Early developmental dysbiosis may disrupt immunological system programming, neurotransmitter homeostasis, and synaptogenesis. Additionally, it can weaken intestinal integrity, causing systemic inflammation and adversely affecting neurological growth and function [15, 22].

6.4.3 Inflammation and Neurotoxicity Mediated by Microbiota

The main characteristics of gut–brain axis dysfunction include microbial metabolite abnormalities and persistent low-grade inflammation, as presented in Fig. 6.5. Endotoxins like LPS can enter the bloodstream through increased intestinal permeability, or "leaky gut," which exacerbates oxidative stress and systemic inflammation. These inflammatory signals have the ability to cross the blood–brain barrier and cause neuronal damage, microglial hyperactivity, and excitotoxicity, all of which are characteristics of a number of neurological conditions, including epilepsy and CP [9, 11]. Furthermore, neurotrophic support and anti-inflammatory responses may be compromised by decreased levels of neuroprotective microbial metabolites, such as SCFAs. These processes collectively reinforce the expanding knowledge that abnormalities in the gut microbiota might influence the development and course of neurological disorders in children by means of neurotoxic and inflammatory pathways [35].

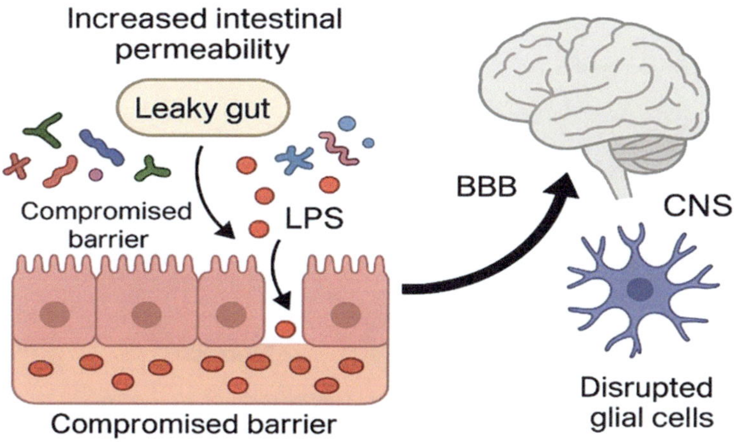

Fig. 6.5 Increased intestinal permeability facilitates LPS translocation and CNS glial disruption. This figure depicts how increased intestinal permeability ("leaky gut") allows harmful molecules like LPS to cross the gut barrier, enter the bloodstream, and reach the brain, where they disrupt glial cells and compromise the BBB, contributing to CNS dysfunction

6.5 Gut Microbiome in Healthy Children

6.5.1 Normal Microbiome Composition in Early Life

Healthy children's gut microbiomes are dynamic, ever-changing ecosystems that start to form at birth and continue to develop throughout early life. *Staphylococcus*, *Streptococcus*, and *Enterobacteriaceae* are among the facultative anaerobes that first invade the stomach of newborns. Strict anaerobes, including *Clostridium*, *Bifidobacterium*, and *Bacteroides*, become increasingly prevalent when intestinal oxygen levels fall [24]. By the age of three, the gut microbiota, which is dominated by the phyla *Firmicutes* and *Bacteroidetes*, usually has diversity and stability similar to that of an adult.

6.5.2 Developmental Stages of Gut Colonization

The evolution of gut microbiota follows a predictable, albeit unique, path. Feeding habits, delivery method, and maternal microbiome all affect the colonization process. Rapid changes in the microbial composition, including an increase in diversity and the growth of beneficial taxa like *Bifidobacterium* in breastfed newborns, are observed throughout the first year of life. Solid food introduction causes a transitional microbiome with a rise in *Firmicutes* and *Bacteroides* between 6 and 24 months. Microbial communities start to settle during the third year, and metabolic processes grow increasingly intricate as the immunological and neurological systems mature [16].

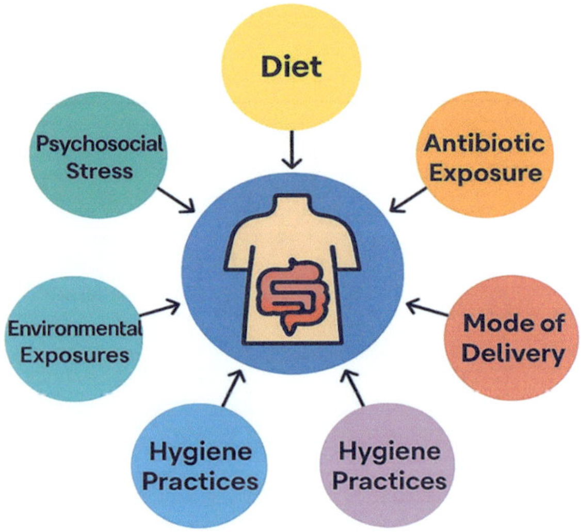

Fig. 6.6 Factors influencing gut microbiome in childhood

6.5.3 Factors Influencing Gut Microbiome in Childhood

The gut microbiota of children is shaped by a variety of host-related and environmental variables, as illustrated in Fig. 6.6. Food is important: formula feeding frequently encourages more microbial diversity but less functional specialization, whereas breast milk encourages the growth of *Bifidobacterium*. The introduction of solid foods promotes microbial maturation, particularly those that are fermented or high in fiber. Early-life antibiotic exposure can significantly alter microbial diversity and postpone microbial succession. Other significant variables include the delivery method (caesarean versus vaginal), personal cleanliness habits, exposure to the environment, and even psychological stress. These early factors may have long-term consequences for metabolic development, immunological programming, and neurocognitive outcomes [32, 33].

6.6 Microbiota Alterations in Cerebral Palsy and Epilepsy

6.6.1 Clinical Studies

Recent developments in metagenomic sequencing have shown that children with neurodevelopmental problems, such as epilepsy and CP, have notable changes in their gut flora. Research shows that gut dysbiosis, which is typified by a loss of beneficial commensals, an overrepresentation of pro-inflammatory taxa, and decreased microbial diversity, is often linked to both disorders. Gut microbiota profiles frequently show decreased populations of *Bifidobacterium* and *Lactobacillus* in children with CP, particularly in those with severe motor and feeding difficulties. These bacteria are essential for preserving the integrity of the gut barrier and

Table 6.1 Summary of clinical studies on gut microbiota alterations in children with CP and epilepsy

Population	Key microbiota findings	Implications	References
Children with spastic CP vs. healthy controls	↓ Diversity, ↑ *Proteobacteria*, ↓ *Bifidobacterium*	Inflammatory gut profile; weakened gut–brain support	[18]
Children with idiopathic epilepsy	↑ Clostrid*ium*, ↑ *Desulfovibrio*, ↓ *Faecalibacterium*	Linked to neurotoxicity and inflammation	[23]
Children on valproate monotherapy (≥6 months)	↓ *Firmicutes*, ↑ *Proteobacteria,* and *Bacteroidetes*	Drug-induced dysbiosis with possible immune activation	[36]
AED-treated children with epilepsy	↓ SCFA-producing bacteria (e.g., *Roseburia*, *Eubacterium*)	Impaired neuroprotection and gastrointestinal health	[8]
Children with CP (various subtypes)	General dysbiosis, ↓ *Lactobacillus*	Suggests the role of the gut–brain axis in CP pathophysiology	[3]

immunological homeostasis [18]. In a pilot investigation, researchers discovered that children with spastic CP had higher abundances of *Proteobacteria*, a phylum linked to inflammatory reactions, and considerably worse microbial richness and evenness when compared to age-matched controls [23].

Likewise, children with epilepsy, especially those with drug-resistant forms, show notable changes in the makeup of their microbiome. According to Zhu et al. [36], *Clostridium* and *Desulfovibrio* species, which are linked to the synthesis of neurotoxic metabolites and pro-inflammatory cytokines, were shown to be substantially more prevalent in juvenile patients with idiopathic epilepsy. Meanwhile, there was a constant decrease in the levels of the anti-inflammatory butyrate-producing bacterium *Faecalibacterium prausnitzii*.

Clinical symptoms commonly observed in CP and epilepsy, such as immunological dysregulation, disturbed sleep, and persistent constipation, are intimately associated with these microbial imbalances. For instance, constipation can both indicate and worsen an imbalance in gut microbes, which can result in a vicious cycle of neurological stress and dysbiosis through the gut–brain axis [3, 8]. Additionally, altered microbial metabolites, such as short-chain fatty acids (SCFAs), may influence neurotransmission, immune responses, and circadian regulation, as illustrated in Table 6.1.

6.6.2 Impact of Medications

The ecology of gut microbes is significantly impacted by pharmacological treatment, especially prolonged use of antiepileptic medications (AEDs). Numerous studies have demonstrated how routinely administered AEDs, like valproic acid, carbamazepine, and levetiracetam, can modify the microbiome. Valproate, which is frequently used to treat pediatric epilepsy, has been demonstrated to change the relative abundance of particular bacterial phyla and drastically limit microbial diversity. A pro-inflammatory shift was suggested by Zhu et al. [36], who showed that children treated with valproate for more than 6 months had higher levels of *Bacteroidetes* and *Proteobacteria* and lower levels of *Firmicutes* and *Actinobacteria*. Increased indicators of systemic inflammation and intestinal permeability accompanied these alterations. Furthermore, reductions in beneficial SCFA-producing bacteria have been linked to long-term antiepileptic medication, which may compromise neurotransmitter production and mucosal immunity. In populations receiving AEDs, there has been evidence of gut microbial depletion of butyrate-producing bacteria such as *Roseburia* and *Eubacterium*, which may be a factor in the neuropsychiatric adverse effects seen in certain patients.

Notably, dietary limitations, gastrointestinal dysfunction, and polypharmacy, all of which are common in children with epilepsy and cerebral palsy, may worsen microbiome disruptions. The necessity of integrative therapeutic approaches that take gut microbiota preservation into account, like probiotic supplements or customized dietary interventions, is highlighted by these findings.

6.6.3 Ketogenic Diet and Microbiota-Mediated Effects

A high-fat, low-carb diet, known as the ketogenic diet (KD), has long been used to treat drug-resistant epilepsy, especially in those in their youth. The KD may have anti-seizure benefits because, according to recent data, it also significantly alters the composition of the gut microbiota in addition to its metabolic effects. Research has indicated that while KD decreases overall microbial diversity, it increases the growth of particular microbial taxa, such as *Parabacteroides* and *Akkermansia muciniphila* [19]. The gut barrier integrity and systemic inflammation, two important variables in seizure susceptibility, are thought to be modulated by these alterations. Additionally, circulating neurotransmitter or metabolite levels may change as a result of microbial remodeling by the KD, establishing a neuroprotective milieu in the brain.

6.6.4 Microbial Influence on Neurotransmitters

The gut microbiota has the ability to produce, alter, or break down a variety of neuroactive substances, which can affect behavior and brain function. For example, it is known that some strains of *Bifidobacterium* and *Lactobacillus* generate GABA, the primary inhibitory neurotransmitter in the central nervous system. Cortical

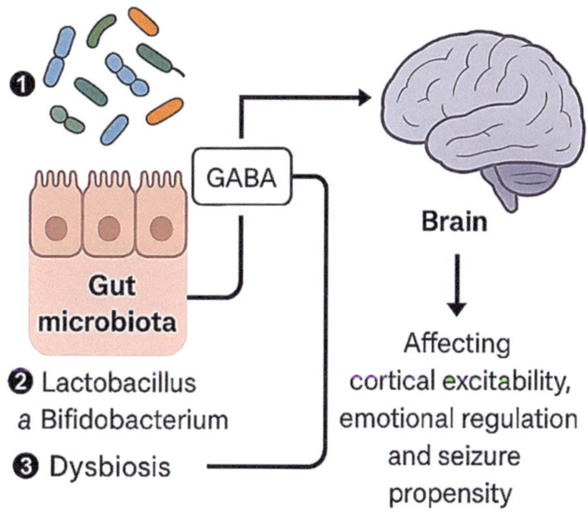

Fig. 6.7 Mechanistic pathway linking gut microbiota to brain function via GABA and neurotransmitter modulation

excitability, emotional modulation, and the likelihood of seizures can all be impacted by dysbiosis, which can result in changed levels of GABA and other important neurotransmitters, including serotonin, dopamine, and acetylcholine [28]. These disruptions are especially significant in epilepsy, where impaired GABAergic signaling, crucial for inhibitory control in the brain, can lower the seizure threshold and exacerbate neuronal excitability, as illustrated in Fig. 6.7.

6.6.5 Microbial Metabolites: Short-Chain Fatty Acids (SCFAs)

When bacteria ferment dietary fibers, they produce important metabolites called SCFAs, especially acetate, propionate, and butyrate, as shown in Fig. 6.8. The functions of SCFAs are varied and include immune response regulation, gut epithelial integrity maintenance, and CNS function modulation. For example, butyrate can affect gene expression by inhibiting histone deacetylase (HDAC) and has anti-inflammatory and neuroprotective qualities. Decreased amounts of bacteria that produce SCFA (such as *Roseburia* and *Eubacterium*) may be a factor in systemic inflammation, microglial activation, and decreased neurotrophic signaling in children with cerebral palsy or epilepsy [1, 29]. All of these processes point to a dynamic and possibly adjustable relationship between neurodevelopmental outcomes and the gut microbiome. They offer a strong case for using the microbiota as a unique supplementary strategy in the clinical treatment of epilepsy and CP.

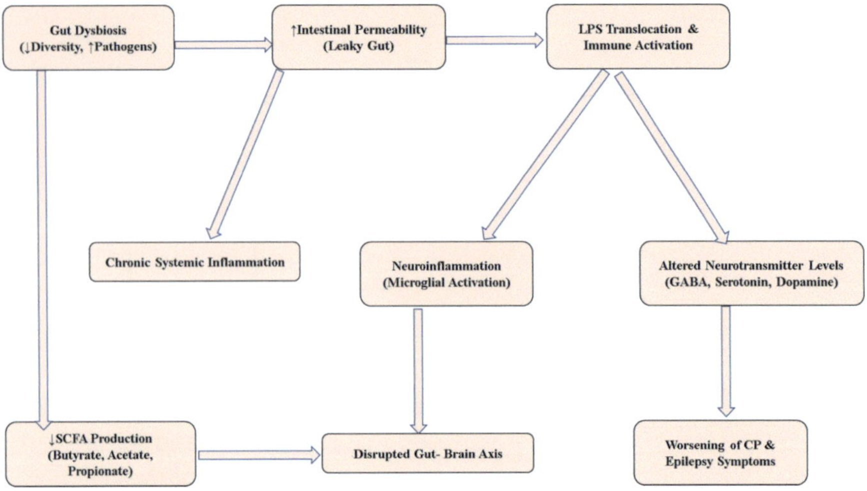

Fig. 6.8 This figure illustrates how gut microbiota dysbiosis may contribute to CP and epilepsy by increasing intestinal permeability ("leaky gut"), allowing bacterial toxins like LPS to trigger systemic inflammation and neuroinflammation. Dysbiosis also reduces beneficial SCFAs and disrupts neurotransmitter regulation (e.g., GABA, serotonin), further aggravating motor dysfunction and seizure activity

6.7 Therapeutic Interventions Targeting the Gut Microbiome

6.7.1 Probiotics and Prebiotics in Pediatric Neurorehabilitation

Prebiotics, which are indigestible food elements that encourage the growth of good bacteria, and probiotics, which are living microorganisms that provide health advantages when given in sufficient proportions, have become viable adjuncts in juvenile neurorehabilitation. Probiotic supplements have been shown in numerous studies to alleviate gastrointestinal symptoms, lower systemic inflammation, and alter neuroactive metabolites in children with neurodevelopmental disorders, such as autism spectrum disorder (ASD) and CP. In animal models, certain strains of *Lactobacillus rhamnosus*, *Bifidobacterium longum*, and *Lactobacillus plantarum* have been associated with decreased anxiety- or seizure-like behaviors and increased GABA synthesis [34]. Prebiotics such as fructooligosaccharides (FOS) and galactooligosaccharides (GOS) may also enhance microbial diversity and intestinal barrier integrity, contributing to systemic immune regulation and improved neurological outcomes [6].

6.7.2 Dietary Strategies

Pediatric patients with neurological diseases may benefit greatly from dietary therapies, which have long been involved in regulating the gut flora. In addition to controlling seizures, the previously mentioned KD also produces a positive microbial profile linked to anti-inflammatory benefits. A high-fiber diet, on the other hand, promotes the growth of bacteria that produce SCFA, such as *Bifidobacterium* species and *Faecalibacterium prausnitzii*, which improve gut barrier function and may have neuroprotective effects. It has also been demonstrated that diets high in polyphenols, fermented foods, and omega-3 fatty acids improve brain function and microbial diversity [13]. Personalized nutritional strategies, tailored to individual microbiome profiles, are increasingly being explored in precision medicine approaches.

6.7.3 Fecal Microbiota Transplantation (FMT): Promise and Limitations

A fast-developing treatment that has demonstrated effectiveness in treating recurrent *Clostridium difficile* infections is fecal microbiota transplantation (FMT), which involves transferring fecal matter from a healthy donor to the recipient's stomach. It is now being studied for its potential in neurological diseases. According to preclinical research, FMT from healthy donors can enhance behavioral outcomes in animal models of autism and epilepsy, lower neuroinflammation, and restore microbial balance [10]. However, safety, ethical, and regulatory issues continue to restrict clinical use in juvenile neurorehabilitation. It is unclear how donor microbiota may affect host immune response and brain development in the long run. Before FMT may be extensively used in pediatric neurology, rigorous randomized controlled trials and standardization of donor selection and administration procedures are required.

6.8 Future Research Directions

6.8.1 Multi-Omics Approaches

Traditional observational research must give way to systems-level analyses in order to advance our knowledge of the gut–brain axis in juvenile neurological diseases. Metagenomics, transcriptomics, proteomics, and metabolomics are examples of multi-omics techniques that provide a thorough understanding of the gut microbiota's taxonomic makeup, functional ability, and metabolic output. Microbial communities can be profiled at high resolution using metagenomic sequencing, and bioactive metabolites, like SCFAs, neurotransmitter precursors, and inflammatory mediators, can be tracked in real time using metabolomics [37]. Finding new microbial or metabolic indicators of illness severity or responsiveness to treatment, as well as defining causal pathways connecting dysbiosis to neurodevelopmental abnormalities, can be accomplished by integrating these datasets.

6.8.2 Personalized Microbiome Therapy

Individual differences in microbiota composition and host–microbe interactions are becoming more widely acknowledged, which has opened the door for customized microbiome therapy. Based on the patient's unique microbial profile and clinical characteristics, this strategy seeks to customize therapies like diet, probiotics, or microbial-derived medicines. Technologies like microbiome-based machine learning models and shotgun metagenomics are being investigated to inform the development of new treatments. Personalized approaches, particularly for adolescents with complex neurological and gastrointestinal comorbidities, could maximize microbial restoration while lowering the dangers associated with one-size-fits-all therapy [7]. However, practical implementation will require scalable diagnostic tools and validated microbial targets.

6.8.3 Longitudinal Studies and Clinical Trials

Randomized controlled trials (RCTs) and well-designed longitudinal studies are desperately needed to determine causation and evaluate the long-term effects of microbiome-targeted treatments. Their translational usefulness is limited because the majority of current research is cross-sectional or based on animal models. From infancy through early childhood, longitudinal cohort studies monitoring microbial, immunological, and neurocognitive changes may offer vital insights into temporal dynamics and windows of therapeutic opportunity. To guarantee reproducibility and therapeutic applicability, clinical trials examining probiotics, dietary changes, and FMT should include multi-omics endpoints, standardized interventions, and age-specific outcome measures [14].

6.9 Conclusion

The evidence presented in this chapter underscores the critical role of the gut microbiota in modulating neurodevelopment and influencing the progression and severity of pediatric neurological disorders, particularly CP and epilepsy. A growing body of research highlights the gut–brain axis as a key communication pathway through which microbial communities interact with the nervous, immune, and endocrine systems. Children with CP and epilepsy often exhibit gut dysbiosis, which may exacerbate neuroinflammation, disrupt neurotransmitter synthesis, and impair intestinal barrier function.

From a clinical perspective, these findings open new avenues for intervention. Modulating the gut microbiota through dietary strategies, probiotics, and potentially FMT holds promise as a supportive therapy to complement conventional neurological care. Therapeutic approaches such as the ketogenic diet have shown dual benefits in both seizure control and reshaping the gut microbial landscape. Moreover, early-life nutritional and microbial interventions may offer preventive strategies during critical windows of brain development.

Looking ahead, integrating microbiome research into pediatric neurocare will require a multidisciplinary effort encompassing clinical neurology, gastroenterology, microbiology, and systems biology. Personalized microbiome therapies, informed by multi-omics data and supported by longitudinal clinical trials, represent a promising frontier in pediatric neurorehabilitation. Incorporating routine microbiome assessments into clinical practice could enhance diagnostic precision and inform individualized treatment strategies, ultimately improving outcomes and quality of life for children with CP and epilepsy.

Declarations

Competing Interests The author(s) has no competing interests to declare that are relevant to the content of this manuscript.

Ethics Approval The author(s) declares that this manuscript does not include primary research with human participants.

The author(s) declares that this manuscript does not include primary studies involving animals.

References

1. Ali Q, Ma S, La S, Guo Z, Liu B, Gao Z, Farooq U, Wang Z, Zhu X, Cui Y, Li D, Shi Y. Microbial short-chain fatty acids: a bridge between dietary fibers and poultry gut health — a review. Anim Biosci. 2022;35(10):1461–78. https://doi.org/10.5713/ab.21.0562.
2. Alyoubi RA, Abu-Zaid A. Epilepsy in cerebral palsy: unraveling prevalence, risk factors, and subtype associations in a large-scale population study. Medicina. 2024;60(11):1809. https://doi.org/10.3390/medicina60111809.
3. Biagi E, Candela M, Fairweather-Tait S, Franceschi C, Brigidi P. Ageing of the human metaorganism: the microbial counterpart. Age. 2012;34(1):247–67. https://doi.org/10.1007/s11357-011-9217-5.
4. Cryan JF, O'Riordan KJ, Cowan CSM, Sandhu KV, Bastiaanssen TFS, Boehme M, Codagnone MG, Cussotto S, Fulling C, Golubeva AV, Guzzetta KE, Jaggar M, Long-Smith CM, Lyte JM, Martin JA, Molinero-Perez A, Moloney G, Morelli E, Morillas E, et al. The microbiota-gut-brain Axis. Physiol Rev. 2019;99(4):1877–2013. https://doi.org/10.1152/physrev.00018.2018.
5. De Filippo C, Cavalieri D, Di Paola M, Ramazzotti M, Poullet JB, Massart S, Collini S, Pieraccini G, Lionetti P. Impact of diet in shaping gut microbiota revealed by a comparative study in children from Europe and rural Africa. Proc Natl Acad Sci. 2010;107(33):14691–6. https://doi.org/10.1073/pnas.1005963107.
6. de Paiva IHR, da Silva RS, Mendonça IP, Duarte-Silva E, Botelho de Souza JR, Peixoto CA. Fructooligosaccharide (FOS) and Galactooligosaccharide (GOS) improve neuroinflammation and cognition by up-regulating IRS/PI3K/AKT signaling pathway in diet-induced obese mice. J Neuroimmune Pharmacol. 2023;18(3):427–47. https://doi.org/10.1007/s11481-023-10069-8.

7. Gilbert JA, Blaser MJ, Caporaso JG, Jansson JK, Lynch SV, Knight R. Current understanding of the human microbiome. Nat Med. 2018;24(4):392–400. https://doi.org/10.1038/nm.4517.
8. Gong X, Liu Y, Liu X, Li AQ, Guo KD, Zhou D, Hong Z. Analysis of gut microbiota in patients with epilepsy treated with valproate: results from a three months observational prospective cohort study. Microb Pathog. 2022;162:105340. https://doi.org/10.1016/j.micpath.2021.105340.
9. Kandpal M, Indari O, Baral B, Jakhmola S, Tiwari D, Bhandari V, Pandey RK, Bala K, Sonawane A, Jha HC. Dysbiosis of gut microbiota from the perspective of the gut–brain axis: role in the provocation of neurological disorders. Metabolites. 2022;12(11):1064. https://doi.org/10.3390/metabo12111064.
10. Kang D-W, Adams JB, Gregory AC, Borody T, Chittick L, Fasano A, Khoruts A, Geis E, Maldonado J, McDonough-Means S, Pollard EL, Roux S, Sadowsky MJ, Lipson KS, Sullivan MB, Caporaso JG, Krajmalnik-Brown R. Microbiota transfer therapy alters gut ecosystem and improves gastrointestinal and autism symptoms: an open-label study. Microbiome. 2017;5(1):10. https://doi.org/10.1186/s40168-016-0225-7.
11. Keshavarzian A, Sisodia SS. Gut microbiota dysbiosis and neurologic diseases: new horizon with potential diagnostic and therapeutic impact. Neurotherapeutics. 2024;21(6):e00478. https://doi.org/10.1016/j.neurot.2024.e00478.
12. Kundu S, Nayak S, Rakshit D, Singh T, Shukla R, Khatri DK, Mishra A. The microbiome–gut–brain axis in epilepsy: pharmacotherapeutic target from bench evidence for potential bedside applications. Eur J Neurol. 2023;30(11):3557–67. https://doi.org/10.1111/ene.15767.
13. Larroya A, Pantoja J, Codoñer-Franch P, Cenit MC. Towards tailored gut microbiome-based and dietary interventions for promoting the development and maintenance of a healthy brain. Front Pediatr. 2021;9:705859. https://doi.org/10.3389/fped.2021.705859.
14. Lloyd-Price J, Abu-Ali G, Huttenhower C. The healthy human microbiome. Genome Med. 2016;8(1):51. https://doi.org/10.1186/s13073-016-0307-y.
15. Maiuolo J, Gliozzi M, Musolino V, Carresi C, Scarano F, Nucera S, Scicchitano M, Oppedisano F, Bosco F, Ruga S, Zito MC, Macri R, Palma E, Muscoli C, Mollace V. The contribution of gut microbiota–brain axis in the development of brain disorders. Front Neurosci. 2021;15:616883. https://doi.org/10.3389/fnins.2021.616883.
16. Milani C, Duranti S, Bottacini F, Casey E, Turroni F, Mahony J, Belzer C, Delgado Palacio S, Arboleya Montes S, Mancabelli L, Lugli GA, Rodriguez JM, Bode L, de Vos W, Gueimonde M, Margolles A, van Sinderen D, Ventura M. The first microbial colonizers of the human gut: composition, activities, and health implications of the infant gut microbiota. Microbiol Mol Biol Rev. 2017;81(4):10–1128. https://doi.org/10.1128/MMBR.00036-17.
17. Morais LH, Schreiber HL, Mazmanian SK. The gut microbiota–brain axis in behaviour and brain disorders. Nat Rev Microbiol. 2021;19(4):241–55. https://doi.org/10.1038/s41579-020-00460-0.
18. Ojeda J, Ávila A, Vidal PM. Gut microbiota interaction with the central nervous system throughout life. J Clin Med. 2021;10(6):1299. https://doi.org/10.3390/jcm10061299.
19. Olson CA, Vuong HE, Yano JM, Liang QY, Nusbaum DJ, Hsiao EY. The gut microbiota mediates the anti-seizure effects of the ketogenic diet. Cell. 2018;173(7):1728–1741.e13. https://doi.org/10.1016/j.cell.2018.04.027.
20. Oroojzadeh P, Bostanabad SY, Lotfi H. Psychobiotics: the influence of gut microbiota on the gut-brain axis in neurological disorders. J Mol Neurosci. 2022;72(9):1952–64. https://doi.org/10.1007/s12031-022-02053-3.
21. Oskoui M, Coutinho F, Dykeman J, Jetté N, Pringsheim T. An update on the prevalence of cerebral palsy: a systematic review and meta-analysis. Dev Med Child Neurol. 2013;55(6):509–19. https://doi.org/10.1111/dmcn.12080.
22. Peng L, Li Z-R, Green RS, Holzmanr IR, Lin J. Butyrate enhances the intestinal barrier by facilitating tight junction assembly via activation of AMP-activated protein kinase in Caco-2 cell monolayers. J Nutr. 2009;139(9):1619–25. https://doi.org/10.3945/jn.109.104638.

23. Peng Y, Chiu ATG, Li VWY, Zhang X, Yeung WL, Chan SHS, Tun HM. The role of the gut-microbiome-brain axis in metabolic remodeling amongst children with cerebral palsy and epilepsy. Front Neurol. 2023;14:1109469. https://doi.org/10.3389/fneur.2023.1109469.
24. Robertson RC, Manges AR, Finlay BB, Prendergast AJ. The human microbiome and child growth – first 1000 days and beyond. Trends Microbiol. 2019;27(2):131–47. https://doi.org/10.1016/j.tim.2018.09.008.
25. Sadowska M, Sarecka-Hujar B, Kopyta I. Cerebral palsy: current opinions on definition, epidemiology, risk factors, classification and treatment options. Neuropsychiatr Dis Treat. 2020;16:1505–18. https://doi.org/10.2147/NDT.S235165.
26. Sampson TR, Mazmanian SK. Control of brain development, function, and behavior by the microbiome. Cell Host Microbe. 2015;17(5):565–76. https://doi.org/10.1016/j.chom.2015.04.011.
27. Sharon G, Cruz NJ, Kang D-W, Gandal MJ, Wang B, Kim Y-M, Zink EM, Casey CP, Taylor BC, Lane CJ, Bramer LM, Isern NG, Hoyt DW, Noecker C, Sweredoski MJ, Moradian A, Borenstein E, Jansson JK, Knight R, et al. Human gut microbiota from autism spectrum disorder promote behavioral symptoms in mice. Cell. 2019;177(6):1600–1618.e17. https://doi.org/10.1016/j.cell.2019.05.004.
28. Sharon G, Sampson TR, Geschwind DH, Mazmanian SK. The central nervous system and the gut microbiome. Cell. 2016;167(4):915–32. https://doi.org/10.1016/j.cell.2016.10.027.
29. Silva YP, Bernardi A, Frozza RL. The role of short-chain fatty acids from gut microbiota in gut-brain communication. Front Endocrinol. 2020;11:508738. https://doi.org/10.3389/fendo.2020.00025.
30. Sorboni SG, Moghaddam HS, Jafarzadeh-Esfehani R, Soleimanpour S. A comprehensive review on the role of the gut microbiome in human neurological disorders. Clin Microbiol Rev. 2022;35(1):e00338–20. https://doi.org/10.1128/CMR.00338-20.
31. Stewart CJ, Ajami NJ, O'Brien JL, Hutchinson DS, Smith DP, Wong MC, Ross MC, Lloyd RE, Doddapaneni H, Metcalf GA, Muzny D, Gibbs RA, Vatanen T, Huttenhower C, Xavier RJ, Rewers M, Hagopian W, Toppari J, Ziegler A-G, et al. Temporal development of the gut microbiome in early childhood from the TEDDY study. Nature. 2018;562(7728):583–8. https://doi.org/10.1038/s41586-018-0617-x.
32. Tamburini S, Shen N, Wu HC, Clemente JC. The microbiome in early life: implications for health outcomes. Nat Med. 2016;22(7):713–22. https://doi.org/10.1038/nm.4142.
33. Turroni F, Ribbera A, Foroni E, van Sinderen D, Ventura M. Human gut microbiota and bifidobacteria: from composition to functionality. Antonie Van Leeuwenhoek. 2008;94(1):35–50. https://doi.org/10.1007/s10482-008-9232-4.
34. Wang Y, Kasper LH. The role of microbiome in central nervous system disorders. Brain Behav Immun. 2014;38:1–12. https://doi.org/10.1016/j.bbi.2013.12.015.
35. Wu H, Huang C, Xiong S. Gut microbiota as a potential therapeutic target for children with cerebral palsy and epilepsy. Brain Dev. 2025;47(1):104286. https://doi.org/10.1016/j.braindev.2024.09.002.
36. Zhu H, Wang W, Li Y. The interplay between microbiota and brain-gut axis in epilepsy treatment. Front Pharmacol. 2024;15:1276551. https://doi.org/10.3389/fphar.2024.1276551.
37. Zierer J, Jackson MA, Kastenmüller G, Mangino M, Long T, Telenti A, Mohney RP, Small KS, Bell JT, Steves CJ, Valdes AM, Spector TD, Menni C. The fecal metabolome as a functional readout of the gut microbiome. Nat Genet. 2018;50(6):790–5. https://doi.org/10.1038/s41588-018-0135-7.

The Microbiome and Neurotransmitter Regulation: A Double-Edged Sword

7

Deepraj Paul and Rohini Agrawal

7.1 Introduction

Human-microbiome interaction begins at fetal life, where bacterial extracellular vesicle from maternal gut starts communicating with the fetus. This prepares the prenatal immune system for after-birth gut colonization [1]. The significance of the finding lies in the fact that gut microbiome is the most significant one in maintaining human health compared to oral, respiratory, skin, and vaginal microbiomes. The beneficial functions played by gut microbiome include fermentation of food, nutrient extraction, and metabolism, stimulating immune response, protection against pathogens, and biosynthesis of bioactive molecules (vitamins, amino acids, and lipids) [2]. The interaction between maternal gut microbiome and fetus is crucial for the development of gut brain axis, which plays a role in human health and disease [3]. Gut microbiome can produce neurotransmitters and communicate with the central nervous system (CNS) [4]. In spite of many positive roles, the maternal gut microbiome may also have pathological implications. Recent study shows that the maternal gut microbiome influences the immune microenvironment in utero during pregnancy and participates in pregnancy complications, such as preeclampsia, preterm birth, recurrent miscarriages, etc. [5]. This makes the enlisting of gut microbiome inevitable to explore the molecular mechanisms involved in health and disease.

Human gut harbors trillions of microbes in the form of bacteria, archaea, fungi, and viruses. The stomach and duodenum are predominantly inhabited

D. Paul (✉)
College of Pharmacy, JSS University, Noida, Uttar Pradesh, India

R. Agrawal
Department of Pharmacy, Maharshi Dayanand Saraswati University, Ajmer, Rajasthan, India

M. H. Rahman, Muhasina K M (eds.), *The Microbiome's Impact on the Brain*,
https://doi.org/10.1007/978-981-95-7494-0_7

by *Lactobacilli*, yeast, and *Streptococci*; the jejunum and ileum are predominantly inhabited by *Lactobacilli*, *Coliform* bacteria, *Streptococci*, *Bifidobacteria*, *Bacteroides*, and *Fusobacteria*; colon is predominantly inhabited by *Streptococci*, *Bifidobacteria*, *Bacteroides*, *Eubacteria*, *Clostridia*, *Fusobacteria*, *Coliform* bacteria, *Veilonella*, *Proteus*, *Lactobacilli*, *Pseudomonades*, *Staphylococci*, protozoa, and yeast. The oral cavity has shown inconsistency in the community type of microbiome in comparison to the gut [6]. Nevertheless, seven hundred kinds of microorganisms have been found inside the oral cavity [7]. Bacteria present inside the oral cavity are identified as gram-positive and gram-negative. They are further categorized as cocci and rods. Gram-positive cocci include *Abiotrophia*, *Peptostreptococcus*, *Stomatococcus*, and *Streptococcus*. The gram-negative cocci include *Moraxella*, *Neisseria*, and *Veillonella*. Gram-positive rods include *Actinomyces*, *Bifidobacterium*, *Eubacterium*, *Lactobacillus*, *Corynebacterium*, *Propionibacterium*, *Rothia*, and *Pseudoramibacter*. Gram-negative rods include *Campylobacter*, *Capnocytophaga*, *Desulfovibrio*, *Eikenella*, *Desulfobacter*, *Fusobacterium*, *Hemophilus*, *Prevotella*, *Selemonas*, *Leptotrichia*, *Simonsiella*, *Wolinella*, and *Treponema*. Some common protozoa found in oral cavity include *Entamoeba gingivalis* and *Trichomonas tenax*. Most prevalent fungi include candida species and others include *Cladosporium*, *Aureobasidium*, *Saccharomycetales*, *Aspergillus*, *Fusarium*, and *Cryptococcus* [8]. Interestingly, the microbiome in oral cavity and gut are quite different in composition. However, oral microbiome can colonize the gut in case of dysbiosis of gut [9]. A recent study shows that the interplay between oral and gut miocrobiomes may contribute to many pathological conditions [10]. Contemporary studies have mentioned that dysbiosis of gut microbiome is associated with many diseases, such as obesity, diabetes, cancer, cardiovascular disorders, hypertension, IBDs [11], depression [12], neurodegenerative disorders such as Alzheimer's disease (AD) and Parkinsonism (PD), amyotrophic lateral sclerosis (ALS), Huntington's disease (HD), etc. [13, 14]. It is noteworthy that alterations in neurotransmitter secretion are associated with type 2 diabetes (T2D) and obesity [15], cancer [16], cardiovascular disorder [17], hypertension [18], irritable bowel syndrome (IBS) [19], depression [12], neurodegenerative disorders [20], and many more diseases. This reflects some kind of correlation between dysbiosis of gut microbiome and alteration in neurotransmitter secretion. This also shows the significance of a balanced gut microbiome in human health and its apparent role in regulating the effects of several neurotransmitters and molecular messengers. This necessitates the study of microbiome in the regulation of neurotransmitters. The significance of the present study can be understood from the establishment of Human Microbiome Project (HMP) by the National Institute of Health (NIH) in 2008 [8], with a specific focus on the role of microbiome in normal physiology and disease [21]. Owing to the significance of microbiome, this book chapter will try to decipher the role of neurotransmitters and their modulation by selected microbiome with a specific focus on the implications of such modulation.

7.2 Microbiome

7.2.1 Lactobacillus and Neurotransmitters

There are more than 170 species and 17 subspecies of *Lactobacillus*. Different strains of *Lactobacillus* found in human gut are *L. fermentum*, *L. plantarum*, *L. casei*, and *L. rhamnosus* [22]. Some varieties, such as *L. acidophilus* [22], *L. gasseri*, *L. crispatus*, *L. plantarum*, *L. brevis*, *L. rhamnosus*, *L. salivarius*, *L. paracasei*, and *L. vaginalis*, are commonly found in the mouth as well as in the gut [23]. Subsequent studies have also confirmed the presence of *L. crispatus*, *L. gasseri* [24], and *L. jensenii* in the gut [25], whereas *L. crispatus*, *L. gasseri*, *L. vaginalis*, *L. jensenii*, and *L. iners* are predominantly isolated from vagina [22]. This shows that a certain bacterial population is common to oral, gut, and vaginal cavities. Knowledge about different types of *Lactobacillus* has led to many researches which have aimed to improve human health [26]. One of the researches finds that oral administration of *L. fermentum* inhibits chronic mild stress-induced reduction of serotonin level in the rat brain. *L. fermentum* also lowers interferon-gamma (IFN-γ) gene expression and corticosterone levels in the brain and plasma. This will reduce inflammation and stress level. Maintenance of serotonin level may alleviate depression and AD [27]. Similar research was conducted to find the effect of *L. fermentum* on acute and subacute stress models. *L. fermentum* has been found to significantly increase serotonin levels in acute and subacute restraint-stressed mice compared to negative control [28]. This increases curiosity about the implications of serotonin on obesity because stress may increase or decrease the craving for food [29, 30]. Craving for food that is rich in sugar and fat increases in the case of increased eating pattern associated with chronic life stress. Stress-induced increased eating of such nutrient-dense foods plays a key role in obesity [29]. A decreased serotonin signaling is responsible for human obesity [31]. The brain plays a significant role in regulating obesity. A decreased synthesis of serotonin in the brain leads to weight gain and obesity, though the brain secretes only 3%–5% of total serotonin secreted by the body [32–34]. Obesity, weight, and weight gain are responsible for 20% of all cancer cases [35]. Obesity increases the risk for age-related cognitive decline, vascular dementia, mild cognitive impairment (MCI), and AD [36] and directly contributes to dyslipidemia, type 2 diabetes, hypertension, sleep disorders, and cardiovascular disease mortality [37]. Thus, decreased serotonin level can be linked with obesity, dementia, MCI, AD, dyslipidemia, type 2 diabetes, hypertension, sleep disorders, cardiovascular mortality, cancer, and many more disorders. A study on mice shows that a decreased level of serotonin transporter (SERT) leads to abnormal fat accumulation in both white and brown adipose tissues, glucose intolerance, and insulin resistance. Estrogen replacement to SERT $^{-/-}$ mice reversed obesity and glucose intolerance [38]. The relevance of estrogen in this context is apparent from the finding that *L. fermentum* impedes the progression of postmenopausal symptoms in estrogen-deficient obese animals [39]. *L. fermentum* has been found to increase the activity of osteoblast and favor mineralization of bones. This may alleviate osteoporosis [40] in postmenopausal women. The effect of *L. fermentum* is similar to the osteoblast-activating property of estrogen [41]. Perhaps these evidences suggest for an indirect estrogenic effect of *L. fermentum*. Studies on rats show that estrogen can desensitize and reduce mRNA expression of serotonin receptor ($5HT_{1A}$) [42, 43]. This raises concern about the actual role of estrogen as well as serotonin. 8-hydroxy-2-(di-n-propylamino)tetralin, a $5HT_{1A}$ agonist (at 100 μg/kg), has been found to increase food intake in 3-month-old lean control rats, but at a 300 μg/kg dose, it has inhibited food intake in obese Zucker rats. Shockingly, 3-month-old obese Zucker rats have shown higher levels of serotonin in the hypothalamus and in the frontal and parietal cortices. Serotonin levels were still higher in the parietal cortex and the hypothalamus of 6-month-old rats [44]. This shows that obesity can prevail despite the presence of higher level of serotonin. Thus, the actual role of serotonin in obesity is obscure (see Fig. 7.1). This perhaps shows the dichotomous effects of serotonin on obesity. This raises concern about the beneficial role of *L. fermentum* in chronic mild stress through serotonin, because psychological stress may be the cause as well as the consequence of obesity [45]. Further concern arises because

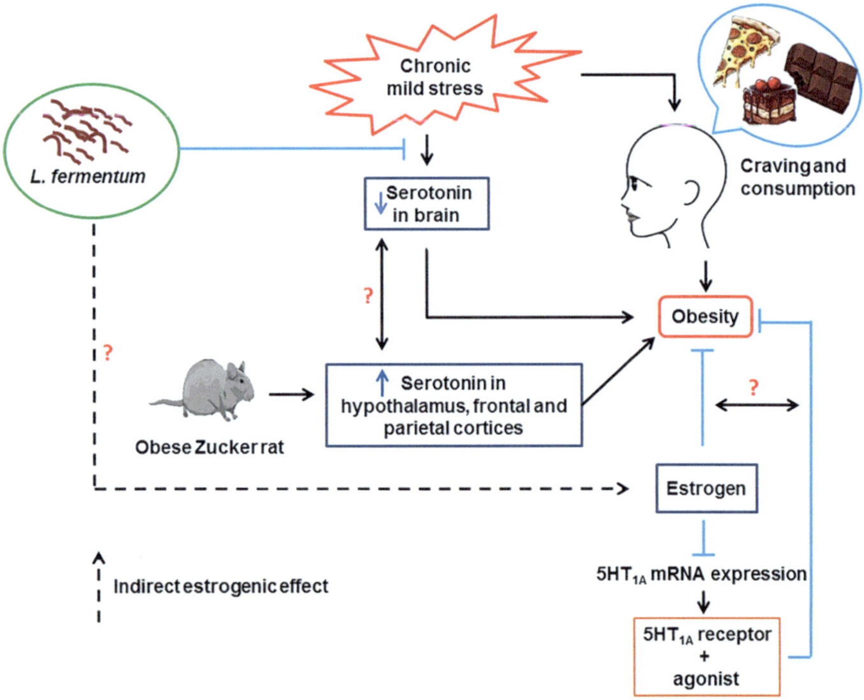

Fig. 7.1 *L. fermentum* prevents chronic stress-induced fall in the brain serotonin level. Chronic stress increases the craving and consumption of high calorie diet and induces obesity. A decreased serotonin level in the brain triggers obesity. Obese Zucker rat shows an increased serotonin level in the brain and raises concern on the beneficial effect of *L. fermentum* in increasing serotonin level in the brain and its actual relation to obesity. *L. fermentum* shows an indirect estrogenic effect. Estrogen prevents obesity but can reduce the expression of $5HT_{1A}$ mRNA. Binding of agonist with $5HT_{1A}$ receptor impeded obesity; shows a complex effect of estrogen and *L. fermentum* on obesity

there is no research that shows the direct effect of *L. fermentum* on the expression pattern of $5HT_{1A}$ receptor. *L. fermentum* is known to protect against diet-induced obesity in mice [46]. Conversely, other species of *Lactobacillus*, such as *L. acidophilus*, have triggered significant weight gain in humans as well as in animals, whereas *L. fermentum* and *L. ingluviei* promote weight gain in animals [47]. This finding contradicts previous assumptions/findings that *L. fermentum* may alleviate obesity by increasing serotonin level or it may protect against diet-induced obesity. Thus, the regulation of serotonin level by *L. fermentum* may or may not protect against obesity.

Similarly, PD shows the presence of depression in around 35% of patients, and PD-associated depression is often persistent [48]. This projects the therapeutic significance of *L. fermentum*, as it can significantly increase central serotonin and norepinephrine levels [28] and alleviate depression [27]. Seventy-five percent PD patients experience tremor [49]. This perhaps shows the limitation of *L. fermentum* therapy in depression among PD patients because serotonin toxicity can produce tremor [50], and norepinephrine may regulate serotonergic neurotransmission. It was found that norepinephrine reuptake inhibitors cause secondary enhancement of central serotonergic activity [51]. Thus, an increase in serotonin level may alleviate depression but aggravate tremor. Another limitation arises due to the metabolic pathways of neurotransmitters. Among many metabolic pathways, norepinephrine and serotonin share a common metabolic pathway, which is mediated by monoamine oxidase (MAO) [52]. MAO-A is a specific isoform of MAO and shows higher affinity to serotonin, followed by norepinephrine. This may divert norepinephrine to other metabolic pathways, such as catechol O-methyltransferase (COMT) [53]. Interestingly, serotonin is known to inhibit COMT [54]. This may increase the concentration of norepinephrine to many folds. *L. fermentum* can significantly increase norepinephrine level [28]. One way, central norepinephrine has an antidepressant effect [55], and the other way, central norepinephrine is associated with rest tremor in PD [56] (see Fig. 7.2). This reflects a complex role of *L. fermentum* in the complications of PD.

Further studies on neurotransmitters reveal that an imbalance in neurotransmitters, such as dopamine and serotonin, may cause tremors or spasms. Serotonin was found to increase motoneuron excitability [57, 58]. Excess release of serotonin leads to the saturation of the serotonin reuptake system. This will increase the availability of serotonin to interact with extrasynaptic 5HT1A receptors, which are present on the axon initial segment of motoneurons. This will inhibit sodium ion channel leading to a subsequent reduction in action potential generation and weaker muscle contraction. Serotonin triggers central fatigue by decreasing the gain of motoneurons [58]. This may aggravate PD, which presents muscle weakness, reduced muscle strength [59], and central fatigue [60]. Central fatigue is responsible for the "impairment in the processing of novel stimuli, which is an essential part of adaptive decision-making behaviour" [60]. The above findings show that despite the beneficial role in depression, serotonin increasing effect of *L. fermentum* may have many negative implications. It may also worsen anxiety disorder. This is due to the fact that anxiety patients represent a higher central

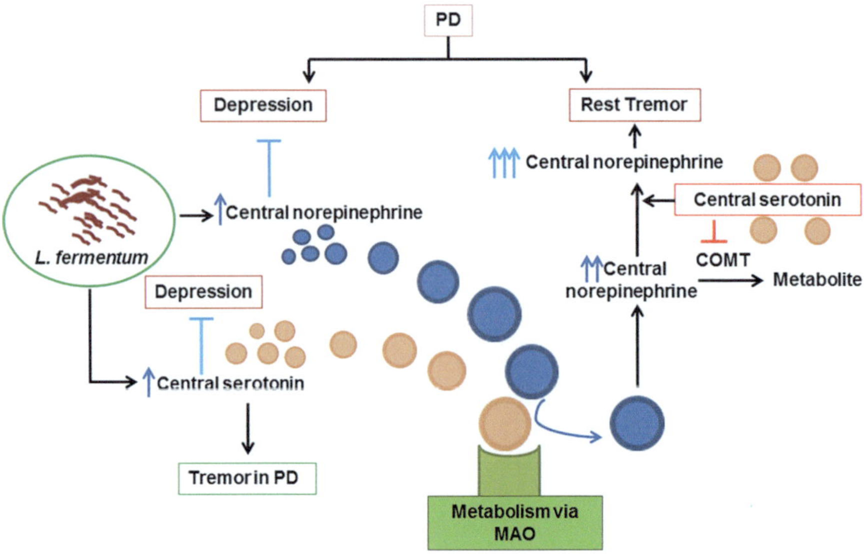

Fig. 7.2 PD represents both depression and tremor. *L. fermentum* can alleviate depression by increasing central norepinephrine and serotonin levels. An excessive increase in central serotonin level is linked with tremor. Serotonin may increase central norepinephrine level by binding with MAO with higher affinity than norepinephrine as well as by blocking the alternative metabolic pathway mediated by COMT. This may trigger rest tremor in PD

serotonin level due to a low serotonin reuptake capacity in the thalamus [61]. This is a critical situation because anxiety has been identified as a neuropsychiatric complication of PD [62]. In fact, PD shows a higher prevalence of anxiety disorders [63]. A study on the patients of PD shows that 40.6% of patients suffer from anxiety disorder, while 31.25% of patients suffer from depression, and 23.4% of patients show an overlap between depression and anxiety [64]. Thus, an increase in serotonin by *L. fermentum* may alleviate depression but aggravate anxiety among patients with PD. This makes the actual role of *L. fermentum* very complex and elusive. As per the revised central fatigue hypothesis, the onset of fatigue is due to an increase in the central ratio of serotonin to dopamine, whereas a low ratio alleviates fatigue [65]. This increases the complexity of the actual impact of *L. fermentum* in PD. This is due to the finding that oral administration of *L. fermentum* can significantly increase dopamine and norepinephrine levels in acute and subacute restraint-stressed mice in comparison to the negative control. mRNA expression of dopamine receptors (D1 and D2) increased significantly along with synaptophysin in the prefrontal cortex and hippocampus of the acute (7 days) and subacute (14 days) restraint-stressed mice. However, there was no significant difference in the expression pattern in comparison to the normal group [28]. The significance of synaptophysin is evident from the finding that loss of synaptophysin protein is observed in many neurodegenerative disorders, including PD and dementia with Lewy bodies [66]. Similarly, the significance of norepinephrine is

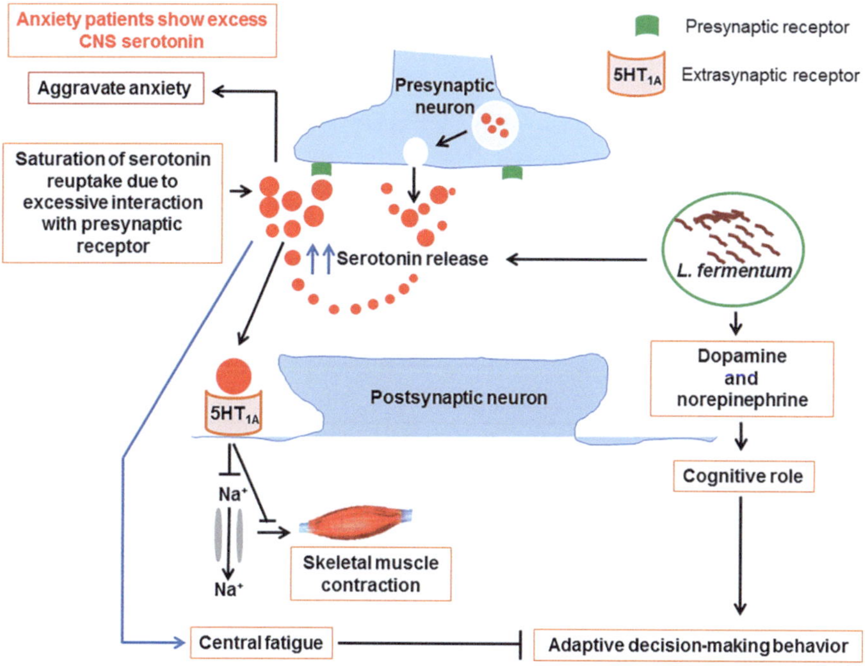

Fig. 7.3 Anxiety patients show an excess CNS serotonin level. *L. fermentum* increases the release of serotonin from presynaptic neuron. This may aggravate anxiety disorders. Excess release of serotonin can saturate the reuptake and favor the interaction of serotonin with extrasynaptic $5HT_{1A}$ receptor. This may lead to weaker skeletal muscle contraction. Excess serotonin may produce central fatigue and affect adaptive decision-making behavior. At the same time, the release of dopamine and norepinephrine will improve cognitive function and adaptive decision-making behavior

evident from the finding that norepinephrine deficiency as a result of selective degeneration of neurons in the locus coeruleus and sympathetic ganglia precedes dopaminergic deficit. Norepinephrine deficiency in the brain and other places is associated with motor, behavioral, cognitive, and autonomic impairments in PD [67]. The significance of norepinephrine is apparent from the finding that it plays a role in cognitive flexibility, which in turn plays a crucial role in adaptive decision-making [68, 69]. The above findings show that by increasing the serotonin level, *L. fermentum* can worsen adaptive decision-making, but by increasing norepinephrine level, it can improve adaptive decision-making (see Fig. 7.3).

L. fermentum increases the production of gamma-aminobutyric acid (GABA) [70, 71]. Other species of *Lactobacillus* that produce GABA are *L. brevis*, *L. buchneri*, *L. delbrueckii* subsp. *bulgaricus*, *L. helveticus*, *L. paracasei*, and *L. plantarum* [72]. GABA-producing activity seems to be neuroprotective among patients with AD because epileptiform activity is observed in 30%–50% of the AD population [73], and epileptiform activity can predispose to seizure [74]. *L. fermentum* may also have a protective role in PD-related epilepsy or epilepsy related PD because

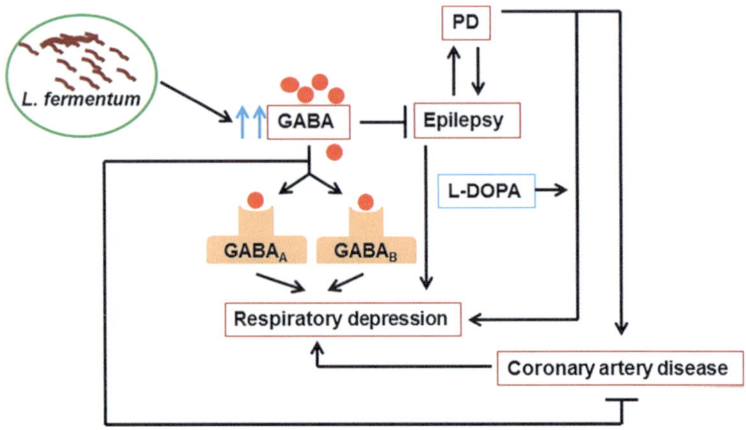

Fig. 7.4 PD increases the chance of epilepsy, and epilepsy in turn can increase the chance of PD. *L. fermentum* increases the concentration of GABA and may prevent/reduce epilepsy, epilepsy-induced PD, and respiratory depression. Patients of PD who are on L-DOPA therapy show respiratory depression. GABA may induce respiratory depression through $GABA_A$ and $GABA_B$ receptors. PD can initiate coronary artery disease (CAD), and CAD can promote respiratory depression, but GABA can suppress CAD and may reduce CAD induced respiratory depression

incident PD increases the risk of incident epileptic seizures [75] and epilepsy increases the risk of PD by 2.19 times [76]. It is noteworthy that epilepsy can suppress breathing and may have a fatal outcome, including death [77]. *L. fermentum* may prevent epilepsy-induced respiratory depression by increasing GABA and preventing epilepsy. These beneficial effects raise other concerns because activation of GABA receptors in Schlaefke's area produces respiratory depression, hypotension, and bradycardia [78]. Contextually, activation of $GABA_A$ and $GABA_B$ receptors leads to respiratory depression [79]. It is worth mentioning that, contrary to previous belief, recent findings suggest that GABA can cross the BBB [80, 81]. Hence, an increase in GABA in the periphery or CNS may have almost similar respiratory effects. The complexity of the situation increases due to the finding that shortness of breath is common among patients of PD who are on L-dopa therapy or patients suffering from pulmonary diseases, coronary artery disease (CAD), and anxiety [82]. Interestingly, higher occurrence of CAD is observed in PD [83], and GABA can prevent CAD by reducing the serum level of triglyceride, cholesterol, and LDL [84]. This makes the role of GABA and *L. fermentum* elusive in the context of respiratory disorder, perhaps representing a dichotomous role of GABA and *L. fermentum* (see Fig. 7.4).

The ability of *L. fermentum* to activate natural killer (NK) cells [85, 86], lower interferon-gamma (IFN-γ) gene expression [27], and increase serotonin level [28] represents a complex phenomenon. This is due to the contradictory finding where NK cells in humans show an increased production of IFN-γ following the activation of 5HT1A receptor by serotonin [87]. Human research shows that intracellular serotonin is required for the optimal production of IFN-γ and interleukin-10 (IL-10).

Notably, the production of IFN-γ may be suppressed by the concentrations of extracellular serotonin at or above the serum values [88]. IFN-γ plays a significant role in neurogenesis through wingless-related integration site (Wnt)/β-catenin cell signaling pathway via promotion of signal transducer and activator of transcript 1 (STAT1) regulation of the β-catenin promoter [89]. Intraperitoneal (i.p.) injection of IFN-γ in APP/PS1 mice can restore microglial autophagy and promote amyloid beta (Aβ) clearance. This leads to a decrease in Aβ plaque load in the cortex and hippocampus of APP/PS1 mice and alleviation of cognitive deficits [90]. This shows the positive role of serotonin. Conversely, IFN-γ induces mRNA and protein expression of leucine-rich repeat kinase 2 (LRRK2) (a Parkinson's disease-associated kinase present in neurons and microglia) in control and dopaminergic neurons. IFN-γ also increases LRRK2 expression in microglia. INF-γ synergizes with LRRK2 to increase the vulnerability of substantia nigra dopaminergic neurons, subsequently leading to inflammation and neurodegeneration in PD. Interestingly, LRRK2 G2019S is associated with PD and represents the most common mutation of LRRK2. IFN-γ can reduce the phosphorylation of AKT serine/threonine kinase 3 (AKT3) in LRRK2 G2019S and control neurons; however, the extent was higher in LRRK2 G2019S neurons compared to control neurons (38.9% vs. 28.8%). There was also a significant reduction in nuclear factor of activated T cells (NFAT3) levels in both LRRK2 G2019S and control neurons following IFN-γ treatment. This is a matter of concern because NFAT transcription complexes are required for neuronal growth and brain circuit formation. The effect of IFN-γ on activated LRRK2 G2019S microglia is observed as neurite shortening, indicating neurotoxicity [91]. This shows the dichotomous effects of IFN-γ and the actual implications of serotonin in neuronal health (see Fig. 7.5) following its modulation by *L. fermentum* in pathological conditions.

The effect of *L. fermentum* on serotonin increases the curiosity about the role of serotonin transporters. Research has shown the presence of 5-HTTLPR polymorphism (a serotonin transporter polymorphism) among patients with AD. Though there are conflicting evidences on the association between 5-HTTLPR polymorphism and AD [92], yet the presence of such a polymorphism among patients with AD may be a matter of concern. This is due to the finding that schizophrenia patients with the 5-HTTLPR ll genotype show significantly higher ratings for psychosis as per the "Brief Psychiatric Rating Scale." There is also a significant increase in the intensity of hallucinations [93]. It is noteworthy that approximately 50% of AD patients show psychotic symptoms, delusions, and hallucinations [94]. This perhaps shows the healing effect of serotonin in depression but the worsening effect in patients with AD or schizophrenia with 5-HTTLPR polymorphism.

Corticosterone in rats and cortisol in human cross the blood-brain barrier (BBB) and interact with neurons and astrocytes. Such interaction produces functional and anatomical changes in those areas [95]. In fact, corticosterone/glucocorticoid can regulate the secretion along the hypothalamic-pituitary-adrenal axis. Cortisone also acts as a signaling molecule during stress and activates the immune system [96]. This reflects the role of corticosterone/corticosteroids as neurotransmitters. Other studies also prove glucocorticoids to be neurotransmitters [97]. In this context,

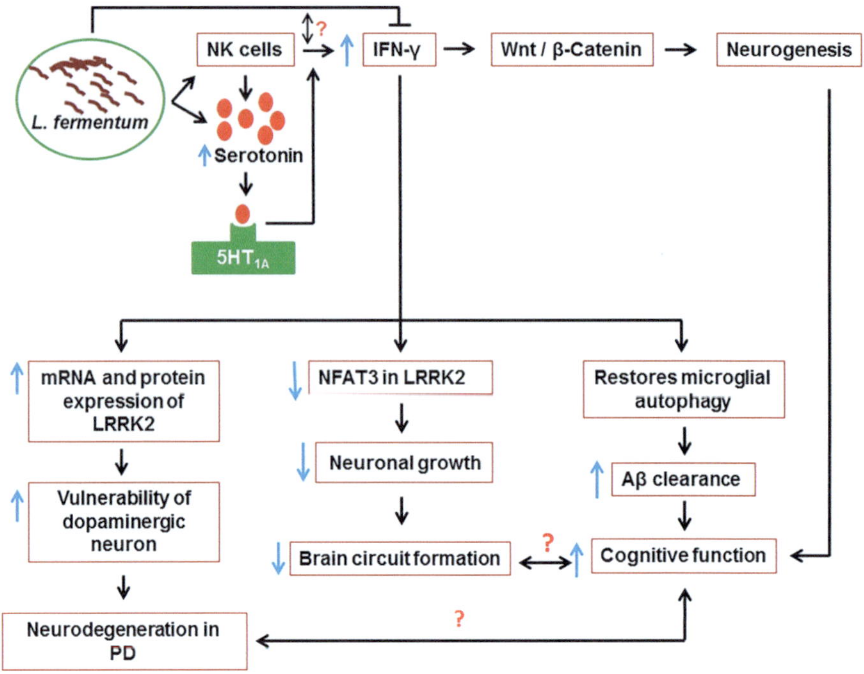

Fig. 7.5 *L. fermentum* activates NK cells and increases serotonin but decreases IFN-γ gene expression. NK cells can increase the expression of IFN-γ. *L. fermentum*-activated NK cells may oppose the decreasing effect of *L. fermentum* on IFN-γ expression. Increased IFN-γ promotes neurogenesis, which in turn may improve cognitive function. Increased expression of IFN-γ can decrease NFAT3 in LRRK2 and subsequently decrease neuronal growth, affecting brain circuit formation. IFN-γ increases cognitive function but decreases neuronal growth and brain circuit formation. IFN-γ can increase LRRK2 mRNA and protein expression and lead to neurodegeneration in PD. This shows the complex role of *L. fermentum* in cognitive function and neurodegeneration

corticosterone lowering effect of *L. fermentum* in the brain and plasma [27] raises concern because the actual role of corticosteroids in the context of AD/ dementia is much more complex. One way, long-term usage of glucocorticoids can cause significant atrophy in the hippocampus and amygdala, significant decline in executive function (working memory); another way, many studies suggest a null to negative association between long-term usage of steroids and most dementia types, including AD. Studies rather suggest a protective role in most dementia types and AD [98]. This finding is corroborated by a recent finding where steroids have been found to delay the progression of dementia [99]. The beneficial role may be contradicted by other research where corticosterone has been found to enhance N-methyl-D-aspartate (NMDA) receptor signaling [100], whereas NMDA receptor antagonists are one of the treatment modalities for AD and dementia [101]. This raises concern on the actual implications of long-term oral administration of *L. fermentum* or its dysbiosis in neurodegenerative condition (see Fig. 7.6). Further complication arises

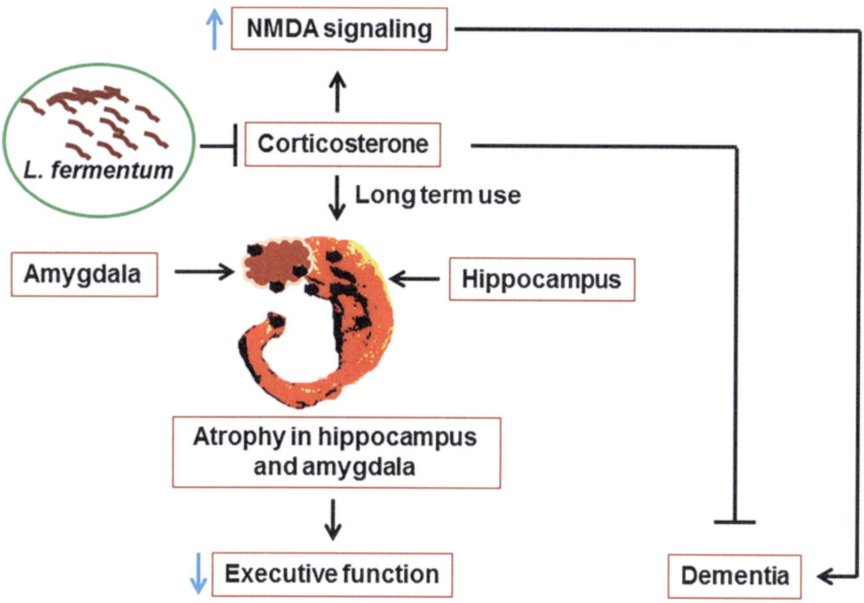

Fig. 7.6 *L. fermentum* lowers corticosterone in the brain. Long-term therapeutic use of corticosterone leads to atrophy of the hippocampus and amygdala, further leading to decreased executive function. Conversely, corticosterone delays the progression of dementia, showing a protective role. Corticosterone enhances NMDA receptor signaling and worsens dementia. Corticosterone represents a dichotomous role in the cognitive domain

from the finding that executive dysfunction may coexist with dementia or may appear at the later stages [102]. This shows the complex role of gut microbiome and the pathological significance of dysbiosis. However, bidirectional communication between gut microbiome and the brain regulates glial functions and protects from neurodegenerative diseases. A balanced gut microbiome maintains the integrity of the gut wall and BBB. This finding shows the significance of probiotics and prebiotics in human health [103, 104]. However, the above findings contradict the beneficial role of corticosterone and raise concerns.

Similarly, the actual role of serotonin in AD is much more complex. Serotonin signaling decreases Aβ accumulation in mice brain interstitial fluid (ISF) and in human [105]. Subsequent study on mice has shown that serotonin reuptake inhibitors can reduce toxic Aβ production and plaque formation. However, constitutive serotonin receptor 7 (5HT7R) interacts with cyclin-dependent kinase 5 (CDK5) and promotes hyperphosphorylation of tau protein independent of the G-protein pathway. Hyperphosphorylation leads to the formation of highly bundled tau structures, eventually leading to neuronal death and a reduction in long-term potentiation (LTP). This will culminate in memory impairment [106]. In fact, 5HT7R recruits CDK5 [107]. This raises a question on the role of serotonin as it activates 5HT7R [108]. Activation of 5HT7R leads to depressive-like behavior through MMP-9 [109], and a depressed individual presents poor memory [110]. Most importantly,

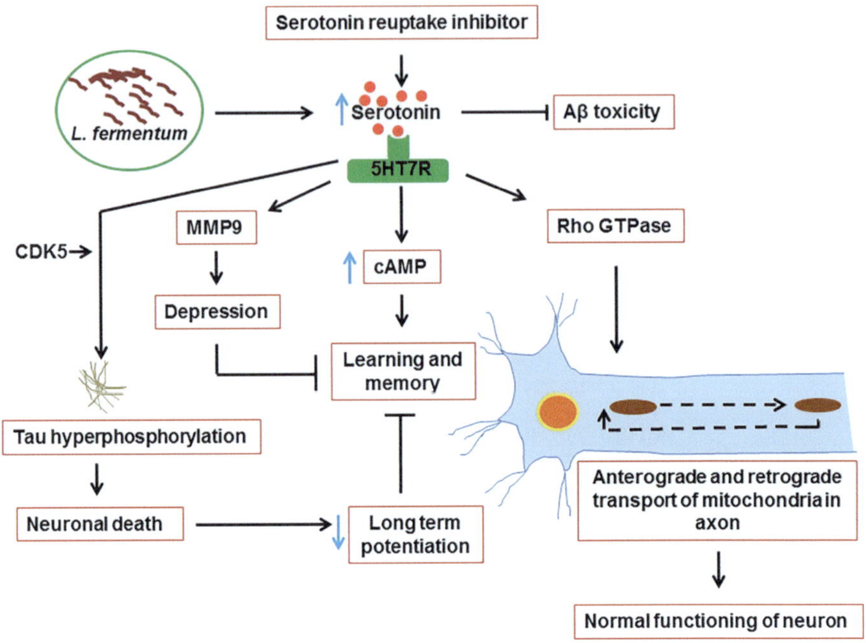

Fig. 7.7 *L. fermentum* increases serotonin concentration, and serotonin decreases Aβ accumulation in brain interstitial fluid, leading to a decrease in Aβ toxicity. Serotonin signaling through 5HT7R in the presence of CDK5 triggers tau hyperphosphorylation, leading to neuronal death. Neuronal death will decrease long-term potentiation, leading to the loss of memory and learning. 5HT7R signaling may trigger depression through MMP9, ultimately affecting memory. 5HT7R signaling increases cAMP and improves learning and memory. 5HT7R signaling improves anterograde and retrograde mitochondrial transport via Rho GTPase and improves neuronal function

increased brain serotonin level shows a positive correlation with MMP-9 activity [111]. Shockingly, this reveals the depressing effect of serotonin through 5HT7R signaling. Conversely, some research shows that 5HT7R is responsible for increasing intracellular cAMP and plays a positive role in thermoregulation, circadian rhythm, sleep, learning, and memory [112]. Contemporary study on human and mouse model of AD has proved that different regions of the brain, including the hippocampus, show a decreased level of cAMP, adenylyl cyclase, and protein kinase A (PKA) [113, 114]. 5HT7R plays a significant role in neuronal function by activating small Rho GTPases such as RhoA and Cdc42 [115]. Rho GTPase plays a significant role in anterograde and retrograde axonal transport of mitochondria. It is also responsible for calcium homeostasis, mitochondrial fusion, and fission. All these phenomena assure normal neuronal functioning [116]. The above findings make the role of serotonin and its receptor ambiguous (see Fig. 7.7) in the context of neurophysiology.

Endothelins are neurotransmitters as well as neuromodulators [117]. *L. fermentum* can increase the serum level of neurotransmitters such as endothelin and vasoactive intestinal peptide (VIP) and decrease the level of somatostatin. It may decrease the level of acetylcholine as it increases the level of serum acetylcholinesterase [118]. Endothelin gives rise to endothelin-1, which has been identified as a neurotransmitter [119]. Endothelin-1 is a potent endogenous vasoconstrictor [120] which can cause hypertension in patients with type 2 diabetes [121]. At the same time, another strain of *Lactobacillus* named *L. helveticus* has been found to increase angiotensin-converting enzyme (ACE) inhibitors in fermented milk. This can reduce vasoconstriction and blood pressure. Increased ACE inhibitors can prevent ACE-mediated breakdown of bradykinin, which has a vasodilatory role [122]. Even studies with another strain of *Lactobacillus* named *L. casei* have confirmed their blood pressure-reducing effect in mice [123]. Thus, different population of bacteria shows opposite effects on blood pressure through different mediators. The net effect on blood pressure will be determined by the balance between different populations of bacteria or by their dysbiosis. To have a complete understanding of the actual effect on blood pressure, it is necessary to decipher the relation between endothelin-1 and angiotensin-II. In vitro studies show that the expression or secretion of endothelin-1 is increased by angiotensin-II. Other studies show that angiotensin-II has been found to upregulate the expression of endothelin A receptor (ET_AR) but not endothelin B receptor (ET_BR). The binding of endothelin-1 with ET_AR produces vasoconstriction but binding with ET_BR produces vasodilation. Angiotensin-II is reported to show synergism with endothelin-1 and produce vasoconstriction and hypertension. Endothelin-1 is the most potent vasoconstrictor [124], and the production or secretion of endothelin-1 can be induced by direct stimulation other than by angiotensin-II [125]. Hence, production of ACE inhibitors by certain types of *Lactobacillus* may not be able to completely nullify the effect of endothelin-1. Research also suggests that endothelin-1 is mainly involved in pathological vasoconstriction (abnormal renal vascular function) [126]. This shifts the focus mainly to the serum concentration of endothelin-1. Increased serum concentration of endothelin-1 in the presence of hypertension leads to atherosclerosis. Endothelin-1 antagonist can protect hypertensive rats from stroke and renal injury. The preclinical findings are alongside the clinical findings, where hypertensive patients show an increased expression of endothelin-1 in the artery, and the plasma also shows an increased level [127]. This shows the cardiovascular implications of endothelin-1. Endothelin-1 also plays a significant role in cancer progression by promoting angiogenesis, cell proliferation, invasion, and metastasis [128]. Signaling between endothelin-1 and peripheral ET_AR is responsible for carcinoma-induced soft tissue pain [129]. The actual influence of *L. fermentum* in blood vessels is ambiguous because endothelin-1 causes coronary vasoconstriction [130], but VIP is known to decrease coronary artery resistance and improve coronary blood flow. It also increases cardiac contraction and heart rate [131]. The ability of *L. fermentum* to increase the expression of endothelin and VIP has gained attention due to the finding that endothelin-1 and vasoactive intestinal contractor (VIC) were found to significantly decrease the colonic output of VIP into portal venous blood [132]. The role of endothelin-1 in hypertension raises other concerns with respect to PD. Hypertension increases the risk of PD by 1.9 times in the Asian population [133], and serum somatostatin has been found to increase in early-stage PD patients [134]. However, somatostatin has been found to

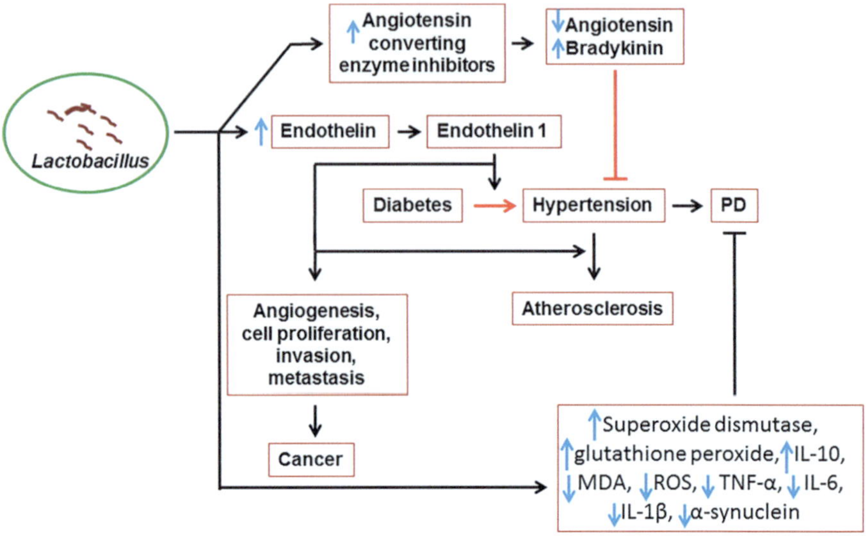

Fig. 7.8 *Lactobacillus* increases the production of ACE inhibitors, leading to a decrease in angiotensin but an increase in bradykinin. This will lead to a decrease in blood pressure. *Lactobacillus* increases serum concentration of endothelin. Endothelin gives rise to endothelin-1. Endothelin-1 can cause hypertension in diabetes patients, and hypertension can cause PD. Endothelin-1 can trigger atherosclerosis among hypertensive patients. Endothelin-1 plays a role in angiogenesis, cell proliferation, invasion, and metastasis. This may lead to cancer. *Lactobacillus* can increase the levels of superoxide dismutase, glutathione peroxide, IL-10, but decrease the levels of MDA, ROS, TNF-α, IL-6, IL-1β, and α-synuclein. This will reduce the progression of PD. This shows the complex role of *Lactobacillus* in hypertension, PD, and associated pathological conditions

lower blood pressure [135]. Another strain of *Lactobacillus*, namely *L. plantarum*, has been found to increase the levels of superoxide dismutase, glutathione peroxide, IL-10, but decrease the levels of malondialdehyde (MDA), reactive oxygen species (ROS), tumor necrosis factor-α (TNF-α), IL-6, and IL-1β. *L. plantarum* also reduces the accumulation of α-synuclein in the substantia nigra of mice treated with 1-methyl-4-phenyl-1,2,3,6-tetrahydropyridine (MPTP) to induce PD [136]. This raises concern about the actual implications of *Lactobacillus* in hypertension and PD (see Fig. 7.8).

7.2.2 Lactobacillus and Other Pathologies

Above evidences suggest for many possibilities of positive and negative outcomes with *Lactobacillus* because of the intricate signaling pathways. Amidst the positive findings, research has confirmed certain negative findings also, which raise pathological concerns. These pathological concerns may be related to diseases that are not of CNS origin but can be linked to CNS disorders. For example, a case study reports that an 82-year-old susceptible female suffered from liver abscess and bacteremia following the consumption of *Lactobacillus*-containing probiotic throughout the treatment of *Clostridium difficile* colitis [137]. The significance of the finding lies in the fact that liver abscess shows many complications, including brain abscess and bacterial meningitis [138], whereas bacteremia can lead to systemic infection in the susceptible population [139]. Certain studies report that increased use of probiotics rich in *Lactobacillus* did not produce bacteremia [140]. Other studies show a positive correlation between *Lactobacillus* and bacteremia. *Lactobacillus* has also been reported to be associated with occasional infective endocarditis, peritonitis, and meningitis [141, 142]. Endocarditis and splenic abscess are observed in people with impaired immunity [143]. This raises concerns because infective endocarditis shows frequent neurological manifestations, such as meningitis, stroke, cerebral hemorrhage, cerebral abscess, encephalopathy, etc. [144].

Some studies have found a positive correlation between *Lactobacillus* overgrowth and tumor cell growth. This is due to the ability of *Lactobacillus* to produce large amounts of N-nitroso compounds. These compounds have been known to favor mutagenesis, angiogenesis, and protooncogene expression, leading to gastric carcinogenesis. Different species of *Lactobacillus* found in gastric carcinoma are *L. paracasei*, *L. fermentum*, *L. rhamosus*, *L. salivarius*, *L. delbreuckii*, and *L. acidophilus* [145]. Conversely, a compelling number of evidences show cancer protective role of *Lactobacillus*. Probiotic supernatant of *L. rhamnosus* has a wide range of effects. In HT-29 cancer cells, it can upregulate proapoptotic genes, such as caspase-3, caspase-9, and Bax. At the same time, it can downregulate Bcl2. *L. rhamnosus* can decrease the expression of cyclin D1, cyclin E, and ERBB2; can prevent colon cancer by arresting the G0/G1 phase of the cell cycle [146]. Another research claims that *L. rhamnosus* free supernatant can exhibit anticancer properties by arresting G2/M phase of cell cycle and reducing cell viability. Decreased cell viability was not related to apoptosis as confirmed by the study conducted in four cell lines, such as HT-29, HCT-116, Caco-2, and A375 [147]. Another strain, namely *L. acidophilus*, shows apoptosis in HT-29 cells through the mitochondrial pathway. *L. acidophilus* upregulates Bax and downregulates Bcl-2, also releases mitochondrial cytochrome C, and eventually activates caspase-9 and caspase-3 [148]. Irrespective of the protective role of *Lactobacillus* in cancer, certain research shows the opposite findings. *Lactobacillus* has been found to convert dietary tryptophan to indoles, a step crucial for activating aryl hydrocarbon receptors on macrophages and promoting pancreatic cancer [149]. The ability of *L. fermentum* to significantly increase dopamine and norepinephrine levels as well as mRNA expression of dopamine receptors (D1 and D2) [28] raises concerns about the actual role of *Lactobacillus* in cancer. This is due to the fact that norepinephrine triggers colorectal cancer via CREB1–miR-373 signaling. This signaling triggers colon cancer cell proliferation and metastasis [150]. Other studies have also proved that norepinephrine can promote metastasis of breast and colon cells through β-adrenergic receptors [151]. Strangely, norepinephrine increases Bax but decreases Bcl-2 expression [152]. This raises curiosity about the actual implications of *Lactobacillus* and norepinephrine in cancer. Similarly, dopamine also increases Bax but decreases Bcl-2 expression

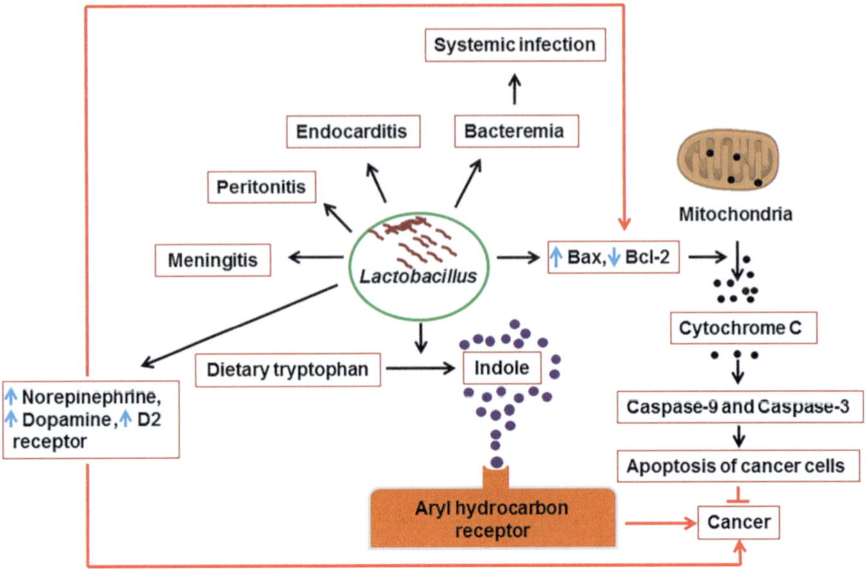

Fig. 7.9 *Lactobacillus* may trigger many pathological conditions, such as meningitis, peritonitis, endocarditis, bacteremia, and systemic infection. *Lactobacillus* can increase Bax but decrease Bcl-2 expression and cause the release of mitochondrial cytochrome C. This will trigger caspase-9 and caspase-3, leading to apoptosis of cancer cells. This will prevent cancer. Conversely, *Lactobacillus* can convert dietary tryptophan into indole. Indole can bind with aryl hydrocarbon receptors and trigger cancer. *Lactobacillus* also increases norepinephrine level, dopamine level, and D2 receptor expression. Increased norepinephrine level and D2 receptor expression may lead to the development of cancer but norepinephrine and dopamine can increase Bax and decrease Bcl-2 expression. This may prevent cancer

[153]. D2 dopamine receptors are upregulated in many cancers [154]. Subsequent research has also confirmed that D2 dopamine receptor expression was very high in certain cancers, such as glioblastoma, whereas the expression levels of D1, D3, D4, and D5 were very low. D2 receptor agonist has been found to increase cell survival and proliferation of glioblastoma cells [155]. The above evidences show many ambiguous mechanistic pathways and contradictory findings pertinent to *Lactobacillus* (see Fig. 7.9). Irrespective of the controversies, the concern of cognitive performance prevails because N-nitroso compounds have been found to reduce cognitive performance in mice, perhaps proving their potential to induce or aggravate AD [156].

7.2.3 Archaea and Neurotransmitters

Archaea are prokaryotic microorganisms [157]. Irrespective of human contact with different types of archaea (hyperthermophiles, methanogens, and halophiles), researchers could identify methanogens as the only archaeal type present in human

[158]. Hence, methanogens deserve special mention, though all the types of archaea have been found to secrete acetylcholine [159]. Secretion of acetylcholine may improve neurological conditions like AD and dementia [160] because AD shows central as well as peripheral cholinergic disturbances. Lymphocytes and platelets from AD patients show 148% and 161% increase in asymmetric (A12) acetylcholine esterase (AChE) forms. Other forms of AChE present in lymphocytes and platelets include dimeric globular (G2) and tetrameric globular (G4) AChE [161]. High serum cholinesterase activities are correlated with inflammation and increased heart rate. It is worth mentioning that peripheral as well as central acetyl choline (ACh) prevents the secretion of IL-1β, IL-6, IL-18, and TNF-α by macrophages [162]. AD and multi-infarct dementia show a significant increase in peripheral IL-1β and TNF-α [163]. Peripheral TNF-α can increase serum and brain levels of TNF-α, IL-6, and MCP-1 in mice. This will induce neuroinflammation [164]. In fact, TNF-α plays a prime role in AD [165]. This shows the involvement of peripheral AChE in neuroinflammation and AD. Research involving rats shows that acetylcholine can induce Janus kinase-2 (JAK-2) [166]. Normally, JAK-2 negatively regulates glycogen synthase kinase 3β (GSK-3β), Fyn (a tyrosine kinase), and cyclin-dependent kinase 5 (CDK5). Negative regulation of GSK-3β/Fyn/CDK5 pathway suppresses neuronal differentiation in mouse embryonic stem cells [167]. This raises concern because patients with AD show a decrease in the differentiation of new neurons [168]. JAK-2 phosphorylation activates signal transducer and activator of transcription 3 (STAT3). JAK-2/STAT3 pathway is linked with phosphoinositide 3-kinase (PI3k)/ protein kinase B (AKT)/ mammalian target of rapamycin (mTOR) pathway. Inhibition of JAK2/STAT3–PI3K/AKT/mTOR pathway has been found to improve spatial learning and memory in female rats. This may alleviate neuroinflammation, AD, and dementia [169]. The actual role of STAT3 is very complex and elusive in the context of AD. Twenty-two natural compounds have been found to inhibit STAT3 in microglia and astrocytes, leading to the alleviation of AD, whereas three natural compounds have been found to activate STAT3 in neurons and alleviate AD [170].

Methanogens are mainly known for producing ATP and methane gas. They convert 60%–99% of substrate into methane [171]. The significance of methane can be understood from the finding that methane-rich saline plays a neuroprotective role in rats by increasing the production of superoxide dismutase and reducing the production of malondialdehyde, 3-nitrotyrosine, 8-hydroxydeoxyguanosine, TNF-α, and IL-1β [172]. Methane-rich saline has been found to downregulate caspase-3 and Bax protein levels but upregulate Bcl-2 protein level. Methane-rich saline blocks the activation of Janus kinase 1 (JAK1)/signal transducer and activator of the transcription 1 (STAT1)/NF-κB-p65 pathway. This will lead to the improvement of cognitive function by decreasing oxidative stress, inflammation, and apoptosis [173]. Methane inhalation has been found to increase large surfactant aggregates in lungs, improve oxygenation, lung compliance, and protect lungs against ischemia-reperfusion injury. Methane upregulates the expression of nuclear factor 2 (Nrf-2) in the nucleus but decreases Nrf-2 expression in the cytosol. Nrf-2 upregulation in the nucleus is responsible for the antioxidant effect

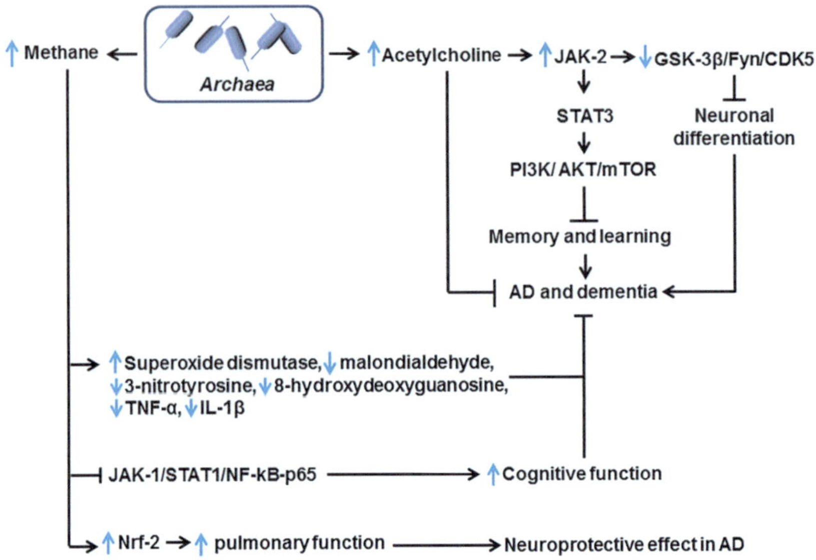

Fig. 7.10 Archaea secrete acetylcholine. Acetylcholine can delay the progression of AD and dementia. Conversely, acetylcholine induces JAK-2. JAK-2 negatively regulates GSK-3β/Fyn/CDK5 and suppresses neuronal differentiation, leading to AD and dementia. JAK-2 activates STAT3. This, along with PI3K/AKT/mTOR, will affect memory and learning, leading to AD and dementia. Archaea produce methane. Methane increases superoxide dismutase but decreases malondialdehyde, 3-nitrotyrosine, 8-hydroxydeoxyguanosine, TNF-α, and IL-1β levels. This will lead to the suppression of AD and dementia. Methane suppresses JAK1/STAT1/NF-κB-p65, leading to improved cognitive function. This will alleviate AD and dementia. Methane also increases Nrf-2. This will improve pulmonary function and will produce a neuroprotective effect in AD

and regulation of pulmonary surfactant [174]. Improved pulmonary function and oxygenation may have a neuroprotective effect in AD [175]. This shows the complex role of archaea in neuronal health (see Fig. 7.10).

7.2.4 Archaea and Other Pathologies

The pathological correlation of archaea with neurodegenerative disorders is corroborated by a study that correlates multiple sclerosis with an increase in gut archaea (Methanobrevibacter aceae) [176]. Increased copies of methanogens (*M. smithii*) have been found in the gut of irritable bowel syndrome (IBS) patients with constipation as compared to healthy control [177]. IBS is associated with an increased risk of AD and dementia. IBS associated risk of dementia is levied only on patients aged 50 years and older [178]. Conversely, *M. smithii* population is significantly higher among healthy people compared to the patients of inflammatory bowel disease (IBD) [179]. This perhaps shows the protective role of *M. smithii* against IBD. A decrease in the population of another methanogen, namely, *Ms. stadtmanae*, in IBD was also confirmed among pediatric patients. *Ms. stadtmanae* has shown the most pronounced positive correlation with fecal concentration of calprotectin (FCP) [180]. FCP is a very sensitive marker indicative of gastrointestinal inflammation. Presence of a higher level of FCP indicates IBD and rules out the chances of IBS [181]. This once again shows the protective role of methanogens in IBD. Contextually, IBD can "exacerbate the progression toward AD" [182]. The above findings show that methanogens may alleviate as well as aggravate AD, showing the dichotomous effect of archaea in human health.

7.2.5 Candida and Neurotransmitters

Candida is mainly known for its virulence. Many species of *Candida* identified include *C. albicans*, *C. tropicalis*, *C. glabrata*, *C. krusei*, and *C. parapsilosis*. *C. albicans* alone accounts for 65.3% of all infections, whereas others account for 2.4%–11.3% of infections [183]. This underscores the need to scan *C. albicans* for its overall implication in human health. There are very few researches showing the impact of *Candida* on neurotransmitter secretion. It was found that glycoproteins from the cell wall of *C. albicans* induce the release of serotonin from isolated platelets [184]. Serotonin released from platelets can degranulate neutrophil and release myeloperoxidase and hydrogen peroxide. There is also an increase in the expression of membrane-bound leukocyte adhesion molecule CD11b on neutrophil. Consequently, there is an increase in inflammation in the infarct area, worsening myocardial infarction (MI) [185]. MI disturbs brain homeostasis, leading to anxiety, depression, cognitive deficits, and stroke [186]. Conversely, a recent study shows that colonization of *C. albicans* in the gut of mice reduces the expression of D1 and D2 classes of dopamine receptors in the dorsal striatum [187]. D1 receptor polymorphism (DRD1A-48G) is associated with MI [188]. Similarly, prolonged activation of the D2 receptor plays a role in the development of MI [189]. Thus, the ability of *C. albicans* to suppress D1 and D2 receptor expressions may suppress MI. The above findings make the role of serotonin and *C. albicans* ambiguous in the context of MI (see Fig. 7.11). Furthermore, serotonin favors platelet aggregation [190]. Though a previous study claims that platelet aggregation plays a minor role in controlling gastric mucosal bleeding [191], yet a recent study shows a decrease in platelet function in acute gastrointestinal bleeding [192]. This raises concerns because gastrointestinal bleeding is associated with MI [193]. As per some studies gastrointestinal bleeding is a complication of MI [194], and MI may lead to cognitive deficits and stroke [186]. This shows that *C. albicans* may suppress gastrointestinal bleeding and may have implications for MI, cognitive deficits, and stroke.

7.2.6 Candida and Other Pathologies

C. albicans produces formyl-methionine and activates intestinal HIF-2α-ceramide signaling, leading to atherosclerosis [195]. Atherosclerosis is the most common cause of myocardial infarction [196] and stroke [197]. Atherosclerosis shows a

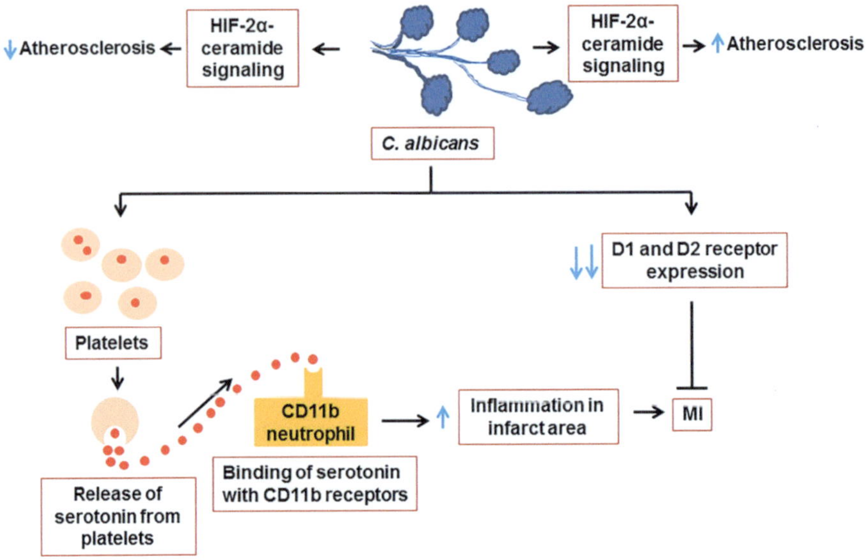

Fig. 7.11 *C. albicans* activates HIF-2α-ceramide signaling. HIF-2α-ceramide signaling has been found to increase or decrease atherosclerosis. *C. albicans* increases the release of serotonin from platelets. Serotonin binds with CD11b receptors on neutrophils and increases inflammation in infarct areas of the heart, aggravating MI. *C. albicans* decreases the expression of D1 and D2 receptors. This may suppress MI

significant association with AD [198]. Surprisingly, another finding shows that HIF-2α promotes adipose ceramide catabolism and suppresses atherosclerosis [199]. This shows the dichotomous effects of *C. albicans*-activated HIF-2α-ceramide signaling (Fig. 7.11). A study shows that HIF-2α-mediated ceramide catabolism plays a role in cognitive impairment [200]. This shows that HIF-2α-mediated ceramide catabolism may alleviate or worsen atherosclerosis but certainly worsens cognitive impairment [199, 200]. If HIF-2α-ceramide signaling alleviates atherosclerosis but worsens cognitive impairment, then this represents a contradiction to the established association between atherosclerosis and cognitive impairment. If HIF-2α-ceramide signaling worsens atherosclerosis as well as cognitive impairment, then this further confirms the association between atherosclerosis and cognitive impairment. This represents many elusive and ambiguous intricate signaling pathways, which need to be ascertained and explained by further research.

7.3 Conclusion

Gut microbiome plays a significant role in the maintenance of health. Despite their presence in the gut, they can communicate with the CNS by producing neurotransmitters. They can also induce the production and release of neurotransmitters in the CNS. They have been found to modulate the expression of many receptors, thereby deciding the course of action of many neurotransmitters. The complex pattern of action of the microbiome is observed in many diseases, including cancer. It becomes very difficult to predict the actual effect of microbiome-induced production/secretion of certain neurotransmitters and expression of their receptors on the course of cancer. The real concern arises from the cross-talk pathways, where modulation of neurotransmitters by microbiome may alleviate certain disease but may aggravate others, representing a double-edged sword. This can be seen with *L. fermentum*. It can improve cognitive outcome but may worsen anxiety. Pinpointing the actual implications of microbiome and microbiome-induced regulation of neurotransmitters becomes difficult in many other disease conditions, such as obesity, PD, AD, dementia, MI, etc. The difficulty in pinpointing the actual implications is further worsened by the dysbiosis of microbiome. These dichotomous effects of microbiome and neurotransmitters pave the path for further research for the betterment of therapeutic outcome.

Declarations

Competing Interests The author(s) has no competing interests to declare that are relevant to the content of this manuscript.

Ethics Approval The author(s) declares that this manuscript does not include primary research with human participants.

The author(s) declares that this manuscript does not include primary studies involving animals.

References

1. Kaisanlahti A, Turunen J, Byts N, Samoylenko A, Bart G, Virtanen N, et al. Maternal microbiota communicates with the fetus through microbiota-derived extracellular vesicles. Microbiome. 2023;11(1):249. https://doi.org/10.1186/s40168-023-01694-9.
2. Hou K, Wu ZX, Chen XY, Wang JQ, Zhang D, Xiao C, et al. Microbiota in health and diseases. Signal Transduct Target Ther. 2022;7(1):135. https://doi.org/10.1038/s41392-022-00974-4.
3. Sajdel-Sulkowska EM. The impact of maternal gut microbiota during pregnancy on fetal gut–brain axis development and life-long health outcomes. Microorganisms. 2023;11(9):2199. https://doi.org/10.3390/microorganisms11092199.
4. Dicks LM. Gut bacteria and neurotransmitters. Microorganisms. 2022;10(9):1838. https://doi.org/10.3390/microorganisms10091838.
5. Lu X, Shi Z, Jiang L, Zhang S. Maternal gut microbiota in the health of mothers and offspring: from the perspective of immunology. Front Immunol. 2024;15:1362784. https://doi.org/10.3389/fimmu.2024.1362784.
6. Cresci GA, Bawden E. Gut microbiome: what we do and don't know. Nutr Clin Pract. 2015;30(6):734–46. https://doi.org/10.1177/0884533615609899.
7. Peng X, Cheng L, You Y, Tang C, Ren B, Li Y, et al. Oral microbiota in human systematic diseases. Int J Oral Sci. 2022;14(1):14. https://doi.org/10.1038/s41368-022-00163-7.
8. Deo PN, Deshmukh R. Oral microbiome: unveiling the fundamentals. J Oral Maxillofac Pathol. 2019;23(1):122–8. https://doi.org/10.4103/jomfp.JOMFP_304_18.
9. Kunath BJ, De Rudder C, Laczny CC, Letellier E, Wilmes P. The oral–gut microbiome axis in health and disease. Nat Rev Microbiol. 2024;22(12):791–805. https://doi.org/10.1038/s41579-024-01075-5.

10. Tan X, Wang Y, Gong T. The interplay between oral microbiota, gut microbiota and systematic diseases. J Oral Microbiol. 2023;15(1):2213112. https://doi.org/10.1080/20002297.2023.2213112.
11. Afzaal M, Saeed F, Shah YA, Hussain M, Rabail R, Socol CT, et al. Human gut microbiota in health and disease: unveiling the relationship. Front Microbiol. 2022;13:999001. https://doi.org/10.3389/fmicb.2022.999001.
12. Zhao T, Liu T, Wang L, Xie K, Tang H, Tang M. Dysfunction of neurotransmitter metabolism is associated with the severity of depression in first-diagnosed, drug-naïve depressed patients. J Affect Disord. 2024;349:332–41. https://doi.org/10.1016/j.jad.2024.01.023.
13. Hashimoto K. Emerging role of the host microbiome in neuropsychiatric disorders: overview and future directions. Mol Psychiatry. 2023;28(9):3625–37. https://doi.org/10.1038/s41380-023-02287-6.
14. Ma YY, Li X, Yu JT, Wang YJ. Therapeutics for neurodegenerative diseases by targeting the gut microbiome: from bench to bedside. Transl Neurodegener. 2024;13(1):12. https://doi.org/10.1186/s40035-024-00404-1.
15. Al-Sayyar A, Hammad MM, Williams MR, Al-Onaizi M, Abubaker J, Alzaid F. Neurotransmitters in type 2 diabetes and the control of systemic and central energy balance. Meta. 2023;13(3):384. https://doi.org/10.3390/metabo13030384.
16. Yang Y, Sun L, Liu X, Liu W, Zhang Z, Zhou X, et al. Neurotransmitters: impressive regulators of tumor progression. Biomed Pharmacother. 2024;176:116844. https://doi.org/10.1016/j.biopha.2024.116844.
17. Elia A, Fossati S. Autonomic nervous system and cardiac neuro-signaling pathway modulation in cardiovascular disorders and Alzheimer's disease. Front Physiol. 2023;14:1060666. https://doi.org/10.3389/fphys.2023.1060666.
18. Tsuda K. Renin-Angiotensin system and sympathetic neurotransmitter release in the central nervous system of hypertension. Int J Hypertens. 2012;2012:474870. https://doi.org/10.1155/2012/474870.
19. Aggarwal S, Ahuja V, Paul J. Dysregulation of GABAergic signalling contributes in the pathogenesis of diarrhea-predominant irritable bowel syndrome. J Neurogastroenterol Motil. 2018;24(3):422. https://doi.org/10.5056/jnm17100.
20. Nimgampalle M, Chakravarthy H, Sharma S, Shree S, Bhat AR, Pradeepkiran JA, et al. Neurotransmitter systems in the etiology of major neurological disorders: emerging insights and therapeutic implications. Ageing Res Rev. 2023;89:101994. https://doi.org/10.1016/j.arr.2023.101994.
21. Turnbaugh PJ, Ley RE, Hamady M, Fraser-Liggett CM, Knight R, Gordon JI. The human microbiome project. Nature. 2007;449(7164):804–10. https://doi.org/10.1038/nature06244.
22. Goldstein EJ, Tyrrell KL, Citron DM. Lactobacillus species: taxonomic complexity and controversial susceptibilities. Clin Infect Dis. 2015;60(Suppl_2):S98–107. https://doi.org/10.1093/cid/civ072.
23. Walter J. Ecological role of lactobacilli in the gastrointestinal tract: implications for fundamental and biomedical research. Appl Environ Microbiol. 2008;74(16):4985–96. https://doi.org/10.1128/AEM.00753-08.
24. Pan M, Hidalgo-Cantabrana C, Goh YJ, Sanozky-Dawes R, Barrangou R. Comparative analysis of *Lactobacillus gasseri* and *Lactobacillus crispatus* isolated from human urogenital and gastrointestinal tracts. Front Microbiol. 2020;10:499288. https://doi.org/10.3389/fmicb.2019.03146.
25. Sun J, Qi C, Zhu H, Zhou Q, Xiao H, Le G, et al. IgA-targeted *Lactobacillus jensenii* modulated gut barrier and microbiota in high-fat diet-fed mice. Front Microbiol. 2019;10:1179. https://doi.org/10.3389/fmicb.2019.01179.
26. Dempsey E, Corr SC. Lactobacillus spp. for gastrointestinal health: current and future perspectives. Front Immunol. 2022;13:840245. https://doi.org/10.3389/fimmu.2022.840245.
27. Liu YW, Ong JS, Gan CY, Khoo BY, Yahaya S, Choi SB, et al. *Lactobacillus fermentum* PS150 showed psychotropic properties by altering serotonergic pathway during stress. J Funct Foods. 2019;59:352–61. https://doi.org/10.1016/j.jff.2019.05.043.

28. Rehman MU, Ghazanfar S, Ul Haq R, Ullah S, Khan S, Wu J, et al. Probiotics (*Bacillus clausii* and *Lactobacillus fermentum* NMCC-14) ameliorate stress behavior in mice by increasing monoamine levels and mRNA expression of dopamine receptors (D1 and D2) and synaptophysin. Front Pharmacol. 2022;13:915595. https://doi.org/10.3389/fphar.2022.915595.
29. Torres SJ, Nowson CA. Relationship between stress, eating behavior, and obesity. Nutrition. 2007;23(11–12):887–94. https://doi.org/10.1016/j.nut.2007.08.008.
30. Ulrich-Lai YM, Fulton S, Wilson M, Petrovich G, Rinaman L. Stress exposure, food intake and emotional state. Stress. 2015;18(4):381–99. https://doi.org/10.3109/10253890.2015.1062981.
31. van Galen KA, Ter Horst KW, Serlie MJ. Serotonin, food intake, and obesity. Obes Rev. 2021;22(7):e13210. https://doi.org/10.1111/obr.13210.
32. Conde K, Fang S, Xu Y. Unraveling the serotonin saga: from discovery to weight regulation and beyond—a comprehensive scientific review. Cell Biosci. 2023;13(1):143. https://doi.org/10.1186/s13578-023-01091-7.
33. Breisch ST, Zemlan FP, Hoebel BG. Hyperphagia and obesity following serotonin depletion by intraventricular p-chlorophenylalanine. Science. 1976;192(4237):382–5. https://doi.org/10.1126/science.130678.
34. Saller CF, Stricker EM. Hyperphagia and increased growth in rats after intraventricular injection of 5, 7-dihydroxytryptamine. Science. 1976;192(4237):385–7. https://doi.org/10.1126/science.1257774.
35. Wolin KY, Carson K, Colditz GA. Obesity and cancer. Oncologist. 2010;15(6):556–65. https://doi.org/10.1634/theoncologist.2009-0285.
36. Flores-Cordero JA, Pérez-Pérez A, Jiménez-Cortegana C, Alba G, Flores-Barragán A, Sánchez-Margalet V. Obesity as a risk factor for dementia and Alzheimer's disease: the role of leptin. Int J Mol Sci. 2022;23(9):5202. https://doi.org/10.3390/ijms23095202.
37. Powell-Wiley TM, Poirier P, Burke LE, Després JP, Gordon-Larsen P, Lavie CJ, et al. Obesity and cardiovascular disease: a scientific statement from the American Heart Association. Circulation. 2021;143(21):e984–1010. https://doi.org/10.1161/CIR.0000000000000973.
38. Zha W, Ho HT, Hu T, Hebert MF, Wang J. Serotonin transporter deficiency drives estrogen-dependent obesity and glucose intolerance. Sci Rep. 2017;7(1):1–14. https://doi.org/10.1038/s41598-017-01291-5.
39. Jeong SY, Kang S, Hua CS, Ting Z, Park S. Synbiotic effects of β-glucans from cauliflower mushroom and *Lactobacillus fermentum* on metabolic changes and gut microbiome in estrogen-deficient rats. Genes Nutr. 2017;12:1–12. https://doi.org/10.1186/s12263-017-0585-z.
40. Liu S, Xie Y, Li C, Zhang G, Wang X, Zhao H, et al. Anti-osteoporosis activity and mechanism analyses of *Limosilactobacillus fermentum* YSC5001 from Chinese traditional dairy products. J Funct Foods. 2025;130:106905.
41. Tobias JH, Compston JE. Does estrogen stimulate osteoblast function in postmenopausal women? Bone. 1999;24(2):121–4. https://doi.org/10.1016/S8756-3282(98)00156-2.
42. Rossi DV, Dai Y, Thomas P, Carrasco GA, Don Carlos LL, Muma NA, et al. Estradiol-induced desensitization of 5-HT1A receptor signaling in the paraventricular nucleus of the hypothalamus is independent of estrogen receptor-beta. Psychoneuroendocrinology. 2010;35(7):1023–33. https://doi.org/10.1016/j.psyneuen.2010.01.003.
43. Osterlund MK, Hurd YL. Acute 17 beta-estradiol treatment down-regulates serotonin 5HT1A receptor mRNA expression in the limbic system of female rats. Brain Res Mol Brain Res. 1998;55(1):169–72. https://doi.org/10.1016/s0169-328x(98)00018-7.
44. Voigt JP, Schade R, Fink H, Hörtnagl H. Role of 5-HT1A receptors in the control of food intake in obese Zucker rats of different ages. Pharmacol Biochem Behav. 2002;72(1–2):403–9. https://doi.org/10.1016/s0091-3057(01)00763-8.
45. Kumar R, Rizvi MR, Saraswat S. Obesity and stress: a contingent paralysis. Int J Prev Med. 2022;13:95. https://doi.org/10.4103/ijpvm.IJPVM_427_20.
46. Yoon Y, Kim G, Noh MG, Park JH, Jang M, Fang S, et al. *Lactobacillus fermentum* promotes adipose tissue oxidative phosphorylation to protect against diet-induced obesity. Exp Mol Med. 2020;52(9):1574–86. https://doi.org/10.1038/s12276-020-00502-w.

47. Million M, Angelakis E, Paul M, Armougom F, Leibovici L, Raoult D. Comparative meta-analysis of the effect of Lactobacillus species on weight gain in humans and animals. Microb Pathog. 2012;53(2):100–8. https://doi.org/10.1016/j.micpath.2012.05.007.
48. Aarsland D, Påhlhagen S, Ballard CG, Ehrt U, Svenningsson P. Depression in Parkinson disease—epidemiology, mechanisms and management. Nat Rev Neurol. 2012;8(1):35–47. https://doi.org/10.1038/nrneurol.2011.189.
49. Abusrair AH, Elsekaily W, Bohlega S. Tremor in Parkinson's disease: from pathophysiology to advanced therapies. Tremor Other Hyperkinet Mov. 2022;12:29. https://doi.org/10.5334/tohm.712.
50. Ibrahim H, Maksod SA, Khorshed M, Rady H, Alsisi A, Mohamed A, et al. Serotonin syndrome: a rare undiagnosed cause of hyperpyrexia. Egypt J Intern Med. 2023;35(1):40. https://doi.org/10.1186/s43162-023-00223-2.
51. Linnér L, Wiker C, Arborelius L, Schalling M, Svensson TH. Selective noradrenaline reuptake inhibition enhances serotonergic neuronal activity and transmitter release in the rat forebrain. J Neural Transm (Vienna). 2004;111(2):127–39. https://doi.org/10.1007/s00702-003-0084-9.
52. Andreou D, Söderman E, Axelsson T, Sedvall GC, Terenius L, Agartz I, et al. Polymorphisms in genes implicated in dopamine, serotonin and noradrenalin metabolism suggest association with cerebrospinal fluid monoamine metabolite concentrations in psychosis. Behav Brain Funct. 2014;10:26. https://doi.org/10.1186/1744-9081-10-26.
53. Naoi M, Wu Y, Maruyama W, Shamoto-Nagai M. Phytochemicals modulate biosynthesis and function of serotonin, dopamine, and norepinephrine for treatment of monoamine neurotransmission-related psychiatric diseases. Int J Mol Sci. 2025;26(7):2916. https://doi.org/10.3390/ijms26072916.
54. Tsao D, Wieskopf JS, Rashid N, Sorge RE, Redler RL, Segall SK, et al. Serotonin-induced hypersensitivity via inhibition of catechol O-methyltransferase activity. Mol Pain. 2012;8:25. https://doi.org/10.1186/1744-8069-8-25.
55. Yang J, Hellerstein DJ, Chen Y, McGrath PJ, Stewart JW, Peterson BS, et al. Serotonin–norepinephrine reuptake inhibitor antidepressant effects on regional connectivity of the thalamus in persistent depressive disorder: evidence from two randomized, double-blind, placebo-controlled clinical trials. Brain Commun. 2022;4(3):fcac 100. https://doi.org/10.1093/braincomms/fcac100.
56. Kinnerup MB, Sommerauer M, Damholdt MF, Schaldemose JL, Ismail R, Terkelsen AJ, et al. Preserved noradrenergic function in Parkinson's disease patients with rest tremor. Neurobiol Dis. 2021;152:105295. https://doi.org/10.1016/j.nbd.2021.105295.
57. Schmidt A, Meindl T, Albu-Schäffer A, Franklin DW, Stratmann P. Influence of serotonin on the long-term muscle contraction of the Kohnstamm phenomenon. Sci Rep. 2025;15(1):1–11. https://doi.org/10.1038/s41598-025-00444-1.
58. Perrier JF. Modulation of motoneuron activity by serotonin. Dan Med J. 2016;63(2):B5204.
59. Monte A, Magris R, Nardello F, Bombieri F, Zamparo P. Muscle shape changes in Parkinson's disease impair function during rapid contractions. Acta Physiol (Oxf). 2023;238(1):e13957. https://doi.org/10.1111/apha.13957.
60. Pauletti C, Mannarelli D, Locuratolo N, Currà A, Marinelli L, Fattapposta F. Central fatigue and attentional processing in Parkinson's disease: an event-related potentials study. Clin Neurophysiol. 2019;130(5):692–700. https://doi.org/10.1016/j.clinph.2019.01.017.
61. Reimold M, Batra A, Knobel A, Smolka MN, Zimmer A, Mann K, et al. Anxiety is associated with reduced central serotonin transporter availability in unmedicated patients with unipolar major depression: a [11C]DASB PET study. Mol Psychiatry. 2008;13(6):606–57. https://doi.org/10.1038/sj.mp.4002149.
62. Khatri DK, Choudhary M, Sood A, Singh SB. Anxiety: an ignored aspect of Parkinson's disease lacking attention. Biomed Pharmacother. 2020;131:110776. https://doi.org/10.1016/j.biopha.2020.110776.
63. Pontone GM, Williams JR, Anderson KE, Chase G, Goldstein SA, Grill S, et al. Prevalence of anxiety disorders and anxiety subtypes in patients with Parkinson's disease. Mov Disord. 2009;24(9):1333–8. https://doi.org/10.1002/mds.22611.

64. Khedr EM, Abdelrahman AA, Elserogy Y, Zaki AF, Gamea A. Depression and anxiety among patients with Parkinson's disease: frequency, risk factors, and impact on quality of life. Egypt J Neurol Psychiatr Neurosurg. 2020;56:1–9. https://doi.org/10.1186/s41983-020-00253-5.
65. Meeusen R, Watson P, Hasegawa H, Roelands B, Piacentini MF. Central fatigue: the serotonin hypothesis and beyond. Sports Med. 2006;36(10):881–909. https://doi.org/10.2165/00007256-200636100-00006.
66. Bai X, Strong R. Expression of synaptophysin protein in different dopaminergic cell lines. J Biochem Pharmacol Res. 2014;2(4):185–90.
67. Espay AJ, LeWitt PA, Kaufmann H. Norepinephrine deficiency in Parkinson's disease: the case for noradrenergic enhancement. Mov Disord. 2014;29(14):1710–9. https://doi.org/10.1002/mds.26048.
68. Borodovitsyna O, Flamini M, Chandler D. Noradrenergic modulation of cognition in health and disease. Neural Plast. 2017;2017:6031478. https://doi.org/10.1155/2017/6031478.
69. Laureiro-Martínez D, Brusoni S. Cognitive flexibility and adaptive decision-making: evidence from a laboratory study of expert decision makers. Strateg Manag J. 2018;39(4):1031–58. https://doi.org/10.1002/smj.2774.
70. Rayavarapu B, Tallapragada P, Usha MS. Statistical optimization of γ-aminobutyric acid production by response surface methodology and artificial neural network models using *Lactobacillus fermentum* isolated from palm wine. Biocatal Agric Biotechnol. 2019;22:101362. https://doi.org/10.1016/j.bcab.2019.101362.
71. Kaur S, Sharma P, Mayer MJ, Neuert S, Narbad A, Kaur S. Beneficial effects of GABA-producing potential probiotic *Limosilactobacillus fermentum* L18 of human origin on intestinal permeability and human gut microbiota. Microb Cell Factories. 2023;22(1):256. https://doi.org/10.1186/s12934-023-02264-2.
72. Cui Y, Miao K, Niyaphorn S, Qu X. Production of gamma-aminobutyric acid from lactic acid bacteria: a systematic review. Int J Mol Sci. 2020;21(3):995. https://doi.org/10.3390/ijms21030995.
73. Kamondi A, Grigg-Damberger M, Löscher W, Tanila H, Horvath AA. Epilepsy and epileptiform activity in late-onset Alzheimer disease: clinical and pathophysiological advances, gaps and conundrums. Nat Rev Neurol. 2024;20(3):162–82. https://doi.org/10.1038/s41582-024-00932-4.
74. Alper GP, Bergman I. Pediatric neurologic assessment and monitoring. In: Pediatric critical care. Mosby; 2006. p. 804–16. https://www.sciencedirect.com/science/article/pii/B9780323018081500535.
75. Gruntz K, Bloechliger M, Becker C, Jick SS, Fuhr P, Meier CR, et al. Parkinson disease and the risk of epileptic seizures. Ann Neurol. 2018;83(2):363–74. https://doi.org/10.1002/ana.25157.
76. Hwang YS, Kang MG, Yeom SW, Jeong CY, Shin BS, Koh J, et al. Increasing incidence of Parkinson's disease in patients with epilepsy: a Nationwide cohort study. J Neurol Sci. 2024;458:122891. https://doi.org/10.1016/j.jns.2024.122891.
77. Teran FA, Bravo E, Richerson GB. Sudden unexpected death in epilepsy: respiratory mechanisms. Handb Clin Neurol. 2022;189:153–76.
78. Yamada KA, Norman WP, Hamosh P, Gillis RA. Medullary ventral surface GABA receptors affect respiratory and cardiovascular function. Brain Res. 1982;248(1):71–8. https://doi.org/10.1016/0006-8993(82)91148-9.
79. Taveira da Silva AM, Hartley B, Hamosh P, Quest JA, Gillis RA. Respiratory depressant effects of GABA alpha- and beta-receptor agonists in the cat. J Appl Physiol. 1987;62(6):2264–72. https://doi.org/10.1152/jappl.1987.62.6.2264.
80. Hepsomali P, Groeger JA, Nishihira J, Scholey A. Effects of oral gamma-aminobutyric acid (GABA) administration on stress and sleep in humans: a systematic review. Front Neurosci. 2020;14:559962. https://doi.org/10.3389/fnins.2020.00923.
81. Boonstra E, de Kleijn R, Colzato LS, Alkemade A, Forstmann BU, Nieuwenhuis S. Neurotransmitters as food supplements: the effects of GABA on brain and behavior. Front Psychol. 2015;6:1520. https://doi.org/10.3389/fpsyg.2015.01520.

82. Khan W, Naz S, Rana AQ. Shortness of breath, a ‘wearing-off’ symptom in Parkinson’s disease. Clin Drug Investig. 2009;29(10):689–91. https://doi.org/10.2165/11315290-000000000-00000.
83. Zhou Z, Zhang M, Fang Q, Huang J. Relationship between Parkinson’s disease and cardio-cerebrovascular diseases: a Mendelian randomized study. Sci Rep. 2023;13(1):20428. https://doi.org/10.1038/s41598-023-47708-2.
84. Hosseini Dastgerdi A, Sharifi M, Soltani N. GABA administration improves liver function and insulin resistance in offspring of type 2 diabetic rats. Sci Rep. 2021;11(1):23155. https://doi.org/10.1038/s41598-021-02324-w.
85. Perez-Cano FJ, Dong H, Yaqoob P. In vitro immunomodulatory activity of *Lactobacillus fermentum* CECT5716 and *Lactobacillus salivarius* CECT 5713: two probiotic strains isolated from human breast milk. Immunobiology. 2010;215(12):996–1004. https://doi.org/10.1016/j.imbio.2010.01.004.
86. Rodríguez-Sojo MJ, Ruiz-Malagón AJ, Rodríguez-Cabezas ME, Gálvez J, Rodríguez-Nogales A. *Limosilactobacillus fermentum* CECT 5716: mechanisms and therapeutic insights. Nutrients. 2021;13(3):1016. https://doi.org/10.3390/nu13031016.
87. Hellstrand K, Czerkinsky C, Ricksten A, Jansson B, Asea A, Kylefjord H, et al. Role of serotonin in the regulation of interferon-γ production by human natural killer cells. J Interf Res. 1993;13(1):33–8. https://doi.org/10.1089/jir.1993.13.33.
88. Kubera M, Kenis G, Bosmans E, Scharpé S, Maes M. Effects of serotonin and serotonergic agonists and antagonists on the production of interferon-γ and interleukin-10. Neuropsychopharmacology. 2000;23(1):89–98. https://doi.org/10.1016/S0893-133X(99)00150-5.
89. Yuan X, He F, Zheng F, Xu Y, Zou J. Interferon-gamma facilitates neurogenesis by activating Wnt/β-catenin cell signaling pathway via promotion of STAT1 regulation of the β-catenin promoter. Neuroscience. 2020;448:219–33. https://doi.org/10.1016/j.neuroscience.2020.08.018.
90. He Z, Yang Y, Xing Z, Zuo Z, Wang R, Gu H, et al. Intraperitoneal injection of IFN-γ restores microglial autophagy, promotes amyloid-β clearance and improves cognition in APP/PS1 mice. Cell Death Dis. 2020;11(6):440. https://doi.org/10.1038/s41419-020-2644-4.
91. Panagiotakopoulou V, Ivanyuk D, De Cicco S, Haq W, Arsić A, Yu C, et al. Interferon-γ signaling synergizes with LRRK2 in neurons and microglia derived from human induced pluripotent stem cells. Nat Commun. 2020;11(1):5163. https://doi.org/10.1038/s41467-020-18755-4.
92. Kunugi H, Ueki A, Otsuka M, Isse K, Hirasawa H, Kato N, et al. Alzheimer’s disease and 5-HTTLPR polymorphism of the serotonin transporter gene: no evidence for an association. Am J Med Genet. 2000;96(3):307–9. https://doi.org/10.1002/1096-8628(20000612)96:3<307::AID-AJMG14>3.0.CO;2-0.
93. Malhotra AK, Goldman D, Mazzanti C, Clifton A, Breier A, Pickar D. A functional serotonin transporter (5-HTT) polymorphism is associated with psychosis in neuroleptic-free schizophrenics. Mol Psychiatry. 1998;3(4):328–32. https://doi.org/10.1038/sj.mp.4000412.
94. Murray PS, Kumar S, DeMichele-Sweet MAA, Sweet RA. Psychosis in Alzheimer’s disease. Biol Psychiatry. 2014;75(7):542–52. https://doi.org/10.1016/j.biopsych.2013.08.020.
95. Mora F, Segovia G, Del Arco A, de Blas M, Garrido P. Stress, neurotransmitters, corticosterone and body–brain integration. Brain Res. 2012;1476:71–85. https://doi.org/10.1016/j.brainres.2011.12.049.
96. Xu J, Wang B, Ao H. Corticosterone effects induced by stress and immunity and inflammation: mechanisms of communication. Front Endocrinol. 2025;16:1448750. https://doi.org/10.3389/fendo.2025.1448750.
97. Rudolph LM, Cornil CA, Mittelman-Smith MA, Rainville JR, Remage-Healey L, Sinchak K, et al. Actions of steroids: new neurotransmitters. J Neurosci. 2016;36(45):11449–58. https://doi.org/10.1523/JNEUROSCI.2473-16.2016.
98. Shorey CL, Mulla RT, Mielke JG. The effects of synthetic glucocorticoid treatment for inflammatory disease on brain structure, function, and dementia outcomes: a systematic review. Brain Res. 2023;1798:148157. https://doi.org/10.1016/j.brainres.2022.148157.

99. Conic JZ, Chetty A, Chen L, Marsh A, Barry S, Pattabhi R, et al. Effect of steroids on the progression of Alzheimer's dementia: a retrospective chart review. Aging Med. 2025;8(1):e70004. https://doi.org/10.1002/agm2.70004.
100. Berry JN, Saunders MA, Sharrett-Field LJ, Reynolds AR, Bardo MT, Pauly JR, et al. Corticosterone enhances N-methyl-D-aspartate receptor signaling to promote isolated ventral tegmental area activity in a reconstituted mesolimbic dopamine pathway. Brain Res Bull. 2016;120:159–65. https://doi.org/10.1016/j.brainresbull.2015.11.018.
101. Companys-Alemany J, Turcu AL, Schneider M, Müller CE, Vázquez S, Griñán-Ferré C, et al. NMDA receptor antagonists reduce amyloid-β deposition by modulating calpain-1 signaling and autophagy, rescuing cognitive impairment in 5XFAD mice. Cell Mol Life Sci. 2022;79(8):408. https://doi.org/10.1007/s00018-022-04438-4.
102. Hoshi H, Hirata Y, Kobayashi M, Sakamoto Y, Fukasawa K, Ichikawa S, et al. Distinctive effects of executive dysfunction and loss of learning/memory abilities on resting-state brain activity. Sci Rep. 2022;12(1):3459. https://doi.org/10.1038/s41598-022-07202-7.
103. Loh JS, Mak WQ, Tan LKS, Ng CX, Chan HH, Yeow SH, et al. Microbiota–gut–brain axis and its therapeutic applications in neurodegenerative diseases. Signal Transduct Target Ther. 2024;9(1):37. https://doi.org/10.1038/s41392-024-01743-1.
104. Singh H, Chopra C, Singh H, Malgotra V, Wani AK, Dhanjal DS, et al. Gut-brain axis and Alzheimer's disease: therapeutic interventions and strategies. J Funct Foods. 2024;112:105915. https://doi.org/10.1016/j.jff.2023.105915.
105. Cirrito JR, Disabato BM, Restivo JL, Verges DK, Goebel WD, Sathyan A, et al. Serotonin signaling is associated with lower amyloid-β levels and plaques in transgenic mice and humans. Proc Natl Acad Sci USA. 2011;108(36):14968–73. https://doi.org/10.1073/pnas.1107411108.
106. Labus J, Röhrs KF, Ackmann J, Varbanov H, Müller FE, Jia S, et al. Amelioration of Tau pathology and memory deficits by targeting 5-HT7 receptor. Prog Neurobiol. 2021;197:101900. https://doi.org/10.1016/j.pneurobio.2020.101900.
107. Ackmann J, Brüge A, Gotina L, Lim S, Jahreis K, Vollbrecht AL, et al. Structural determinants for activation of the Tau kinase CDK5 by the serotonin receptor 5-HT7R. Cell Commun Signal. 2024;22(1):233. https://doi.org/10.1186/s12964-024-01612-y.
108. Nikiforuk A. Targeting the serotonin 5-HT7 receptor in the search for treatments for CNS disorders: rationale and Progress to date. CNS Drugs. 2015;29(4):265–75. https://doi.org/10.1007/s40263-015-0236-0.
109. Bijata M, Bączyńska E, Müller FE, Bijata K, Masternak J, Krzystyniak A, et al. Activation of the 5-HT7 receptor and MMP-9 signaling module in the hippocampal CA1 region is necessary for the development of depressive-like behavior. Cell Rep. 2022;38(11):110532. https://doi.org/10.1016/j.celrep.2022.110532.
110. Dillon DG, Pizzagalli DA. Mechanisms of memory disruption in depression. Trends Neurosci. 2018;41(3):137–49. https://doi.org/10.1016/j.tins.2017.12.006.
111. Hasan M, Seo JE, Rahaman KA, Kang MJ, Jung BH, Kwon OS. Increased levels of brain serotonin correlated with MMP-9 activity and IL-4 levels resulted in severe experimental autoimmune encephalomyelitis (EAE) in obese mice. Neuroscience. 2016;319:168–82. https://doi.org/10.1016/j.neuroscience.2016.01.045.
112. Carbone C, Lo Russo SLM, Lacivita E, Frank A, Alleva E, Stark H, et al. Prior activation of 5-HT7 receptors modulates the conditioned place preference with methylphenidate. Front Behav Neurosci. 2019;13:208. https://doi.org/10.3389/fnbeh.2019.00208.
113. Gopalakrishna R, Oh A, Bhat NR, Mack WJ. Cyclic adenosine monophosphate-elevating agents inhibit amyloid-beta internalization and neurotoxicity: their action in Alzheimer's disease prevention. Neural Regen Res. 2023;18(12):2675–6. https://doi.org/10.4103/1673-5374.373664.
114. Kelly MP. Cyclic nucleotide signaling changes associated with normal aging and age-related diseases of the brain. Cell Signal. 2018;42:281–91. https://doi.org/10.1016/j.cellsig.2017.11.004.

115. De Filippis B, Nativio P, Fabbri A, Ricceri L, Adriani W, Lacivita E, et al. Pharmacological stimulation of the brain serotonin receptor 7 as a novel therapeutic approach for Rett syndrome. Neuropsychopharmacology. 2014;39(11):2506–18. https://doi.org/10.1038/npp.2014.105.
116. Panchal K, Tiwari AK. Miro (Mitochondrial Rho GTPase), a key player of mitochondrial axonal transport and mitochondrial dynamics in neurodegenerative diseases. Mitochondrion. 2021;56:118–35. https://doi.org/10.1016/j.mito.2020.10.005.
117. Carreño Gutiérrez H, Colanesi S, Cooper B, Reichmann F, Young AM, Kelsh RN, et al. Endothelin neurotransmitter signalling controls zebrafish social behaviour. Sci Rep. 2019;9(1):3040. https://doi.org/10.1038/s41598-019-39907-7.
118. Zhao X, Qian Y, Suo H, Du M, Li G, Liu Z, et al. Preventive effect of *Lactobacillus fermentum* Zhao on activated carbon-induced constipation in mice. J Nutr Sci Vitaminol. 2015;61(2):131–7. https://doi.org/10.3177/jnsv.61.131.
119. Dashwood MR, Loesch A. Endothelin-1 as a neuropeptide: neurotransmitter or neurovascular effects? J Cell Commun Signal. 2010;4:51–62. https://doi.org/10.1007/s12079-009-0073-3.
120. Kowalczyk A, Kleniewska P, Kolodziejczyk M, Skibska B, Goraca A. The role of endothelin-1 and endothelin receptor antagonists in inflammatory response and sepsis. Arch Immunol Ther Exp. 2015;63(1):41–52. https://doi.org/10.1007/s00005-014-0310-1.
121. Kostov K, Blazhev A, Atanasova M, Dimitrova A. Serum concentrations of endothelin-1 and matrix metalloproteinases-2, −9 in pre-hypertensive and hypertensive patients with type 2 diabetes. Int J Mol Sci. 2016;17(8):1182. https://doi.org/10.3390/ijms17081182.
122. Fuglsang A, Nilsson D, Nyborg NC. Cardiovascular effects of fermented milk containing angiotensin-converting enzyme inhibitors evaluated in permanently catheterized, spontaneously hypertensive rats. Appl Environ Microbiol. 2002;68(7):3566–9. https://doi.org/10.1128/AEM.68.7.3566-3569.2002.
123. Fujimura Y, Shimura M, Nagai H, Hamada-Sato N. Evaluation of angiotensin-converting enzyme-inhibitory activity in abalone viscera fermented by *Lactobacillus casei* 001. J Funct Foods. 2021;82:104474. https://doi.org/10.1016/j.jff.2021.104474.
124. Lin YJ, Kwok CF, Juan CC, Hsu YP, Shih KC, Chen CC, et al. Angiotensin II enhances endothelin-1-induced vasoconstriction through upregulating endothelin type a receptor. Biochem Biophys Res Commun. 2014;451(2):263–9.
125. Drawnel FM, Archer CR, Roderick HL. The role of the paracrine/autocrine mediator endothelin-1 in regulation of cardiac contractility and growth. Br J Pharmacol. 2013;168(2):296–317. https://doi.org/10.1111/j.1476-5381.2012.02195.x.
126. Takeya K, Wang X, Kathol I, Loutzenhiser K, Loutzenhiser R, Walsh MP. Endothelin-1, but not angiotensin II, induces afferent arteriolar myosin diphosphorylation as a potential contributor to prolonged vasoconstriction. Kidney Int. 2015;87(2):370–81. https://doi.org/10.1038/ki.2014.284.
127. Schiffrin EL. Role of endothelin-1 in hypertension. Hypertension. 1999;34(4):876–81. https://doi.org/10.1161/01.HYP.34.4.876.
128. Harrison M, Zinovkin D, Pranjol MZI. Endothelin-1 and its role in cancer and potential therapeutic opportunities. Biomedicine. 2024;12(3):511. https://doi.org/10.3390/biomedicines12030511.
129. Schmidt BL, Pickering V, Liu S, Quang P, Dolan J, Connelly ST, et al. Peripheral endothelin A receptor antagonism attenuates carcinoma-induced pain. Eur J Pain. 2007;11(4):406–14. https://doi.org/10.1016/j.ejpain.2006.05.007.
130. Hasdai D, Holmes DR, Garratt KN, Edwards WD, Lerman A. Mechanical pressure and stretch release endothelin-1 from human atherosclerotic coronary arteries in vivo. Circulation. 1997;95(2):357–62. https://doi.org/10.1161/01.CIR.95.2.357.
131. Henning RJ, Sawmiller DR. Vasoactive intestinal peptide: cardiovascular effects. Cardiovasc Res. 2001;49(1):27–37. https://doi.org/10.1016/S0008-6363(00)00229-7.
132. Blank MA, Fuortes M, Nyrén O, Jaffe BM. Effect of endothelin-1 and vasoactive intestinal contractor on blood flow and output of vasoactive intestinal polypeptide in the feline colon. Life Sci. 1991;48(20):1937–44. https://doi.org/10.1016/0024-3205(91)90226-2.

133. Ng YF, Ng E, Lim EW, Prakash KM, Tan LC, Tan EK. Case-control study of hypertension and Parkinson's disease. npj Parkinsons Dis. 2021;7(1):63. https://doi.org/10.1038/s41531-021-00202-w.
134. Shiraishi M, Kobayashi T, Watanabe H, Kamo T, Hasegawa Y. Serum somatostatin in early-stage Parkinson's disease. Acta Neurol Scand. 2010;121(4):225–9. https://doi.org/10.1111/j.1600-0404.2009.01178.x.
135. Moulopoulos SD, Stamatelopoulos SF, Vita L. Somatostatin prolongation of the blood pressure lowering effect of diazoxide. Metabolism. 1978;27(9):1365–7. https://doi.org/10.1016/0026-0495(78)90076-8.
136. Wang L, Zhao Z, Zhao L, Zhao Y, Yang G, Wang C, et al. *Lactobacillus plantarum* DP189 reduces α-SYN aggravation in MPTP-induced Parkinson's disease mice via regulating oxidative damage, inflammation, and gut microbiota disorder. J Agric Food Chem. 2022;70(4):1163–73. https://doi.org/10.1021/acs.jafc.1c07711.
137. Sherid M, Samo S, Sulaiman S, Husein H, Sifuentes H, Sridhar S. Liver abscess and bacteremia caused by lactobacillus: role of probiotics? Case report and review of the literature. BMC Gastroenterol. 2016;16:138. https://doi.org/10.1186/s12876-016-0552-y.
138. Chen YC, Lin CH, Chang SN, Shi ZY. Epidemiology and clinical outcome of pyogenic liver abscess: an analysis from the National Health Insurance Research Database of Taiwan, 2000–2011. J Microbiol Immunol Infect. 2016;49(5):646–53.
139. Haziri D, Prechter F, Stallmach A. Yoghurt-induced *Lactobacillus bacteremia* in a patient with Crohn's disease on therapy with ustekinumab and concomitant HIV-infection. Z Gastroenterol. 2020;59(4):317–20. https://doi.org/10.1055/a-1168-7577.
140. Salminen MK, Tynkkynen S, Rautelin H, Saxelin M, Vaara M, Ruutu P, et al. *Lactobacillus bacteremia* during a rapid increase in probiotic use of *Lactobacillus rhamnosus* GG in Finland. Clin Infect Dis. 2002;35(10):1155–60. https://doi.org/10.1086/342912.
141. Ioannou P, Ziogou A, Giannakodimos I, Giannakodimos A, Baliou S, Samonis G. Infective endocarditis by Lactobacillus species—a narrative review. Antibiotics. 2024;13(1):53. https://doi.org/10.3390/antibiotics13010053.
142. Antoun M, Hattab Y, Akhrass FA, Hamilton LD. Uncommon pathogen, lactobacillus, causing infective endocarditis: case report and review. Case Rep Infect Dis. 2020;2020:8833948.
143. Ozer M, Goksu SY, Shahverdiani A, Mustafa M. *Lactobacillus acidophilus*-induced endocarditis and associated splenic abscess. Case Rep Infect Dis. 2020;2020:1382709. https://doi.org/10.1155/2020/1382709.
144. Novy E, Sonneville R, Mazighi M, Klein IF, Mariotte E, Mourvillier B, et al. Neurological complications of infective endocarditis: new breakthroughs in diagnosis and management. Med Mal Infect. 2013;43(11–12):443–50. https://doi.org/10.1016/j.medmal.2013.09.010.
145. Li ZP, Liu JX, Lu LL, Wang LL, Xu L, Guo ZH, et al. Overgrowth of Lactobacillus in gastric cancer. World J Gastrointest Oncol. 2021;13(9):1099–108. https://doi.org/10.4251/wjgo.v13.i9.1099.
146. Dehghani N, Tafvizi F, Jafari P. Cell cycle arrest and anti-cancer potential of probiotic *Lactobacillus rhamnosus* against HT-29 cancer cells. Bioimpacts. 2021;11(4):245–52. https://doi.org/10.34172/bi.2021.32
147. Salemi R, Vivarelli S, Ricci D, Scillato M, Santagati M, Gattuso G, et al. *Lactobacillus rhamnosus* GG cell-free supernatant as a novel anti-cancer adjuvant. J Transl Med. 2023;21(1):195. https://doi.org/10.1186/s12967-023-04036-3.
148. Guo Y, Zhang T, Gao J, Jiang X, Tao M, Zeng X, et al. *Lactobacillus acidophilus* CICC 6074 inhibits growth and induces apoptosis in colorectal cancer cells in vitro and in HT-29 cells induced-mouse model. J Funct Foods. 2020;75:104290. https://doi.org/10.1016/j.jff.2020.104290.
149. Hezaveh K, Shinde RS, Klötgen A, Halaby MJ, Lamorte S, Quevedo R, et al. Tryptophan-derived microbial metabolites activate the aryl hydrocarbon receptor in tumor-associated macrophages to suppress anti-tumor immunity. Immunity. 2022;55(2):324–40.

150. Han J, Jiang Q, Ma R, Zhang H, Tong D, Tang K, et al. Norepinephrine-CREB1-miR-373 axis promotes progression of colon cancer. Mol Oncol. 2020;14(5):1059–73. https://doi.org/10.1002/1878-0261.12657.
151. Barbieri A, Bimonte S, Palma G, Luciano A, Rea D, Giudice A, et al. The stress hormone norepinephrine increases migration of prostate cancer cells in vitro and in vivo. Int J Oncol. 2015;47(2):527–34. https://doi.org/10.3892/ijo.2015.3038.
152. Ma YX, Guo Z, Sun T. CGRP inhibits norepinephrine induced apoptosis with restoration of Bcl-2/Bax in cultured cardiomyocytes of rat. Neurosci Lett. 2013;549:130–4. https://doi.org/10.1016/j.neulet.2013.05.028.
153. Lan YL, Wang X, Xing JS, Yu ZL, Lou JC, Ma XC, et al. Anti-cancer effects of dopamine in human glioma: involvement of mitochondrial apoptotic and anti-inflammatory pathways. Oncotarget. 2017;8(51):88488–500. https://doi.org/10.18632/oncotarget.19691
154. Weissenrieder JS, Neighbors JD, Mailman RB, Hohl RJ. Cancer and the dopamine D2 receptor: a pharmacological perspective. J Pharmacol Exp Ther. 2019;370(1):111–26. https://doi.org/10.1124/jpet.119.256818.
155. Jeon HM, Oh YT, Shin YJ, Chang N, Kim D, Woo D, et al. Dopamine receptor D2 regulates glioblastoma survival and death through MET and death receptor 4/5. Neoplasia. 2023;39:100894. https://doi.org/10.1016/j.neo.2023.100894.
156. Liu W, Huang J, Yan Z, Lin Y, Huang G, Chen X, et al. Association of N-nitrosodimethylamine exposure with cognitive impairment based on the clues of mice and humans. Front Aging Neurosci. 2023;15:1137164. https://doi.org/10.3389/fnagi.2023.1137164.
157. DasSarma S, Coker JA, DasSarma P. Archaea-overview. In: Desk encyclopedia of microbiology. 2nd ed. Elsevier; 2010. p. 118–39. https://doi.org/10.1016/B978-012373944-5.00108-5.
158. Eckburg PB, Lepp PW, Relman DA. Archaea and their potential role in human disease. Infect Immun. 2003;71(2):591–6. https://doi.org/10.1128/IAI.71.2.591-596.2003.
159. Yamada T, Fujii T, Kanai T, Amo T, Imanaka T, Nishimasu H, et al. Expression of acetylcholine (ACh) and ACh-synthesizing activity in Archaea. Life Sci. 2005;77(16):1935–44. https://doi.org/10.1016/j.lfs.2005.01.026.
160. Chen ZR, Huang JB, Yang SL, Hong FF. Role of cholinergic Signaling in Alzheimer's disease. Molecules. 2022;27(6):1816. https://doi.org/10.3390/molecules27061816.
161. Rakonczay Z, Horváth Z, Juhász A, Kálmán J. Peripheral cholinergic disturbances in Alzheimer's disease. Chem Biol Interact. 2005;157:233–8. https://doi.org/10.1016/j.cbi.2005.10.034.
162. Waiskopf N, Shenhar-Tsarfaty S, Soreq H. Serum cholinesterase activities as biomarkers of cardiac malfunctioning. In: Biomarkers in cardiovascular disease. Dordrecht: Springer; 2016. p. 197–218. https://doi.org/10.1007/978-94-007-7741-5_10-1.
163. De Luigi A, Pizzimenti S, Quadri P, Lucca U, Tettamanti M, Fragiacomo C, et al. Peripheral inflammatory response in Alzheimer's disease and multiinfarct dementia. Neurobiol Dis. 2002;11(2):308–14. https://doi.org/10.1006/nbdi.2002.0556.
164. Biesmans S, Bouwknecht JA, Ver Donck L, Langlois X, Acton PD, De Haes P, et al. Peripheral administration of tumor necrosis factor-alpha induces neuroinflammation and sickness but not depressive-like behavior in mice. Biomed Res Int. 2015;2015:716920.
165. Chang R, Yee KL, Sumbria RK. Tumor necrosis factor α inhibition for Alzheimer's disease. J Cent Nerv Syst Dis. 2017;9:1179573517709278. https://doi.org/10.1177/1179573517709278.
166. Zecchin HG, Priviero FB, Souza CT, Zecchin KG, Prada PO, Carvalheira JB, et al. Defective insulin and acetylcholine induction of endothelial cell–nitric oxide synthase through insulin receptor substrate/Akt signaling pathway in aorta of obese rats. Diabetes. 2007;56(4):1014–24. https://doi.org/10.2337/db05-1147.
167. Oh M, Kim SY, Byun JS, Lee S, Kim WK, Oh KJ, et al. Depletion of Janus kinase-2 promotes neuronal differentiation of mouse embryonic stem cells. BMB Rep. 2021;54(12):626. https://doi.org/10.5483/BMBRep.2021.54.12.154.
168. Terreros-Roncal J, Moreno-Jiménez EP, Flor-García M, Llorens-Martin M. The differentiation of new neurons is impaired in Alzheimer's disease patients: developing topics. Alzheimers Dement. 2020;16:e047383. https://doi.org/10.1002/alz.047383.

169. Hindam MO, Ahmed LA, Sayed NSE, Khattab M, Sallam NA. Repositioning of baricitinib for management of memory impairment in ovariectomized/D-galactose treated rats: a potential role of JAK2/STAT3-PI3K/AKT/mTOR signaling pathway. Life Sci. 2024;351:122838. https://doi.org/10.1016/j.lfs.2024.122838.
170. Wen X, Hu J. Targeting STAT3 signaling pathway in the treatment of Alzheimer's disease with compounds from natural products. Int Immunopharmacol. 2024;141:112936. https://doi.org/10.1016/j.intimp.2024.112936.
171. Buan NR. Methanogens: pushing the boundaries of biology. Emerg Top Life Sci. 2018;2(4):629–46. https://doi.org/10.1042/ETLS20180031.
172. Shen M, Fan D, Zang Y, Chen Y, Zhu K, Cai Z, et al. Neuroprotective effects of methane-rich saline on experimental acute carbon monoxide toxicity. J Neurol Sci. 2016;369:361–7. https://doi.org/10.1016/j.jns.2016.08.055.
173. Wang FD, Li J, Zhai X, Chen R, Wang F. Methane-rich saline restores brain SOD activity and alleviates cognitive impairment in rats with traumatic brain injury. Food Sci Technol. 2021;42:e54921. https://doi.org/10.1590/fst.54921.
174. Zhang B, Tian X, Li G, Zhao H, Wang X, Yin Y, et al. Methane inhalation protects against lung ischemia-reperfusion injury in rats by regulating pulmonary surfactant via the Nrf 2 pathway. Front Physiol. 2021;12:615974. https://doi.org/10.3389/fphys.2021.615974.
175. Yang C, Yang Q, Xiang Y, Zeng XR, Xiao J, Le WD. The neuroprotective effects of oxygen therapy in Alzheimer's disease: a narrative review. Neural Regen Res. 2023;18(1):57–63. https://doi.org/10.4103/1673-5374.343897.
176. Jhangi S, Gandhi R, Glanz B, Cook S, Nejad P, Ward D, et al. Increased Archaea species and changes with therapy in gut microbiome of multiple sclerosis subjects (S24.001). Neurology. 2014;82(10 suppl):S24–001. https://doi.org/10.1212/WNL.82.10_supplement.S24.001.
177. Ghoshal U, Shukla R, Srivastava D, Ghoshal UC. Irritable bowel syndrome, particularly the constipation-predominant form, involves an increase in *Methanobrevibacter smithii*, which is associated with higher methane production. Gut Liver. 2016;10(6):932–8. https://doi.org/10.5009/gnl15588.
178. Chen CH, Lin CL, Kao CH. Irritable bowel syndrome is associated with an increased risk of dementia: a Nationwide population-based study. PLoS One. 2016;11(1):e0144589. https://doi.org/10.1371/journal.pone.0144589.
179. Ghavami SB, Rostami E, Sephay AA, Shahrokh S, Balaii H, Aghdaei HA, et al. Alterations of the human gut *Methanobrevibacter smithii* as a biomarker for inflammatory bowel diseases. Microb Pathog. 2018;117:285–9. https://doi.org/10.1016/j.micpath.2018.01.029.
180. Cisek AA, Szymańska E, Wierzbicka-Rucińska A, Aleksandrzak-Piekarczyk T, Cukrowska B. Methanogenic archaea in the pediatric inflammatory bowel disease in relation to disease type and activity. Int J Mol Sci. 2024;25(1):673. https://doi.org/10.3390/ijms25010673.
181. Pathirana WGW, Chubb SP, Gillett MJ, Vasikaran SD. Faecal calprotectin. Clin Biochem Rev. 2018;39(3):77–90.
182. Heston MB, Hanslik KL, Zarbock KR, Harding SJ, Davenport-Sis NJ, Kerby RL, et al. Gut inflammation associated with age and Alzheimer's disease pathology: a human cohort study. Sci Rep. 2023;13(1):18924. https://doi.org/10.1038/s41598-023-45929-z.
183. Turner SA, Butler G. The Candida pathogenic species complex. Cold Spring Harb Perspect Med. 2014;4(9):a019778. https://doi.org/10.1101/cshperspect.a019778.
184. Nosal R, Menyhardtová Z. The effect of glycoprotein from *Candida albicans* on functions of rat platelets. Toxicon. 1976;14(4):313–8. https://doi.org/10.1016/0041-0101(76)90028-3.
185. Mauler M, Herr N, Schoenichen C, Witsch T, Marchini T, Härdtner C, et al. Platelet serotonin aggravates myocardial ischemia/reperfusion injury via neutrophil degranulation. Circulation. 2019;139(7):918–31. https://doi.org/10.1161/CIRCULATIONAHA.118.033942.
186. Gelosa P, Castiglioni L, Rzemieniec J, Muluhie M, Camera M, Sironi L. Cerebral derailment after myocardial infarct: mechanisms and effects of the signaling from the ischemic heart to brain. J Mol Med. 2022;100(1):23–41. https://doi.org/10.1007/s00109-021-02154-3.
187. Day AW, Perez-Lozada J, DiLeo A, Blandino K, Maguire J, Kumamoto CA. *Candida albicans* colonization modulates murine ethanol consumption and Behavioral responses through

elevation of serum prostaglandin E2 and impact on the striatal dopamine system. bioRxiv. 2025. https://doi.org/10.1101/2025.02.25.640044.

188. Zhao Z, Li S, Zhang L, Deng X, Chen T, Zeng K, et al. Dopamine D1 receptor gene polymorphism is associated with myocardial infarction. DNA Cell Biol. 2012;31(6):1010–4. https://doi.org/10.1089/dna.2011.1466.
189. Fouad Shalaby MA, Abd El Latif HA, El Yamani M, Galal MA, Kamal S, Sindi I, et al. Therapeutic activity of sarpogrelate and dopamine D2 receptor agonists on cardiovascular and renal systems in rats with alloxan-induced diabetes. BMC. Pharmacol Toxicol. 2021;22(1):64. https://doi.org/10.1186/s40360-021-00526-6.
190. Rieder M, Gauchel N, Bode C, Duerschmied D. Serotonin: a platelet hormone modulating cardiovascular disease. J Thromb Thrombolysis. 2021;52(1):42–7. https://doi.org/10.1007/s11239-020-02331-0.
191. Whittle BJ, Kauffman GL Jr, Moncada S. Hemostatic mechanisms, independent of platelet aggregation, arrest gastric mucosal bleeding. Proc Natl Acad Sci USA. 1986;83(15):5683–7.
192. Kringen MK, Narum S, Lygren I, Seljeflot I, Sandset PM, Trøseid AM, et al. Reduced platelet function and role of drugs in acute gastrointestinal bleeding. Basic Clin Pharmacol Toxicol. 2011;108(3):194–201. https://doi.org/10.1111/j.1742-7843.2010.00643.x.
193. Cappell MS. Gastrointestinal bleeding associated with myocardial infarction. Gastroenterol Clin N Am. 2000;29(2):423–44. https://doi.org/10.1016/S0889-8553(05)70121-7.
194. Zhong L, Quan X, Dang P, Tang M, Yu H, Guo F. Clinical characteristics and risk factors of in-hospital gastrointestinal bleeding in patients with acute myocardial infarction. Front Cardiovasc Med. 2022;9:933597. https://doi.org/10.3389/fcvm.2022.933597.
195. Wang X, Zhou S, Hu X, Ye C, Nie Q, Wang K, et al. Candida albicans accelerates atherosclerosis by activating intestinal hypoxia-inducible factor2α signaling. Cell Host Microbe. 2024;32(6):964–79. https://doi.org/10.1016/j.chom.2024.04.017.
196. Młynarska E, Czarnik W, Fularski P, Hajdys J, Majchrowicz G, Stabrawa M, et al. From atherosclerotic plaque to myocardial infarction—the leading cause of coronary artery occlusion. Int J Mol Sci. 2024;25(13):7295. https://doi.org/10.3390/ijms25137295.
197. Banerjee C, Chimowitz MI. Stroke caused by atherosclerosis of the major intracranial arteries. Circ Res. 2017;120(3):502–13. https://doi.org/10.1161/CIRCRESAHA.116.308441.
198. Xie B, Shi X, Xing Y, Tang Y. Association between atherosclerosis and Alzheimer's disease: a systematic review and meta-analysis. Brain Behav. 2020;10(4):e01601. https://doi.org/10.1002/brb3.1601.
199. Zhang X, Zhang Y, Wang P, Zhang SY, Dong Y, Zeng G, et al. Adipocyte hypoxia-inducible factor 2α suppresses atherosclerosis by promoting adipose ceramide catabolism. Cell Metab. 2019;30(5):937–51. https://doi.org/10.1016/j.cmet.2019.09.016.
200. Li G, Liu J, Guo M, Gu Y, Guan Y, Shao Q, et al. Chronic hypoxia leads to cognitive impairment by promoting HIF-2α-mediated ceramide catabolism and alpha-synuclein hyperphosphorylation. Cell Death Discov. 2022;8(1):473. https://doi.org/10.1038/s41420-022-01260-6.

Therapeutic Interventions: Modulating the Microbiome to Combat Neurodegeneration

8

Aparna Anandan, A. K. Mohammed Unais, A. Vishnu Prabhu, Mariya K. Saju, Kowsalya Sonaimuthu, Harinivashini Murugesan, Felicita Danielraj, Dhayalan Marimuthu, Marthandam Asokan Shibu, and Vijaya Padma Viswanadha

8.1 Introduction

Neurodegenerative diseases (NDs) such as Alzheimer's disease (AD), Parkinson's disease (PD), amyotrophic lateral sclerosis (ALS), and Huntington's disease (HD) are chronic, progressive disorders characterized by the selective loss of neuronal populations, leading to motor, cognitive, and behavioral impairments. While aging, genetic predisposition, and environmental factors are established contributors to ND pathogenesis, increasing evidence underscores the significant role of peripheral systems—particularly the gut microbiota—in influencing disease onset and progression [42].

The human gastrointestinal tract harbors trillions of symbiotic microorganisms collectively termed the gut microbiota, whose collective genome and metabolites form the gut microbiome [63]. These microbes exert a profound impact on host physiology, including metabolism, immune modulation, and central nervous system (CNS) development and function [65]. The gut microbiota's influence on neurodevelopment, cognition, and behavior has earned it the title of the body's "second

A. Anandan · A. K. Mohammed Unais · A. Vishnu Prabhu · M. K. Saju · K. Sonaimuthu
F. Danielraj · D. Marimuthu · M. A. Shibu · V. P. Viswanadha (✉)
Department of Biotechnology, Bharathiar University, Coimbatore, Tamil Nadu, India
e-mail: shibu@buc.edu.in; aparna.biotech@buc.edu.in; mohammed.biotech@buc.edu.in; vishnuprabhu.biotech@buc.edu.in; mariya.biotech@buc.edu.in; kowsalya.biotech@buc.edu.in; dhayalan.biotech@buc.edu.in

H. Murugesan
Department of Foods and Nutrition, PSG College of Arts and Science, Coimbatore, Tamil Nadu, India

M. H. Rahman, Muhasina K M (eds.), *The Microbiome's Impact on the Brain*,
https://doi.org/10.1007/978-981-95-7494-0_8

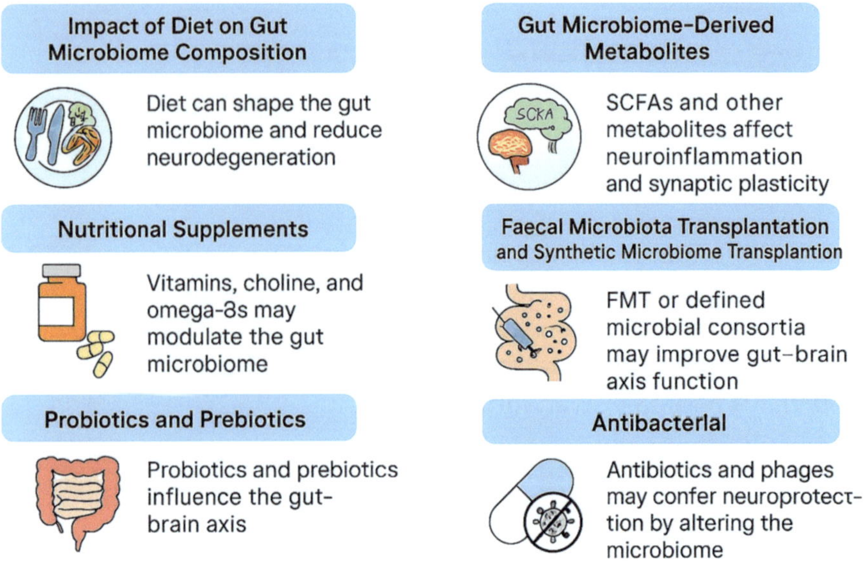

Fig. 8.1 Therapeutic approaches targeting the microbiome regulating the gut-brain axis to combat NDs

brain" [8]. Through mechanisms involving the vagus nerve, immune pathways, endocrine signaling (e.g., hypothalamic-pituitary-adrenal axis), and metabolic intermediates like short-chain fatty acids, the microbiota-gut-brain (MGB) axis facilitates bidirectional communication between the gut and brain [5, 68].

Notably, alterations in gut microbial composition—termed dysbiosis—have been consistently reported in clinical studies of NDs and are thought to exacerbate disease pathology by promoting neuroinflammation, impairing blood-brain barrier integrity, and disrupting neurotransmitter synthesis [93]. Experimental models, such as germ-free or antibiotic-treated animals, exhibit ND-like behavioral and neuropathological changes, further reinforcing the role of the gut microbiota in CNS disorders [43]. With the global burden of NDs projected to rise sharply due to population aging, there is a growing interest in exploring gut microbiota as a modifiable therapeutic target.

This chapter aims to summarize the current understanding of the gut microbiota's role in neurodegeneration and evaluate emerging therapeutic strategies (Fig. 8.1) such as dietary interventions, probiotics, prebiotics, and microbiota transplants that seek to modulate the microbiome for neuroprotection. In doing so, we explore both mechanistic insights and clinical advances, highlighting the opportunities and challenges in translating microbiota-based interventions into effective therapies for NDs.

8.2 Therapeutic Approaches Targeting the Microbiome in NDs

8.2.1 Impact of Diet on Gut Microbiome Composition

Diet is a pivotal modifiable factor that shapes the gut microbiome, which in turn influences neurodevelopment, immunity, and the pathophysiology of NDs such as AD and PD [26, 49]. Recent studies highlight that dietary modulation of gut microbes can reduce disease risk, modulate symptoms, and mitigate neurodegeneration by fostering a beneficial gut environment. This relationship is bidirectional; diet influences both the immune system and the microbiota, which interact through signaling pathways and microbial metabolites [25].

Western dietary patterns—rich in fats and sugars but low in fiber—promote gut dysbiosis, reduce microbial diversity, and drive low-grade systemic inflammation associated with metabolic syndrome (MetS), thereby contributing to ND progression [9, 11, 17]. Evidence suggests that anti-inflammatory diets appear beneficial, as higher Dietary Inflammatory Index scores are linked to greater AD and PD risk, and gut microbiota composition reflects dietary inflammatory potential. Antiinflammatory diets enrich, *Akkermansia muciniphila*, while pro-inflammatory diets increase *Ruminococcus torques* and *Clostridium leptum* [4, 95].

Specific dietary regimens like the Mediterranean (MeDi), MIND, and ketogenic diets have demonstrated neuroprotective potential through microbiome modulation. These diets improve cognition and reduce ND symptoms by promoting the growth of beneficial taxa such as *Christensenellaceae* and *Bilophila*, enhancing short-chain fatty acid (SCFA) production, and maintaining gut-brain axis integrity [47, 50, 55, 57, 59].

Moreover, gut microbes synthesize critical neuroactive compounds like GABA and serotonin and modulate immune responses and blood-brain barrier (BBB) function. Dysbiosis compromises these functions, increasing vulnerability to neurodegeneration [70]. Probiotics have been shown to reinforce intestinal and BBB integrity, while dietary interventions involving prebiotics and postbiotics aim to restore healthy microbial balance [96].

Together, these findings underscore the therapeutic potential of diet-driven gut microbiota manipulation in NDs. Targeted nutritional strategies that enhance the production of neuroprotective metabolites and reduce inflammation could be integrated with pharmacological treatments to yield more effective interventions. Future research should focus on identifying optimal dietary compositions and microbial targets to harness the full benefits of the gut-brain axis in ND management [78].

8.2.2 Nutritional Supplements

Diet significantly influences NDs through its interaction with the gut microbiome. A cohort study in Eastern Finland reported that higher dietary choline intake was associated with improved cognitive performance. In contrast, findings from the Framingham Heart Study suggested a potential increase in the risk of dementia and AD with high choline intake, though without statistical significance [87]. Choline serves as a precursor for trimethylamine (TMA), a microbial metabolite subsequently converted to trimethylamine N-oxide (TMAO) in the liver. Elevated TMAO levels have been linked to cognitive decline and dementia. This discrepancy underscores the influence of gut microbiome composition and function on the metabolic fate of dietary choline and its neurological outcomes, highlighting the need for further mechanistic studies to define optimal intake [43].

Vitamins also play a neuroprotective role in NDs. Vitamins A, B, D, and E have shown protective effects against AD and PD in multiple studies. The gut microbiome is both a regulator and a beneficiary of vitamin metabolism. It contributes significantly to the host's vitamin B supply—up to 30%—through microbial synthesis [7, 27, 33, 76]. Conversely, vitamins modulate gut microbial composition, for example, vitamin D levels are positively associated with gut microbial diversity. Vitamin A deficiency has been shown to exacerbate cognitive deficits and amyloid burden in APP/PS1 AD mice, which correlated with altered gut microbial diversity and decreased *Lactobacillus* abundance [77, 96]. These findings emphasize the bidirectional influence between vitamins and the microbiome in NDs.

Polyunsaturated fatty acids (PUFAs), particularly omega-3 fatty acids, have been linked to slowed cognitive decline in AD and reduced motor symptoms in PD through both observational and interventional studies. Clinical trials indicate that omega-3 supplementation enhances the abundance of Bacteroidetes and butyrate-producing *Lachnospiraceae*, taxa associated with milder ND phenotypes [43].

In conclusion, diet-mediated modulation of the gut microbiome plays a crucial role in the pathophysiology of NDs. However, most existing evidence is derived from observational studies. Randomized controlled trials are essential to establish causality and to define dietary strategies that beneficially remodel the gut-brain axis in neurodegenerative contexts.

8.2.3 Probiotics and Prebiotics

The gut microbiota plays a pivotal role in modulating neuroinflammation, synaptic plasticity, and gut-brain signaling, making it a promising target for therapeutic interventions in NDs. Probiotics (live beneficial microbes), prebiotics (nondigestible substrates that nourish specific microbes), and synbiotics (combinations of both) have been investigated for their potential to mitigate symptoms and slow progression in disorders like AD, PD, and HD [16, 28, 53, 56].

Among probiotics, *Lactobacillus* and *Bifidobacterium* are the most extensively studied genera due to their anti-inflammatory and gut-stabilizing effects. A randomized, double-blind trial showed that supplementation with *L. acidophilus*, *B. bifidum*, and *B. longum* (2×10^9 CFU/day) combined with selenium for 12 weeks significantly improved cognitive performance and metabolic markers in AD patients compared to selenium or placebo alone [1]. In PD, these strains alleviated gastrointestinal symptoms such as constipation and improved motor function [83]. However, their therapeutic efficacy remains inconsistent. A meta-analysis reported no significant cognitive benefit in AD patients following probiotic intervention with these genera. Additionally, elevated fecal levels of *Lactobacillus* and *Bifidobacterium* have been observed in both AD and PD, suggesting a possible link to disease progression. These conflicting findings highlight the importance of strain-specific effects, necessitating high-resolution metagenomic and functional analyses to identify truly beneficial strains.

Next-generation probiotics like *Clostridium butyricum* have shown promising results. In APP/PS1 AD mouse models, oral administration of *C. butyricum* (5×10^8 CFU/day) for 4 weeks improved cognition, reduced Aβ deposition, restored butyrate levels, and modulated microglial activation [22]. In MPTP-induced PD models, it enhanced motor function, preserved dopaminergic neurons, and improved synaptic plasticity through colonic GLP-1 and GPR41/GPR43 signaling, along with increased cerebral GLP-1 receptor expression [90].

Akkermansia muciniphila, a mucin-degrading bacterium associated with gut barrier integrity and metabolic health, also shows neuroprotective potential. In APP/PS1 mice, it reduced Aβ40/Aβ42 burden, improved cognition, restored goblet cell function, and normalized glucose metabolism [90]. Similar cognitive and systemic benefits were reported in $AlCl_3$/D-galactose-induced AD rats [15], while in SOD1-Tg ALS mice it elevated central nicotinamide levels and improved motor outcomes [29]. However, its role in PD remains debated—some studies report increased abundance in patients, raising concerns about mucin depletion and compromised gut barrier function [41].

Prebiotics—nondigestible dietary fibers that promote the growth of beneficial gut microbes—have shown promising neuroprotective effects in both preclinical and clinical studies. In APP/PS1 mouse models of AD, a 2-month soluble fiber-rich diet reduced astrocyte activation and improved cognition by increasing SCFA production, particularly butyrate and propionate; these effects were abolished by antibiotic treatment, highlighting their microbiota dependence [45]. In PD models overexpressing α-synuclein, a 16-week high-fiber diet improved motor function and reduced α-synuclein aggregation in the substantia nigra by mitigating pathogenic microglial activation. Similarly, in HD models, fiber supplementation improved cognitive and emotional behaviors and alleviated gastrointestinal dysfunction, likely via microbiota modulation [72].

Clinical evidence also supports the therapeutic potential of prebiotics. A longitudinal cohort study found that higher intake of dietary fructans—a class of fermentable prebiotics—was associated with a reduced risk of developing AD in elderly individuals. In PD patients with constipation, an insoluble fiber-rich diet enhanced gastrointestinal transit, improved motor performance, and increased plasma L-DOPA levels, potentially boosting drug bioavailability and gut-brain communication [86].

In summary, while prebiotics and probiotics offer potential as adjunctive therapies for NDs, successful clinical translation depends on a nuanced understanding of host-microbe interactions. Given the variability in gut microbiota influenced by factors such as age, sex, and disease stage, personalized interventions guided by metagenomic and multi-omics approaches are essential to achieve consistent therapeutic outcomes.

8.2.4 Gut Microbiome-Derived Metabolites

Gut microbiome-derived metabolites play critical roles in NDs by modulating neuroinflammation, synaptic plasticity, and gut-brain communication. SCFAs—acetate, propionate, and butyrate produced from dietary fiber fermentation—can cross the BBB and influence brain function. Sodium butyrate improves synaptic plasticity and cognition in AD mouse models by enhancing dendritic spine development and promoting neuroprotective astrocyte subtypes [32, 83]. However, the effects of acetate and propionate remain inconsistent, with studies reporting both anti- and pro-inflammatory outcomes in AD models [39, 41, 52]. In PD, butyrate worsened motor deficits and neuroinflammation [61], while SCFA mixtures induced microglial activation in α-synuclein-overexpressing mice. Conversely, in ALS, butyrate restored gut homeostasis and extended lifespan [89], highlighting the dose- and disease-specific actions of SCFAs. Elevated secondary bile acids—deoxycholic and glycodeoxycholic acids—are associated with cognitive decline and PD status, correlating with pro-inflammatory gut bacteria [12, 44]. These act via receptors such as FXR, LXRα/LXRβ, and S1PR2, offering potential therapeutic targets. TMAO, another microbial metabolite, promotes neurodegeneration, while its inhibition improves cognition in AD models [12, 62, 91]. Indole-3-propionic acid and polyphenol-derived metabolites like S-equol and phenyl-γ-valerolactones show neuroprotective potential by restoring gut balance and reducing Aβ aggregation [23, 67]. Overall, these findings highlight the complex, disease-specific impact of microbial metabolites in NDs and their therapeutic promise.

8.2.5 Fecal Microbiota Transplantation and Synthetic Microbiome Transplantation

Fecal microbiota transplantation (FMT) is gaining recognition as a promising strategy to restore gut microbial balance and modulate host immunity in NDs. By introducing diverse microbial communities and metabolites from healthy donors, FMT enhances microbial diversity, shifts host metabolite profiles, and reduces inflammation, contributing to improved gut-brain axis health [92].

In AD models, FMT from wild-type mice—either with or without antibiotic pretreatment—reversed gut immune dysregulation, reduced amyloid-β (Aβ) deposition, attenuated tau phosphorylation, and improved cognitive outcomes [21, 71]. Similarly, in rotenone-induced PD mice, FMT without antibiotic priming alleviated motor deficits and neuroinflammation via suppression of the LPS-TLR4/MyD88/NF-κB pathway and restoration of BBB integrity [94]. Preliminary clinical findings support these results. AD patients with *Clostridioides difficile* infection showed cognitive improvement 1 month after FMT via colonoscopy [58]. In PD, oral FMT capsules improved autonomic and gastrointestinal symptoms and increased gut microbiome complexity without prior antibiotic use [13]. Another study found that colonic FMT provided more sustained relief of motor and non-motor symptoms than nasointestinal delivery, though some patients reported mild gastrointestinal side effects [85].

Together, these findings highlight FMT as a potential adjunct therapy for NDs, though further studies are needed to optimize delivery methods, timing, donor selection, and long-term safety.

While traditional FMT is generally safe, it remains akin to a "black box," inadvertently transferring an undefined mix of bacteria, fungi, parasites, and viruses, which may pose risks to recipients. To mitigate these uncertainties, researchers have developed a synthetic microbiome with a well-defined composition. Cheng et al. designed a consortium of 104—or in later refinement 119—human gut bacterial species, termed hCom2, that can stably engraft in germ-free mice. Remarkably, mice colonized with hCom2 demonstrated immune cell profiles and metabolic outputs closely mirroring those of mice colonized with native human feces (https://news.stanford.edu/stories/2022/09/stanford-researchers-construct-complex-complete-synthetic-microbiome). This approach introduces the intriguing possibility of designing synthetic microbial communities that replicate the beneficial functions of healthy human microbiomes while eliminating the unpredictability of donor-derived FMT. By selectively incorporating species with proven neuroprotective or anti-inflammatory effects, such as those linked to cognitive preservation or immune modulation, synthetic consortia offer a targeted and safer strategy for microbiota-based interventions in neurodegenerative diseases.

8.2.6 Antibacterial

Beyond their antimicrobial role, antibiotics can modulate the gut microbiota and offer neuroprotective benefits. Certain classes, such as tetracyclines, exhibit anti-inflammatory and antioxidant properties, making them attractive candidates for NDD therapy. However, early-life antibiotic exposure, such as low-dose penicillin during gestation, has been shown to disrupt gut microbiota, impair BBB integrity, alter brain cytokines, and affect behavior in offspring, potentially increasing neuropsychiatric risk [40]. In PD models, antibiotics suppress α-synuclein aggregation and neuroinflammation [81, 84], while in AD mice long-term antibiotic use altered microbial profiles, reduced glial activation, and modulated Aβ deposition dynamics [48]. Doxycycline also reduced tau aggregation and toxicity in tauopathy models [46].

Clinically, *H. pylori* eradication in PD patients improves motor symptoms, especially bradykinesia. Antibiotics have also reduced neuroinflammation and motor deficits in PD and delayed disease onset in MS models [14, 37]. However, nonabsorbable antibiotics worsened demyelination in MS models, an effect mitigated by butyrate, highlighting the complex microbiota-immune interplay.

Overall, antibiotics may influence ND progression via gut-brain axis modulation, though long-term risks and variable responses necessitate cautious, targeted application [54].

Emerging evidence also points to bacteriophages as modulators of gut-brain interactions. In aged mice, accumulation of isoamylamine—a metabolite from *Ruminococcaceae*—induces neuroinflammation via S100A8-driven microglial apoptosis [75]. Aging reduces gut *Myoviridae* phages, which target *Ruminococcaceae*. Oral phage therapy restored phage levels, reduced isoamylamine, protected microglia, and alleviated neuroinflammation [75]. These findings highlight phages as novel, targeted tools to reshape microbial metabolism and mitigate age-related cognitive decline, warranting further investigation [19, 75, 79].

8.3 Modulating the Microbiome in Neurodegenerative Diseases

8.3.1 Parkinson's Disease

Current therapeutic approaches in PD primarily aim to alleviate symptoms and slow disease progression. Pharmacologically, dopaminergic therapies remain central. Levodopa (L-DOPA), administered alongside dopa decarboxylase inhibitors like carbidopa or benserazide, enhances dopamine availability in the brain. Additional agents, such as monoamine oxidase-B (MAO-B) inhibitors like safinamide and adenosine A2A receptor antagonists, are used to manage motor complications and delay dopaminergic neuron degeneration [31, 60] (Table 8.1).

Growing evidence highlights the gut microbiome's role in PD pathogenesis, offering opportunities for microbiota-targeted interventions. Dietary modifications have shown beneficial effects: high-fiber diets suppress pro-inflammatory gut bacteria like *Bacteroides* and promote the growth of butyrate-producing taxa such as *Blautia*, *Roseburia*, and *Coprococcus*, which support gut barrier integrity and immune regulation. In contrast, high intake of added sugars may enhance the abundance of *Klebsiella*, a bacterium capable of amyloid production that exacerbates α-synuclein aggregation and inflammation [36, 38].

Probiotic supplementation has demonstrated clinical utility. Multi-strain formulations containing *Lactobacillus*, *Bifidobacterium*, and *Enterococcus* have shown improvements in gastrointestinal symptoms, particularly constipation, in PD patients over short treatment periods. Similarly, prebiotics—nondigestible food components that stimulate beneficial gut bacteria—have been associated with improved non-motor symptoms and modulation of fecal biomarkers in early-stage PD [74].

Collectively, these therapeutic strategies, ranging from dopamine replacement and microbiota modulation to dietary interventions and regenerative medicine, highlight the multifaceted management of PD and underscore the importance of integrated, patient-tailored treatment approaches.

Table 8.1 Neurodegenerative disorders and their therapeutic interventions targeting microbiome till date

NDD	Research subjects	Intervention	Outcome
Antibiotics			
AD	A53T α-synuclein transgenic mice	Intranasal rifampicin	Lower α-synuclein oligomers in the hippocampus and improved spatial memory
AD	APPSWE/ PS1ΔE9 mice	Oral broad-spectrum antibiotics (gentamicin, vancomycin, metronidazole, etc.) and cefoperazone	Altered gut microbiota and inflammatory mediators, decreased Aβ plaques, increased soluble Aβ, and reduced glial activation
PD	Patients	Triple therapy with omeprazole, amoxicillin, and clarithromycin	Eradication of *H. pylori* improved motor retardation symptoms
PD	6-OHDA rat model	Long-term nonabsorbable antibiotics (neomycin, vancomycin)	Decreased striatal inflammation and improved motor function
Dietary interventions			
AD	Patients	Mediterranean diet with coconut oil	Improved memory and orientation scores
AD	Patients	Ketogenic diet with sustained ketosis	Better daily function and quality of life compared to standard diet
PD	Patients	Mediterranean diet	Improved executive function, memory, and overall cognitive score
PD	Patients	Ketogenic diet	Improved both motor and non-motor symptoms
MS	EAE mice	Fasting-mimicking or ketogenic diet	Reduced inflammation and demyelination severity
ALS	Patients	High-calorie fatty diet (HCFD)	Survival benefit in patients with rapid disease progression
Probiotics			
AD	Patients	Selenium + probiotics (*L. acidophilus*, *B. bifidum*, *B. longum*)	Enhanced cognition and metabolic health
AD	APP/PS1 mice	*Clostridium butyricum* gavage	Improved cognition, reduced Aβ plaques, microglial activation, and pro-inflammatory cytokines
AD	3xTg-AD mice	Mixed probiotics (lactic acid bacteria and bifidobacteria)	Reduced plaque size/number and elevated SCFAs
AD	SAMP8 mice	ProBiotic-4 (Lactobaci*llus*/*Bifidobacterium* mix)	Improved cognition through suppression of NF-κB pathways

(continued)

Table 8.1 (continued)

NDD	Research subjects	Intervention	Outcome
PD	6-OHDA rats	SLAB51 probiotic blend	Reduced neuroinflammation and motor deficits
PD	Patients	Oral probiotics (*Lactobacillus*, *Bifidobacterium*, *L. reuteri*, *L. fermentum*)	Improved motor scores (MDS-UPDRS) and metabolic profiles
MS	Patients	Probiotics (*L. acidophilus*, *B. bifidum*, *L. fermentum*, *L. casei*)	Reduced IL-8 and TNF-α in PBMCs
MS	Patients	Probiotic mix (*Lactobacillus*, *Bifidobacterium*, *Streptococcus*)	Decreased dendritic cell activation and MS risk allele expression
MS	EAE mice	*Lactibiane iki* or Vivomixx	*Lactibiane iki* improved demyelination and symptoms, Vivomixx ineffective
Fecal microbiota transplantation (FMT)			
PD	C57BL/6 mice	FMT via oral gavage	Corrected dysbiosis, restored SCFAs, improved DA/5-HT levels, and reduced neuroinflammation
PD	Patients with constipation	FMT via nasoduodenal tube	Increased beneficial microbes (*Blautia*, *Prevotella*), resolved SIBO, and improved symptoms
AD	Mn-exposed rats	FMT from healthy donors	Reduced Aβ, Tau, and neuroinflammatory markers
AD	ADLP APT mice	FMT from wild-type donors	Improved plaques, tangles, glial reactivity, and cognition, normalized immune-related gene expression
MS	EAE mice	Oral synthetic miR-30d	Increased *Akkermansia*, boosted Tregs, and alleviated symptoms
ALS	Patients	Fresh stool FMT via nasal delivery	Increased Tregs and induced anti-inflammatory immune status

8.3.2 Alzheimer's Disease

Therapeutic strategies targeting the gut microbiota have gained significant attention in AD. Probiotics, particularly strains of *Lactobacillus* (e.g., *L. plantarum*, *L. rhamnosus*) and *Bifidobacterium* (e.g., *B. longum*, *B. breve*), enhance gut barrier integrity, reduce systemic inflammation, and modulate neurotransmitter activity, particularly by producing gamma-aminobutyric acid (GABA) and SCFAs. Clinical studies have demonstrated improvements in gastrointestinal symptoms and modest gains in motor and non-motor functions following probiotic administration in PD patients [2]. Prebiotics, such as inulin and oligosaccharides, promote the growth of beneficial microbes and elevate SCFA production, thereby supporting the gut-brain axis and modulating inflammation [35]. Synbiotics, a combination of probiotics and prebiotics, have demonstrated synergistic benefits in animal models, including memory enhancement and upregulation of neurotrophic factors such as BDNF[18] (Table 8.1).

FMT from healthy donors has shown the ability to restore microbial balance, reduce amyloid-β (Aβ) plaque deposition, and improve cognitive function in AD mouse models [34, 71].

Dietary interventions, including the Mediterranean and ketogenic diets, are also effective in enhancing microbial diversity and enriching neuroprotective taxa like *Akkermansia muciniphila* and *Bifidobacteria*. These diets are associated with reduced inflammation and improved cognitive outcomes [3].

Additionally, lifestyle modifications—such as regular physical activity and stress reduction—complement these interventions by supporting microbiota stability and promoting neuroimmune balance. Despite these promising strategies, challenges remain, including individual variability in microbiome composition, probiotic strain specificity, and long-term safety concerns. Nonetheless, gut microbiota modulation represents a safe, noninvasive, and multi-targeted approach that holds substantial promise for the management of AD [10, 24, 69, 82].

8.3.3 Amyotrophic Lateral Sclerosis

Emerging evidence highlights gut microbiota modulation as a promising therapeutic approach for amyotrophic lateral sclerosis (ALS). In preclinical models, dysbiosis has been shown to disrupt the gut-brain axis, impair intestinal barrier function, increase systemic inflammation, and contribute to motor neuron degeneration [30]. Therapeutic strategies targeting these pathways include probiotics, phage therapy, antibiotics, and microbiota-derived metabolites, which aim to restore microbial balance and reduce neuroinflammation (Table 8.1).

Supplementation with *Lacticaseibacillus rhamnosus* has shown neuroprotective effects, partly due to its unique fatty acid profile, which helped restore energy metabolism independent of the carnitine shuttle [73]. Similarly, administration of *Akkermansia muciniphila* in SOD1-G93A mouse models improved mitochondrial function and alleviated motor symptoms, underscoring the potential of specific taxa in disease modulation [20].

Metabolites—such as SCFAs, bile acids, and nicotinamide produced by gut bacteria—can regulate microglial activity, mitochondrial health, and Treg/Th17 immune balance, all of which are implicated in ALS pathophysiology [88]. Notably, ALS patients often exhibit decreased levels of SCFA-producing bacteria like *Faecalibacterium prausnitzii* and reduced microbial diversity, which may exacerbate inflammation and metabolic dysfunction. Although further clinical validation is required, these findings suggest that correcting gut microbial imbalances may slow disease progression and improve symptom management in ALS.

8.3.4 Other Neurodegenerative Disorders (Huntington's and Multiple Sclerosis)

Targeting the gut microbiota has emerged as a promising therapeutic strategy for several neurodegenerative disorders beyond AD, PD, and ALS, particularly in Huntington's disease (HD) and multiple sclerosis (MS). In HD gene expansion carriers, gut microbial imbalances—including reduced *Firmicutes*, *Lachnospiraceae*, and *Akkermansiaceae*—have been linked to immune dysregulation. Notably, *Bilophila* abundance negatively correlates with pro-inflammatory IL-6, while *Intestinimonas* shows a positive correlation with anti-inflammatory IL-4 levels, suggesting a gut-immune interaction in HD pathophysiology. In HD mouse models, FMT from healthy donors significantly improved cognitive function in females, possibly due to sex-specific microbial and immune responses [43] (Table 8.1).

In MS, therapeutic targeting of the gut microbiome aims to regulate systemic and neuroinflammation through microbial metabolites such as SCFAs, butyrate, propionate, and acetate, produced by *Firmicutes* and *Bacteroidetes*. These SCFAs modulate T-cell responses and cytokine production and may impact demyelination processes. Gut-derived molecules such as lipopolysaccharides and polysaccharide A can also influence immune function and CNS inflammation. Dysbiosis has been implicated in altering hypothalamic-pituitary-adrenal (HPA) axis function and disrupting immunoregulation, making microbiota restoration a key therapeutic goal in MS [80].

Probiotic interventions have shown efficacy in experimental autoimmune encephalomyelitis (EAE), the animal model of MS. Strains of *Lactobacillus*, *Escherichia coli*, and *Prevotella* reduced disease incidence, delayed symptom onset, and attenuated disease severity when administered orally or intraperitoneally. In marmoset models, a yoghurt-based diet reduced demyelination and pro-inflammatory immune responses, indicating that diet-induced microbiota changes influence neuroimmune outcomes [51].

Emerging research also suggests the potential of precision therapies using prebiotics, probiotics, and synbiotics tailored to an individual's microbiome profile. Metagenomic and metabolomic profiling could help identify strain-specific markers relevant to neurodegenerative progression and therapy. Such integrative strategies, when combined with existing pharmacological or dietary interventions, may open new avenues for personalized treatment approaches in HD, MS, and other less-studied neurodegenerative disorders [6, 64, 66].

8.4 Conclusion

Intriguing preclinical and clinical evidence has established a strong link between gut microbiome alterations and the development and progression of NDDs. Therapeutic strategies aimed at modulating the gut microbiome through dietary interventions, probiotics, prebiotics, synbiotics, antibacterials, microbial metabolite supplementation, and FMT have shown promise in alleviating neuroinflammation, neuropathology, and behavioral symptoms associated with AD, PD, ALS, HD, and MS. These approaches either target the microbial composition directly or modify the metabolite environment to restore gut-brain axis homeostasis.

Despite these advances, several challenges remain. The gut microbiome is highly diverse, with significant interindividual variability influenced by genetic, environmental, and lifestyle factors. High-resolution metagenomic and metabolomic studies across large, longitudinal cohorts are required to define disease-specific microbial signatures and identify reliable biomarkers for early diagnosis and therapeutic targeting. Furthermore, mechanistic insights into how gut microbes and their metabolites influence key ND pathways, such as immune cell activation, amyloid deposition, tau aggregation, and barrier integrity, are still limited.

The current therapeutic options also face practical limitations. Long-term effectiveness and in vivo survival of probiotic preparations remain uncertain, the use of antibiotics and FMT carries risks of disrupting microbial equilibrium, and dietary interventions often require prolonged adherence to achieve measurable effects. Additionally, standardized methodologies for sample processing, microbial characterization, and animal modeling are needed to ensure reproducibility and comparability across studies. Looking forward, advances in high-throughput sequencing, droplet-based microbial culture, and mass spectrometry-based metabolomics are expected to refine our understanding of host-microbiome interactions. Personalized microbiome-based therapies tailored to individual microbial profiles could enhance efficacy, while engineered probiotics, synthetic microbiome communities, and targeted microbial metabolite analogues hold promise as next-generation interventions.

In conclusion, modulation of the gut microbiome represents a compelling, multi-targeted strategy for combating neurodegenerative diseases. Achieving clinical translation will require an integrated approach that combines rigorous mechanistic studies, standardized methodologies, and precision medicine frameworks. By maintaining or restoring gut microbial equilibrium, microbiome-targeted therapies have the potential to revolutionize the prevention and management of NDs in the near future.

Acknowledgments Aparna Anandan sincerely acknowledges the DST-INSPIRE Program, Government of India (DST/INSPIRE Fellowship/2019/IF190185). Felicita Danielraj and Dhayalan Marimuthu acknowledge the CSIR-UGC-NET, Government of India (CSIR-UGC-NET/UGC-NET Fellowship/10/06/2022/221610051932 and CSIR-UGC-NET/UGC-NET Fellowship/10/05/2024/231620054741, respectively). Harinivashini Murugesan acknowledges NFOBC-NET(NBCFDC/E-70291/28/03/2023/220510006190). Vijaya Padma Viswanadha acknowledge Tamil Nadu State Council for Higher Education-RGP (RGP/2019-20/BU/HECP-0005) no. (C3/CRTD/995/2021).

Author Contributions All authors had full access and are responsible for the integrity of the content in the chapter.

Declarations

Competing Interests The author(s) has no competing interests to declare that are relevant to the content of this manuscript.

Ethics Approval The author(s) declares that this manuscript does not include primary research with human participants.

The author(s) declares that this manuscript does not include primary studies involving animals.

References

1. Abdel-Haq R, Schlachetzki JCM, Boktor JC, Cantu-Jungles TM, Thron T, Zhang M, Bostick JW, Khazaei T, Chilakala S, Morais LH, Humphrey G, Keshavarzian A, Katz JE, Thomson M, Knight R, Gradinaru V, Hamaker BR, Glass CK, Mazmanian SK. A prebiotic diet modulates microglial states and motor deficits in α-synuclein overexpressing mice. elife. 2022;11:e81453. https://doi.org/10.7554/eLife.81453.
2. Akbari E, Asemi Z, Daneshvar Kakhaki R, Bahmani F, Kouchaki E, Tamtaji OR, Hamidi GA, Salami M. Effect of probiotic supplementation on cognitive function and metabolic status in Alzheimer's disease: a randomized, double-blind and controlled trial. Front Aging Neurosci. 2016;8:229544. https://doi.org/10.3389/fnagi.2016.00256.
3. Askarova S, Umbayev B, Masoud A-R, Kaiyrlykyzy A, Safarova Y, Tsoy A, Olzhayev F, Kushugulova A. The links between the gut microbiome, aging, modern lifestyle and Alzheimer's disease. Front Cell Infect Microbiol. 2020;10:104. https://doi.org/10.3389/fcimb.2020.00104.
4. Balomenos V, Bounou L, Charisis S, Stamelou M, Ntanasi E, Georgiadi K, Mourtzinos I, Tzima K, Anastasiou CA, Xiromerisiou G, Maraki M, Yannakoulia M, Kosmidis MH, Dardiotis E, Hadjigeorgiou G, Sakka P, Stefanis L, Scarmeas N. Dietary inflammatory index score and prodromal Parkinson's disease incidence: the HELIAD study. J Nutr Biochem. 2022;105:108994. https://doi.org/10.1016/j.jnutbio.2022.108994.
5. Bianchimano P, Britton GJ, Wallach DS, Smith EM, Cox LM, Liu S, Iwanowski K, Weiner HL, Faith JJ, Clemente JC, Tankou SK. Mining the microbiota to identify gut commensals modulating neuroinflammation in a mouse model of multiple sclerosis. Microbiome. 2022;10(1):174. https://doi.org/10.1186/s40168-022-01364-2.
6. Bonfili L, Cecarini V, Gogoi O, Gong C, Cuccioloni M, Angeletti M, Rossi G, Eleuteri AM. Microbiota modulation as preventative and therapeutic approach in Alzheimer's disease. FEBS J. 2021;288(9):2836–55. https://doi.org/10.1111/febs.15571.
7. Brauer-Nikonow A, Zimmermann M. How the gut microbiota helps keep us vitaminized. Cell Host Microbe. 2022;30(8):1063–6. https://doi.org/10.1016/j.chom.2022.07.010.
8. Brown K, Thomson CA, Wacker S, Drikic M, Groves R, Fan V, Lewis IA, McCoy KD. Microbiota alters the metabolome in an age- and sex-dependent manner in mice. Nat Commun. 2023;14(1):1348. https://doi.org/10.1038/s41467-023-37055-1.
9. Calvo-Ochoa E, Hernández-Ortega K, Ferrera P, Morimoto S, Arias C. Short-term high-fat-and-fructose feeding produces insulin signaling alterations accompanied by neurite and synaptic reduction and astroglial activation in the rat hippocampus. J Cereb Blood Flow Metab. 2014;34(6):1001–8. https://doi.org/10.1038/jcbfm.2014.48.
10. Chandra S, Sisodia SS, Vassar RJ. The gut microbiome in Alzheimer's disease: what we know and what remains to be explored. Mol Neurodegener. 2023;18(1):9. https://doi.org/10.1186/s13024-023-00595-7.
11. Chen X, Devaraj S. Gut microbiome in obesity, metabolic syndrome, and diabetes. Curr Diab Rep. 2018;18(12):129. https://doi.org/10.1007/s11892-018-1104-3.
12. Chen S-J, Chen C-C, Liao H-Y, Wu Y-W, Liou J-M, Wu M-S, Kuo C-H, Lin C-H. Alteration of gut microbial metabolites in the systemic circulation of patients with Parkinson's disease. J Parkinsons Dis. 2022;12(4):1219–30. https://doi.org/10.3233/JPD-223179.

13. Cheng Y, Tan G, Zhu Q, Wang C, Ruan G, Ying S, Qie J, Hu X, Xiao Z, Xu F, Chen L, Chen M, Pei Y, Zhang H, Tian Y, Chen D, Liu X, Huang H, Wei Y. Efficacy of fecal microbiota transplantation in patients with Parkinson's disease: clinical trial results from a randomized, placebo-controlled design. Gut Microbes. 2023;15(2):2284247. https://doi.org/10.1080/19490976.2023.2284247.
14. Colpitts SL, Kasper EJ, Keever A, Liljenberg C, Kirby T, Magori K, Kasper LH, Ochoa-Repáraz J. A bidirectional association between the gut microbiota and CNS disease in a biphasic murine model of multiple sclerosis. Gut Microbes. 2017;8(6):561–73. https://doi.org/10.1080/19490976.2017.1353843.
15. Cristofori F, Dargenio VN, Dargenio C, Miniello VL, Barone M, Francavilla R. Anti-inflammatory and immunomodulatory effects of probiotics in gut inflammation: A door to the body. Front Immunol. 2021;12:578386. https://doi.org/10.3389/fimmu.2021.578386.
16. Cuervo-Zanatta D, Syeda T, Sánchez-Valle V, Irene-Fierro M, Torres-Aguilar P, Torres-Ramos MA, Shibayama-Salas M, Silva-Olivares A, Noriega LG, Torres N, Tovar AR, Ruminot I, Barros LF, García-Mena J, Perez-Cruz C. Dietary Fiber modulates the release of gut bacterial products preventing cognitive decline in an Alzheimer's mouse model. Cell Mol Neurobiol. 2023;43(4):1595–618. https://doi.org/10.1007/s10571-022-01268-7.
17. Dabke K, Hendrick G, Devkota S. The gut microbiome and metabolic syndrome. J Clin Invest. 2019;129(10):4050–7. https://doi.org/10.1172/JCI129194.
18. Denman CR, Park SM, Jo J. Gut-brain axis: gut dysbiosis and psychiatric disorders in Alzheimer's and Parkinson's disease. Front Neurosci. 2023;17:1268419. https://doi.org/10.3389/fnins.2023.1268419.
19. Duan Y, Young R, Schnabl B. Bacteriophages and their potential for treatment of gastrointestinal diseases. Nat Rev Gastroenterol Hepatol. 2022;19(2):135–44. https://doi.org/10.1038/s41575-021-00536-z.
20. Eddin AN, Alfuwais M, Eddin RN, Alkattan K, Yaqinuddin A. Gut-modulating agents and amyotrophic lateral sclerosis: current evidence and future perspectives. Nutrients. 2024;16(5):590. https://doi.org/10.3390/nu16050590.
21. Elangovan S, Borody TJ, Holsinger RMD. Fecal microbiota transplantation reduces pathology and improves cognition in a mouse model of Alzheimer's disease. Cells. 2022;12(1):119. https://doi.org/10.3390/cells12010119.
22. Fakhri S, Yarmohammadi A, Yarmohammadi M, Farzaei MH, Echeverria J. Marine natural products: promising candidates in the modulation of gut-brain Axis towards neuroprotection. Mar Drugs. 2021;19(3):165. https://doi.org/10.3390/md19030165.
23. Fang H, Fang M, Wang Y, Zhang H, Li J, Chen J, Wu Q, He L, Xu J, Deng J, Liu M, Deng Y, Chen C. Indole-3-propionic acid as a potential therapeutic agent for sepsis-induced gut microbiota disturbance. Microbiol Spectrum. 2022;10(3):e0012522. https://doi.org/10.1128/spectrum.00125-22.
24. Ferreiro AL, Choi J, Ryou J, Newcomer EP, Thompson R, Bollinger RM, Hall-Moore C, Ndao IM, Sax L, Benzinger TLS, Stark SL, Holtzman DM, Fagan AM, Schindler SE, Cruchaga C, Butt OH, Morris JC, Tarr PI, Ances BM, Dantas G. Gut microbiome composition may be an indicator of preclinical Alzheimer's disease. Sci Transl Med. 2023;15(700):eabo2984. https://doi.org/10.1126/scitranslmed.abo2984.
25. García-Montero C, Fraile-Martínez O, Gómez-Lahoz AM, Pekarek L, Castellanos AJ, Noguerales-Fraguas F, Coca S, Guijarro LG, García-Honduvilla N, Asúnsolo A, Sanchez-Trujillo L, Lahera G, Bujan J, Monserrat J, Álvarez-Mon M, Álvarez-Mon MA, Ortega MA. Nutritional components in Western diet versus Mediterranean diet at the gut microbiota–immune system interplay. Implications for health and disease. Nutrients. 2021;13(2):699. https://doi.org/10.3390/nu13020699.
26. Glans I, Sonestedt E, Nägga K, Gustavsson A-M, González-Padilla E, Borne Y, Stomrud E, Melander O, Nilsson PM, Palmqvist S, Hansson O. Association between dietary habits in midlife with dementia incidence over a 20-year period. Neurology. 2023;100(1):e28–37. https://doi.org/10.1212/WNL.0000000000201336.

27. Gong X, Shi L, Wu Y, Luo Y, Kwok T. B vitamin supplementation slows cognitive decline in mild cognitive impairment patients with frontal lobe atrophy. J Alzheimer's Dis. 2022;89(4):1453–61. https://doi.org/10.3233/JAD-220685.
28. He X, Yan C, Zhao S, Zhao Y, Huang R, Li Y. The preventive effects of probiotic *Akkermansia muciniphila* on D-galactose/AlCl3 mediated Alzheimer's disease-like rats. Exp Gerontol. 2022;170:111959. https://doi.org/10.1016/j.exger.2022.111959.
29. Hill C, Guarner F, Reid G, Gibson GR, Merenstein DJ, Pot B, Morelli L, Canani RB, Flint HJ, Salminen S, Calder PC, Sanders ME. The International Scientific Association for Probiotics and Prebiotics consensus statement on the scope and appropriate use of the term probiotic. Nat Rev Gastroenterol Hepatol. 2014;11(8):506–14. https://doi.org/10.1038/nrgastro.2014.66.
30. Hong D, Zhang C, Wu W, Lu X, Zhang L. Modulation of the gut–brain axis via the gut microbiota: a new era in treatment of amyotrophic lateral sclerosis. Front Neurol. 2023;14:1133546. https://doi.org/10.3389/fneur.2023.1133546.
31. Jankovic J. Current approaches to the treatment of Parkinson's disease. Neuropsychiatr Dis Treat. 2008;4:743. https://doi.org/10.2147/NDT.S2006.
32. Jiang Y, Li K, Li X, Xu L, Yang Z. Sodium butyrate ameliorates the impairment of synaptic plasticity by inhibiting the neuroinflammation in 5XFAD mice. Chem Biol Interact. 2021;341:109452. https://doi.org/10.1016/j.cbi.2021.109452.
33. Jiang Q, Lin L, Xie F, Jin W, Zhu W, Wang M, Qiu Q, Li Z, Liu J, Mao S. Metagenomic insights into the microbe-mediated B and K2 vitamin biosynthesis in the gastrointestinal microbiome of ruminants. Microbiome. 2022;10(1):109. https://doi.org/10.1186/s40168-022-01298-9.
34. Jiang X, Zheng Y, Sun H, Dang Y, Yin M, Xiao M, Wu T. Fecal microbiota transplantation improves cognitive function of a mouse model of Alzheimer's disease. CNS Neurosci Ther. 2025;31(2):e70259. https://doi.org/10.1111/cns.70259.
35. Kang JW, Zivkovic AM. The potential utility of prebiotics to modulate Alzheimer's disease: A review of the evidence. Microorganisms. 2021;9(11):2310. https://doi.org/10.3390/microorganisms9112310.
36. Keshavarzian A, Green SJ, Engen PA, Voigt RM, Naqib A, Forsyth CB, Mutlu E, Shannon KM. Colonic bacterial composition in Parkinson's disease. Mov Disord. 2015;30(10):1351–60. https://doi.org/10.1002/mds.26307.
37. Koutzoumis DN, Vergara M, Pino J, Buddendorff J, Khoshbouei H, Mandel RJ, Torres GE. Alterations of the gut microbiota with antibiotics protects dopamine neuron loss and improve motor deficits in a pharmacological rodent model of Parkinson's disease. Exp Neurol. 2020;325:113159. https://doi.org/10.1016/j.expneurol.2019.113159.
38. Kwon D, Zhang K, Paul KC, Folle AD, Del Rosario I, Jacobs JP, Keener AM, Bronstein JM, Ritz B. Diet and the gut microbiome in patients with Parkinson's disease. npj Parkinsons Disease. 2024;10(1):89. https://doi.org/10.1038/s41531-024-00681-7.
39. Lang W, Li X, Wang Y, Duan Y, Wang Y, Wei P, Li X, Zhu Q, Cui Y. Sodium propionate improves cognitive and memory function in mouse models of Alzheimer's disease. Neurosci Lett. 2022;791:136887. https://doi.org/10.1016/j.neulet.2022.136887.
40. Leclercq S, Mian FM, Stanisz AM, Bindels LB, Cambier E, Ben-Amram H, Koren O, Forsythe P, Bienenstock J. Low-dose penicillin in early life induces long-term changes in murine gut microbiota, brain cytokines and behavior. Nat Commun. 2017;8(1):15062. https://doi.org/10.1038/ncomms15062.
41. Li C, Niu Z, Zou M, Liu S, Wang M, Gu X, Lu H, Tian H, Jha R. Probiotics, prebiotics, and synbiotics regulate the intestinal microbiota differentially and restore the relative abundance of specific gut microorganisms. J Dairy Sci. 2020;103(7):5816–29. https://doi.org/10.3168/jds.2019-18003.
42. Li S, Zhao L, Xiao J, Guo Y, Fu R, Zhang Y, Xu S. The gut microbiome: an important role in neurodegenerative diseases and their therapeutic advances. Mol Cell Biochem. 2024;479(9):2217–43. https://doi.org/10.1007/s11010-023-04853-6.

43. Ma Y-Y, Li X, Yu J-T, Wang Y-J. Therapeutics for neurodegenerative diseases by targeting the gut microbiome: from bench to bedside. Transl Neurodegener. 2024;13(1):12. https://doi.org/10.1186/s40035-024-00404-1.
44. MahmoudianDehkordi S, Arnold M, Nho K, Ahmad S, Jia W, Xie G, Louie G, Kueider-Paisley A, Moseley MA, Thompson JW, Williams LSJ, Tenenbaum JD, Blach C, Baillie R, Han X, Bhattacharyya S, Toledo JB, Schafferer S, Klein S, et al. Altered bile acid profile associates with cognitive impairment in Alzheimer's disease—an emerging role for gut microbiome. Alzheimers Dement. 2019;15(1):76–92. https://doi.org/10.1016/j.jalz.2018.07.217.
45. Markowiak P, Śliżewska K. Effects of probiotics, prebiotics, and synbiotics on human health. Nutrients. 2017;9(9):1021. https://doi.org/10.3390/nu9091021.
46. Medina L, González-Lizárraga F, Dominguez-Meijide A, Ploper D, Parrales V, Sequeira S, Cima-Omori M-S, Zweckstetter M, Del Bel E, Michel PP, Outeiro TF, Raisman-Vozari R, Chehín R, Socias SB. Doxycycline interferes with tau aggregation and reduces its neuronal toxicity. Front Aging Neurosci. 2021;13:635760. https://doi.org/10.3389/fnagi.2021.635760.
47. Metcalfe-Roach A, Yu AC, Golz E, Cirstea M, Sundvick K, Kliger D, Foulger LH, Mackenzie M, Finlay BB, Appel-Cresswell S. <scp>MIND</scp> and Mediterranean diets associated with later onset of Parkinson's disease. Mov Disord. 2021;36(4):977–84. https://doi.org/10.1002/mds.28464.
48. Minter MR, Zhang C, Leone V, Ringus DL, Zhang X, Oyler-Castrillo P, Musch MW, Liao F, Ward JF, Holtzman DM, Chang EB, Tanzi RE, Sisodia SS. Antibiotic-induced perturbations in gut microbial diversity influences neuro-inflammation and amyloidosis in a murine model of Alzheimer's disease. Sci Rep. 2016;6(1):30028. https://doi.org/10.1038/srep30028.
49. Molsberry S, Bjornevik K, Hughes KC, Healy B, Schwarzschild M, Ascherio A. Diet pattern and prodromal features of Parkinson disease. Neurology. 2020;95(15):e2095–108. https://doi.org/10.1212/WNL.0000000000010523.
50. Moustafa B, Trifan G, Isasi CR, Lipton RB, Sotres-Alvarez D, Cai J, Tarraf W, Stickel A, Mattei J, Talavera GA, Daviglus ML, González HM, Testai FD. Association of Mediterranean Diet with cognitive decline among Diverse Hispanic or Latino adults from the Hispanic Community Health Study/Study of Latinos. JAMA Netw Open. 2022;5(7):e2221982. https://doi.org/10.1001/jamanetworkopen.2022.21982.
51. Nemati MH, Yazdanpanah E, Kazemi R, Orooji N, Dadfar S, Oksenych V, Haghmorad D. Microbiota-driven mechanisms in multiple sclerosis: pathogenesis, therapeutic strategies, and biomarker potential. Biology. 2025;14(4):435. https://doi.org/10.3390/biology14040435.
52. Neuffer J, González-Domínguez R, Lefèvre-Arbogast S, Low DY, Driollet B, Helmer C, Du Preez A, de Lucia C, Ruigrok SR, Altendorfer B, Aigner L, Lucassen PJ, Korosi A, Thuret S, Manach C, Pallàs M, Urpi-Sardà M, Sánchez-Pla A, Andres-Lacueva C, Samieri C. Exploration of the gut–brain axis through metabolomics identifies serum propionic acid associated with higher cognitive decline in older persons. Nutrients. 2022;14(21):4688. https://doi.org/10.3390/nu14214688.
53. Nishiwaki H, Ito M, Hamaguchi T, Maeda T, Kashihara K, Tsuboi Y, Ueyama J, Yoshida T, Hanada H, Takeuchi I, Katsuno M, Hirayama M, Ohno K. Short chain fatty acids-producing and mucin-degrading intestinal bacteria predict the progression of early Parkinson's disease. npj Parkinsons Disease. 2022;8(1):65. https://doi.org/10.1038/s41531-022-00328-5.
54. Obrenovich M, Jaworski H, Tadimalla T, Mistry A, Sykes L, Perry G, Bonomo R. The role of the microbiota–gut–brain axis and antibiotics in ALS and neurodegenerative diseases. Microorganisms. 2020;8(5):784. https://doi.org/10.3390/microorganisms8050784.
55. Olson CA, Vuong HE, Yano JM, Liang QY, Nusbaum DJ, Hsiao EY. The gut microbiota mediates the anti-seizure effects of the ketogenic diet. Cell. 2018;173(7):1728–1741.e13. https://doi.org/10.1016/j.cell.2018.04.027.
56. Ou Z, Deng L, Lu Z, Wu F, Liu W, Huang D, Peng Y. Protective effects of *Akkermansia muciniphila* on cognitive deficits and amyloid pathology in a mouse model of Alzheimer's disease. Nutr Diabetes. 2020;10(1):12. https://doi.org/10.1038/s41387-020-0115-8.
57. Paknahad Z, Sheklabadi E, Derakhshan Y, Bagherniya M, Chitsaz A. The effect of the Mediterranean diet on cognitive function in patients with Parkinson's disease: a randomized

clinical controlled trial. Complement Ther Med. 2020;50:102366. https://doi.org/10.1016/j.ctim.2020.102366.
58. Park S-H, Lee J-H, Kim J-S, Kim TJ, Shin J, Im JH, Cha B, Lee S, Kwon KS, Shin YW, Ko S-B, Choi SH. Fecal microbiota transplantation can improve cognition in patients with cognitive decline and Clostridioides difficile infection. Aging. 2022;14(16):6449–66. https://doi.org/10.18632/aging.204230
59. Phillips MCL, Deprez LM, Mortimer GMN, Murtagh DKJ, McCoy S, Mylchreest R, Gilbertson LJ, Clark KM, Simpson PV, McManus EJ, Oh J-E, Yadavaraj S, King VM, Pillai A, Romero-Ferrando B, Brinkhuis M, Copeland BM, Samad S, Liao S, Schepel JAC. Randomized crossover trial of a modified ketogenic diet in Alzheimer's disease. Alzheimers Res Ther. 2021;13(1):51. https://doi.org/10.1186/s13195-021-00783-x.
60. Poewe W, Seppi K, Tanner CM, Halliday GM, Brundin P, Volkmann J, Schrag A-E, Lang AE. Parkinson disease. Nat Rev Dis Primers. 2017;3(1):17013. https://doi.org/10.1038/nrdp.2017.13.
61. Qiao C-M, Sun M-F, Jia X-B, Li Y, Zhang B-P, Zhao L-P, Shi Y, Zhou Z L, Zhu Y-L, Cui C, Shen Y-Q. Sodium butyrate exacerbates Parkinson's disease by aggravating neuroinflammation and colonic inflammation in MPTP-induced mice model. Neurochem Res. 2020;45(9):2128–42. https://doi.org/10.1007/s11064-020-03074-3.
62. Quan W, Qiao C-M, Niu G-Y, Wu J, Zhao L-P, Cui C, Zhao W-J, Shen Y-Q. Trimethylamine N-oxide exacerbates neuroinflammation and motor dysfunction in an acute MPTP mice model of Parkinson's disease. Brain Sci. 2023;13(5):790. https://doi.org/10.3390/brainsci13050790.
63. Ritz NL, Brocka M, Butler MI, Cowan CSM, Barrera-Bugueño C, Turkington CJR, Draper LA, Bastiaanssen TFS, Turpin V, Morales L, Campos D, Gheorghe CE, Ratsika A, Sharma V, Golubeva AV, Aburto MR, Shkoporov AN, Moloney GM, Hill C, et al. Social anxiety disorder-associated gut microbiota increases social fear. Proc Natl Acad Sci. 2024;121(1):e2308706120. https://doi.org/10.1073/pnas.2308706120.
64. Rob M, Yousef M, Lakshmanan AP, Mahboob A, Terranegra A, Chaari A. Microbial signatures and therapeutic strategies in neurodegenerative diseases. Biomed Pharmacother. 2025;184:117905. https://doi.org/10.1016/j.biopha.2025.117905.
65. Ross FC, Mayer DE, Gupta A, Gill CIR, Del Rio D, Cryan JF, Lavelle A, Ross RP, Stanton C, Mayer EA. Existing and future strategies to manipulate the gut microbiota with diet as a potential adjuvant treatment for psychiatric disorders. Biol Psychiatry. 2024;95(4):348–60. https://doi.org/10.1016/j.biopsych.2023.10.018.
66. Ruwatia R, Sarkar SR, Panda S, Banerjee S. Gut microbiome and neurodegenerative disorders: therapeutic implications. Aging Pathobiol Ther. 2025;7(1):05–24. https://doi.org/10.31491/APT.2025.03.164
67. Sekikawa A, Wharton W, Butts B, Veliky CV, Garfein J, Li J, Goon S, Fort A, Li M, Hughes TM. Potential protective mechanisms of S-equol, a metabolite of soy isoflavone by the gut microbiome, on cognitive decline and dementia. Int J Mol Sci. 2022;23(19):11921. https://doi.org/10.3390/ijms231911921.
68. Shandilya S, Kumar S, Jha NK, Kesari KK, Ruokolainen J. Interplay of gut microbiota and oxidative stress: perspective on neurodegeneration and neuroprotection. J Adv Res. 2022;38:223–44. https://doi.org/10.1016/j.jare.2021.09.005.
69. Sochocka M, Donskow-Łysoniewska K, Diniz BS, Kurpas D, Brzozowska E, Leszek J. The gut microbiome alterations and inflammation-driven pathogenesis of Alzheimer's disease—a critical review. Mol Neurobiol. 2019;56(3):1841–51. https://doi.org/10.1007/s12035-018-1188-4.
70. Strandwitz P. Neurotransmitter modulation by the gut microbiota. Brain Res. 2018;1693:128–33. https://doi.org/10.1016/j.brainres.2018.03.015.
71. Sun J, Xu J, Ling Y, Wang F, Gong T, Yang C, Ye S, Ye K, Wei D, Song Z, Chen D, Liu J. Fecal microbiota transplantation alleviated Alzheimer's disease-like pathogenesis in APP/PS1 transgenic mice. Transl Psychiatry. 2019;9(1):189. https://doi.org/10.1038/s41398-019-0525-3.
72. Sun J, Xu J, Yang B, Chen K, Kong Y, Fang N, Gong T, Wang F, Ling Z, Liu J. Effect of *Clostridium butyricum* against microglia-mediated neuroinflammation in Alzheimer's

disease via regulating gut microbiota and metabolites butyrate. Mol Nutr Food Res. 2020;64(2):1900636. https://doi.org/10.1002/mnfr.201900636.
73. Sun J, Huang T, Debelius JW, Fang F. Gut microbiome and amyotrophic lateral sclerosis: A systematic review of current evidence. J Intern Med. 2021;290(4):758–88. https://doi.org/10.1111/joim.13336.
74. Tan AH, Lim SY, Lang AE. The microbiome–gut–brain axis in Parkinson disease—from basic research to the clinic. Nat Rev Neurol. 2022;18(8):476–95. https://doi.org/10.1038/s41582-022-00681-2.
75. Teng Y, Mu J, Xu F, Zhang X, Sriwastva MK, Liu QM, Li X, Lei C, Sundaram K, Hu X, Zhang L, Park JW, Hwang JY, Rouchka EC, Zhang X, Yan J, Merchant ML, Zhang H-G. Gut bacterial isoamylamine promotes age-related cognitive dysfunction by promoting microglial cell death. Cell Host Microbe. 2022;30(7):944–960.e8. https://doi.org/10.1016/j.chom.2022.05.005.
76. Thiel A, Hermanns C, Lauer AA, Reichrath J, Erhardt T, Hartmann T, Grimm MOW, Grimm HS. Vitamin D and its analogues: from differences in molecular mechanisms to potential benefits of adapted use in the treatment of Alzheimer's disease. Nutrients. 2023;15(7):1684. https://doi.org/10.3390/nu15071684.
77. Thomas RL, Jiang L, Adams JS, Xu ZZ, Shen J, Janssen S, Ackermann G, Vanderschueren D, Pauwels S, Knight R, Orwoll ES, Kado DM. Vitamin D metabolites and the gut microbiome in older men. Nat Commun. 2020;11(1):5997. https://doi.org/10.1038/s41467-020-19793-8.
78. Tran SM-S, Mohajeri MH. The role of gut bacterial metabolites in brain development, aging and disease. Nutrients. 2021;13(3):732. https://doi.org/10.3390/nu13030732.
79. Twort FW. An investigation on the nature of ultra-microscopic viruses. Lancet. 1915;186(4814):1241–3. https://doi.org/10.1016/S0140-6736(01)20383-3.
80. Ullah H, Tovchiga O, Daglia M, Khan H. Modulating gut microbiota: an emerging approach in the prevention and treatment of multiple sclerosis. Curr Neuropharmacol. 2021;19(11):1966–83. https://doi.org/10.2174/1570159X19666210217084827.
81. Umeda T, Hatanaka Y, Sakai A, Tomiyama T. Nasal rifampicin improves cognition in a mouse model of dementia with Lewy bodies by reducing α-synuclein oligomers. Int J Mol Sci. 2021;22(16):8453. https://doi.org/10.3390/ijms22168453.
82. Vogt NM, Kerby RL, Dill-McFarland KA, Harding SJ, Merluzzi AP, Johnson SC, Carlsson CM, Asthana S, Zetterberg H, Blennow K, Bendlin BB, Rey FE. Gut microbiome alterations in Alzheimer's disease. Sci Rep. 2017;7(1):13537. https://doi.org/10.1038/s41598-017-13601-y.
83. Wang C, Zheng D, Weng F, Jin Y, He L. Sodium butyrate ameliorates the cognitive impairment of Alzheimer's disease by regulating the metabolism of astrocytes. Psychopharmacology. 2022;239(1):215–27. https://doi.org/10.1007/s00213-021-06025-0.
84. Wu L-H, Huang B-R, Lai S-W, Lin C, Lin H-Y, Yang L-Y, Lu D-Y. SIRT1 activation by minocycline on regulation of microglial polarization homeostasis. Aging. 2020;12(18):17990–8007. https://doi.org/10.18632/aging.103542
85. Xue L-J, Yang X-Z, Tong Q, Shen P, Ma S-J, Wu S-N, Zheng J-L, Wang H-G. Fecal microbiota transplantation therapy for Parkinson's disease. Medicine. 2020;99(35):e22035. https://doi.org/10.1097/MD.0000000000022035.
86. Yang X, Yu D, Xue L, Li H, Du J. Probiotics modulate the microbiota–gut–brain axis and improve memory deficits in aged SAMP8 mice. Acta Pharm Sin B. 2020;10(3):475–87. https://doi.org/10.1016/j.apsb.2019.07.001.
87. Ylilauri MPT, Voutilainen S, Lönnroos E, Virtanen HEK, Tuomainen T-P, Salonen JT, Virtanen JK. Associations of dietary choline intake with risk of incident dementia and with cognitive performance: the Kuopio Ischaemic Heart Disease Risk Factor Study. Am J Clin Nutr. 2019;110(6):1416–23. https://doi.org/10.1093/ajcn/nqz148.
88. Zeng Q, Shen J, Chen K, Zhou J, Liao Q, Lu K, Yuan J, Bi F. The alteration of gut microbiome and metabolism in amyotrophic lateral sclerosis patients. Sci Rep. 2020;10(1):12998. https://doi.org/10.1038/s41598-020-69845-8.
89. Zhang Y, Wu S, Yi J, Xia Y, Jin D, Zhou J, Sun J. Target intestinal microbiota to alleviate disease progression in amyotrophic lateral sclerosis. Clin Ther. 2017;39(2):322–36. https://doi.org/10.1016/j.clinthera.2016.12.014.

90. Zhang C, Wang H, Chen T. Interactions between intestinal microflora/probiotics and the immune system. Biomed Res Int. 2019;2019:1–8. https://doi.org/10.1155/2019/6764919.
91. Zhang K, Paul KC, Jacobs JP, Chou H-C, Duarte Folle A, Del Rosario I, Yu Y, Bronstein JM, Keener AM, Ritz B. Parkinson's disease and the gut microbiome in rural California. J Parkinsons Dis. 2022;12(8):2441–52. https://doi.org/10.3233/JPD-223500.
92. Zhang X, Luo X, Tian L, Yue P, Li M, Liu K, Zhu D, Huang C, Shi Q, Yang L, Xia Z, Zhao J, Ma Z, Li J, Leung JW, Lin Y, Yuan J, Meng W, Li X, Chen Y. The gut microbiome dysbiosis and regulation by fecal microbiota transplantation: umbrella review. Front Microbiol. 2023;14:1286429. https://doi.org/10.3389/fmicb.2023.1286429.
93. Zhang X, Tang B, Guo J. Parkinson's disease and gut microbiota: from clinical to mechanistic and therapeutic studies. Transl Neurodegener. 2023;12(1):59. https://doi.org/10.1186/s40035-023-00392-8.
94. Zhao Z, Ning J, Bao X, Shang M, Ma J, Li G, Zhang D. Fecal microbiota transplantation protects rotenone-induced Parkinson's disease mice via suppressing inflammation mediated by the lipopolysaccharide-TLR4 signaling pathway through the microbiota-gut-brain axis. Microbiome. 2021;9(1):226. https://doi.org/10.1186/s40168-021-01107-9.
95. Zheng J, Hoffman KL, Chen J-S, Shivappa N, Sood A, Browman GJ, Dirba DD, Hanash S, Wei P, Hebert JR, Petrosino JF, Schembre SM, Daniel CR. Dietary inflammatory potential in relation to the gut microbiome: results from a cross-sectional study. Br J Nutr. 2020;124(9):931–42. https://doi.org/10.1017/S0007114520001853.
96. Zhu X, Li B, Lou P, Dai T, Chen Y, Zhuge A, Yuan Y, Li L. The relationship between the gut microbiome and neurodegenerative diseases. Neurosci Bull. 2021;37(10):1510–22. https://doi.org/10.1007/s12264-021-00730-8.

9 The Role of Inflammation in the Brain: A Gut-Driven Phenomenon

Nasir Uddin Mahbub, Md Minarul Islam, and Seong Tshool Hong

9.1 Introduction

The human brain, long regarded as an isolated command center, is now increasingly recognized as part of a complex, dynamic network of interorgan communication—most notably with the gut [1–4]. The concept of the gut-brain axis has revolutionized our understanding of brain function and dysfunction, providing a unifying framework that bridges neuroscience, immunology, microbiology, and endocrinology [2–5]. In recent years, inflammation in the brain has emerged as a central feature of many neurological and neuropsychiatric disorders [6]. Traditionally, neuroinflammation was considered a localized response to injury or infection [7, 8]. However, accumulating evidence now suggests that the origins of neuroinflammation often stem from the gut, where microbial imbalances, or gut dysbiosis, can trigger systemic immune responses that extend to and influence the CNS [9–13]. This has profound implications for how we understand, diagnose, and treat brain-related conditions.

The vagus nerve serves as a critical bidirectional conduit within the gut-brain axis, facilitating the transmission of neurophysiological signals between the gut and the CNS (gut-brain axis) [14–18]. Moreover, metabolites produced by microbial

N. U. Mahbub · S. T. Hong (✉)
Department of Biomedical Sciences and Institute for Medical Science, Jeonbuk National University Medical School, Jeonju, Republic of Korea
e-mail: nasir.mahbub@jbnu.ac.kr; seonghong@jbnu.ac.kr

M. M. Islam (✉)
Department of Food Science and Biotechnology, Gachon University, Seongnam, Gyeonggi-do, Republic of Korea
e-mail: minarul2026@gachon.ac.kr

M. H. Rahman, Muhasina K M (eds.), *The Microbiome's Impact on the Brain*, https://doi.org/10.1007/978-981-95-7494-0_9

metabolites—particularly short-chain fatty acids (SCFAs), neurotransmitter precursors, and other bioactive compounds—play a pivotal role in modulating immune homeostasis, maintaining the integrity of the blood-brain barrier (BBB), and regulating neuronal function [19–25]. Perturbations in these microbiota-derived signaling pathways have been implicated in the initiation and perpetuation of chronic inflammation, thereby contributing to the pathogenesis of various CNS disorders [26–28].

Consider the case of Sarah, a 45-year-old educator who has been previously recognized for her exceptional memory and cognitive agility. Over the span of a year, she began experiencing subtle yet progressively debilitating cognitive symptoms, including episodic forgetfulness—such as difficulty recalling students' names—diminished concentration during instructional periods, and heightened levels of anxiety. These impairments began to significantly interfere with both her professional responsibilities and personal well-being, ultimately motivating her to seek medical evaluation. Despite undergoing comprehensive neurological assessments, no definitive etiology for her cognitive decline was identified.

In pursuit of alternative explanations, Sarah consulted an integrative medicine specialist who posed a seemingly unrelated but insightful question: "How is your gastrointestinal health?" She reported frequent episodes of bloating, irregular bowel habits, and intermittent abdominal discomfort—symptoms she had historically regarded as inconsequential. The clinician proposed that a dysbiotic gut microbiota could be contributing to her neurocognitive symptoms, initiating a pivotal shift in her clinical trajectory [29]. Through the implementation of targeted nutritional interventions, including the integration of anti-inflammatory foods and probiotic supplementation, Sarah experienced marked improvements in both gastrointestinal function and cognitive clarity. Her case underscores a growing recognition of the gut-brain axis as a critical determinant of neurological health [30–32].

Sarah's story is not unique. The connection between the global population experiencing neuropsychiatric and neurodegenerative symptoms and the role of chronic, low-grade systemic inflammation and neuroinflammation lies in the gut-brain axis, where gastrointestinal dysfunction serves as a key driver of inflammation-mediated neurological disorders [33]. However, when inflammation becomes chronic—persisting long after its initial trigger—it can become destructive [34–36]. The gut, home to approximately 100 trillion microbes collectively known as the microbiome, is a major source of this chronic inflammation [37, 38]. Factors such as poor diet, chronic stress, or antibiotic overuse can disrupt the balance of gut microbes [39–42]. This imbalance can lead to a "leaky gut," allowing inflammatory molecules like cytokines to enter the bloodstream and reach the brain [33, 43, 44]. Once in the brain, these molecules can activate immune cells such as microglia and astrocytes, triggering neuroinflammation that disrupts neuronal function and contributes to disorders like Alzheimer's, Parkinson's, depression, and anxiety [45–49]. Traditional treatments for these conditions often focus solely on the brain, targeting symptoms with medications or therapies that may not address underlying causes. However, emerging research highlights the gut-brain axis as a critical pathway, suggesting that gut-driven inflammation may precede and exacerbate neurological symptoms [9, 13, 50–53].

9.2 The Gut-Brain Axis as a Central Node in Health and Disease

The human body is a marvel of interconnected systems, and one of the most fascinating connections is the gut-brain axis—a dynamic, bidirectional communication network linking the gastrointestinal tract to the brain [50, 54–57]. This axis is not just a passive conduit but an active player in our physical and mental well-being, influencing everything from mood to memory [58, 59]. The gut-brain axis is a sophisticated communication system involving multiple pathways: neural, hormonal, immune, and metabolic [60, 61]. Each component plays a distinct role in ensuring the gut and brain work in harmony [60]. At its core, the gut-brain axis involves a complex interplay of neural pathways, microbial activity, and immune signaling, with inflammation emerging as a critical factor when this system goes awry [2, 52, 62, 63]. In this section, we unravel the anatomy of the gut-brain axis, explore the gut microbiome's role in producing key neurotransmitters and metabolites, and examine how an imbalanced gut can trigger inflammation that affects the brain.

The vagus nerve represents a pivotal element of the gut-brain axis, serving as a principal bidirectional communication pathway linking the gastrointestinal tract with the CNS [15, 64, 65]. It is a mixed nerve, with about 80% of its fibers being afferent (sensory, carrying signals from the gut to the brain) and 20% efferent (motor, carrying signals from the brain to the gut) [15, 64]. This means most vagal communication is gut to brain, allowing the CNS to constantly monitor the internal state of the GI tract. Increasingly, the vagus nerve is recognized for its pivotal role in neuroimmune interactions and its capacity to influence central inflammatory processes [66–68]. Approximately 80% of vagal fibers are afferent, transmitting diverse sensory inputs—including mechanical stimuli, nutrient-derived chemical signals, and immunological cues—to the nucleus tractus solitarius (NTS) in the brain stem [14, 69, 70]. The vagus nerve senses changes in the gut environment—such as the presence of microbial metabolites, hormones, and inflammatory signals—and transmits this information to the CNS, influencing mood, stress responses, and even cognitive functions [64, 65, 71]. In turn, the CNS can adjust gut function via vagal efferents, affecting motility, secretion, and immune activity. This bidirectional loop ensures that psychological states (e.g., stress, anxiety) can directly impact gut physiology, and vice versa [15]. The vagus nerve is activated by neurotransmitters and metabolites produced by gut microbiota, and its integrity is necessary for microbiota-induced changes in brain function and behavior [71, 72]. Disruption of the vagus (e.g., by vagotomy) can block the effects of gut microbiota on brain plasticity and behavior [71]. The vagus nerve mediates a cholinergic anti-inflammatory pathway, dampening peripheral inflammation and reducing intestinal permeability, which can influence microbiota composition and systemic immune responses [64,

72]. Chronic stress suppresses vagal tone, thereby amplifying inflammatory responses and facilitating the development of gastrointestinal pathologies such as irritable bowel syndrome (IBS) and inflammatory bowel disease (IBD) [64, 73]. Low vagal tone is associated with increased peripheral inflammation and worse outcomes in these conditions. Consequently, the vagus nerve is a critical conduit in the gut-brain axis, enabling continuous, bidirectional communication that integrates neural, microbial, and immune signals. Its function is essential for maintaining gastrointestinal and mental health, and its modulation offers promising therapeutic avenues for a range of disorders [64].

The gut houses approximately 70% of the body's immune system, primarily in the gut-associated lymphoid tissue (GALT) [74–77]. Immune cells in the gut produce cytokines—small proteins that regulate inflammation—and these molecules can travel to the brain, influencing its immune environment [62, 77, 78]. When the gut microbiome is healthy, it promotes anti-inflammatory signals [79–83]. However, dysbiosis can trigger the release of proinflammatory cytokines, including tumor necrosis factor-alpha (TNF-α) and interleukin-6 (IL-6), which cross the BBB and activate microglia, the brain's immune cells, leading to neuroinflammation [84–86].

The gut microbiota profoundly influences the endocrine system by regulating the release of gut-derived hormones such as ghrelin, leptin, and peptide YY (PYY), which can affect mood, cognition, and appetite [87, 88]. Additionally, the hypothalamic-pituitary-adrenal (HPA) axis connects stress responses with gut function, establishing another feedback loop that links psychological states with gastrointestinal health [89, 90]. When the brain perceives a stressor, the hypothalamus initiates the stress response by secreting corticotropin-releasing hormone (CRH), which subsequently stimulates the anterior pituitary gland to release adrenocorticotropic hormone (ACTH). ACTH then acts on the adrenal cortex to promote the synthesis and release of cortisol, the principal glucocorticoid involved in the stress response [91]. Cortisol exerts its effects via glucocorticoid receptors expressed on various cell types, including intestinal epithelial cells, immune cells, and enteroendocrine cells, thereby influencing gastrointestinal motility, secretory activity, and epithelial barrier integrity [89].

Prolonged activation of the hypothalamic-pituitary-adrenal (HPA) axis and sustained elevation of circulating cortisol levels are associated with increased intestinal permeability (commonly referred to as "leaky gut"), dysbiosis of the gut microbiota, and the initiation or exacerbation of mucosal and systemic inflammation [89, 90]. These perturbations may contribute to the pathogenesis of functional gastrointestinal disorders and chronic inflammatory conditions. Moreover, stress-induced changes in gut physiology can alter microbial composition, while microbial metabolites, neuroactive compounds, and signaling through the vagus nerve and enteric nervous system modulate central stress pathways, underscoring the bidirectional dynamics of the gut-brain axis. The ENS influences the HPA axis through the secretion of gut hormones, neuropeptides, and cytokines, while cortisol and other components of the HPA axis, in turn, modulate ENS function, impacting gastrointestinal motility and secretory responses. This reciprocal communication network suggests that dysregulation in either the central or the enteric component may propagate pathophysiological effects throughout the system, with implications for both mental and gastrointestinal health [89, 90].

Chronic hyperactivation of the HPA axis is implicated in the etiology of mood disorders such as depression and anxiety, partly through mechanisms involving increased intestinal permeability, heightened systemic inflammation, and disruptions in neurotransmitter biosynthesis (e.g., serotonin, a substantial proportion of which is synthesized in the gut) [90]. Dysbiosis and HPA axis dysregulation have also been observed in neurodevelopmental and neurodegenerative disorders, including ASD and PD, as well as in functional bowel disorders like IBS [89].

Under physiological conditions, the gut-brain-microbiota axis contributes to the maintenance of systemic homeostasis. However, chronic psychological stress can perturb this dynamic equilibrium, leading to persistent dysregulation of the HPA axis, compromised gut barrier function, microbial imbalance, and aberrant neuroimmune signaling—thereby perpetuating a pathological feedback loop with broad implications for both somatic and neuropsychiatric disease states [89, 90]. Understanding the gut-brain-microbiota axis offers promising avenues for therapeutic interventions aimed at restoring homeostasis and treating complex, multifactorial diseases.

9.3 Gut Dysbiosis and Its Role in Neuroinflammation

Gut dysbiosis has emerged as a pivotal contributor to systemic and neuroinflammatory processes through its modulation of the gut-brain axis [92, 93]. This disruption of microbial homeostasis perturbs the intestinal milieu, compromises epithelial barrier integrity, and initiates a cascade of immune responses that extend beyond the gastrointestinal tract to affect systemic physiology and CNS function [22, 94]. The disturbance of the gut-brain axis by dysbiosis contributes to neuroinflammation via several interconnected mechanisms.

One primary consequence of dysbiosis is increased intestinal permeability, commonly referred to as "leaky gut," which permits translocation of microbial products such as lipopolysaccharide (LPS), bacterial amyloids, and pro-inflammatory cytokines into the systemic circulation [44, 95, 96]. These circulating molecules can impair the structural and functional integrity of the BBB, a critical interface safeguarding the CNS from peripheral insults [97]. Enhanced BBB permeability allows these pro-inflammatory mediators to access the brain parenchyma, where they initiate and propagate inflammatory responses [98, 99], for instance, LPS interacts with Toll-like receptor 4 (TLR4) expressed on brain endothelial cells, promoting the release of cytokines and chemokines that further compromise BBB function and facilitate neuroinflammatory cascades [100].

Microglia, the innate immune cells resident in the CNS, are highly responsive to peripheral immune signals [45, 101, 102]. Microbial-derived metabolites, endotoxins, and circulating cytokines that breach a compromised BBB can activate microglia, inducing a shift from a homeostatic to a pro-inflammatory phenotype [103, 104]. Activated microglia secrete pro-inflammatory cytokines (e.g., interleukin-1β, tumor necrosis factor-α), chemokines, and reactive oxygen species, contributing to neuronal injury and chronic neuroinflammation [22, 105, 106]. Similarly, astrocytes—essential for neuronal support and BBB maintenance—can undergo reactive transformation in response to dysbiosis-associated signals, thereby exacerbating inflammatory signaling within the CNS [107, 108]. Persistent activation of glial cells is a defining feature of many neurodegenerative pathologies [109].

Systemic inflammation resulting from microbial imbalance (dysbiosis) in the gut leads to elevated levels of circulating cytokines, including tumor necrosis factor-alpha (TNF-α), interleukin-1 beta (IL-1β), and interleukin-6 (IL-6), which can impact the brain both directly and indirectly. Chronic systemic inflammation undermines the structural and functional integrity of the BBB, rendering it more permeable to cytokines, immune cells, and other pro-inflammatory mediators [10, 110, 111]. Endothelial cells in the BBB respond to these circulating cytokines by altering tight junction proteins, thereby increasing BBB "leakiness" [112]. Even without directly crossing the BBB, cytokines can transmit inflammatory signals to the brain via endothelial and neural routes, triggering a broader neuroinflammatory response [10, 111]. While cytokines play essential roles in maintaining neuronal and glial homeostasis through autocrine and paracrine signaling, elevated peripheral cytokine levels can overwhelm these regulatory mechanisms [113, 114]. This disruption leads to neuronal dysfunction and, in severe cases, cell death. Furthermore, elevated cytokines—particularly IL-1β and TNF-α—interfere with synaptic plasticity, disrupting both Hebbian and homeostatic mechanisms that underpin learning, memory, and cognitive function [113, 115]. These effects manifest as impaired long-term potentiation (LTP) and abnormal synaptic scaling, both of which are hallmarks of neurodegenerative processes [113]. Additionally, cytokine-induced activation of microglia results in excessive synaptic pruning, disrupting neural circuit connectivity and further contributing to cognitive deficits [113].

Moreover, certain commensal bacteria produce amyloidogenic proteins (e.g., curli fibers) and other neurotoxic metabolites capable of entering systemic circulation and accessing the CNS [116]. These microbial amyloids may act as nucleation sites for endogenous protein misfolding and aggregation—a central feature in the etiologies of Alzheimer's disease (AD) and Parkinson's disease (PD) [117, 118], for example, bacterial amyloids may potentiate the aggregation of amyloid-beta in AD or α-synuclein in PD [116, 117]. This "cross-seeding" effect has been demonstrated in mouse, rat, and *C. elegans* models, as well as in vitro [116].

Concurrently, dysbiosis leads to a reduction in the production of neuroprotective SCFAs, such as butyrate, which are known to support neuronal viability, facilitate amyloid clearance, and exert anti-inflammatory effects within the CNS. The resultant imbalance between neurotoxic and neuroprotective microbial metabolites further amplifies neuroinflammatory and neurodegenerative processes [118, 119]. Advancing our understanding of host-microbiota interactions is thus essential for developing novel strategies to prevent and treat CNS inflammatory and degenerative diseases.

9.4 Neurological Disorders Linked to Gut Microbiota-Induced Inflammation

The gut microbiota, a key regulator of host immunity, metabolism, and neural function, critically influences neurological health via the gut-brain axis. Dysbiosis compromises intestinal barrier integrity, drives systemic immune activation, and induces neuroinflammation, BBB disruption, and neurotransmitter imbalance. These processes contribute to neurological disorders such as AD, PD, multiple sclerosis (MS), autism spectrum disorder (ASD), depression, and anxiety. Despite clinical heterogeneity, these conditions converge on shared pathways of microbial dysregulation, immune perturbation, and neural dysfunction, highlighting inflammation as a central link between the gut microbiota and neurological pathology.

9.4.1 Alzheimer's Disease (AD)

AD is a progressive, irreversible neurodegenerative disorder and the most prevalent cause of dementia worldwide, with incidence strongly associated with aging [120–123]. Clinically, AD manifests as a gradual and relentless decline in cognitive domains such as episodic memory, executive function, visuospatial skills, and language, ultimately impairing daily living and resulting in complete dependency [124–126]. The neuropathological features of AD are multifactorial and include the accumulation of extracellular β-amyloid (Aβ) plaques and intracellular neurofibrillary tangles composed of hyperphosphorylated tau protein [127–134]. Aβ is generated via abnormal proteolytic cleavage of amyloid precursor protein (APP) mediated by β-secretase (BACE1) and γ-secretase [135–141]. The aggregation of Aβ into oligomers and fibrils is neurotoxic, impairing synaptic function, disrupting calcium homeostasis, and triggering a cascade of oxidative and inflammatory processes [132, 133, 142–146]. In AD, tau undergoes abnormal chemical modifications (notably hyperphosphorylation), which cause it to detach from microtubules and aggregate into neurofibrillary tangles within neurons. These tangles disrupt the internal structure and function of neurons, leading to cell damage, synaptic loss, and, ultimately, neuronal death [147]. Tau pathology arises when tau, a microtubule-associated protein, becomes hyperphosphorylated and dissociates from microtubules, leading to the formation of insoluble paired helical filaments that aggregate into tangles, destabilize cytoskeletal integrity, and promote neuronal dysfunction and apoptosis [148–153]. Tau protein dysfunction, tangles, and their spread underpin much of the neurodegeneration and cognitive impairment seen in AD and are a stronger predictor of clinical symptoms than amyloid-beta [147, 154]. Primed microglia release pro-inflammatory cytokines such as IL-1β, which can trigger tau protein hyperphosphorylation via pathways like p38-MAPK and GSK-3β, ultimately contributing to neurofibrillary tangle formation characteristic of AD and tauopathies [155–158]. This cycle can be self-perpetuating: increased inflammation exacerbates tau pathology, while tau pathology further activates microglia, locking the CNS in a state of ongoing neuroinflammation and neurodegeneration [156, 158, 159].

In addition to the classical amyloid and tau hypotheses, AD involves extensive neuroinflammation, mitochondrial dysfunction, impaired autophagy, and cerebrovascular abnormalities [142, 160–162]. Activated microglia and astrocytes in the AD brain release pro-inflammatory cytokines, including interleukin-1β (IL-1β), interleukin-6 (IL-6), and tumor necrosis factor-α (TNF-α), chemokines, and reactive oxygen species (ROS), establishing a chronic inflammatory milieu that exacerbates neuronal damage and promotes further Aβ and tau pathology [163–168]. Oxidative stress, a prominent feature in AD, leads to lipid peroxidation, DNA damage, and protein oxidation, which impair neuronal metabolism and synaptic plasticity [169–174]. Mitochondrial dysfunction contributes to bioenergetic failure, increased ROS generation, and release of pro-apoptotic factors [161, 175–177]. Synaptic degeneration—particularly in the hippocampus and cortex—is an early event in AD, closely correlating with cognitive decline and mediated by both Aβ toxicity and tau-induced axonal transport disruption [127, 142, 178, 179]. Furthermore, neurotransmitter systems are significantly affected: cholinergic deficits underlie memory impairments, glutamatergic dysregulation contributes to excitotoxicity, and impairments in serotonergic, dopaminergic, and GABAergic signaling contribute to behavioral and mood disturbances [180–185]. Emerging research highlights the crucial role of the gut-brain axis, particularly gut microbiota dysbiosis, in modulating key aspects of AD pathophysiology [94, 186, 187]. LPS has been shown to enhance amyloidogenesis by upregulating BACE1 expression and increasing tau phosphorylation via MAPK signaling pathways [188–190].

In AD, a reduction in SCFA-producing taxa such as *Faecalibacterium prausnitzii* and *Lactobacillus* is frequently observed, impairing the regulatory function of SCFAs and exacerbating neuroinflammation [81, 191, 192]. A study published in *Gut Microbes* found that patients with AD had lower levels of *Bifidobacterium*, which correlated with reduced SCFA production and heightened neuroinflammation [193–197]. Additionally, bacterial species such as *Escherichia coli*, *Bacillus subtilis*, *Salmonella*, and *Staphylococcus aureus* produce functional amyloids (e.g., curli fibers) that are structurally and immunologically similar to human Aβ peptides. These bacterial amyloids can potentially cross-seed with Aβ, promoting its aggregation and the formation of amyloid plaques in the brain—a hallmark of AD [198]. Pistollato et al. (2016) highlight that the gut microbiota can release significant amounts of amyloids and LPS, which may reach the brain through the gut-brain axis, contributing to neuroinflammatory and amyloidogenic pathways implicated in AD [199]. Seminal work by Harach et al. (2017) demonstrated that germ-free (GF) transgenic mouse models of AD show a dramatic reduction in cerebral Aβ (amyloid-beta) amyloid pathology compared to control AD mice with normal intestinal microbiota [200]. Moreover, the persistent secretion of IL-6 and other neurotoxic substances by activated microglia leads to an amplified neuroinflammatory response and increased oxidative damage, both of which are fundamental driving forces in the neurodegenerative process seen in AD [163, 201–203].

Animal studies show that AD mouse models raised in germ-free conditions or treated with antibiotics demonstrate a significant reduction in amyloid-β (Aβ) plaque formation and neuroinflammatory responses. This suggests a critical role of gut microbiota in AD pathogenesis, particularly concerning Aβ accumulation and inflammatory processes within the brain [204]. Transplanting fecal microbiota from healthy donors to AD models restores a more balanced and diverse gut microbial community. This intervention reverses dysbiosis—a condition associated with neurodegenerative progression in AD [204–206]. FMT treatment is associated with decreased deposition of Aβ plaques in the brain—a hallmark of AD. This reduction is often paralleled by improvements in cognition and behavioral outcomes, highlighting a strong link between gut microbiota composition and AD neuropathology [206–208]. FMT enriches beneficial gut bacteria (e.g., *Bacteroides*, Lachnospiraceae) and increases anti-inflammatory metabolites, suppressing microglial activation and pro-inflammatory cytokine production [206, 209]. Reports indicate not only improvements in cognitive and pathological features in mouse models but also case observations of AD patients showing symptom relief after FMT [204, 205].

Adopting dietary patterns like the Mediterranean or MIND diet—rich in polyphenols, omega-3 fatty acids, and fermentable fibers—not only fosters a beneficial gut microbiota but is also robustly associated with a lower risk for AD and slower cognitive decline [210–214]. Consequently, interventions aimed at restoring microbial homeostasis—via dietary modulation, microbial-based therapeutics, and precision medicine approaches—represent a promising frontier in the prevention and treatment of AD.

9.4.2 Parkinson's Disease (PD)

PD is a progressive neurodegenerative condition marked by a constellation of motor and non-motor symptoms. Its pathophysiology is characterized by the degeneration of dopaminergic neurons in the substantia nigra pars compacta and the aberrant aggregation of α-synuclein into Lewy bodies and Lewy neurites [215]. Clinically, PD presents with hallmark motor symptoms—resting tremor, bradykinesia, rigidity, and postural instability—as well as a spectrum of non-motor manifestations such as constipation, hyposmia, depression, cognitive impairment, and sleep disturbances [216, 217]. The characteristic lesion involves progressive degeneration of dopaminergic neurons in the substantia nigra pars compacta, culminating in significant striatal dopamine depletion [215, 218]. These are intracellular inclusions composed primarily of aggregated, misfolded α-synuclein. They represent a key neuropathological feature and propagate in the nervous system in a prion-like, spreading manner [215, 219]. Lewy pathology and neuron loss extend beyond the substantia nigra, affecting cortical, limbic, and peripheral autonomic regions, accounting for both motor and non-motor manifestations [217].

In PD, dopaminergic neurons in the nigrostriatal pathway progressively degenerate, causing disruptions in basal ganglia function and motor coordination. This degeneration triggers compensatory alterations in glutamatergic, GABAergic, and cholinergic signaling, contributing to both motor and non-motor symptoms [220]. The pathogenesis of PD involves synergistic cellular stressors, including mitochondrial dysfunction, oxidative stress, impaired autophagy-lysosomal clearance, and chronic neuroinflammation [221, 222]. Key genetic contributors—such as mutations in SNCA, LRRK2, PINK1, PARK2, and DJ-1—disrupt protein homeostasis, mitochondrial quality control, and redox balance [223, 224].

Gut-brain and gut-microbiome interactions play a pivotal role in the initiation and progression of PD. Gut dysbiosis is not just a secondary feature but appears to actively contribute to disease pathogenesis through molecular, inflammatory, and neurodegenerative pathways [225–227]. Emerging evidence suggests that gut dysbiosis plays a central role in the pathogenesis of PD through multiple interconnected mechanisms. Dysbiosis compromises intestinal barrier integrity, resulting in increased intestinal permeability ("leaky gut") and translocation of bacterial endotoxins such as LPS into systemic circulation. LPS stimulates Toll-like receptor 4 (TLR4) signaling and systemic inflammation, which can cross the BBB, activate microglia, and promote neuroinflammatory cascades that exacerbate dopaminergic neurodegeneration. LPS and other metabolites trigger systemic and neuronal inflammation, notably by promoting the misfolding and aggregation of alpha-synuclein—a hallmark pathology of PD [225, 228]. PD patients exhibit altered gut microbiome profiles, with consistent increases in genera such as *Lactobacillus*, *Akkermansia*, and *Bifidobacterium*, and depletion of SCFA-producing bacteria like those from the *Lachnospiraceae* family and the genus *Faecalibacterium* [229]. These microbiome shifts occur even before the onset of motor symptoms and are associated with more severe gastrointestinal manifestations in PD [230].

Accumulation of alpha-synuclein may first occur in the enteric nervous system as a response to toxins or metabolites from dysbiotic microbiota [225]. An altered gut microbiota may promote the misfolding and aggregation of α-synuclein in enteric neurons, which can then propagate to the brain via retrograde transsynaptic transmission along the vagus nerve [228]—a hypothesis supported by studies showing reduced PD risk in patients who have undergone vagotomy [231, 232]. In addition to dysbiosis, which also alters the metabolism of bile acids and neurotransmitters, notably, gut microbes synthesize and degrade dopamine precursors and influence serotonin and gamma-aminobutyric acid (GABA) signaling, contributing to the non-motor features of PD such as depression and sleep disorders [9, 233–236]. In PD, loss of antioxidant-producing microbes and increased gut-derived ROS may exacerbate mitochondrial dysfunction in vulnerable neurons [237–239]. Collectively, these mechanisms illustrate how disturbances in the gut microbiome can initiate and exacerbate neurodegenerative processes in PD [225, 228, 230].

Preclinical studies provide compelling evidence for these mechanisms. Germ-free or antibiotic-treated mouse models overexpressing α-synuclein exhibit attenuated motor deficits and reduced neuroinflammation compared to conventionally colonized counterparts, whereas colonization with PD patient-derived microbiota induces motor dysfunction and microglial activation [240, 241]. In clinical settings, multiple studies have identified consistent microbial signatures in PD patients, including a decrease in anti-inflammatory genera (*Prevotella*, *Bifidobacterium*,

Blautia) [225, 229, 242, 243] and an increase in pro-inflammatory taxa (*Proteobacteria*, *Enterobacteriaceae*, *Akkermansia*) [244–248]. These microbial alterations correlate with disease severity, α-synuclein burden, and inflammatory markers [225, 233]. The gut microbiota also affects the bioavailability and efficacy of medications, including levodopa, the mainstay pharmacological treatment for PD [249, 250]. FMT is under investigation as a strategy to restore microbial homeostasis and mitigate PD progression [251]. Moreover, dietary interventions—such as the Mediterranean diet and ketogenic diet—promote microbial diversity and reduce systemic inflammation, potentially conferring neuroprotection in PD [252–255]. Continued interdisciplinary research into the gut-brain axis will be essential for developing microbiome-targeted interventions in the context of PD.

9.4.3 Multiple Sclerosis (MS)

MS is a chronic, immune-mediated, demyelinating disorder of the CNS, characterized by neuroinflammation, axonal damage, and progressive neurological disability [256–259]. The pathology of MS includes widespread abnormalities in both white and grey matter regions of the CNS [260, 261]. A hallmark of MS lesions is perivascular lymphocytic infiltration, where clusters of T and B lymphocytes accumulate around small veins, driving localized immune responses in both white matter and cortical grey matter [262, 263]. Microglial activation is another central feature, with these resident immune cells becoming activated and proliferating in response to injury, thereby contributing to neuroinflammation and neuronal damage [262, 264, 265]. Astrocytic gliosis, marked by the proliferation of reactive astrocytes, leads to scar-like tissue formation within lesion sites, especially in chronic stages [262, 264]. Over time, chronic plaques emerge—demyelinated areas that persist and expand in both white and grey matter, often evolving from earlier acute lesions rich in inflammatory cells to more inactive forms characterized by persistent gliosis and low-grade microglial activity [257, 262, 264].

The underlying autoimmune processes in MS are driven by several pro-inflammatory cytokines. Interferon gamma (IFN-γ) promotes Th1 responses and activates macrophages and microglia, contributing to demyelination [266]. Interleukin-17 (IL-17), produced by Th17 cells, plays a pivotal role by disrupting the BBB and amplifying neuroinflammation; its levels often correlate with disease activity as seen in MRI findings [266]. Tumor necrosis factor-alpha (TNF-α) has dual roles: while its membrane-bound form may support neuroprotection, its soluble form exacerbates inflammation and tissue injury [266]. Dysregulation of immune tolerance is also a key factor, as regulatory T-cell (Treg) dysfunction impairs the suppression of autoreactive lymphocytes, thereby sustaining chronic inflammation and CNS damage [266].

MS susceptibility is shaped by a complex interaction between genetic and environmental factors. Genetically, certain HLA haplotypes and cytokine gene polymorphisms are associated with increased risk, highlighting the role of immune-related genes [266]. Among environmental factors, Epstein-Barr virus (EBV) infection is strongly implicated in MS pathogenesis, possibly through its effects on B-cell function and immune regulation [257]. Vitamin D deficiency is another significant contributor, associated with both heightened MS risk and worsened disease progression, likely due to its immunomodulatory roles [257]. Smoking has been consistently linked to increased susceptibility and faster disease progression [257]. Other factors, such as childhood obesity, geographic latitude, and certain infections, also influence MS onset and course [257]. Collectively, the pathogenesis of MS reflects a dynamic interplay between immune dysregulation, genetic susceptibility, and modifiable environmental exposures, each contributing to disease development and progression [257, 262–264, 266, 267].

Dysbiosis has emerged as a significant factor in the pathophysiology of MS, implicating the gut-brain axis and microbiota-immune interactions in both the onset and progression of the disease. MS patients typically exhibit a reduced abundance of SCFA-producing bacteria such as *Faecalibacterium*, *Roseburia*, *Coprococcus*, and *Bifidobacterium*, alongside an increased presence of potentially pro-inflammatory species like *Akkermansia*, *Methanobrevibacter*, and *Blautia* [268, 269]. This altered microbial composition contributes to a shift in the balance of T-cell populations—specifically, a reduction in regulatory T cells (Tregs) that normally promote immune tolerance and a concomitant increase in pro-inflammatory Th1 and Th17 cells, both of which are implicated in MS-related neuroinflammation [270, 271]. Concurrently, dysbiosis promotes chronic systemic inflammation through the overgrowth of pathogenic bacteria such as *Proteobacteria*, leading to heightened production of pro-inflammatory cytokines like IFN-γ, TNF-α, and IL-1β [272, 273]. Dysbiotic microbiota can compromise intestinal tight junctions, increasing gut permeability. This allows bacterial products (e.g., lipopolysaccharides) to translocate into circulation, activating the immune system and contributing to a pro-inflammatory milieu [272, 273]. Molecules like LPS stimulate Toll-like receptor pathways, fueling systemic and central inflammation directly linked to MS pathogenesis [272]

Lower levels of SCFAs (especially butyrate) seen in MS dampen anti-inflammatory signaling and BBB, increasing vulnerability to CNS inflammation [269, 273]. Fewer gut bacteria in MS patients are coated with immunoglobulin A (IgA), indicating a breakdown in normal immune surveillance and regulation at the mucosal surface. This impairs the clearance of harmful microbes, exacerbating inflammation [274, 275]. Animal and human studies suggest gut microbiota from MS patients can promote CNS-specific autoimmunity when transferred to susceptible hosts, further reinforcing the gut's role in disease pathogenesis [271]. There is also evidence of disrupted bidirectional communication between the gut and CNS. Pro-inflammatory signals originating in the gut can weaken the BBB and facilitate the migration of immune cells and inflammatory mediators into the CNS, exacerbating neuroinflammation and demyelination [273, 276, 277]

The gut microbiome further mediates the effects of environmental risk factors such as diet, infections, and smoking, which may influence MS risk and course by altering microbial composition and host immune responses [278]. Importantly, this growing understanding of gut dysbiosis in MS opens the door to new therapeutic opportunities. Interventions targeting the gut microbiota—such as probiotics, prebiotics, or dietary modulation—show promise as adjunctive treatments. Moreover, specific microbial profiles may serve as biomarkers for diagnosis, prognosis, or therapeutic response [269, 279]. Ongoing research into gut-derived immunomodulators and microbial biomarkers holds potential for developing personalized, microbiome-based therapies for this debilitating condition.

9.4.4 Autism Spectrum Disorder (ASD)

ASD is a multifaceted neurodevelopmental disorder marked by impairments in social communication, the presence of restricted or repetitive behaviors, and atypical sensory processing [280–282]. Symptoms typically emerge in early childhood and persist throughout life, though severity and functional impact vary widely [283, 284]. Common features include impaired verbal and nonverbal communication, poor eye contact, difficulty in forming peer relationships, sensory sensitivities, and insistence on routines [285]. ASD is highly heterogeneous, encompassing a broad spectrum of intellectual and behavioral profiles [282, 286]. The etiology is multifactorial, involving a complex interplay between genetic susceptibility (e.g., mutations in SHANK3, CHD8, SCN2A, MECP2) and environmental influences such as prenatal infections, maternal immune activation, toxic exposures, and perinatal complications [287–291]. Neurobiologically, ASD is associated with atypical synaptic connectivity, impaired synaptogenesis and pruning, altered excitation/inhibition (E/I) balance, neuroinflammation, and abnormal neurodevelopmental trajectories affecting regions such as the prefrontal cortex, amygdala, and cerebellum [288, 292–295]. Emerging evidence implicates gastrointestinal dysbiosis and systemic inflammation as central pathophysiological contributors to ASD, mediated through the gut-brain axis [296–298]. Dysbiosis disrupts the integrity of the intestinal mucosal barrier, increasing intestinal permeability. This allows bacterial toxins (e.g., LPS), harmful metabolites, and dietary peptides to enter the bloodstream [297, 299, 300]. LPS, a potent pro-inflammatory endotoxin, can cross the BBB and modulate brain regions involved in emotion and behavior, such as the amygdala [300]. This breach in gut and BBB integrity contributes to systemic and neuroinflammation, characterized by elevated cytokines like IL-6 and TNF-α, which activate microglia and disrupt normal brain development and function [297, 300]. Microbial metabolites also play a pivotal role; SCFAs such as butyrate and propionate, produced by gut bacteria, can influence gene expression, neurodevelopment, and behavior via epigenetic mechanisms. However, some metabolites, including ammonia and hydrogen sulfide, may exert neurotoxic effects [297, 301]. The gut microbiota synthesizes and modulates neurotransmitters such as serotonin, dopamine, and GABA, for example, gut bacteria influence serotonin levels through the tryptophan metabolic pathway, which is crucial for mood and behavior regulation [297]. Dysbiosis can activate innate immune pathways such as TLR/NF-κB and disrupt metabolic signaling pathways like PI3K/Akt/mTOR, linking gut health to both metabolic disturbances (e.g., insulin resistance) and neurodevelopmental outcomes [297, 301]. These inflammatory and metabolic signals can also drive epigenetic modifications in the brain, affecting gene expression and neural outcomes. Multiple studies confirm that children with ASD exhibit significant gut microbiome dysbiosis—characterized by decreased levels of beneficial bacteria, such as *Bifidobacterium*, *Prevotella*, and *Lactobacillus*, and increased levels of potentially pathogenic species, like *Clostridium*, *Desulfovibrio*, and *Sutterella*—which correlate with both gastrointestinal (GI) and behavioral symptom severity [302–304]. Although supported by extensive mechanistic and animal model evidence, translation to human studies remains challenging due to the clinical and microbial heterogeneity of ASD, and the precise causal relationships continue to be elucidated [299, 305].

Animal studies have shown that fecal FMT from individuals with ASD can induce ASD-like behaviors in mice, supporting a potential causal role of gut microbiota in the disorder [297]. Genetic and environmental factors—including mode of delivery, antibiotic use, early diet, and infections—also shape the gut microbiome and may modulate ASD risk and severity [298, 306]. In response, interventions targeting the GBA, such as probiotics, prebiotics, dietary modifications, and FMT, are being explored to restore microbial balance and reduce inflammation, with some reports indicating improvements in gastrointestinal and behavioral symptoms [298, 306]. However, responses vary among individuals, likely reflecting the substantial heterogeneity of both ASD and microbiota profiles. Despite compelling associations, causality and precise mechanisms remain elusive, and inconsistencies across studies underscore the need for further investigation. To advance this field, research using multi-omics approaches—such as metagenomics, metabolomics, and transcriptomics—is essential to uncover mechanistic insights, identify reliable biomarkers, and guide the development of effective, personalized therapies [296, 298, 307]. In sum, gastrointestinal dysbiosis and systemic inflammation appear to play important roles in ASD pathogenesis via the gut-brain axis, influencing brain function through microbial metabolites, immune activation, and barrier dysfunction; however, further rigorous studies are required to establish causality and therapeutic relevance.

9.4.5 Other Psychiatric Disorders

Mounting evidence suggests that gut-driven inflammation, mediated by dysbiosis of the gut microbiome, plays a critical role in the pathogenesis of major psychiatric conditions such as depression, anxiety, and schizophrenia [308–310]. Dysbiosis is associated with reduced levels of beneficial, anti-inflammatory bacteria (e.g., butyrate-producing *Faecalibacterium* and *Coprococcus*) and increased levels of pro-inflammatory bacteria (e.g., *Eggerthella*) [308]. This imbalance can heighten gut permeability, allowing bacterial products to enter systemic circulation, triggering immune activation and promoting both systemic and neuroinflammation—key contributors to psychiatric pathology [308, 310]. The gut microbiome directly influences the production, metabolism, and signaling of neurotransmitters such as serotonin, dopamine, GABA, and glutamate, for instance, certain bacteria are involved in glutamate and GABA metabolism, and dysbiosis can disrupt the balance of these crucial neurotransmitters, which are central to mood regulation and cognitive function [308]. Animal studies demonstrate that FMT from humans with depression into germ-free mice induces depressive-like behaviors, while colonization with "healthy" microbiota does not [309]. These effects are mediated by disturbances in gut microbial genes and host metabolites, particularly those involved in carbohydrate and amino acid metabolism, further linking gut dysbiosis to behavioral changes [309]. The gut-brain axis is bidirectional: not only does the microbiome shape brain function and behavior, but psychological stress and psychiatric symptoms can also alter the composition of the gut microbiome, creating a feedback loop that may exacerbate and perpetuate mental health disorders [309].

Altered gut microbiota also affect the synthesis and availability of key neurotransmitters, including serotonin, dopamine, and GABA, by modulating the tryptophan-kynurenine pathway and producing neuroactive metabolites [311–313], for instance, certain *Clostridium* species can deconjugate bile acids and generate neuroactive compounds, while increased levels of harmful microbial metabolites like p-cresol and 4-ethylphenyl sulfate (4EPS) have been associated with anxiety-like behaviors in animal models [311, 312]. These changes are reversible in part; restoration of beneficial bacterial populations—such as *Lactobacillus* and *Bifidobacterium*—has been shown to decrease pathogenic metabolite production and improve behavioral outcomes [311, 312]. Early life stress—such as maternal separation, poor diet, and early antibiotic exposure—can also program the gut microbiome and immune system toward a pro-inflammatory state, increasing long-term vulnerability to psychiatric disorders [313].

Therapeutically, microbiome-targeted strategies show promise. Probiotics (e.g., *Lactobacillus casei*, *Bifidobacterium longum*), prebiotics, synbiotics, and even FMT have demonstrated beneficial effects in preclinical studies, improving mood,

Table 9.1 Gut microbiome dysbiosis and psychiatric disorders

Disorder	Key microbiome alterations	Inflammatory/ neurochemical effects	Behavioral outcomes
Depression	Reduced *Faecalibacterium*, *Coprococcus*, increased *Eggerthella*	Increased inflammation, altered serotonin, dopamine	Depressive-like behaviors
Anxiety	Similar to depression, reduced diversity	Increased pro-inflammatory cytokines	Anxiety-like behaviors
Schizophrenia	Reduced SCFA producers, increased pro-inflammatory taxa	Neuroinflammation, glutamate/GABA imbalance	Cognitive/behavioral symptoms

reducing anxiety-like behaviors, and modulating inflammatory markers [312, 314, 315]. Some small-scale clinical trials have reported reduced depressive and anxiety symptoms following probiotic supplementation, though results are heterogeneous and often lack mechanistic clarity [316, 317]. Because persistent inflammation underlies many psychiatric symptoms, anti-inflammatory interventions—whether through dietary modulation, microbiota restoration, or pharmacological means—may represent a novel therapeutic frontier (Table 9.1) [318–320].

These findings highlight the gut-brain axis as a critical determinant of mental health and underscore the therapeutic potential of microbiome-based interventions. Nevertheless, clinical translation remains challenging due to individual variability, microbial complexity, and limited understanding of causal mechanisms. Future research integrating multi-omics analyses, longitudinal human cohorts, and well-controlled interventional trials will be essential to validate microbiome-targeted strategies and develop personalized treatments for psychiatric disorders [321–323].

9.5 Therapeutic Opportunities: Targeting the Gut to Heal the Brain

The gut-brain axis is increasingly recognized as a critical therapeutic target for mitigating neuroinflammation and improving clinical outcomes in neuropsychiatric and neurodegenerative disorders [22, 324, 325]. Modulating the gut microbiota and its associated metabolites offers a novel avenue to reduce both systemic and CNS inflammation, restore neuroimmune homeostasis, and potentially decelerate the progression of cognitive and affective pathologies [22, 62, 326, 327]. These current and emerging therapeutic strategies—encompassing microbial modulation, pharmacological interventions, nutritional and lifestyle modifications, and biotechnological innovations—highlight their underlying mechanisms, empirical support, and potential for personalized medicine.

Probiotics—defined as live microorganisms that confer health benefits to the host when administered in adequate amounts—are foundational to gut-targeted therapeutic interventions. Specific strains, including *Lactobacillus rhamnosus*, *Bifidobacterium longum*, and *Faecalibacterium prausnitzii*, demonstrate notable anti-inflammatory effects through the suppression of pro-inflammatory cytokines (e.g., IL-6, TNF-α) and the enhancement of intestinal epithelial barrier integrity [328]. Clinical investigations in major depressive disorder, generalized anxiety disorder, and neurodegenerative diseases such as Alzheimer's and Parkinson's have shown that probiotic supplementation may alleviate mood disturbances, improve cognitive performance, and reduce systemic inflammatory biomarkers [209, 329–332], for example, a 2023 meta-analysis in PD reported high-quality evidence that probiotics significantly improve motor and non-motor symptoms, including depression, and moderate-quality evidence for reduced anxiety and improved quality of life [333]. These effects are attributed, in part, to modulation of the serotonergic system and the hypothalamic-pituitary-adrenal (HPA) axis, though the underlying mechanisms remain an active area of research and show considerable variability depending on strain, dose, and individual response [334]. Probiotics exhibit potential across a range of neuropsychiatric and neurodegenerative conditions, mediated by gut-barrier reinforcement, anti-inflammatory activity, and gut-brain axis modulation. However, their benefits are not universal, and optimal outcomes depend on careful selection of strain, dose, and individual patient factors.

Prebiotics are nondigestible food components, primarily certain oligosaccharides such as inulin, fructooligosaccharides (FOS), and galactooligosaccharides (GOS), that reach the colon intact and serve as nourishment for beneficial gut microbes [335]. Found naturally in fruits, vegetables, and grains or taken as dietary supplements, these compounds selectively stimulate the growth and metabolic activity of beneficial bacteria, especially *Bifidobacterium* and *Lactobacillus* [336–340]. Empirical evidence supports the role of SCFAs in modulating stress reactivity and enhancing cognitive function, particularly in the context of mood and anxiety disorders [58, 341, 342]. Their mechanisms of action include serving as fermentation substrates for these bacteria, producing SCFAs that lower intestinal pH, thereby inhibiting pathogenic microbes and promoting a favorable gut environment [343]. Additionally, they enhance the colonization of beneficial bacteria by outcompeting pathogens. In summary, prebiotics like inulin, FOS, and GOS play a crucial role in maintaining gut health by selectively supporting beneficial bacterial populations and contributing to overall well-being [340].

FMT—the transfer of stool from screened healthy donors to recipients with disrupted microbiota—aims to restore microbial diversity and function. Preclinical models of AD, PD, and ASD have demonstrated that FMT can attenuate neuroinflammation, improve behavioral outcomes, and rebalance immune signaling, likely through enriching SCFA-producing taxa and reducing pro-inflammatory bacteria such as *Clostridium* and *Proteobacteria* [344, 345]. While clinical applications remain investigational, early-phase trials and case reports suggest benefits, including improved behavioral regulation in ASD and reduced inflammatory biomarkers in PD [344]. Challenges include the need for rigorous donor screening, standardization of procedures, and long-term safety evaluations, especially regarding potential transmission of pathogens or undesirable microbial traits [345].

Broad-spectrum antibiotics can rapidly alter gut microbial composition, often reducing diversity and potentially leading to dysbiosis, increased susceptibility to infections, and the emergence of antimicrobial resistance due to selective pressure on both pathogens and commensals [346, 347]. These nonspecific effects and the risk of long-term ecological disruption make broad-spectrum antibiotics unsuitable for sustained modulation of the gut microbiome, especially given that some compositional changes can persist for months to years [347]. In contrast, rifaximin—a minimally absorbed, gut-targeted antibiotic—has shown efficacy in reducing gut-derived endotoxins (such as LPS) and ameliorating symptoms in conditions like hepatic encephalopathy (HE), with emerging relevance to neuropsychiatric manifestations associated with gut-brain axis dysfunction [348–350]. Its benefits may be mediated more by functional modulation (e.g., reducing bacterial translocation, endotoxemia, and mucin degradation) than by broad taxonomic disruption [350].

Pharmacological agents that modulate microbial-derived metabolites represent a promising direction for gut-brain axis therapeutics [351], for example, inhibitors of the kynurenine pathway aim to reduce the production of neurotoxic metabolites such as quinolinic acid, implicated in depression and schizophrenia [33]. Agonists of bile acid receptors, including obeticholic acid, have been shown to reinforce gut barrier integrity and suppress systemic inflammation [351]. Additionally, inflammasome inhibitors targeting the NLRP3 pathway have exhibited preclinical efficacy in reducing microglial activation, a key feature in neurodegenerative pathology [352–354]. These agents suppress the release of pro-inflammatory cytokines (IL-1β, IL-18) and may attenuate neuronal pyroptosis and cognitive deficits [352, 353], for example, CY-09 directly inhibits NLRP3, alleviates amyloid pathology in Alzheimer's models, and shows antidepressant-like effects by reducing neuroinflammation in microglia [353]. Inhibition of the NLRP3 pathway can also shift microglial phenotype from pro-inflammatory (M1) to anti-inflammatory (M2), further mitigating neural damage [355]. These targeted interventions remain in early clinical development but offer mechanistic precision in modulating gut-derived neuroimmune processes.

VNS—both implantable and noninvasive (transcutaneous) forms—modulates the gut-brain axis primarily by activating the cholinergic anti-inflammatory pathway (CAP), a neuroimmune circuit that suppresses systemic inflammation [356–358]. This pathway is triggered when the vagus nerve is stimulated, leading to the release of acetylcholine, which binds to receptors on immune cells (especially macrophages) and inhibits the production of pro-inflammatory cytokines such as IL-6, TNF-α, IL-1β, and IL-8 [356, 357]. Importantly, anti-inflammatory cytokines like IL-10 are not suppressed, maintaining a balanced immune response [356]. Exploratory studies suggest potential benefits in schizophrenia and neurodegenerative diseases, with indirect effects on gut microbial composition, but these applications require further validation [73, 358, 359]. Noninvasive VNS is an area of active development, offering the possibility of broader clinical application [360, 361].

Diet is a principal determinant of gut microbial composition and function [362, 363]. Diets rich in fermentable fiber (e.g., whole grains, legumes), polyphenols (e.g., berries, cocoa), and omega-3 polyunsaturated fatty acids (e.g., EPA/DHA from fatty fish) are associated with increased abundance of SCFA-producing taxa and lower systemic inflammation [363–365]. The Mediterranean diet, in particular, has been linked to improvements in mood and cognition, likely due to its anti-inflammatory and neuroprotective properties; human and animal studies show that such dietary patterns enhance microbiota diversity, increase SCFA production, reduce neuroinflammation and brain insulin resistance, and correlate with better cognitive performance [363–365]. In contrast, Western dietary patterns—characterized by high intake of saturated fats and refined sugars—are associated with microbial dysbiosis, increased gut permeability, and exacerbated neuropsychiatric symptoms, including poorer memory and elevated neural inflammation in both animal models and some human studies [363–365]. These associations are mediated by multiple pathways, including microbial metabolite production (e.g., SCFAs), immune signaling, and gut barrier integrity, though further clinical research is needed to establish causal mechanisms in humans [362, 363].

Postbiotic interventions have been shown to restore gut microbial balance, reduce pathological amyloid-beta deposition in the brain, and improve memory and learning in rodent models of AD [366, 367]. Mechanistically, postbiotics may enhance gut barrier integrity, regulate neurotransmitter systems, and support synaptic plasticity—key processes in combating cognitive decline [366]. Their benefits likely arise both from direct actions on neural cells and from indirect effects via the gut-brain axis, modulating local and systemic inflammation and oxidative stress [366]. Due to their chemical stability, scalability, and defined composition, postbiotics represent a promising next-generation intervention for neurological disorders [368, 369]. However, while preclinical data are encouraging, further clinical studies are needed to firmly establish efficacy, optimal dosing, and long-term safety in humans [368, 370]. Although research into the underlying mechanisms and clinical validation is still ongoing, the microbiota-gut-brain axis presents a significant opportunity for developing innovative treatments for neurological and psychiatric disorders.

9.6 Conclusion

The intricate relationship between the gut and the brain—mediated through the gut-brain axis—has emerged as a transformative paradigm in understanding neuroinflammation and its broader implications for neurological and psychiatric health [4, 371]. At the heart of this axis lies the gut microbiome, a dynamic and metabolically active ecosystem whose composition and function have far-reaching effects on brain development, function, and disease [372]. Dysbiosis disrupts the integrity of the intestinal barrier, promotes peripheral inflammation, and alters central immune responses—ultimately compromising brain health. The vagus nerve, acting as a neuroimmune conduit, and microbiota-derived metabolites, such as SCFAs, tryptophan catabolites, and secondary bile acids, have emerged as key mediators in these processes [373]. Disruptions in gut microbiota have been implicated in a spectrum of neurological disorders, such as AD, PD, MS, ASD, and various neuropsychiatric conditions [9, 50, 240, 374–376]. Despite their distinct pathophysiological features, these disorders often converge on a shared axis of neuroinflammation, frequently rooted in or amplified by microbial imbalances in the gut.

Crucially, this emerging understanding opens new therapeutic frontiers. Interventions aimed at modulating the gut microbiota—such as probiotics, prebiotics, dietary changes, FMT, and vagus nerve stimulation—offer promising strategies to mitigate neuroinflammation and restore neurological function [62, 205, 377]. These approaches mark a shift from a traditionally brain-centric view of neurological disease to a more holistic model that integrates gut health as a critical determinant of brain health.

As the field of gut-brain research advances, future studies must continue to decipher the molecular mechanisms underlying this axis, identify microbial biomarkers for early disease detection, and develop precision microbiome-targeted therapies [378]. A deeper understanding of the gut-brain axis will not only illuminate the pathogenesis of complex neurological disorders but also pave the way for innovative, noninvasive interventions that heal the brain from the gut outward.

In conclusion, the gut-brain axis is far more than a physiological novelty—it is a pivotal component of the neuroimmune interface [379, 380]. Unlocking its mechanisms and harnessing its therapeutic potential could drive a transformative shift across neuroscience, neurology, and psychiatry, placing the gut at the forefront as both guardian and architect of brain health.

Declarations

Competing Interests The author(s) has no competing interests to declare that are relevant to the content of this manuscript.

Ethics Approval The author(s) declares that this manuscript does not include primary research with human participants.

The author(s) declares that this manuscript does not include primary studies involving animals.

References

1. Shigehara A, et al. Gut microbiota and interorgan communication among the gut, brain, and beyond. Intern Med. 2025;65(2):203–13.
2. Carabotti M, et al. The gut-brain axis: interactions between enteric microbiota, central and enteric nervous systems. Ann Gastroenterol. 2015;28(2):203–9.
3. Martin CR, et al. The brain-gut-microbiome axis. Cell Mol Gastroenterol Hepatol. 2018;6(2):133–48.
4. Cryan JF, et al. The microbiota-gut-brain axis. Physiol Rev. 2019;99(4):1877–2013.
5. Dash S, Syed YA, Khan MR. Understanding the role of the gut microbiome in brain development and its association with neurodevelopmental psychiatric disorders. Front Cell Deve Biol. 2022;10:880544.
6. Cervellati C, et al. Inflammation in neurological disorders: the thin boundary between brain and periphery. Antioxid Redox Signal. 2020;33(3):191–210.
7. Bentivoglio M, Mariotti R, Bertini G. Neuroinflammation and brain infections: historical context and current perspectives. Brain Res Rev. 2011;66(1–2):152–73.
8. Shabab T, et al. Neuroinflammation pathways: a general review. Int J Neurosci. 2017;127(7):624–33.
9. Ullah H, et al. The gut microbiota–brain axis in neurological disorder. Front Neurosci. 2023;17:1225875.
10. Mou Y, et al. Gut microbiota interact with the brain through systemic chronic inflammation: implications on neuroinflammation, neurodegeneration, and aging. Front Immunol. 2022;13:796288.
11. Chakraborty P, et al. Gut microbiota nexus: exploring the interactions with the brain, heart, lungs, and skin axes and their effects on health. Med Microecol. 2024;20:100104.
12. Montagnani M, et al. The crosstalk between gut microbiota and nervous system: a bidirectional interaction between microorganisms and metabolome. Int J Mol Sci. 2023;24(12):10322.
13. Ramadan YN, et al. Microbiome gut-brain-axis: impact on brain development and mental health. Mol Neurobiol. 2025;62(8):10813–33.

14. Hwang YK, Oh JS. Interaction of the vagus nerve and serotonin in the gut–brain axis. Int J Mol Sci. 2025;26(3):1160.
15. Breit S, et al. Vagus nerve as modulator of the brain–gut axis in psychiatric and inflammatory disorders. Front Psych. 2018;9:298797.
16. Ma L, Wang H-B, Hashimoto K. The vagus nerve: an old but new player in brain–body communication. Brain Behav Immun. 2025;124:28–39.
17. Petrut S-M, et al. Gut over mind: exploring the powerful gut–brain axis. Nutrients. 2025;17(5):842.
18. Dos Santos JCC, et al. Gut-microbiome-brain axis: the crosstalk between the vagus nerve, alpha-synuclein and the brain in Parkinson's disease. Neural Regen Res. 2023;18(12):2611–4.
19. O'Riordan KJ, et al. Short chain fatty acids: microbial metabolites for gut-brain axis signalling. Mol Cell Endocrinol. 2022;546:111572.
20. Liu J, et al. Functions of gut microbiota metabolites, current status and future perspectives. Aging Dis. 2022;13(4):1106.
21. Lee B, et al. Gut microbiota metabolite messengers in brain function and pathology at a view of cell type-based receptor and enzyme reaction. Biomol Ther. 2024;32(4):403.
22. Loh JS, et al. Microbiota–gut–brain axis and its therapeutic applications in neurodegenerative diseases. Signal Transduct Target Ther. 2024;9(1):37.
23. Abou Izzeddine N, et al. The microbial guardians: unveiling the role of gut microbiota in shaping neurodegenerative disease. IBRO Neurosci Rep. 2025;19:17.
24. Jung YH, Chae CW, Han HJ. The potential role of gut microbiota-derived metabolites as regulators of metabolic syndrome-associated mitochondrial and endolysosomal dysfunction in Alzheimer's disease. Exp Mol Med. 2024;56(8):1691–702.
25. Abildinova GZ, et al. The gut-brain-metabolic axis: exploring the role of microbiota in insulin resistance and cognitive function. Front Microbiol. 2024;15:1463958.
26. Vargas A, et al. Gut microbiota-derived metabolites and chronic inflammatory diseases. Explor Med. 2025;6:1001275.
27. Wang X, et al. Gut-brain axis in the pathogenesis of sepsis-associated encephalopathy. Neurobiol Dis. 2024;195:106499.
28. Camberos-Barraza J, Guadrón-Llanos AM, De la Herrán-Arita AK. The gut microbiome-neuroglia axis: implications for brain health, inflammation, and disease. Neuroglia. 2024;5(3):254–73.
29. Jain A, Madkan S, Patil P. The role of gut microbiota in neurodegenerative diseases: current insights and therapeutic implications. Cureus. 2023;15(10):e47861.
30. Rogers G, et al. From gut dysbiosis to altered brain function and mental illness: mechanisms and pathways. Mol Psychiatry. 2016;21(6):738–48.
31. Anderson, S.C., J.F. Cryan, and T. Dinan, The psychobiotic revolution: mood, food, and the new science of the gut-brain connection. 2017: National Geographic Books.
32. Montiel-Castro AJ, et al. The microbiota-gut-brain axis: neurobehavioral correlates, health and sociality. Front Integr Neurosci. 2013;7:70.
33. Kearns R. Gut–brain axis and neuroinflammation: the role of gut permeability and the kynurenine pathway in neurological disorders. Cell Mol Neurobiol. 2024;44(1):64.
34. Nigam M, et al. Evaluation of the association of chronic inflammation and cancer: insights and implications. Biomed Pharmacother. 2023;164:115015.
35. Maddipati KR. Distinct etiology of chronic inflammation–implications on degenerative diseases and cancer therapy. Front Immunol. 2024;15:1460302.
36. Van Dyke TE, Bartold PM, Reynolds EC. The nexus between periodontal inflammation and dysbiosis. Front Immunol. 2020;11:511.
37. Guinane CM, Cotter PD. Role of the gut microbiota in health and chronic gastrointestinal disease: understanding a hidden metabolic organ. Ther Adv Gastroenterol. 2013;6(4):295–308.
38. Afzaal M, et al. Human gut microbiota in health and disease: unveiling the relationship. Front Microbiol. 2022;13:999001.
39. Mansour S, et al. Impact of diet on human gut microbiome and disease risk. New Microbes New Infect. 2021;41:100845.

40. Ramirez J, et al. Antibiotics as major disruptors of gut microbiota. Front Cell Infect Microbiol. 2020;10:572912.
41. Patangia DV, et al. Impact of antibiotics on the human microbiome and consequences for host health. Microbiology. 2022;11(1):e1260.
42. Madison A, Kiecolt-Glaser JK. Stress, depression, diet, and the gut microbiota: human–bacteria interactions at the core of psychoneuroimmunology and nutrition. Curr Opin Behav Sci. 2019;28:105–10.
43. Tang W, et al. The impact of gut microbiota disorders on the blood–brain barrier. Infect Drug Resist. 2020;13:3351–63.
44. Carloni S, Rescigno M. The gut-brain vascular axis in neuroinflammation. Semin Immunol. 2023;69:101802.
45. Gao C, et al. Microglia in neurodegenerative diseases: mechanism and potential therapeutic targets. Signal Transduct Target Ther. 2023;8(1):359.
46. Sălcudean A, et al. Neuroinflammation—a crucial factor in the pathophysiology of depression—a comprehensive review. Biomolecules. 2025;15(4):502.
47. Thomas SD, et al. Targeting microglia in neuroinflammation: H3 receptor antagonists as a novel therapeutic approach for Alzheimer's disease, Parkinson's disease, and autism spectrum disorder. Pharmaceuticals. 2024;17(7):831.
48. Müller L, Di Benedetto S. Neuroimmune crosstalk in chronic neuroinflammation: microglial interactions and immune modulation. Front Cell Neurosci. 2025;19:1575022.
49. Houser MC, Tansey MG. The gut-brain axis: is intestinal inflammation a silent driver of Parkinson's disease pathogenesis? NPJ Parkinson's Dis. 2017;3(1):3.
50. Suganya K, Koo B-S. Gut–brain axis: role of gut microbiota on neurological disorders and how probiotics/prebiotics beneficially modulate microbial and immune pathways to improve brain functions. Int J Mol Sci. 2020;21(20):7551.
51. Kurhaluk N, et al. Role of antioxidants in modulating the microbiota–gut–brain axis and their impact on neurodegenerative diseases. Int J Mol Sci. 2025;26(8):3658.
52. Agirman G, Yu KB, Hsiao EY. Signaling inflammation across the gut-brain axis. Science. 2021;374(6571):1087–92.
53. Hasan A, et al. An updated and comprehensive review exploring the gut–brain axis in neurodegenerative disorders and neurotraumas: implications for therapeutic strategies. Brain Sci. 2025;15(6):654.
54. Carabotti M, et al. The gut-brain axis: interactions between enteric microbiota, central and enteric nervous systems. Ann Gastroenterol Q Publ Hellenic Soc Gastroenterol. 2015;28(2):203.
55. Wang H-X, Wang Y-P, Chen X. Gut microbiota-brain axis. Chin Med J. 2016;129(19):2373–80.
56. Gwak M-G, Chang S-Y. Gut-brain connection: microbiome, gut barrier, and environmental sensors. Immune Netw. 2021;21(3):e20.
57. Kasarello K, Cudnoch-Jedrzejewska A, Czarzasta K. Communication of gut microbiota and brain via immune and neuroendocrine signaling. Front Microbiol. 2023;14:1118529.
58. Appleton J. The gut-brain Axis: influence of microbiota on mood and mental health. Integr Med (Encinitas). 2018;17(4):28–32.
59. Clapp M, et al. Gut microbiota's effect on mental health: the gut-brain axis. Clin Pract. 2017;7(4):987.
60. Post Z, Manfready RA, Keshavarzian A. Overview of the gut-brain axis: from gut to brain and back again. Semin Neurol. 2023;43(4):506–17.
61. Hattori N, Yamashiro Y. The gut-brain axis. Ann Nutr Metab. 2021;77(Suppl 2):1–3.
62. O'Riordan KJ, et al. The gut microbiota-immune-brain axis: therapeutic implications. Cell Rep Med. 2025;6(3):101982.
63. Lu S, et al. The communication mechanism of the gut-brain axis and its effect on central nervous system diseases: a systematic review. Biomed Pharmacother. 2024;178:117207.
64. Bonaz B, Bazin T, Pellissier S. The vagus nerve at the Interface of the microbiota-gut-brain axis. Front Neurosci. 2018;12:49.

65. Kanai T, Teratani T. Role of the vagus nerve in the gut-brain axis: development and maintenance of gut regulatory T cells via the liver-brain-gut vago-vagal reflex. Brain Nerve. 2022;74(8):971–7.
66. Liu L, et al. Neuro-immune interactions: exploring the anti-inflammatory role of the vagus nerve. Int Immunopharmacol. 2025;159:114941.
67. Rangon C-M, Niezgoda A. Understanding the pivotal role of the vagus nerve in health from pandemics. Bioengineering. 2022;9(8):352.
68. Futoh Y, et al. Vagus nerve signal has an inhibitory influence on the development of peritoneal metastasis in murine gastric cancer. Sci Rep. 2024;14(1):7832.
69. Yu CD, Xu QJ, Chang RB. Vagal sensory neurons and gut-brain signaling. Curr Opin Neurobiol. 2020;62:133–40.
70. Shaffer C, Barrett LF, Quigley KS. Signal processing in the vagus nerve: hypotheses based on new genetic and anatomical evidence. Biol Psychol. 2023;182:108626.
71. Siopi E, et al. Gut microbiota changes require vagus nerve integrity to promote depressive-like behaviors in mice. Mol Psychiatry. 2023;28(7):3002–12.
72. Mörkl S, Butler MI, Wagner-Skacel J. Gut-brain-crosstalk- the vagus nerve and the microbiota-gut-brain axis in depression. A narrative review. J Affect Disord Rep. 2023;13:100607.
73. Breit S, et al. Vagus nerve as modulator of the brain–gut axis in psychiatric and inflammatory disorders. Front Psychiatry. 2018;9:44.
74. Bemark M, et al. Gut-associated lymphoid tissue: a microbiota-driven hub of B cell immunity. Trends Immunol. 2024;45(3):211–23.
75. Morbe UM, et al. Human gut-associated lymphoid tissues (GALT); diversity, structure, and function. Mucosal Immunol. 2021;14(4):793–802.
76. Dupont HL, et al. The intestinal microbiome in human health and disease. Trans Am Clin Climatol Assoc. 2020;131:178–97.
77. Xue J, Ajuwon KM, Fang R. Mechanistic insight into the gut microbiome and its interaction with host immunity and inflammation. Anim Nutr. 2020;6(4):421–8.
78. Sittipo P, et al. The function of gut microbiota in immune-related neurological disorders: a review. J Neuroinflammation. 2022;19(1):154.
79. Zheng D, Liwinski T, Elinav E. Interaction between microbiota and immunity in health and disease. Cell Res. 2020;30(6):492–506.
80. Zhao M, et al. Immunological mechanisms of inflammatory diseases caused by gut microbiota dysbiosis: a review. Biomed Pharmacother. 2023;164:114985.
81. Hou K, et al. Microbiota in health and diseases. Signal Transduct Target Ther. 2022;7(1):135.
82. Belkaid Y, Hand TW. Role of the microbiota in immunity and inflammation. Cell. 2014;157(1):121–41.
83. Bagheri S, Zolghadri S, Stanek A. Beneficial effects of anti-inflammatory diet in modulating gut microbiota and controlling obesity. Nutrients. 2022;14(19):3985.
84. Chidambaram SB, et al. Gut dysbiosis, defective autophagy and altered immune responses in neurodegenerative diseases: Tales of a vicious cycle. Pharmacol Ther. 2022;231:107988.
85. Anand N, Gorantla VR, Chidambaram SB. The role of gut dysbiosis in the pathophysiology of neuropsychiatric disorders. Cells. 2022;12(1):54.
86. Abdel-Haq R, et al. Microbiome-microglia connections via the gut-brain axis. J Exp Med. 2019;216(1):41–59.
87. Clemente-Suarez VJ, et al. Microbiota implications in endocrine-related diseases: from development to novel therapeutic approaches. Biomedicine. 2024;12(1):221.
88. Yu M, Yu B, Chen D. The effects of gut microbiota on appetite regulation and the underlying mechanisms. Gut Microbes. 2024;16(1):2414796.
89. Rusch JA, Layden BT, Dugas LR. Signalling cognition: the gut microbiota and hypothalamic-pituitary-adrenal axis. Front Endocrinol (Lausanne). 2023;14:1130689.
90. Bertollo AG, et al. Hypothalamus-pituitary-adrenal and gut-brain axes in biological interaction pathway of the depression. Front Neurosci. 2025;19:1541075.

91. Kageyama K, Iwasaki Y, Daimon M. Hypothalamic regulation of Corticotropin-releasing factor under stress and stress resilience. Int J Mol Sci. 2021;22(22):12242.
92. Kandpal M, et al. Dysbiosis of gut microbiota from the perspective of the gut-brain axis: role in the provocation of neurological disorders. Meta. 2022;12(11):1064.
93. Zhang R, et al. The gut microbiome, immune modulation, and cognitive decline: insights on the gut-brain axis. Front Immunol. 2025;16:1529958.
94. Solanki R, Karande A, Ranganathan P. Emerging role of gut microbiota dysbiosis in neuroinflammation and neurodegeneration. Front Neurol. 2023;14:1149618.
95. Sahu KK, et al. Pharmacological insights into gut microbiota modulation in systemic lupus erythematosus: mechanisms, treatment strategies, and clinical implications. J Pharmacol Exp Ther. 2025;392:103659.
96. Liang L, Saunders C, Sanossian N. Food, gut barrier dysfunction, and related diseases: a new target for future individualized disease prevention and management. Food Sci Nutr. 2023;11(4):1671–704.
97. Jaeger LB, et al. Lipopolysaccharide alters the blood-brain barrier transport of amyloid beta protein: a mechanism for inflammation in the progression of Alzheimer's disease. Brain Behav Immun. 2009;23(4):507–17.
98. Mayer MG, Fischer T. Shared mechanisms of blood-brain barrier dysfunction and neuroinflammation in coronavirus disease 2019 and Alzheimer disease. Am J Pathol. 2025;195:2075.
99. Takata F, et al. Blood-brain barrier dysfunction amplifies the development of neuroinflammation: understanding of cellular events in brain microvascular endothelial cells for prevention and treatment of BBB dysfunction. Front Cell Neurosci. 2021;15:661838.
100. Guijarro-Muñoz I, et al. Lipopolysaccharide activates toll-like receptor 4 (TLR4)-mediated NF-κB signaling pathway and proinflammatory response in human pericytes. J Biol Chem. 2014;289(4):2457–68.
101. Lenz KM, Nelson LH. Microglia and beyond: innate immune cells as regulators of brain development and behavioral function. Front Immunol. 2018;9:698.
102. Colonna M, Butovsky O. Microglia function in the central nervous system during health and neurodegeneration. Annu Rev Immunol. 2017;35:441–68.
103. Woodburn SC, Bollinger JL, Wohleb ES. The semantics of microglia activation: neuroinflammation, homeostasis, and stress. J Neuroinflammation. 2021;18(1):258.
104. Mayer MG, Fischer T. Microglia at the blood brain barrier in health and disease. Front Cell Neurosci. 2024;18:1360195.
105. Kaur R, Kumar S, Singh L. A comprehensive review: neuroinflammation and immune communication between the central nervous system and the periphery. Cytokine. 2025;192:156974.
106. Swer NM, et al. Gut microbiota-derived metabolites and their importance in neurological disorders. Mol Biol Rep. 2023;50(2):1663–75.
107. Zißler J, Rothhammer V, Linnerbauer M. Gut-brain interactions and their impact on astrocytes in the context of multiple sclerosis and beyond. Cells. 2024;13(6):497.
108. Ye Q, et al. Astrocytic Slc4a4 regulates blood-brain barrier integrity in healthy and stroke brains via a CCL2-CCR2 pathway and NO dysregulation. Cell Rep. 2024;43(5):114193.
109. Manu DR, et al. Astrocyte involvement in blood-brain barrier function: a critical update highlighting novel, complex, neurovascular interactions. Int J Mol Sci. 2023;24(24):17146.
110. Takeda S, Sato N, Morishita R. Systemic inflammation, blood-brain barrier vulnerability and cognitive/non-cognitive symptoms in Alzheimer disease: relevance to pathogenesis and therapy. Front Aging Neurosci. 2014;6:171.
111. Galea I. The blood–brain barrier in systemic infection and inflammation. Cell Mol Immunol. 2021;18(11):2489–501.
112. Yang J, et al. New insight into neurological degeneration: inflammatory cytokines and blood–brain barrier. Front Mol Neurosci. 2022;15:1013933.
113. Zipp F, Bittner S, Schafer DP. Cytokines as emerging regulators of central nervous system synapses. Immunity. 2023;56(5):914–25.
114. Sei Y, Vitković L, Yokoyama MM. Cytokines in the central nervous system: regulatory roles in neuronal function, cell death and repair. Neuroimmunomodulation. 1995;2(3):121–33.

115. Rizzo FR, et al. Tumor necrosis factor and interleukin-1β modulate synaptic plasticity during neuroinflammation. Neural Plast. 2018;2018(1):8430123.
116. Sampson T. Microbial amyloids in neurodegenerative amyloid diseases. FEBS J. 2025;292(6):1265–81.
117. Wang C, et al. Genome-wide screen identifies curli amyloid fibril as a bacterial component promoting host neurodegeneration. Proc Natl Acad Sci USA. 2021;118(34):e2106504118.
118. Das TK, et al. Bacterial amyloid Curli associated gut epithelial neuroendocrine activation predominantly observed in Alzheimer's disease mice with central amyloid-β pathology. J Alzheimer's Dis. 2022;88(1):191–205.
119. Miller AL, et al. Microbiome or infections: amyloid-containing biofilms as a trigger for complex human diseases. Front Immunol. 2021;12:638867.
120. Kamatham PT, et al. Pathogenesis, diagnostics, and therapeutics for Alzheimer's disease: breaking the memory barrier. Ageing Res Rev. 2024;101:102481.
121. Breijyeh Z, Karaman R. Comprehensive review on Alzheimer's disease: causes and treatment. Molecules. 2020;25(24):5789.
122. Zhang J, et al. Recent advances in Alzheimer's disease: mechanisms, clinical trials and new drug development strategies. Signal Transduct Target Ther. 2024;9(1):211.
123. Safiri S, et al. Alzheimer's disease: a comprehensive review of epidemiology, risk factors, symptoms diagnosis, management, caregiving, advanced treatments and associated challenges. Front Med. 2024;11:1474043.
124. Tarawneh R, Holtzman DM. The clinical problem of symptomatic Alzheimer disease and mild cognitive impairment. Cold Spring Harb Perspect Med. 2012;2(5):a006148.
125. Emilien G, et al. Alzheimer disease: neuropsychology and pharmacology. 2003: Springer Science & Business Media.
126. Guarino A, et al. Executive functions in Alzheimer disease: a systematic review. Front Aging Neurosci. 2019;10:437.
127. Zhang H, et al. Interaction between Aβ and Tau in the pathogenesis of Alzheimer's disease. Int J Biol Sci. 2021;17(9):2181–92.
128. Bloom GS. Amyloid-β and tau: the trigger and bullet in Alzheimer disease pathogenesis. JAMA Neurol. 2014;71(4):505–8.
129. Hampel H, et al. The Amyloid-β pathway in Alzheimer's Disease. Mol Psychiatry. 2021;26(10):5481–503.
130. Chen Y, Yu Y. Tau and neuroinflammation in Alzheimer's disease: interplay mechanisms and clinical translation. J Neuroinflammation. 2023;20(1):165.
131. Zheng Q, Wang X. Alzheimer's disease: insights into pathology, molecular mechanisms, and therapy. Protein Cell. 2024;16(2):83–120.
132. Azargoonjahromi A. The duality of amyloid-β: its role in normal and Alzheimer's disease states. Mol Brain. 2024;17(1):44.
133. Roda AR, et al. Amyloid-beta peptide and tau protein crosstalk in Alzheimer's disease. Neural Regen Res. 2022;17(8):1666–74.
134. Bi C, Bi S, Li B. Processing of mutant β-Amyloid precursor protein and the clinicopathological features of familial Alzheimer's disease. Aging Dis. 2019;10(2):383–403.
135. Nhan HS, Chiang K, Koo EH. The multifaceted nature of amyloid precursor protein and its proteolytic fragments: friends and foes. Acta Neuropathol. 2015;129(1):1–19.
136. Hampel H, et al. The β-secretase BACE1 in Alzheimer's disease. Biol Psychiatry. 2021;89(8):745–56.
137. Chen G-F, et al. Amyloid beta: structure, biology and structure-based therapeutic development. Acta Pharmacol Sin. 2017;38(9):1205–35.
138. Durrant CS, et al. Beta secretase 1-dependent amyloid precursor protein processing promotes excessive vascular sprouting through NOTCH3 signalling. Cell Death Dis. 2020;11(2):98.
139. Yamakawa H, et al. B-Secretase inhibitor potency is decreased by Aberrant β-Cleavage location of the "Swedish mutant" amyloid precursor protein. J Biol Chem. 2010;285(3):1634–42.

140. Mamada N, et al. Amyloid β-protein oligomers upregulate the β-secretase, BACE1, through a post-translational mechanism involving its altered subcellular distribution in neurons. Mol Brain. 2015;8(1):73.
141. Roßner S, Lichtenthaler SF. Alzheimer's disease BACE proteases. In: Lajtha A, Perez-Polo JR, Rossner S, editors. Handbook of neurochemistry and molecular neurobiology: development and aging changes in the nervous system. Boston: Springer; 2008. p. 261–84.
142. Zhang Y, et al. Amyloid β-based therapy for Alzheimer's disease: challenges, successes and future. Signal Transduct Target Ther. 2023;8(1):248.
143. Cardaci V, et al. Characterizing oxidative stress induced by Aβ oligomers and the protective role of carnosine in primary mixed glia cultures. Free Radic Biol Med. 2025;229:213–24.
144. Rajmohan R, Reddy PH. Amyloid-Beta and phosphorylated Tau accumulations cause abnormalities at synapses of Alzheimer's disease neurons. J Alzheimer's Dis. 2017;57(4):975–99.
145. Tang H, et al. Emerging biophysical origins and pathogenic implications of amyloid oligomers. Nat Commun. 2025;16(1):2937.
146. Fanlo-Ucar H, et al. The dual role of amyloid Beta-peptide in oxidative stress and inflammation: unveiling their connections in Alzheimer's disease etiopathology. Antioxidants. 2024;13(10):1208.
147. Thal DR, Tomé SO. The central role of tau in Alzheimer's disease: from neurofibrillary tangle maturation to the induction of cell death. Brain Res Bull. 2022;190:204–17.
148. Rawat P, et al. Phosphorylated Tau in Alzheimer's disease and other Tauopathies. Int J Mol Sci. 2022;23(21):12841.
149. Himmelstein DS, et al. Tau as a therapeutic target in neurodegenerative disease. Pharmacol Ther. 2012;136(1):8–22.
150. Brandt R, Hundelt M, Shahani N. Tau alteration and neuronal degeneration in tauopathies: mechanisms and models. Biochim Biophys Acta (BBA) – Mol Basis Dis. 2005;1739(2):331–54.
151. Hong X, et al. The role and pathogenesis of tau protein in Alzheimer's disease. Biomolecules. 2025;15(6):824.
152. Gendron TF, Petrucelli L. The role of tau in neurodegeneration. Mol Neurodegener. 2009;4(1):13.
153. Di Lorenzo D. Tau protein and tauopathies: exploring tau protein–protein and microtubule interactions, cross-interactions and therapeutic strategies. ChemMedChem. 2024;19(21):e202400180.
154. Abyadeh M, et al. Amyloid-beta and tau protein beyond Alzheimer's disease. Neural Regen Res. 2024;19(6):1262–76.
155. Maphis N, et al. Reactive microglia drive tau pathology and contribute to the spreading of pathological tau in the brain. Brain. 2015;138(Pt 6):1738–55.
156. Barron M, et al. A state of delirium: deciphering the effect of inflammation on tau pathology in Alzheimer's disease. Exp Gerontol. 2017;94:103–7.
157. Li JW, et al. Microglial priming in Alzheimer's disease. Ann Transl Med. 2018;6(10):176.
158. Bivona G, et al. Microglial activation and priming in Alzheimer's disease: state of the art and future perspectives. Int J Mol Sci. 2023;24(1):884.
159. Lima MN, Barbosa-Silva MC, Maron-Gutierrez T. Microglial priming in infections and its risk to neurodegenerative diseases. Front Cell Neurosci. 2022;16:878987.
160. Fišar Z. Linking the amyloid, tau, and mitochondrial hypotheses of Alzheimer's disease and identifying promising drug targets. Biomolecules. 2022;12(11):1676.
161. D'Alessandro MCB, et al. Mitochondrial dysfunction in Alzheimer's disease. Ageing Res Rev. 2025;107:102713.
162. Dominguez-Gortaire J, et al. Alzheimer's disease: exploring pathophysiological hypotheses and the role of machine learning in drug discovery. Int J Mol Sci. 2025;26(3):1004.
163. Wang WY, et al. Role of pro-inflammatory cytokines released from microglia in Alzheimer's disease. Ann Transl Med. 2015;3(10):136.
164. Wu Y, Eisel ULM. Microglia-astrocyte communication in Alzheimer's disease. J Alzheimer's Dis. 2023;95(3):785–803.

165. Calsolaro V, Edison P. Neuroinflammation in Alzheimer's disease: current evidence and future directions. Alzheimers Dement. 2016;12(6):719–32.
166. Dias D, Socodato R. Beyond amyloid and tau: the critical role of microglia in Alzheimer's disease therapeutics. Biomedicine. 2025;13(2):279.
167. Zhang H, et al. Astrocyte-mediated inflammatory responses in traumatic brain injury: mechanisms and potential interventions. Front Immunol. 2025;16:1584577.
168. Deng Q, et al. Microglia and astrocytes in Alzheimer's disease: significance and summary of recent advances. Aging Dis. 2024;15(4):1537–64.
169. Dash UC, et al. Oxidative stress and inflammation in the pathogenesis of neurological disorders: mechanisms and implications. Acta Pharm Sin B. 2025;15(1):15–34.
170. Cheignon C, et al. Oxidative stress and the amyloid beta peptide in Alzheimer's disease. Redox Biol. 2018;14:450–64.
171. Trofin D-M, et al. Oxidative stress in brain function. Antioxidants. 2025;14(3):297.
172. Zhao Y, Zhao B. Oxidative stress and the pathogenesis of Alzheimer's disease. Oxidative Med Cell Longev. 2013;2013(1):316523.
173. Sharma C, Kim SR. Oxidative stress: culprit or consequence in Alzheimer's amyloidopathy. Neural Regen Res. 2023;18(9):1948–9.
174. Foret MK, et al. Early oxidative stress and DNA damage in Aβ-burdened hippocampal neurons in an Alzheimer's-like transgenic rat model. Commun Biol. 2024;7(1):861.
175. Meng K, et al. Mitochondrial dysfunction in neurodegenerative diseases: mechanisms and corresponding therapeutic strategies. Biomedicine. 2025;13(2):327.
176. Xu X, Pang Y, Fan X. Mitochondria in oxidative stress, inflammation and aging: from mechanisms to therapeutic advances. Signal Transduct Target Ther. 2025;10(1):190.
177. Ježek J, Cooper KF, Strich R. Reactive oxygen species and mitochondrial dynamics: the yin and Yang of mitochondrial dysfunction and cancer progression. Antioxidants. 2018;7(1):13.
178. Shankar GM, Walsh DM. Alzheimer's disease: synaptic dysfunction and Abeta. Mol Neurodegener. 2009;4:48.
179. Spires-Jones TL, Hyman BT. The intersection of Amyloid Beta and Tau at synapses in Alzheimer's disease. Neuron. 2014;82(4):756–71.
180. Kandimalla R, Reddy PH. Therapeutics of neurotransmitters in Alzheimer's disease. J Alzheimer's Dis. 2017;57(4):1049–69.
181. Gasiorowska A, et al. The biology and pathobiology of glutamatergic, cholinergic, and dopaminergic signaling in the aging brain. Front Aging Neurosci. 2021;13:654931.
182. Teleanu RI, et al. Neurotransmitters—key factors in neurological and neurodegenerative disorders of the central nervous system. Int J Mol Sci. 2022;23(11):5954.
183. Kaczmarski P, et al. Influence of glutamatergic and GABAergic neurotransmission on obstructive sleep apnea. Front Neurosci. 2023;17:1213971.
184. Beretta E, Cuboni G, Deidda G. Unveiling GABA and serotonin interactions during neurodevelopment to re-open adult critical periods for neuropsychiatric disorders. Int J Mol Sci. 2025;26(12):5508.
185. Cutler AJ, Mattingly GW, Maletic V. Understanding the mechanism of action and clinical effects of neuroactive steroids and GABAergic compounds in major depressive disorder. Transl Psychiatry. 2023;13(1):228.
186. Kowalski K, Mulak A. Brain-gut-microbiota axis in Alzheimer's disease. J Neurogastroenterol Motil. 2019;25(1):48–60.
187. Zhong SR, et al. Functional roles of the microbiota-gut-brain axis in Alzheimer's disease: implications of gut microbiota-targeted therapy. Transl Neurosci. 2021;12(1):581–600.
188. Wang J, et al. Supplementation of lycopene attenuates lipopolysaccharide-induced amyloidogenesis and cognitive impairments via mediating neuroinflammation and oxidative stress. J Nutr Biochem. 2018;56:16–25.
189. Chen Y, et al. The multifaceted roles of apolipoprotein E4 in Alzheimer's disease pathology and potential therapeutic strategies. Cell Death Discov. 2025;11(1):312.
190. Du Y, et al. Inhibition of PKCδ reduces amyloid-β levels and reverses Alzheimer disease phenotypes. J Exp Med. 2018;215(6):1665–77.

191. Escorcia Mora P, Valbuena D, Diez-Juan A. The role of the gut microbiota in female reproductive and gynecological health: insights into endometrial signaling pathways. Life. 2025;15(5):762.
192. Ney LM, et al. Short chain fatty acids: key regulators of the local and systemic immune response in inflammatory diseases and infections. Open Biol. 2023;13(3):230014.
193. Chandra S, Sisodia SS, Vassar RJ. The gut microbiome in Alzheimer's disease: what we know and what remains to be explored. Mol Neurodegener. 2023;18(1):9.
194. Liang Y, et al. The link between gut microbiome and Alzheimer's disease: from the perspective of new revised criteria for diagnosis and staging of Alzheimer's disease. Alzheimers Dement. 2024;20(8):5771–88.
195. Qian XH, et al. Mechanisms of short-chain fatty acids derived from gut microbiota in Alzheimer's disease. Aging Dis. 2022;13(4):1252–66.
196. Lei W, et al. Gut microbiota-driven neuroinflammation in Alzheimer's disease: from mechanisms to therapeutic opportunities. Front Immunol. 2025;16:1582119.
197. Huang Y, et al. Short-chain fatty acids: important components of the gut-brain axis against AD. Biomed Pharmacother. 2024;175:116601.
198. Wu S, et al. Roles and mechanisms of gut microbiota in patients with Alzheimer's disease. Front Aging Neurosci. 2021;13:650047.
199. Pistollato F, et al. Role of gut microbiota and nutrients in amyloid formation and pathogenesis of Alzheimer disease. Nutr Rev. 2016;74(10):624–34.
200. Harach T, et al. Reduction of Abeta amyloid pathology in APPPS1 transgenic mice in the absence of gut microbiota. Sci Rep. 2017;7(1):41802.
201. Simpson DSA, Oliver PL. ROS generation in microglia: understanding oxidative stress and inflammation in neurodegenerative disease. Antioxidants (Basel). 2020;9(8):743.
202. Adamu A, et al. The role of neuroinflammation in neurodegenerative diseases: current understanding and future therapeutic targets. Front Aging Neurosci. 2024;16:1347987.
203. Chen H, et al. Neuroinflammation of microglial regulation in Alzheimer's disease: therapeutic approaches. Molecules. 2024;29(7):1478.
204. Seo D-O, Holtzman DM. Current understanding of the Alzheimer's disease-associated microbiome and therapeutic strategies. Exp Mol Med. 2024;56(1):86–94.
205. Eslami M, et al. A novel frontier in gut-brain axis research: the transplantation of fecal microbiota in neurodegenerative disorders. Biomedicine. 2025;13(4):915.
206. Zhou C, et al. Bidirectional modulation of Alzheimer's disease via gut microbiota: rescue by fecal transplantation from healthy donors and aggravation by colitis-associated dysbiosis. Front Neurosci. 2025;19:1593854.
207. Xiang W, et al. Fecal microbiota transplantation: a novel strategy for treating Alzheimer's disease. Front Microbiol. 2023;14:1281233.
208. Sun J, et al. Fecal microbiota transplantation alleviated Alzheimer's disease-like pathogenesis in APP/PS1 transgenic mice. Transl Psychiatry. 2019;9(1):189.
209. Zhou X-P, et al. The complex relationship between gut microbiota and Alzheimer's disease: a systematic review. Ageing Res Rev. 2025;104:102637.
210. Barber TM, et al. The effects of the Mediterranean diet on health and gut Microbiota. Nutrients. 2023;15(9):2150.
211. Frausto DM, et al. Dietary regulation of gut-brain axis in Alzheimer's disease: importance of microbiota metabolites. Front Neurosci. 2021;15:736814.
212. Merra G, et al. Influence of Mediterranean diet on human gut microbiota. Nutrients. 2020;13(1):7.
213. Ellouze I, et al. Dietary patterns and Alzheimer's disease: an updated review linking nutrition to neuroscience. Nutrients. 2023;15(14):3204.
214. Mazzola G, et al. Sustainable plant-based diets and food allergies: a scoping review inspired by EAT-lancet. Appl Sci. 2025;15(13):7296.
215. Yang K, et al. White matter changes in Parkinson's disease. Npj Parkinson's Disease. 2023;9(1):150.

216. Gómez-Benito M, et al. Modeling Parkinson's disease with the alpha-Synuclein protein. Front Pharmacol. 2020;11:356.
217. Radad K, et al. Pathophysiology of non-motor signs in Parkinson's disease: some recent updating with brief presentation. Explor Neuroprot Ther. 2023;3(1):24–46.
218. Guatteo E, et al. Pathophysiological features of Nigral dopaminergic neurons in animal models of Parkinson's disease. Int J Mol Sci. 2022;23(9):4508.
219. Calabresi P, et al. Alpha-synuclein in Parkinson's disease and other synucleinopathies: from overt neurodegeneration back to early synaptic dysfunction. Cell Death Dis. 2023;14(3):176.
220. Błaszczyk J. Nigrostriatal interaction in the aging brain: new therapeutic target for Parkinson's disease. Acta Neurobiol Exp. 2017;77(1):106–12.
221. Chinta SJ, et al. Inducible alterations of glutathione levels in adult dopaminergic midbrain neurons result in nigrostriatal degeneration. J Neurosci. 2007;27(51):13997–4006.
222. Gubellini P, et al. Downstream mechanisms triggered by mitochondrial dysfunction in the basal ganglia: from experimental models to neurodegenerative diseases. Biochim Biophys Acta. 2010;1802(1):151–61.
223. Larsen S, Hanss Z, Krüger R. The genetic architecture of mitochondrial dysfunction in Parkinson's disease. Cell Tissue Res. 2018;373(1):21–37.
224. van der Vlag M, Havekes R, Heckman PR. The contribution of Parkin, PINK1 and DJ-1 genes to selective neuronal degeneration in Parkinson's disease. Eur J Neurosci. 2020;52(4):3256–68.
225. Zhu M, et al. Gut microbiota: a novel therapeutic target for Parkinson's disease. Front Immunol. 2022;13:937555.
226. Huang B, et al. Gut microbiome dysbiosis across early Parkinson's disease, REM sleep behavior disorder and their first-degree relatives. Nat Commun. 2023;14(1):2501.
227. DeWeerdt S. Parkinson's gut-microbiota links raise treatment possibilities. Nature. 2025;
228. Mahbub NU, et al. Dysbiosis of the gut microbiota and its effect on α-synuclein and prion protein misfolding: consequences for neurodegeneration. Front Cell Infect Microbiol. 2024;14:1348279.
229. Romano S, et al. Meta-analysis of the Parkinson's disease gut microbiome suggests alterations linked to intestinal inflammation. Npj Parkinson's Disease. 2021;7(1):27.
230. Yang D, et al. The role of the gut microbiota in the pathogenesis of Parkinson's disease. Front Neurol. 2019;10:1155.
231. Liu B, et al. Vagotomy and Parkinson disease: a Swedish register-based matched-cohort study. Neurology. 2017;88(21):1996–2002.
232. Svensson E, et al. Vagotomy and subsequent risk of Parkinson's disease. Ann Neurol. 2015;78(4):522–9.
233. Metcalfe-Roach A, Finlay BB. The role of the gut microbiome in Parkinson's disease. Future Neurol. 2025;20(1):2494981.
234. Kollaparampil Kishanchand D, et al. The intricate interplay: microbial metabolites and the gut-liver-brain axis in Parkinson's disease. J Neurosci Res. 2025;103(1):e70016.
235. Nakhal MM, et al. The microbiota–gut–brain axis and neurological disorders: a comprehensive review. Life. 2024;14(10):1234.
236. Wang Q, et al. The role of gut dysbiosis in Parkinson's disease: mechanistic insights and therapeutic options. Brain. 2021;144(9):2571–93.
237. Henrich MT, et al. Mitochondrial dysfunction in Parkinson's disease – a key disease hallmark with therapeutic potential. Mol Neurodegener. 2023;18(1):83.
238. Liang Y, et al. Gut microbial metabolites in Parkinson's disease: implications of mitochondrial dysfunction in the pathogenesis and treatment. Mol Neurobiol. 2021;58(8):3745–58.
239. Ebadpour N, et al. From mitochondrial dysfunction to neuroinflammation in Parkinson's disease: pathogenesis and mitochondrial therapeutic approaches. Int Immunopharmacol. 2024;142:113015.
240. Sampson TR, et al. Gut microbiota regulate motor deficits and neuroinflammation in a model of parkinson's disease. Cell. 2016;167(6):1469–1480.e12.

241. Kim N, et al. Transplantation of gut microbiota derived from Alzheimer's disease mouse model impairs memory function and neurogenesis in C57BL/6 mice. Brain Behav Immun. 2021:98, 357–365.
242. Wang L, et al. Gut microbiota and Parkinson's disease. Chin Med J. 2025;138(3):289–97.
243. Heravi FS, Naseri K, Hu H. Gut microbiota composition in patients with neurodegenerative disorders (Parkinson's and Alzheimer's) and healthy controls: a systematic review. Nutrients. 2023;15(20):4365.
244. Proano AC, et al. Gut microbiota and its repercussion in Parkinson's disease: a systematic review in occidental patients. Neurol Int. 2023;15(2):750–63.
245. Hu Y, et al. Retrospective analysis of diet and gut microbiota diversity and clinical pharmacology outcomes in patients with parkinsonism syndrome. Heliyon. 2024;10(21):e38645.
246. Amorim Neto DP, et al. Akkermansia muciniphila induces mitochondrial calcium overload and α -synuclein aggregation in an enteroendocrine cell line. iScience. 2022;25(3):103908.
247. Bullich C, et al. Gut vibes in Parkinson's disease: the microbiota-gut-brain axis. Mov Disord Clin Pract. 2019;6(8):639–51.
248. Marzouk NH, et al. Proinflammatory and GABA eating bacteria in Parkinson's disease gut microbiome from a meta-analysis perspective. Npj Parkinson's Dis. 2025;11(1):145.
249. Menozzi E, Schapira AHV. The gut microbiota in Parkinson disease: interactions with drugs and potential for therapeutic applications. CNS Drugs. 2024;38(5):315–31.
250. Jameson KG, Hsiao EY. A novel pathway for microbial metabolism of levodopa. Nat Med. 2019;25(8):1195–7.
251. Shekar S, et al. Fecal microbiota transplantation for Parkinson's disease: a systematic review of clinical evidence. Med Microecol. 2025;25:100128.
252. Jackson A, et al. Diet in Parkinson's disease: critical role for the microbiome. Front Neurol. 2019;10:1245.
253. Santos S, et al. The role of microbiota dysbiosis in Parkinson's disease: pathophysiology and therapeutic opportunities. Eng Microbiol. 2025;5(3):100222.
254. Salim S, et al. Gut microbiome and Parkinson's disease: perspective on pathogenesis and treatment. J Adv Res. 2023;50:83–105.
255. Bisaglia M. Mediterranean diet and Parkinson's disease. Int J Mol Sci. 2022;24:42.
256. Papiri G, et al. Multiple sclerosis: inflammatory and neuroglial aspects. Curr Issues Mol Biol. 2023;45(2):1443–70.
257. Filippi M, et al. Multiple sclerosis. Nat Rev Dis Primers. 2018;4(1):43.
258. Ismael R. From gut to brain: exploring the role of extracellular vesicles in multiple sclerosis. Mol Neurobiol. 2024;63(1):349.
259. Mey GM, Mahajan KR, DeSilva TM. Neurodegeneration in multiple sclerosis. WIREs Mech Dis. 2023;15(1):e1583.
260. Ghasemi N, Razavi S, Nikzad E. Multiple sclerosis: pathogenesis, symptoms, diagnoses and cell-based therapy. Cell J. 2017;19(1):1–10.
261. Geurts JJG, Barkhof F. Grey matter pathology in multiple sclerosis. Lancet Neurol. 2008;7(9):841–51.
262. Lassmann H. Multiple sclerosis pathology. Cold Spring Harb Perspect Med. 2018;8(3):a028936.
263. Jeon YH, et al. Sudden death associated with possible flare-ups of multiple sclerosis after COVID-19 vaccination and infection: a case report and literature review. J Korean Med Sci. 2023;38(10):e78–0.
264. O'Loughlin E, et al. Microglial phenotypes and functions in multiple sclerosis. Cold Spring Harb Perspect Med. 2018;8(2):a028993.
265. Vila A, et al. Multiple sclerosis: enigmatic factors and new controversies. Clin Case Rep Rev. 2016;2:1–9.
266. Meyer-Arndt L, et al. Inflammatory cytokines associated with multiple sclerosis directly induce alterations of neuronal cytoarchitecture in human neurons. J Neuroimmune Pharmacol. 2023;18(1–2):145–59.

267. Pukoli D, Vécsei L. Smouldering lesion in MS: microglia, lymphocytes and pathobiochemical mechanisms. Int J Mol Sci. 2023;24(16):12631.
268. Jangi S, et al. Alterations of the human gut microbiome in multiple sclerosis. Nat Commun. 2016;7(1):12015.
269. Ordoñez-Rodriguez A, et al. Changes in gut microbiota and multiple sclerosis: a systematic review. Int J Environ Res Public Health. 2023;20(5):4624.
270. Nemati MH, et al. Microbiota-driven mechanisms in multiple sclerosis: pathogenesis, therapeutic strategies, and biomarker potential. Biology. 2025;14(4):435.
271. Altieri C, et al. Gut-microbiota, and multiple sclerosis: background, evidence, and perspectives. Nutrients. 2023;15(4):942.
272. Mohsen E, et al. Multiple sclerosis: a story of the interaction between gut microbiome and components of the immune system. Mol Neurobiol. 2025;62(6):7762–75.
273. Parodi B, de Rosbo NK. The gut-brain axis in multiple sclerosis. Is its dysfunction a pathological trigger or a consequence of the disease? Front Immunol. 2021;12:718220.
274. Buscarinu MC, et al. The contribution of gut barrier changes to multiple sclerosis pathophysiology. Front Immunol. 2019;10:1916.
275. Camara-Lemarroy CR, et al. The intestinal barrier in multiple sclerosis: implications for pathophysiology and therapeutics. Brain. 2018;141(7):1900–16.
276. Doroszkiewicz J, Groblewska M, Mroczko B. The role of gut microbiota and gut–brain interplay in selected diseases of the central nervous system. Int J Mol Sci. 2021;22(18):10028.
277. Oshaghi M, Kourosh-Arami M, Roozbehkia M. Role of neurotransmitters in immune-mediated inflammatory disorders: a crosstalk between the nervous and immune systems. Neurol Sci. 2023;44(1):99–113.
278. Correale J, Hohlfeld R, Baranzini SE. The role of the gut microbiota in multiple sclerosis. Nat Rev Neurol. 2022;18(9):544–58.
279. Dunalska A, Saramak K, Szejko N. The role of gut microbiome in the pathogenesis of multiple sclerosis and related disorders. Cells. 2023;12(13):1760.
280. Hodges H, Fealko C, Soares N. Autism spectrum disorder: definition, epidemiology, causes, and clinical evaluation. Transl Pediatr. 2020;9(Suppl 1):S55–s65.
281. Thye MD, et al. The impact of atypical sensory processing on social impairments in autism spectrum disorder. Dev Cogn Neurosci. 2018;29:151–67.
282. Lord C, et al. Autism spectrum disorder. Nat Rev Dis Prim. 2020;6(1):5.
283. Seltzer MM, et al. The symptoms of autism spectrum disorders in adolescence and adulthood. J Autism Dev Disord. 2003;33(6):565–81.
284. Lord C, et al. Autism spectrum disorder. Lancet. 2018;392(10146):508–20.
285. Fakhoury M. Autistic spectrum disorders: a review of clinical features, theories and diagnosis. Int J Dev Neurosci. 2015;43:70–7.
286. Nazeer A, Ghaziuddin M. Autism spectrum disorders: clinical features and diagnosis. Pediatr Clin. 2012;59(1):19–25.
287. Al-Beltagi M, et al. Decoding the genetic landscape of autism: a comprehensive review. World J Clin Pediatr. 2024;13(3):98468.
288. Jiang C-C, et al. Signalling pathways in autism spectrum disorder: mechanisms and therapeutic implications. Signal Transduct Target Ther. 2022;7(1):229.
289. Srinath S, et al. Small SNPs, big effects: a review of single nucleotide variations and polymorphisms in key genes associated with autism spectrum disorder. Int J Dev Neurosci. 2025;85(2):e70016.
290. Masini E, et al. An overview of the main genetic, epigenetic and environmental factors involved in autism spectrum disorder focusing on synaptic activity. Int J Mol Sci. 2020;21(21):8290.
291. Rylaarsdam L, Guemez-Gamboa A. Genetic causes and modifiers of autism spectrum disorder. Front Cell Neurosci. 2019;13:385.
292. Toscano CV, et al. Neuroinflammation in autism spectrum disorders: exercise as a "pharmacological" tool. Neurosci Biobehav Rev. 2021;129:63–74.

293. Canitano R, Palumbi R. Excitation/inhibition modulators in autism spectrum disorder: current clinical research. Front Neurosci. 2021;15:753274.
294. Ibrahim K, et al. Reduced amygdala-prefrontal functional connectivity in children with autism Spectrum disorder and co-occurring disruptive behavior. Biol Psychiatry Cogn Neurosci Neuroimaging. 2019;4(12):1031–41.
295. Mediane DH, et al. Medial prefrontal cortex circuitry and social behaviour in autism. Neuropharmacology. 2024;260:110101.
296. Morton JT, et al. Multi-level analysis of the gut–brain axis shows autism spectrum disorder-associated molecular and microbial profiles. Nat Neurosci. 2023;26(7):1208–17.
297. Zhou M, et al. Intervention and research progress of gut microbiota-immune-nervous system in autism spectrum disorders among students. Front Microbiol. 2025;16:1535455.
298. Gonçalves CL, et al. The many faces of microbiota-gut-brain axis in autism spectrum disorder. Life Sci. 2024;337:122357.
299. Fattorusso A, et al. Autism spectrum disorders and the gut microbiota. Nutrients. 2019;11(3):521.
300. Mehra A, et al. Gut microbiota and autism spectrum disorder: from pathogenesis to potential therapeutic perspectives. J Tradit Complement Med. 2023;13(2):135–49.
301. Guevara-Ramírez P, et al. Mechanistic links between gut dysbiosis, insulin resistance, and autism spectrum disorder. Int J Mol Sci. 2025;26(13):6537.
302. Moreno RJ, Ashwood P. An update on microbial interventions in autism spectrum disorder with gastrointestinal symptoms. Int J Mol Sci. 2024;25(23):13078.
303. Nguyen HD, Kim W-K. Disrupted microbial cross-feeding and altered L-phenylalanine consumption in people living with HIV. Brief Bioinform. 2025;26(2):bbaf111.
304. Kumar A, et al. Probiotics as modulators of gut-brain axis for cognitive development. Front Pharmacol. 2024;15:1348297.
305. Fang Z, et al. Gut microbiota and autism spectrum disorder: advances in dietary intervention strategies based on the microbiota-gut-brain axis mechanism. Front Neurosci. 2025;19:1587818.
306. Taniya MA, et al. Role of gut microbiome in autism spectrum disorder and its therapeutic regulation. Front Cell Infect Microbiol. 2022;12:915701.
307. Zarimeidani F, et al. Gut Microbiota and autism spectrum disorder: a neuroinflammatory mediated mechanism of pathogenesis? Inflammation. 2025;48(2):501–19.
308. Hashimoto K. Emerging role of the host microbiome in neuropsychiatric disorders: overview and future directions. Mol Psychiatry. 2023;28(9):3625–37.
309. Rogers GB, et al. From gut dysbiosis to altered brain function and mental illness: mechanisms and pathways. Mol Psychiatry. 2016;21(6):738–48.
310. Denman CR, Park SM, Jo J. Gut-brain axis: gut dysbiosis and psychiatric disorders in Alzheimer's and Parkinson's disease. Front Neurosci. 2023;17:1268419.
311. Marwaha K, et al. Exploring the complex relationship between psychosocial stress and the gut microbiome: implications for inflammation and immune modulation. J Appl Physiol. 2025;138(2):518–35.
312. Kumar A, et al. Gut Microbiota in anxiety and depression: unveiling the relationships and management options. Pharmaceuticals (Basel). 2023;16(4):565.
313. Warren A, et al. Dangers of the chronic stress response in the context of the microbiota-gut-immune-brain axis and mental health: a narrative review. Front Immunol. 2024;15:1365871.
314. Mallick K, et al. Gut microbes: therapeutic target for neuropsychiatric disorders. J Psychiatr Res. 2025;184:27–38.
315. Kalaga P, Ray SK. Mental health disorders due to gut microbiome alteration and NLRP3 Inflammasome activation after spinal cord injury: molecular mechanisms, promising treatments, and aids from artificial intelligence. Brain Sci. 2025;15(2):197.
316. Noonan S, et al. Food & mood: a review of supplementary prebiotic and probiotic interventions in the treatment of anxiety and depression in adults. BMJ Nutr Prev Health. 2020;3(2):351–62.

317. Chao L, et al. Effects of probiotics on depressive or anxiety variables in healthy participants under stress conditions or with a depressive or anxiety diagnosis: a meta-analysis of randomized controlled trials. Front Neurol. 2020;11:421.
318. Krsek A, et al. Navigating the neuroimmunomodulation frontier: pioneering approaches and promising horizons—a comprehensive review. Int J Mol Sci. 2024;25(17):9695.
319. Miller AH, Raison CL. Are anti-inflammatory therapies viable treatments for psychiatric disorders?: where the rubber meets the road. JAMA Psychiatry. 2015;72(6):527–8.
320. Maia JM, et al. Exploring the links between pro-inflammatory diets, gut dysbiosis, serotonin, and their implications for psychiatric and neurodegenerative disorders. Brain Behav Immun Integr. 2024;8:100097.
321. Liu Q, et al. Multi-kingdom microbiota analysis reveals bacteria-viral interplay in IBS with depression and anxiety. npj Biofilms Microbiomes. 2025;11(1):129.
322. Schmidt TSB, Raes J, Bork P. The human gut microbiome: from association to modulation. Cell. 2018;172(6):1198–215.
323. Goldiş A, et al. Personalized microbiome modulation to improve clinical outcomes in pediatric inflammatory bowel disease: a multi-omics and interventional approach. Microorganisms. 2025;13(5):1047.
324. Ashique S, et al. Gut-brain axis: a cutting-edge approach to target neurological disorders and potential synbiotic application. Heliyon. 2024;10(13):e34092.
325. Dandamudi BJ, et al. Neurodegenerative disorders and the gut-microbiome-brain axis: a literature review. Cureus. 2024;16(10):e72427.
326. Che Mohd Nassir CMN, et al. The microbiota–gut–brain axis: key mechanisms driving glymphopathy and cerebral small vessel disease. Life. 2025;15(1):3.
327. Tang W, et al. The gut microbiota–oligodendrocyte axis: a promising pathway for modulating oligodendrocyte homeostasis and demyelination-associated disorders. Life Sci. 2024;354:122952.
328. Zheng Y, et al. Probiotics fortify intestinal barrier function: a systematic review and meta-analysis of randomized trials. Front Immunol. 2023;14:1143548.
329. Chakravarty K, et al. Exploring the multifaceted therapeutic potential of probiotics: a review of current insights and applications. Probiotics Antimicrob Proteins. 2025;17(1):341–63.
330. Kim JH, et al. Probiotics as potential treatments for neurodegenerative diseases: a review of the evidence from in vivo to clinical trial. Biomol Ther (Seoul). 2025;33(1):54–74.
331. Zhu F, et al. Effects of probiotics on neurodegenerative disease-related symptoms and systemic inflammation: a systematic review. Int J Gen Med. 2024;17:5941–58.
332. Renesteen E, et al. Microbiome engineering for biotherapeutic in Alzheimer's disease Through the gut–brain axis: potentials and limitations. Int J Mol Sci. 2025;26(11):5351.
333. Park JM, et al. Effect of probiotic supplementation on gastrointestinal motility, inflammation, motor, non-motor symptoms and mental health in Parkinson's disease: a meta-analysis of randomized controlled trials. Gut Pathog. 2023;15(1):9.
334. Crocetta A, et al. From gut to brain: unveiling probiotic effects through a neuroimaging perspective—a systematic review of randomized controlled trials. Front Nutr. 2024;11:1446854.
335. Gibson GR, et al. Expert consensus document: The International Scientific Association for Probiotics and Prebiotics (ISAPP) consensus statement on the definition and scope of prebiotics. Nat Rev Gastroenterol Hepatol. 2017;14(8):491–502.
336. Kaur AP, et al. Plant prebiotics and their role in the amelioration of diseases. Biomolecules. 2021;11(3):440.
337. Azad MAK, et al. Opportunities of prebiotics for the intestinal health of monogastric animals. Anim Nutr. 2020;6(4):379–88.
338. Macfarlane GT, Steed H, Macfarlane S. Bacterial metabolism and health-related effects of galacto-oligosaccharides and other prebiotics. J Appl Microbiol. 2008;104(2):305–44.
339. Roy S, Dhaneshwar S. Role of prebiotics, probiotics, and synbiotics in management of inflammatory bowel disease: current perspectives. World J Gastroenterol. 2023;29(14):2078–100.

340. Smolinska S, Popescu F-D, Zemelka-Wiacek M. A review of the influence of prebiotics, probiotics, synbiotics, and postbiotics on the human gut microbiome and intestinal integrity. J Clin Med. 2025;14(11):3673.
341. Quagebeur R, et al. The role of short-chain fatty acids (SCFAs) in regulating stress responses, eating behavior, and nutritional state in anorexia nervosa: protocol for a randomized controlled trial. J Eat Disord. 2023;11(1):191.
342. Lee S-H, Han C, Shin C. IUPHAR review: microbiota-gut-brain axis and its role in neuropsychiatric disorders. Pharmacol Res. 2025;216:107749.
343. You S, et al. The promotion mechanism of prebiotics for probiotics: a review. Front Nutr. 2022;9:1000517.
344. Vendrik KEW, et al. Fecal microbiota transplantation in neurological disorders. Front Cell Infect Microbiol. 2020;10:98.
345. Matheson JT, Holsinger RMD. The role of fecal Microbiota transplantation in the treatment of neurodegenerative diseases: a review. Int J Mol Sci. 2023;24(2):1001.
346. Fishbein SRS, Mahmud B, Dantas G. Antibiotic perturbations to the gut microbiome. Nat Rev Microbiol. 2023;21(12):772–88.
347. Pilmis B, Le Monnier A, Zahar JR. Gut Microbiota, antibiotic therapy and antimicrobial resistance: a narrative review. Microorganisms. 2020;8(2):269.
348. Kaji K, et al. Rifaximin ameliorates hepatic encephalopathy and endotoxemia without affecting the gut microbiome diversity. World J Gastroenterol. 2017;23(47):8355–66.
349. Ponziani FR, et al. Effect of rifaximin on gut microbiota composition in advanced liver disease and its complications. World J Gastroenterol. 2015;21(43):12322–33.
350. Patel VC, et al. Rifaximin-α reduces gut-derived inflammation and mucin degradation in cirrhosis and encephalopathy: RIFSYS randomised controlled trial. J Hepatol. 2022;76(2):332–42.
351. Park KJ, Gao Y. Gut-brain axis and neurodegeneration: mechanisms and therapeutic potentials. Front Neurosci. 2024;18:1481390.
352. Han Q, et al. Microglial NLRP3 inflammasome-mediated neuroinflammation and therapeutic strategies in depression. Neural Regen Res. 2024;19(9):1890–8.
353. Xu W, Huang Y, Zhou R. NLRP3 inflammasome in neuroinflammation and central nervous system diseases. Cell Mol Immunol. 2025;22(4):341–55.
354. Hanslik KL, Ulland TK. The role of microglia and the Nlrp3 Inflammasome in Alzheimer's disease. Front Neurol. 2020;11:570711.
355. Zhang X, et al. Inhibitors of the NLRP3 inflammasome pathway as promising therapeutic candidates for inflammatory diseases (review). Int J Mol Med. 2023;51(4):35.
356. Wang W, et al. Advances in VNS efficiency and mechanisms of action on cognitive functions. Front Physiol. 2024;15:1452490.
357. Goggins E, Mitani S, Tanaka S. Clinical perspectives on vagus nerve stimulation: present and future. Clin Sci (Lond). 2022;136(9):695–709.
358. Chen Z, Liu K. Mechanism and applications of vagus nerve stimulation. Curr Issues Mol Biol. 2025;47(2):122.
359. Faraji N, et al. Vagus nerve stimulation and gut microbiota interactions: a novel therapeutic avenue for neuropsychiatric disorders. Neurosci Biobehav Rev. 2025;169:105990.
360. Hesampour F, Bernstein CN, Ghia JE. Brain-gut Axis: invasive and noninvasive Vagus nerve stimulation, limitations, and potential therapeutic approaches. Inflamm Bowel Dis. 2024;30(3):482–95.
361. Liu F-J, et al. Non-invasive vagus nerve stimulation in anti-inflammatory therapy: mechanistic insights and future perspectives. Front Neurosci. 2024;18:1490300.
362. Berding K, et al. Diet and the microbiota-gut-brain axis: sowing the seeds of good mental health. Adv Nutr. 2021;12(4):1239–85.
363. Horn J, et al. Role of diet and its effects on the gut microbiome in the pathophysiology of mental disorders. Transl Psychiatry. 2022;12(1):164.
364. Koblinsky ND, et al. The role of the gut microbiome in diet and exercise effects on cognition: a review of the intervention literature. J Gerontol A Biol Sci Med Sci. 2023;78(2):195–205.

365. Koblinsky ND, et al. The role of the gut microbiome in diet and exercise effects on cognition: a review of the intervention literature. J Gerontol Series A. 2022;78(2):195–205.
366. Aran KR, et al. Postbiotics as a therapeutic tool in Alzheimer's disease: insights into molecular pathways and neuroprotective effects. Ageing Res Rev. 2025;106:102685.
367. Xiao L, et al. Postbiotics regulate intestinal Microbiota and reduce amyloid β-protein deposition in the brain, thereby ameliorating cognitive impairment in rats with Alzheimer's disease. Food Sci. 2025;46(5):182–93.
368. Gupta MK, et al. Innovative interventions: postbiotics and psychobiotics in neurodegenerative disease treatment. Probiotics Antimicr Prot. 2025;
369. Głowacka P, et al. Postbiotics as molecules targeting cellular events of aging brain-the role in pathogenesis, prophylaxis and treatment of neurodegenerative diseases. Nutrients. 2024;16(14):2244.
370. Mohammadpour D, et al. Application of probiotics and postbiotics in neurological disorders. Curr Pharm Biotechnol. 2025;27
371. Wadan A-HS, El-Aziz MKA, Ellakwa DE-S. The microbiota-gut-brain-axis theory: role of gut microbiota modulators (GMMs) in gastrointestinal, neurological, and mental health disorders. Naunyn Schmiedeberg's Arch Pharmacol. 2025;398:13397.
372. Hill C, et al. The International Scientific Association for probiotics and prebiotics consensus statement on the scope and appropriate use of the term probiotic. Nat Rev Gastroenterol Hepatol. 2014;11(8):506–14.
373. Dalile B, et al. The role of short-chain fatty acids in microbiota–gut–brain communication. Nat Rev Gastroenterol Hepatol. 2019;16(8):461–78.
374. Dinan TG, Cryan JF. Brain–gut–microbiota axis — mood, metabolism and behaviour. Nat Rev Gastroenterol Hepatol. 2017;14(2):69–70.
375. Cryan JF, et al. The gut microbiome in neurological disorders. Lancet Neurol. 2020;19(2):179–94.
376. Charitos IA, et al. The gut Microbiota's role in neurological, psychiatric, and neurodevelopmental disorders. Nutrients. 2024;16(24):4404.
377. Galland L. The gut microbiome and the brain. J Med Food. 2014;17(12):1261–72.
378. Fang X, et al. Precision microbiome: a new era of targeted therapy with core probiotics. Research (Wash D C). 2025;8:0658.
379. Dinan TG, Cryan JF. Gut instincts: microbiota as a key regulator of brain development, ageing and neurodegeneration. J Physiol. 2017;595(2):489–503.
380. Rutsch A, Kantsjö JB, Ronchi F. The gut-brain axis: how microbiota and host inflammasome influence brain physiology and pathology. Front Immunol. 2020;11:604179.

Microbiota-Gut-Brain Axis and Its Potential Implications on Neuropsychiatric Disorders with a Special Emphasis on Bile Acid Signaling

10

Muhasina K M, Puja Ghosh, Deepika N P, Krishnaveni Nagappan, Dhanabal S. Palaniswamy, Md Ekhtiar Uddin, and Md Abul Hassan

10.1 Introduction

The gut microbiota is a large and varied group of micro-sized organisms that live inside the gut or gastrointestinal organs. This intricate ecosystem, comprising bacteria, viruses, archaea, fungi, and protozoa, is essential for preserving host homeostasis. The microbes that reside inside the body, referred to as microbiota, are more dominant than the human cells or genomes. The microbial genomes, also known as the microbiome, consist of features that confer features that significantly impact human evolution. Humans can be considered as an amalgamation of human as well as microbial cells, depicting the fact that humans are a "supraorganism" [1]. In

Muhasina K M (✉)
Department of Pharmaceutical Analysis, Prime College of Pharmacy, Palakkad, Kerala, India

P. Ghosh · Deepika N P · D. S. Palaniswamy
Department of Pharmacognosy, JSS College of Pharmacy, Ooty, Tamil Nadu, India
e-mail: spdhanabal@jssuni.edu.in

K. Nagappan
Department of Pharmaceutical Analysis, JSS College of Pharmacy, JSS Academy of Higher Education & Research, Ooty, The Nilgiris, Tamil Nadu, India
e-mail: krisath@jssuni.edu.in

M. E. Uddin
The School of Medicine, Medical Sciences and Nutrition, University of Aberdeen, Aberdeen, UK

M. A. Hassan
Obihiro University of Agriculture and Veterinary Medicine, Obihiro, Japan

M. H. Rahman, Muhasina K M (eds.), *The Microbiome's Impact on the Brain*,
https://doi.org/10.1007/978-981-95-7494-0_10

addition to that, they have well-known roles in digestion, nutrient absorption, and immune modulation, and accumulating evidence indicates that the gut microbiome significantly impacts the CNS (central nervous system) through a two-way communication network known as the microbiota-gut-brain axis (MGBA) [2]. This axis combines neural, immune, endocrine, and metabolic pathways, allowing the gut microbiota to change the thinking inside the brain, how people will react, and also how they feel.

Numerous distinct processes within our body may be associated with the gut microbiota, particularly the development and maturation of the CNS and the formation and regulation of the immune response. It is a significant component of the gut-brain axis, interacting with the brain through the microbiota-gut-brain axis [3]. It may regulate intestinal homeostasis, human metabolism, and immunological development [4], as well as neurological processes and behavior [5]. A diverse and stable gut microbiota, essential for maintaining health, produces chemicals that facilitate metabolic and physiological functioning. This organism performs different functions, including the metabolism of medicine and food, preservation of gut barrier integrity, immunomodulation, and infection defense, therefore contributing to a healthy gut ecology [6]. Neuropsychiatric disorders, including depression, bipolar disorder, schizophrenia, anxiety, and autism spectrum disorders (ASD), have historically been regarded as predominantly brain-centric conditions. However, recent studies indicate that disturbances in gut microbial composition (dysbiosis) and the changes in microbial production of metabolites may play a role in their pathogenesis and severity [7]. This paradigm shift emphasizes the necessity of investigating peripheral factors, especially the gut microbiota, as potential modulators and therapeutic targets in mental health.

Dysbiosis, characterized by alterations in the structure of the gut microbiota, is associated with a diverse array of conditions, including inflammatory bowel disease [8], coeliac disease [9], metabolic syndrome [10], diabetes mellitus, depression, autism, anxiety, colon cancer, and neurodegenerative disorders [11, 12]. This has led to a significant focus on potential changes in microbiota associated with inflammation and the development of neurological and cognitive diseases. The pathways connecting gut microbiomes and neurocognitive disorders remain unclear. Research on the gut-brain axis has yielded crucial insights into human perception and its connection to neuroendocrinology and gastrointestinal (GI) disorders [12]. This specific axis includes a chain of action that integrates the GI tract, CNS, and the enteric nervous system, encompassing the brain. While the precise mechanisms linking gut microbiota and brain function require further clarification, multiple potential pathways exist through which gut microbiota may influence brain activity [13]. Microbiota-gut-brain connection may transpire via several systems that constitute the gut-brain axis, including the CNS, enteric nervous system, autonomic nervous system, neuroendocrine systems, and immune system. Microorganisms can affect CNS processes through the vagus nerve [14] and through the immunomodulation [15], the hypothalamic-pituitary-adrenal (HPA) axis [16], and tryptophan metabolism. They also possess the capability to synthesize various neurotransmitters [17] and produce metabolites, such as short-chain fatty acids (SCFAs).

These SCFAs are believed to play a crucial role in the interaction between the gut, microbiota, and the brain [18]. Besides the established functions of the colon in energy provision and trophic factor [19], SCFAs will manage the T-regulatory (Treg) cell [20]. Additionally, increasing evidence indicates that SCFAs have significant physiological impacts on the brain cells. SCFAs pass from the gut to the blood and blood to the brain barriers, get into the CNS, and accumulate within the brain cells. They may promote acidification intracellular matrix [21] and possibly influence calcium signaling, neurotransmitter release, function of mitochondria, activation of natural immunity, and expression of genes, so altering the communications in the neurons and host behavior. Depression and anxiety disorders are the most prevalent psychiatric illnesses among adults. Disruption of HPA axis signaling is related to mood disorders, often characterized by elevated inflammatory mediators and cortisol levels that contribute to a persistent pro-inflammatory state [16]. The amygdala, a brain neuronal structure, is essential for the creation, processing, and experience of emotional information [22]. The influence of microbes on the morphology and physiology of the amygdala may elucidate several emotional and stress-related variables associated with the microbiome [23]. Serotonin is connected to the etiology of different illnesses, including depression and anxiety. Studies indicate that certain treatments involving prebiotics and probiotics may modify the gut microbiome composition and enhance the peripheral levels of the serotonin precursor, tryptophan, in the brains of animal models demonstrating depression and chronic stress [24].

The relation between AD and gut microbiota has also been investigated. The association between environmental and genetic factors is thought to affect the genesis of Alzheimer's disease, with the gut microbiota being one such factor. Studies utilizing animal models have demonstrated that gut dysbiosis has a role in the initiation, progression, and exacerbation of Alzheimer's disease, encompassing persistent neuroinflammation and oxidative stress, among other elements. The relationship between cognitive and motor impairment in Parkinson's disease and alterations in gut-brain axis symptoms has gained substantial academic interest in recent years. Gastrointestinal symptoms may come into effect before the emergence of cognitive and motor problems by several years [3, 25]. The influence of gut microbiota on the CNS is being examined in ADHD, a neurodevelopmental disease marked by impulsivity, lack of attention, and hyperactivity disorders [26]. ADHD may be linked to "unhealthy" diets characterized by inadequate amounts of zinc, magnesium, and iron, and dietary practices significantly influence the makeup of gut microbiota [27]. Consequently, they have been assigned to play a possible role in the pathogenesis of ADHD. Comprehending the initial interplay between gut microbiota and the development of the specified

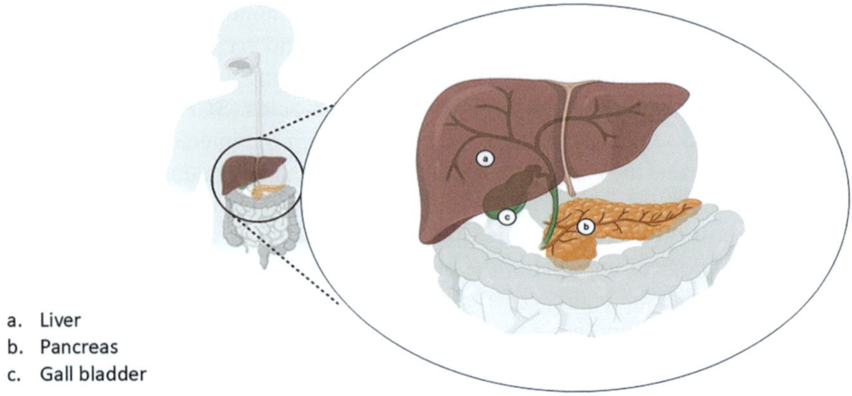

Fig. 10.1 Bile acid release mechanism

illnesses may provide novel intervention strategies, especially for early diagnosis and treatment.

Bile acids (BAs) have become more important as one of the many microbial-derived metabolites that affect the MGBA. Bile acids are known for the breakdown of fats and the processing of cholesterol a long ago. Now, however, they are also known to act as signaling molecules with effects that go beyond the digestive system. They interact with nuclear receptors like the farnesoid X receptor (FXR) [7] and membrane-bound receptors like the Takeda G protein-coupled receptor 5 (TGR5) [28]. This affects not only metabolic pathways but also immune responses and neuroinflammation. Gut microbiota significantly influences bile acid composition via deconjugation, dihydroxylation, and epimerization reactions, resulting in secondary bile acids with unique biological activities [29]. Recent findings suggest that bile acid signaling may directly or indirectly influence neuronal excitability, neurotransmitter synthesis, blood-brain barrier (BBB) integrity, and neuroinflammatory pathways [30]. Dysregulation in bile acid metabolism (Fig. 10.1), frequently resulting from gut dysbiosis, has been linked to modified neurobehavioral outcomes in both animal models and human studies. This connection establishes bile acids as essential mediators within the MGBA and as prospective biomarkers or therapeutic targets in neuropsychiatric disorders [31]. This chapter aims to investigate the link between gut microbiota and its role in various illnesses, and a comprehensive knowledge of the pathophysiology of many illnesses may facilitate early diagnosis, improved symptom progression, and enhanced therapy choices. We are also discussing the chemical interplay between the GB axis, giving significant attention toward bile acid regulation and its potential implications on gut microbiota.

10.2 Signaling Cascade in Neuroendocrine Organs

The hypothalamic-pituitary-adrenal (HPA) axis is the principal endocrine system that connects gut activity to the CNS. Signals from the gut, such as inflammatory mediators and microbial metabolites, can turn on the HPA axis, which causes the liberation of corticotropin-releasing hormone (CRH), cortisol, and adrenocorticotropic hormone (ACTH) [32]. Chronic activation of the HPA axis, frequently seen in stress-related disorders, has been associated with gut dysbiosis, modified bile acid metabolism, and heightened gut permeability. Endocrine communication is two-way because stress hormones can change how the gut moves, how well it acts as a barrier, and the types of bacteria that live there. This creates a feedback loop that can make neuropsychiatric symptoms worse. The gut-associated lymphoid tissue (GALT) is the largest immune compartment in the body. It keeps a delicate balance between tolerance and immune activation [33]. Microbial dysbiosis can upset this balance, causing low-grade systemic inflammation. Cytokines (e.g., IL-6, IL-1β, TNF-α) released in the periphery can traverse or signal across the BBB, affecting microglial activation and neuroinflammation [34]. Bile acids can regulate immune function by binding to receptors such as FXR and TGR5 on immune cells, thereby influencing inflammatory cascades pertinent to neuropsychiatric disorders.

Microbiota generates various bioactive metabolites that function as chemical messengers in the MGBA because of the processing of short-chain fatty acids, after the addition of fiber to the diet. Some important examples are short-chain fatty acids (SCFAs) like acetate, butyrate, and propionate, which change gene expression, neurotransmitter production, and immune function. Tryptophan metabolites, such as serotonin and kynurenine, affect mood and cognitive processes [35]. Bile acids, in addition to their digestive functions, serve as signaling molecules capable of influencing metabolic, immune, and neural pathways (Fig. 10.2). They are a type of metabolite that has two functions such as involvement in digestion and acting like hormones to connect gut bacteria with the body's physiology [36]. Their capacity to engage with nuclear and membrane-bound receptors positions them centrally to affect neuropsychiatric outcomes. They are also amphipathic steroid molecules in the liver from cholesterol and are crucial for lipid digestion and absorption [37]. BAs are well known for their classical role in emulsifying fats, but new research shows that they are also bioactive signaling molecules that have effects throughout the body, including on the CNS. The gut microbiota is significant for changing the composition, diversity, and signaling potential of the bile acid pool. This makes BA metabolism a key part of the MGBA.

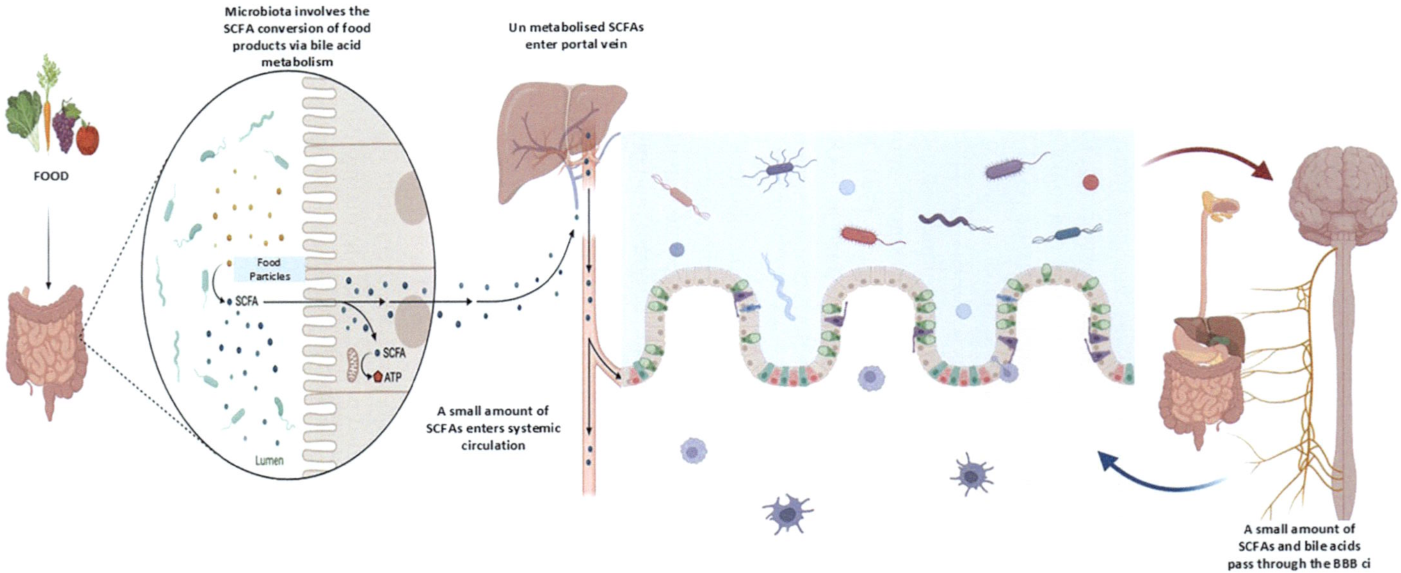

Fig. 10.2 Activity of food on bile acid metabolism and SCFs that have been influenced by gut microbiome and further enter the molecules into brain cells

10.3 Potential Implications of Bile Acid Regulation on Gut Microbiota

The MGBA is a two-way communication network that links the GI tract and the microorganisms that live there with the CNS. This axis brings together neural, endocrine, immune, and metabolic signaling pathways, making sure that peripheral and central systems are always communicating with each other. Changes in any part of this network can have big effects on the host's body, including mental health and brain function. The vagus nerve acts as a direct pathway between the brain and the gut. It sends sensory information about the contents of the gut, how it moves, and how microbes are doing their work. The ENS is also called the "second brain" because it works on its own to control gastrointestinal motility, secretion, and barrier function [38]. It also sends information to the CNS. Microbial metabolites, such as SCFAs and bile acids, can affect neuronal activity via these neural pathways, modifying vagal tone and sensory neuron excitability.

Bile acids are signaling hormones that regulate several metabolic processes. The gut microbiota modulates the capacity of the bile acids to combine and communicate through their receptors, as it influences the regulation, production, and physicochemical characteristics of bile acids. In humans, hepatocytes synthesize primary bile acids such as chenodeoxycholic acid (CDCA) and cholic acid (CA) through the classical (CYP7A1-mediated) and alternative (CYP27A1-mediated) pathways [39]. These acids are conjugated with glycine or taurine before being secreted into the bile. This makes them more soluble and stops them from being passively reabsorbed in the proximal intestine. Conjugated bile acids are released into the duodenum, where they help the small intestine break down and absorb fats [40]. Bacteria in the gut transmit messages to the CNS through their metabolites, so influencing brain function and the behavior of the host. Recent years have elucidated the impact of bile acids on cognitive activity in both normal physiological and pathological contexts. The primary source of bile acids present in the brain may be their absorption from the systemic circulation, although they may also be synthesized locally in the brain. The bacteria in the gut may change the makeup of the bile acid pool in the brain. This suggests that bile acids communicate between the gut microbiome and the brain [41].

A significant portion of bile acids reenter into the terminal ileum through active transport and then come back to the hepatic system through the enterohepatic circulation [42]. The rest of the bile acids go to the colon, where they meet the gut flora. They may alter bile acids via biotransformation reactions that significantly modify their chemical structure and signaling characteristics, such as deconjugation, 7α-dehydroxylation, epimerization, and oxidation. This has been done by microbial bile salt hydrolases (BSHs), which mostly come from the generated *Bacteroides*, *Lactobacillus*, *Bifidobacterium*, and *Clostridium*. Alterations in hydroxyl groups into different stereoisomers (like ursodeoxycholic acid (UDCA)) have different anti-inflammatory and neuroprotective effects [31]. Anaerobic bacteria (like *Clostridium scindens*) do this to turn CA into deoxycholic acid (DCA) and CDCA into lithocholic acid (LCA). These changes not only add to the variety of bile acids but also create molecules that bind to new receptors, which can alter metabolism of the host, immunity, and possibly even the function of the central nervous system [43]. Several enquiries persist about the selection pressures that regulate the development of a stable microbiota, the factors influencing its transfer to the host's system, and the components that distinguish the evolution of the human gut microbiota from that of other mammals [44].

Conjugated bile acids (CBAs) [45, 46] improve lipid absorption and function as signaling molecules that modulate systemic endocrine functions, including the regulation of triglycerides, cholesterol, and potentially glucose homeostasis [45, 47, 48]. Moreover, CBAs are proposed to inhibit bacterial proliferation in the small intestine via direct antibacterial properties, enhancement of human mucosal defenses, or a combination of both methods [45, 49–51]. The principal "gateway" event in the metabolism of CBAs of bacteria is mediated by bile salt hydrolase (BSH), often referred to as choloylglycine hydrolase, which catalyzes the deconjugation process. The principal "gateway" process in bacterial metabolism of conjugated bile acids (CBAs) is facilitated by choloylglycine hydrolase, also known as bile salt hydrolase (BSH). Then the deconjugation of CBAs will be catalyzed by BSH to release free primary bile acids (BAs), such as chenodeoxycholic acid or cholic acid, along with amino acids. Unconjugated primary bile acids are then extensively modified by the gut flora, via dehydrogenation, dihydroxylation, and sulfation and form secondary and tertiary bile acids [45, 46, 52, 53].

Deconjugated and modified bile acids have been connected to the development of gallstones and colorectal cancer, while current research suggests that the gut microbiota may alter the host's bile acid production and metabolism, potentially affecting the lipid metabolism of the organism. The activity of BSH and consequent BA alteration in this community may substantially affect host physiology by disrupting BA-regulated endocrine activities and altering the risk of metabolic disorders, including obesity, diabetes, and atherosclerosis [47, 48]. This study employs metagenomics to find functional anchors, akin to phylogenetic anchors such as 16S rRNA, within a group of microorganisms and utilizes these genes within a phylogenetic framework to assess the gut flora based on gastrointestinal tract survival activities. This approach evaluates functional distribution and redundancy within a community, while functional anchoring will facilitate sequence-based metagenomic analysis. To establish a healthy microbiota (symbiotic, commensal, or mutualistic relationships with the human host), bacteria will surmount numerous challenges when colonizing the mammalian gastrointestinal tract without jeopardizing host fitness. The primary mechanisms for preventing small intestine bacterial overgrowth are the antibacterial capabilities of bile acids and the activation of mucosal defenses. Our findings indicate that BSH activity enhances bacterial resistance to CBA and promotes their survival in the digestive tract, facilitating colonization and the evolution of gut flora. The host will indirectly benefit from the establishment of a gut microbial group that safeguards against gastrointestinal diseases and enhances caloric absorption. The preservation and enhancement of BSH activity in the microbiota metagenomes of healthy individuals indicate that BSH may assist the small intestine in combating CBA without causing bacterial overgrowth or injury to the host. Local mucosal defenses and the suppressive effects of BA may restrict bacterial colonization and suppress BSH synthesis in the small intestine [9, 13–15].

The broad phylogenetic dissemination and notable redundancy of BSH activity in the human gut flora imply host-level choice for this function, and recent investigations demonstrate direct gain of BSH activity on host health. Investigations into the intestinal pathogen *Clostridium difficile* revealed that metabolites derived from microbial bile acid (BA) metabolism inhibit the in vitro growth of vegetative cells, implying a direct role of modified BA in protecting the host from *C. difficile* and other gastrointestinal infections. To establish a healthy microbiota, bacteria must surmount numerous challenges when colonizing the mammalian GI tract without jeopardizing host fitness. The primary mechanisms for preventing small intestine bacterial overgrowth are the antibacterial capabilities of bile acids and the activation of mucosal defenses. Our findings indicate that BSH activity enhances bacterial resistance to CBA and promotes their survival in the gastrointestinal tract, facilitating colonization and the evolution of gut microbiota. The host will indirectly benefit from the establishment of a gut microbial community that safeguards against gastrointestinal diseases and enhances caloric absorption. The possible advantageous effects of BSH activity and BA alteration suggest that BSH activity may have been favored at the host level due to its good impact on host fitness. Conversely, changed bile acids have been associated with carcinogenic and cocarcinogenic properties, and substantial evidence supports their involvement in colorectal cancer (CRC) [19, 20].

Regarding the process of counterselection against the short-term advantages of BSH activity and the eventual adjustment of BA metabolism a hypothesis suggests that the activity of BSH may safeguard the colonic epithelium from the genotoxic implications of CBA that can permeate this recess [30]. A disruption of CBA has been suggested as a means of preventing the accumulation of these compounds in colonic epithelial cells and the subsequent damage to DNA [30]. Additional research is required to identify the specific species and elucidate the mechanisms by which they mitigate the detrimental impacts of CBA or promote BA transformation. The activation of mucosal defenses and various signaling functions of bile acids is facilitated by several receptors, with bile acids acting as natural ligands, including TGR5 and the farnesoid X receptor (FXR). The induction of FXR, located in the liver and intestine, is essential for mucosal defense against bacterial invasion in the SI and for sustaining homeostasis of bile acids, cholesterol, and triglycerides [54].

TGR5 incitement is associated with energy and may be homeostasis of glucose metabolism. The influence of altered bile acids on the activation of these receptors is unclear; nevertheless, the gut microbiota may substantially impact the physiology of the host and the development of metabolic disorders through the regulation of the bile acid signaling. Additional inquiry into the significant impacts of microbial bile acid alteration on the metabolic pathway of the host is essential. The research should elaborate on BSH activity and BA modification by the gut microbiota, along with its implications for host health and physiology, which might elucidate the coevolution of humans and their gut flora, perhaps resulting in approaches to influence this community for illness treatment and the improvement of human well-being.

A healthy gut microbiota keeps the ratio of primary to secondary bile acids stable, which is important for digestion and signaling. But dysbiosis, which can be caused by diet, antibiotics, long-term inflammation, or disease, can change this balance. Lower BSH activity could lead to higher levels of conjugated bile acids, which would change how FXR and TGR5 are activated. FXR activation regulates inflammatory signaling and metabolic equilibrium, both of which affect neuroinflammatory pathways associated with depression, Alzheimer's disease, and Parkinson's disease. These nuclear receptors, which can also be activated by some bile acid derivatives, help the body get rid of toxins and change how the immune system works. Their role in keeping the gut barrier intact indirectly protects against systemic inflammation, which is known to cause mood disorders and cognitive decline.

Ligands are activated primarily by secondary bile acids, especially lithocholic acid (LCA) and deoxycholic acid (DCA). It changes how much energy is used, how many anti-inflammatory cytokines are released, and how much GLP-1 is secreted. CNS impacts TGR5 activation and this can affect the activity in the hypothalamus and limbic system by having anti-inflammatory effects on the periphery and stimulating the vagus nerve. This could change how people feel and how they react to stress. Sphingosine-1-phosphate receptors (S1PRs) and bile acid derivatives may indirectly influence sphingolipid metabolism, impacting neuroinflammatory pathways as well as modulating muscarinic and ion channel. Some bile acids can directly change how neurons respond by changing how ion channels work in membranes.

The overproduction of certain secondary bile acids (e.g., DCA, LCA) may have pro-inflammatory and cytotoxic effects, which could affect the integrity of the gut barrier and systemic inflammation. The decreased production of beneficial bile acids like UDCA may lower neuroprotective signaling. Changes in the configuration of the bile acids have been observed in neuropsychiatric disorders, indicating a mechanistic connection between microbial bile acid metabolism and mental health outcomes. BAs are not only important for digestive activity; they also work as powerful signaling molecules that affect metabolic, immune, and neural processes. A network of membrane-bound G protein-coupled receptors (GPCRs) and intracellular signaling cascades is what causes these effects. BAs play a pivotal role in the microbiota-gut-brain axis because they can cross or indirectly affect the blood-brain barrier (BBB). They act as a biochemical bridge between gut microbial activity and central nervous system (CNS) function.

The BBB restricts the passage of numerous circulating molecules; however, evidence indicates that unconjugated bile acids and specific conjugated variants may traverse it under pathological or inflammatory circumstances. Moreover, peripheral effects (immune modulation, endocrine signaling) may indirectly affect CNS function without direct bile acid entry. Increased bile acid concentrations in cerebrospinal fluid (CSF) have been documented in neurodegenerative diseases, indicating their possible involvement in CNS pathophysiology. Bile acids act as links between gut microbial activity, systemic metabolism, immune tone, and brain function through FXR, TGR5, and other receptor-mediated pathways. Disruptions in these signaling networks, frequently stemming from gut dysbiosis, can alter the equilibrium toward neuroinflammation, oxidative stress, and modified neurotransmission, all of which are pivotal to neuropsychiatric disorders.

During years of bile acid research, different biological functions for bile acids have been revealed lately, owing to breakthroughs in understanding the human microbiome. As more bacterial enzymes are characterized and the methods for identifying new bile acids become increasingly sensitive, the variety of bile acids metabolized and/or synthesized by bacteria continues to expand. Bile acids also influence the composition and functionality of the microbiome community. This chapter emphasizes the manipulation of the bile acid mechanism by the gut flora, its dependence on the metabolic activity of the microorganisms, and the influence of other factors, such as diet and antibiotics, on the composition of the bile acid. Understanding the influence of bile acid signaling networks in various organs with differing bile acid compositions is becoming increasingly crucial, particularly regarding their impact on infectious, neoplastic, and metabolic diseases. These advancements have enabled the creation of therapies that address abnormalities in bile acid profiles connected with microbiota.

Bile acids, a significant category of microbially generated metabolites, are endogenous compounds originating from cholesterol in the liver and then metabolized by microbiota in the gut [55]. These chemicals enhanced receptors in the liver, stomach, and peripheral tissues, regulating several host activities, including metabolic functions. BAs are associated with the amino acid's glycine and, to a lesser extent, taurine in humans before excretion into bile and subsequent transit to the duodenum. BAs are secreted into the duodenum after consuming food, and conjugated bile acids are then reabsorbed in the ileum and transported back to the liver through portal blood in a process called enterohepatic circulation, which saves more than 95% of the BA pool. We tried to provide a comprehensive overview of bile acid composition in this chapter and its relation with gut microbiota. This viewpoint summarizes the current understanding of how microbiota affects bile acids and how this modulation may contribute to metabolic diseases [55].

10.4 Microbial Modulation of Bile Acid Biosynthesis

Changes in bile acid metabolism have become a common occurrence in various neuropsychiatric disorders, indicating that gut microbiota-mediated bile acid signaling may have a pathophysiological function. Dysbiosis, changes in the ratios of primary to secondary bile acids, and problems with bile acid receptor signaling can all affect neuroinflammation, neurotransmitter balance, and neuronal metabolism, which are all important factors in mental health disorders. Chronic stress in rodents causes gut dysbiosis, lowers the variety of secondary bile acids, and raises behaviors that are similar to depression. These effects can be undone by adding bile acids or restoring the microbiota. Clinical metabolomics studies have indicated diminished levels of specific secondary bile acids and modified conjugation patterns in individuals with major depressive disorder (MDD) [56]. Children with ASD frequently exhibit unique gut microbiota profiles characterized by diminished bile salt hydrolase (BSH)-producing bacteria, resulting in modified primary bile acid accumulation. Certain studies have identified correlations between BA composition and the severity of ASD symptoms, although the causal relationship is still being examined. Individuals with schizophrenia and bipolar disorder frequently exhibit metabolic comorbidities, including obesity and insulin resistance, both of which are affected by bile acid metabolism [57]. In small groups of patients, lower levels of circulating secondary bile acids and changes in the expression patterns of BA receptors have been seen. Dysregulated BA signaling may worsen systemic inflammation and mitochondrial dysfunction, leading to cognitive and mood instability [58].

Initial research indicates that the modulation of gut microbiota or bile acid pathways may enhance metabolic and behavioral outcomes in pertinent animal models. Neurodegenerative disorders, including Alzheimer's and Parkinson's disease, frequently manifest with psychiatric symptoms such as psychosis, anxiety, and depression, despite not being classified as psychiatric conditions. Increased cytotoxic bile acids (e.g., deoxycholic acid) and diminished neuroprotective UDCA have been observed in patients with AD and PD. BA-induced mitochondrial dysfunction and oxidative stress may expedite neuronal degeneration while facilitating psychiatric comorbidity. The synthesis of BAs is intricate and comprises numerous reaction steps, requiring different enzymes [59]. BA synthesis transpires in the hepatic system and can be accomplished through two separate pathways. The classical pathway accounts for at least three quarters of BA synthesis under typical conditions, initiated by the 7α-hydroxylation of cholesterol, catalyzed by cholesterol 7α-hydroxylase (CYP7A1) [60]. CYP7A1 is the rate-limiting enzyme that governs bile acid production. The alternative (or acidic) pathway commences with sterol-27-hydroxylase (CYP27A1) [59]. The 27-hydroxycholesterol generated is then hydroxylated by oxysterol 7α-hydroxylase (CYP7B1). The 7α-hydroxylated intermediates originating from cholesterol and oxysterols then experience modifications to the sterol ring, alongside oxidation and truncation of the side chain, through multiple reaction steps [60]. The gut microbiota modulates the expression of several enzymes, such as CYP7A1, CYP7B1, and CYP27A1 [61]. Microbial deconjugation, which involves the elimination of the glycine or taurine conjugate, inhibits active reuptake from the small intestine via the ASBT. Bile acid deconjugation is performed by bacteria with BSH activity. There are different types of bacteria and archaea in the human gut. These include *Bifidobacteria*, *Bacteroides*, *Lactobacilli*, and *Clostridium*. BSH is more common in the gut microbiota than in other microbial communities. It is also linked to higher susceptibility to bile toxicity. These oxidation events are reversible and may eventually lead to epimerization. The chemicals generated are often overlooked in both standard and sophisticated bile acid studies due to the scarcity of suitable reference molecules. Iso-bile acids are present in human serum and urine, especially in the colon and feces.

10.5 Conclusion

Over the last 10 years, we have learned a lot about how the brain and gut work together. The microbiota-gut-brain axis (MGBA) is a complex and changing system that connects the two. The MGBA is a two-way communication system that connects the microbial ecosystem in the GI tract to the CNS. This has important effects on neurological and psychiatric health. Bile acid signaling is a metabolite-centered mechanism that connects gut microbiota and brain function. It is at the forefront of current research and points to promising new ways to understand and treat neuropsychiatric disorders. The MGBA includes many different ways for the body to communicate, such as neural, endocrine, immune, and metabolic pathways. Its main parts are gut microbiota: there are trillions of microorganisms in the human gastrointestinal tract, mostly bacteria. These microorganisms have an unmediated effect on the host's metabolism, immunity, and even behavior. The enteric nervous system (ENS) communicates with the CNS through autonomic nerves and the blood-brain barrier. Signaling molecules and microbial metabolites, including short-chain fatty acids (SCFAs), neurotransmitters (e.g., serotonin, dopamine), and bile acids, function as signaling molecules along this axis. The hypothalamic-pituitary-adrenal (HPA) axis and the autonomic nervous system let the brain control gut functions, while the gut sends signals to the brain through metabolic, immunological, and hormonal mediators. Disruption of this axis, commonly referred to as dysbiosis, has been associated with numerous neuropsychiatric disorders, including depression, anxiety, autism, Parkinson's disease, and Alzheimer's disease.

Declarations

Competing Interests The author(s) has no competing interests to declare that are relevant to the content of this manuscript.

Ethics Approval The author(s) declares that this manuscript does not include primary research with human participants.

The author(s) declares that this manuscript does not include primary studies involving animals.

References

1. Turnbaugh PJ, et al. The human microbiome project. Nature. 2007;449(7164):804–10.
2. Muhasina KM, Ghosh P, Nagappan K, Palaniswamy DS, Begum R, Islam MR, Tagde P, Shaikh NK, Farahim F, Mondal TK. From gut microbiomes to infectious pathogens: neurological disease game changers. Mol Neurobiol. 2025;62(1):1184–204.
3. Liang S, Wu X, Jin F. Gut-brain psychology: rethinking psychology from the microbiota–gut–brain axis. Front Integr Neurosci. 2018;12:33.
4. Lynch SV, Pedersen O. The human intestinal microbiome in health and disease. N Engl J Med. 2016;375(24):2369–79.
5. Mayer EA, Tillisch K, Gupta A. Gut/brain axis and the microbiota. J Clin Invest. 2015;125(3):926–38.

6. Ahern PP, Maloy KJ. Understanding immune–microbiota interactions in the intestine. Immunology. 2020;159(1):4–14.
7. Mori H, et al. Farnesoid X receptor, bile acid metabolism, and gut microbiota. Metabolites. 2022;12(7):647.
8. Fasano A. All disease begins in the (leaky) gut: role of zonulin-mediated gut permeability in the pathogenesis of some chronic inflammatory diseases. F1000Res. 2020;9:F1000-Faculty.
9. Rossi RE, et al. Interaction between gut microbiota and celiac disease: from pathogenesis to treatment. Cells. 2023;12(6):823.
10. Fan Y, Pedersen O. Gut microbiota in human metabolic health and disease. Nat Rev Microbiol. 2021;19(1):55–71.
11. Rutsch A, Kantsjö JB, Ronchi F. The gut-brain axis: how microbiota and host inflammasome influence brain physiology and pathology. Front Immunol. 2020;11:604179.
12. Maiuolo J, et al. The contribution of gut microbiota–brain axis in the development of brain disorders. Front Neurosci. 2021;15:616883.
13. Borre YE, et al. Microbiota and neurodevelopmental windows: implications for brain disorders. Trends Mol Med. 2014;20(9):509–18.
14. Bonaz B, Sinniger V, Pellissier S. The vagus nerve in the neuro-immune axis: implications in the pathology of the gastrointestinal tract. Front Immunol. 2017;8:1452.
15. Erny D, et al. Host microbiota constantly control maturation and function of microglia in the CNS. Nat Neurosci. 2015;18(7):965–77.
16. Keller J, Gomez R, Williams G, Lembke A, Lazzeroni L, Murphy GM, Schatzberg AF. HPA axis in major depression: cortisol, clinical symptomatology and genetic variation predict cognition. Mol Psychiatry. 2017;22(4):527–36.
17. O'Mahony SM, Clarke G, Borre YE, Dinan TG, Cryan JF. Serotonin, tryptophan metabolism and the brain-gut-microbiome axis. Behav Brain Res. 2015;277:32–48.
18. Dalile B, Van Oudenhove L, Vervliet B, Verbeke K. The role of short-chain fatty acids in microbiota–gut–brain communication. Nat Rev Gastroenterol Hepatol. 2019;16(8):461–78.
19. Pascale A, et al. Microbiota and metabolic diseases. Endocrine. 2018;61(3):357–71.
20. Arpaia N, et al. Metabolites produced by commensal bacteria promote peripheral regulatory T-cell generation. Nature. 2013;504(7480):451–5.
21. Bonnet U, Bingmann D, Wiemann M. Intracellular pH modulates spontaneous and epileptiform bioelectric activity of hippocampal CA3-neurones. Eur Neuropsychopharmacol. 2000;10(2):97–103.
22. Phelps EA, LeDoux JE. Contributions of the amygdala to emotion processing: from animal models to human behavior. Neuron. 2005;48(2):175–87.
23. Sarkar A, Harty S, Lehto SM, Moeller AH, Dinan TG, Dunbar RI, Cryan JF, Burnet PW. The microbiome in psychology and cognitive neuroscience. Trends Cogn Sci. 2018;22(7):611–36.
24. Desbonnet L, Garrett L, Clarke G, Bienenstock J, Dinan TG. The probiotic Bifidobacteria infantis: an assessment of potential antidepressant properties in the rat. J Psychiatr Res. 2008;43(2):164–74.
25. Chapelet G, Leclair-Visonneau L, Clairembault T, Neunlist M, Derkinderen P. Can the gut be the missing piece in uncovering PD pathogenesis? Park Relat Disord. 2019;59:26–31.
26. Boonchooduang N, Louthrenoo O, Chattipakorn N, Chattipakorn SC. Possible links between gut–microbiota and attention-deficit/hyperactivity disorders in children and adolescents. Eur J Nutr. 2020;59(8):3391–403.
27. Sinn N. Nutritional and dietary influences on attention deficit hyperactivity disorder. Nutr Rev. 2008;66(10):558–68.
28. Baars A, et al. The gut microbiota as a therapeutic target in IBD and metabolic disease: a role for the bile acid receptors FXR and TGR5. Microorganisms. 2015;3(4):641–66.
29. Guzior DV, Quinn RA. Microbial transformations of human bile acids. Microbiome. 2021;9(1):140.
30. Xing C, et al. Roles of bile acids signaling in neuromodulation under physiological and pathological conditions. Cell Biosci. 2023;13(1):106.

31. Wang S, et al. Connecting the gut microbiota and neurodegenerative diseases: the role of bile acids. Mol Neurobiol. 2023;60(8):4618–40.
32. DeMorrow S. Role of the hypothalamic–pituitary–adrenal axis in health and disease. Int J Mol Sci. 2018;19(4):986.
33. Gaboriau-Routhiau M-C, Moreau V. Influence of resident intestinal microflora on the development and functions of the gut-associated lymphoid tissue. Microb Ecol Health Dis. 2001;13(2):65–86.
34. Yang J, et al. New insight into neurological degeneration: inflammatory cytokines and blood–brain barrier. Front Mol Neurosci. 2022;15:1013933.
35. Correia AS, Vale N. Tryptophan metabolism in depression: a narrative review with a focus on serotonin and kynurenine pathways. Int J Mol Sci. 2022;23(15):8493.
36. Marin JG, Jose, et al. Bile acids in physiology, pathology and pharmacology. Curr Drug Metab. 2016;17(1):4–29.
37. Šarenac TM, Mikov M. Bile acid synthesis: from nature to the chemical modification and synthesis and their applications as drugs and nutrients. Front Pharmacol. 2018;9:939.
38. Prasad SD. A neurochemical perspective on brain-gut communication and its gastrointestinal regulation: a ENS as the "second brain". J Adv Immunol. 2025;2(2). https://doi.org/10.5281/zenodo.15600765.
39. Ng DZW, et al. Ex vivo metabolism kinetics of primary to secondary bile acids via a physiologically relevant human faecal microbiota model. Chem Biol Interact. 2024;399:111140.
40. Hofmann AF, Eckmann L. How bile acids confer gut mucosal protection against bacteria. Proc Natl Acad Sci. 2006;103(12):4333–4.
41. Monteiro-Cardoso VF, Corlianò M, Singaraja RR. Bile acids: a communication channel in the gut-brain axis. NeuroMolecular Med. 2021;23:99–117. https://doi.org/10.1007/s12017-020-08625-z.
42. Alrefai WA, Gill RK. Bile acid transporters: structure, function, regulation and pathophysiological implications. Pharm Res. 2007;24(10):1803–23.
43. Mohanty I, et al. The changing metabolic landscape of bile acids–keys to metabolism and immune regulation. Nat Rev Gastroenterol Hepatol. 2024;21(7):493–516.
44. Ley RE, Peterson DA, Gordon JI. Ecological and evolutionary forces shapingmicrobial diversity in the human intestine. Cell. 2006;124:837–48.
45. Begley M, Hill C, Gahan CG. Bile salt hydrolase activity in probiotics. Appl Environ Microbiol. 2006;72:1729–38.
46. Ridlon JM, Kang DJ, Hylemon PB. Bile salt biotransformations by human intes-tinal bacteria. J Lipid Res. 2006;47:241–59.
47. Houten SM, Watanabe M, Auwerx J. Endocrine functions of bile acids. EMBO J. 2006;25:1419–25.
48. Watanabe M, et al. Bile acids induce energy expenditure by promoting intra-cellular thyroid hormone activation. Nature. 2006;439:484–9.
49. Hofmann AF, Eckmann L. How bile acids confer gut mucosal protection againstbacteria. Proc Natl Acad Sci USA. 2006;103:4333–4.
50. Inagaki T, et al. Regulation of antibacterial defense in the small intestine by the nuclear bile acid receptor. Proc Natl Acad Sci USA. 2006;103:3920–5.
51. Begley M, Sleator RD, Gahan CG, Hill C. Contribution of three bile-associated loci, bsh, pva, and. btlB, to gastrointestinal persistence and bile tolerance of Listeriamonocytogenes. Infect Immun. 2005;73:894–904.
52. Batta AK, et al. Side chain conjugation prevents bacterial 7-dehydroxylation ofbile acids. J Biol Chem. 1990;265:10925–8.
53. van Eldere J, Celis P, De Pauw G, Lesaffre E, Eyssen H. Tauroconjugation of cholic acid stimulates 7 alpha-dehydroxylation by fecal bacteria. Appl Environ Microbiol. 1996;62:656–61.
54. Ticho AL, et al. Bile acid receptors and gastrointestinal functions. Liver Res. 2019;3(1):31–9.
55. De Aguiar Vallim TQ, Tarling EJ, Edwards PA. Pleiotropic roles of bile acids in metabolism. Cell Metab. 2013;17(5):657–69.

56. Huang Y, et al. Possible association of Firmicutes in the gut microbiota of patients with major depressive disorder. Neuropsychiatr Dis Treat. 2018;Volume 14:3329–37.
57. Skonieczna-Żydecka K, et al. Second-generation antipsychotics and metabolism alterations: a systematic review of the role of the gut microbiome. Psychopharmacology. 2019;236(5):1491–512.
58. Büttiker P, et al. Dysfunctional mitochondrial processes contribute to energy perturbations in the brain and neuropsychiatric symptoms. Front Pharmacol. 2023;13:1095923.
59. Russell DW. The enzymes, regulation, and genetics of bile acid synthesis. Annu Rev Biochem. 2003;72:137–74.
60. Thomas C, Gioiello A, Noriega L, Strehle A, Oury L, Rizzo G, et al. TGR5-mediated bile acid sensing controls glucose homeostasis. Cell Metab. 2009;10(3):167–77.
61. Sayin SI, Wahlström A, Felin J, Jäntti S, Marschall HU, Bamberg K, et al. Gut microbiota regulates bile acid metabolism by reducing the levels of tauro-beta-muricholic acid, a naturally occurring FXR antagonist. Cell Metab. 2013;17(2):225–35.

Gut Dysbiosis in Alzheimer's Disease: Insights from the Microbiome

11

Neelakanta Sarvashiva Kiran, Chandrashekar Yashaswini, Ankita Chatterjee, and Bhupendra Prajapati

11.1 Introduction

The most prevalent kind of cognitive impairment, Alzheimer's disease (AD), is typified by neurological conditions, memory loss, and gradual loss of cognitive ability. Extracellular amyloid-beta (Aβ) plaque buildup, neurofibrillary tangles within the cell made of hyperphosphorylated tau, synaptic breakdown, and the death of neurons, especially in the hippocampus along with cortex, are destructive characteristics of AD [1]. These neurodegenerative changes lead to deterioration in thinking, behavior, and the ability to perform daily activities [2]. The precise cause of AD is still unknown after decades of research, and existing therapies only slightly reduce symptoms without slowing the course of the illness. This emphasizes how vital it is to investigate new pathogenic factors and therapeutic techniques [3].

The gut microbiome, a diverse and dynamic community of microbes found in the human gut, is essential for preserving host equilibrium [4]. Comprising trillions of microorganisms, including bacteria, viruses, fungi, and archaea, the gut microbiota influences nutrient metabolism, immune modulation, mucosal barrier integrity, and

N. S. Kiran · C. Yashaswini · A. Chatterjee
Department of Biotechnology, School of Applied Sciences, REVA University, Bengaluru, Karnataka, India

Center for Biotechnology and Molecular Biology, REVA Research Center, REVA University, Bengaluru, Karnataka, India

B. Prajapati (✉)
Department of Pharmaceutics, Parul Institute of Pharmacy, Faculty of Pharmacy, Parul University, Waghodia, Vadodara, Gujarat, India

M. H. Rahman, Muhasina K M (eds.), *The Microbiome's Impact on the Brain*,
https://doi.org/10.1007/978-981-95-7494-0_11

even brain function [5]. Advances in metagenomic sequencing have revealed that microbial composition varies significantly between healthy individuals and those with various metabolic, inflammatory, and neurodegenerative conditions [6].

A crucial foundation for comprehending how the brain's central nervous system (CNS) interacts through the gastrointestinal surroundings is the idea of the relationship between the gut and the brain [7]. This bidirectional signaling system involves neural, hormonal, immune, and metabolic pathways, allowing gut microbes to influence brain development, behavior, and neuroinflammation. Disruptions in this axis have been implicated in mood disorders, autism spectrum disorders, Parkinson's disease, and, more recently, Alzheimer's disease [8].

A disparity in microbiological makeup or activity, known as intestinal dysbiosis, is becoming more widely acknowledged as a risk factor for neurological along with metabolic illnesses. Dysbiosis may result in the overgrowth of pathogenic microbes, reduced microbial diversity, compromised intestinal barrier function, and excessive production of pro-inflammatory metabolites—all of which can affect brain health [9].

Emerging evidence supports a hypothesis that gut dysbiosis may play a pivotal role in the onset and progression of AD [10]. Increased permeation of the intestinal as well as blood-brain barriers, decreased amyloid elimination, and heightened neuroinflammatory reactions have all been linked to changed microbial profiles [11]. This chapter explores the growing body of research linking gut microbiome disturbances with AD pathogenesis, offering new avenues for understanding disease mechanisms and developing microbiota-centered therapeutic strategies [12].

11.2 The Healthy Gut Microbiome: Composition and Function

The gut microbiome, a complex and ever-changing microbial ecology found in the human gastrointestinal tract, is made up of billions of species, including viruses, fungi, bacteria, and archaea [13]. This microbial consortium is predominantly composed of a few major bacterial phyla—Proteobacteria, Verrucomicrobia, Actinobacteria, Firmicutes, and Bacteroidetes—each of which plays specialized and often overlapping roles in host physiology (Table 11.1). Firmicutes, which include genera such as *Ruminococcus*, *Lactobacillus*, *Clostridium*, and *Faecalibacterium*, are heavily involved in the fermentation of dietary fibers, leading to the production of critical metabolites such as short-chain fatty acids (SCFAs) [14]. *Bacteroidetes*, including *Bacteroides* and *Prevotella*, excel in polysaccharide degradation and mucin turnover, contributing to both nutrient availability and mucosal health [15]. Actinobacteria, particularly the genus *Bifidobacterium*, are key colonizers of the early-life microbiome and play significant roles in immune maturation, pathogen resistance, and maintenance of epithelial integrity [16]. Proteobacteria, although a minor component in healthy individuals, include many facultative anaerobes and opportunistic pathogens, such as *Escherichia* and *Salmonella*, and are often considered markers of dysbiosis when disproportionately expanded [17].

Table 11.1 Major components and functions of a healthy gut microbiome

Bacterial phylum	Key genera (examples)	Primary functions	Microbiota-derived metabolites	References
Firmicutes	*Clostridium*, *Lactobacillus*, *Faecalibacterium*	SCFA (esp. butyrate) production, energy harvest, immune regulation	Butyrate, acetate, lactate	[18]
Bacteroidetes	*Bacteroides*, *Prevotella*	Polysaccharide breakdown, bile acid metabolism, modulating T-cell response	Propionate, acetate, succinate	[19]
Actinobacteria	*Bifidobacterium*	Protection against pathogens, fermentation of oligosaccharides, vitamin production	Acetate, lactate, folate	[20]
Proteobacteria	*Escherichia*, *Enterobacter*	Typically low in abundance, opportunistic under dysbiosis	LPS (under pathological expansion)	[21]
Verrucomicrobia	*Akkermansia muciniphila*	Mucin degradation, gut barrier maintenance, immune homeostasis	SCFAs, mucin-derived sugars	[22]

In addition to its taxonomic makeup, the microbiota in the gastrointestinal tract carries out a number of physiological tasks that are essential for host growth and survival [23]. It plays a fundamental role in nutrient metabolism, including the breakdown of indigestible polysaccharides into absorbable monosaccharides; the synthesis of essential vitamins such as biotin, folate, and vitamin K; and the regulation of lipid metabolism and bile acid recirculation. These metabolic functions significantly impact host energy homeostasis and are closely linked to systemic metabolic health [24]. The immunological interactions between the host and its microbiota are equally significant. Microbial-associated molecular patterns (MAMPs) influence immunological tolerance, inhibit aberrant inflammation, as well as stimulate the growth and proliferation of regulatory T cells along with different immunological lineages via receptors that detect patterns (e.g., TLRs, NOD-like receptors) [25]. Additionally, the gut microbiota strengthens the integrity of the gastrointestinal tract by promoting the creation of mucus layers through goblet cells, triggering the development of tightly coupled proteins, and preventing the colonization of harmful microorganisms through the generation of peptides that are antimicrobial and competitive elimination [26].

The creation of bioactive substances by the gut microbiota, many of which have systemic effects, is arguably one of among the most fundamental ways that it affects the well-being of the host. Short-chain fatty acids (SCFAs) such as acetate, propionate, and butyrate are the primary end products of microbial fermentation of dietary fibers [27]. These SCFAs serve as energy substrates for colonocytes, modulate gene expression via histone deacetylase (HDAC) inhibition, and engage G-protein-coupled receptors (e.g., GPR41, GPR43) to regulate immune and inflammatory responses [28]. Through the farnesoid X receptor (FXR) and Takeda G-protein receptor 5 (TGR5), bile acids—which are produced in the liver and transformed by microbes in the intestinal tract into secondary liver acids like deoxycholic acid and lithocholic acid—act as signaling molecules that affect immune cell distinctiveness, intestinal motility, as well the breakdown of fatty acids [29]. Another critical metabolic pathway is tryptophan catabolism, whereby microbial species convert dietary tryptophan into a range of neuroactive compounds including indoles, serotonin, and kynurenine derivatives [30]. These metabolites can cross the intestinal and blood-brain barriers, modulate central nervous system activity, and influence neuroinflammation and oxidative stress pathways—mechanisms increasingly implicated in the pathogenesis of neurodegenerative disorders [31].

11.3 Gut-Brain Axis: Mechanistic Communication

Through neurological, immunological, endocrine, as well as biological processes, the gut-brain axis (GBA) is a complex, interconnected network that unifies the enteric nervous system (ENS) along with the central nervous system (CNS) [32]. This multifaceted cross talk enables gut microbes to influence brain development, cognition, mood, and behavior, while the brain can simultaneously modulate gastrointestinal function and microbial composition [33]. The GBA represents a dynamic physiological interface, wherein microbial signals are translated into host neurochemical responses, and vice versa (Fig. 11.1).

The neurological pathway is one of the main channels of this interaction, and the vagus nervous system is an essential link between the brain stem and the stomach [34]. This cranial nerve transmits afferent sensory signals from the gut to the CNS, responding to microbial metabolites and hormones, while efferent vagal signaling regulates intestinal motility, mucus secretion, and local immune responses [35]. Experimental studies have shown that vagotomy abolishes certain microbiota-induced behavioral and inflammatory effects, underscoring its indispensable role in gut-brain communication [36].

The immune pathway is equally vital, with the gut microbiota modulating both local and systemic immune responses [37]. Lipopolysaccharides (LPS) and other microbial-associated molecular patterns (MAMPs) may translocate into the circulatory system as a result of dysbiotic microbial communities impairing the effectiveness of the intestinal barrier [38]. These molecules activate peripheral immune cells

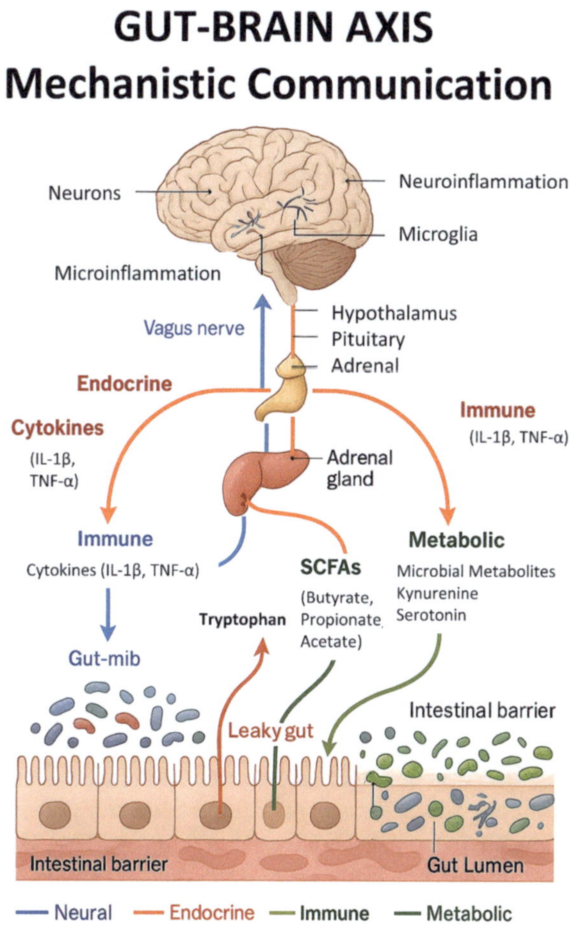

Fig. 11.1 Mechanistic representation of the gut-brain axis

and provoke the release of pro-inflammatory cytokines (e.g., IL-1β, IL-6, TNF-α), which can cross the blood-brain barrier (BBB) or influence its permeability, thereby initiating neuroinflammation and microglial activation—hallmarks of neurodegenerative diseases like Alzheimer's [39].

Interactions between the microbiota and the brain are also significantly influenced by the endocrine system, namely the hypothalamic-pituitary-adrenal (HPA) axis [40]. The HPA axis regulates the stress response through the secretion of glucocorticoids such as cortisol, which can alter gut permeability, microbial composition, and immune tone. Conversely, gut microbes can influence HPA axis activation through the production of neuroactive compounds and modulation of peripheral inflammation, linking chronic stress to both gut and brain dysfunction [41].

The metabolic pathway involves microbial production of neuroactive metabolites, which can directly or indirectly affect neuronal signaling and brain function

[42]. Key metabolites include short-chain fatty acids (SCFAs) like butyrate, which can enhance BBB integrity, suppress neuroinflammation, and influence epigenetic regulation of gene expression in neuronal cells. Furthermore, chemicals including serotonin, kynurenine, as well as indoles are produced by microbes' metabolism of tryptophan and are involved in immunological signaling, neurological regulation, and the control of oxidative damage [43].

Collectively, these pathways enable microbial products to exert far-reaching effects on the brain, influencing neurodevelopment, synaptic plasticity, and neurodegenerative processes [44]. The gut-brain axis can become unstable and accelerate the development of neurological diseases like Alzheimer's disease if any of these signaling pathways are disrupted by gut dysbiosis, prolonged inflammation, or triggers from the environment [45]. Understanding the precise mechanisms of GBA communication holds profound implications for developing microbiome-targeted therapeutics aimed at preserving cognitive health and preventing neurodegeneration [46].

11.4 Gut Dysbiosis in Alzheimer's Disease

A paradigm shift from a disease that is solely brain-centric to one which could originate, or potentially be influenced by, external systems like the digestive system has been brought about by emerging research that has highlighted the vital function of the microbiome of the digestive tract in the pathological process of Alzheimer's disease (AD) [47]. Gut dysbiosis, known as a disruption in the equilibrium of microbial composition, diversity, and function, has been strongly correlated with the progression of AD through its influence on neuroinflammatory pathways, systemic immune activation, and neurodegenerative cascades [48]. These alterations are not merely correlative but are now being explored as causal contributors to AD pathogenesis [49].

The gut microbiota identities of AD patients and intellectually healthy, age-matched counterparts differ significantly, according to several human cohort studies [50]. These studies have consistently demonstrated a reduction in microbial alpha-diversity, a hallmark of ecological instability, and an overrepresentation of pro-inflammatory bacterial taxa [51], for example, a groundbreaking study found that while anti-inflammatory properties *Bifidobacterium* as well as *Eubacterium rectale* levels were down, *Escherichia/Shigella* along with other *Proteobacteria*—a phylum frequently linked to intestinal inflammation along with the generation of endotoxins—were up. This shift is paralleled by elevated serum levels of pro-inflammatory cytokines (IL-6, IL-1β, TNF-α), suggesting a mechanistic link between microbial dysbiosis and systemic immune activation [52].

Preclinical studies using transgenic mouse models of AD, such as APP/PS1, 3xTg-AD, and 5xFAD, have provided deeper insights into causal relationships [53]. These models exhibit age-dependent dysbiosis that precedes or accompanies the onset of amyloid-beta (Aβ) deposition and tau pathology. Notably, germ-free APP/PS1 mice exhibit significantly reduced amyloid burden, which is exacerbated upon colonization with microbiota from conventionally housed AD mice—indicating a direct microbiota-driven modulation of Aβ pathology. Additionally, it has been demonstrated that fecal microbiome transplantation (FMT) from individuals who are healthy into dysbiotic AD mice can decrease neuroinflammatory markers (including IL-1β, GFAP, and Iba1), restore synaptic expression of proteins (like PSD-95 and synaptophysin), and ameliorate cognitive impairment. These results imply that altering the gut microbiota can prevent or reverse the neurodegeneration linked to AD [54].

By increasing intestinal permeability (also known as "leaky gut"), gut dysbiosis facilitates the relocation of microbial-associated molecular patterns (MAMPs) into the systemic blood flow, including microbial amyloids and lipopolysaccharides (LPS) [55]. Once in circulation, these molecules can breach the blood-brain barrier (BBB), particularly when it is already compromised by age or inflammation, and trigger microglial activation through pattern recognition receptors (e.g., TLR4, NLRP3 inflammasome). This, in turn, amplifies neuroinflammation, promotes the misfolding and aggregation of endogenous Aβ, and impairs its clearance via glymphatic and microglial pathways [56].

In addition to immunological perturbations, dysbiosis significantly alters the gut-derived metabolic landscape. One prominent instance is the disruption of tryptophan metabolic processes, in which microbes interact with the serotonin as well as kynurenine pathways, increasing the synthesis of neurotoxic intermediates including 3-hydroxykynurenine along with quinolinic acid [57]. The pathogenic characteristics of AD, including excitotoxicity, dysfunction of mitochondria, and chronic oxidative stress have all been linked to these compounds. Additionally, both AD patients and models have shown reduced synthesis of neuroprotective short-chain fatty acids (SCFAs), particularly butyrate, which compromises the integrity of the blood-brain barrier, inhibits the activity of histone deacetylase (HDAC), and lowers the level of expression of anti-inflammatory genes [58, 59].

Collectively, these converging lines of evidence from clinical and experimental models reinforce the concept that the gut microbiome is not a passive bystander but an active regulator of neurodegenerative processes [60]. The microbial shifts observed in AD—marked by decreased beneficial commensals and enrichment of pro-inflammatory, proteolytic, and endotoxin-producing species—serve as both upstream drivers and amplifiers of amyloid pathology, tau hyperphosphorylation, synaptic dysfunction, and neuronal loss [61]. As such, gut dysbiosis represents a novel and potentially modifiable target for early diagnosis, biomarker development, and therapeutic intervention in Alzheimer's disease [62].

11.5 Microbial Metabolites and Neurotoxicity in AD

The gut microbiota exerts its systemic influence not only through its taxonomic composition but also via a vast repertoire of bioactive metabolites, many of which possess profound neuromodulatory, immunological, and neurotoxic potential [63]. Short-chain fatty acids (SCFAs), lipopolysaccharides (LPS), and tryptophan catabolites, along with bile acid derivatives, are examples of microbial-derived metabolites whose altered fabrication and modified circulation throughout the system have been linked to the onset or aggravation of neurodegenerative processes in Alzheimer's disease (AD) [64]. These molecules interact with the host through a variety of mechanisms—crossing biological barriers, modulating epigenetic machinery, activating immune receptors, and disrupting neurotransmitter networks—thus serving as critical mediators in the gut-brain axis and potential effectors of neurotoxicity in AD [65].

The short-chain fatty acids (SCFAs), which are mainly acetate, propionate, and butyrate, are one of the most researched classes of microbial compounds in AD. These are created when anaerobic gut bacteria like *Faecalibacterium prausnitzii*, *Roseburia* spp., and *Clostridium* cluster IV and XIVa ferment dietary fibers. These SCFAs function not only as energy substrates for colonocytes but also as epigenetic modulators via histone deacetylase (HDAC) inhibition, thereby influencing the transcription of genes related to inflammation, synaptic plasticity, and blood-brain barrier (BBB) integrity [66]. In AD, reduced fecal and systemic SCFA levels—particularly butyrate—have been reported, correlating with increased microglial activation, peripheral inflammation, and cognitive impairment. Lack of butyrate reduces the production of tight junction proteins in the BBB's endothelial cells, making the membrane more susceptible to inflammatory chemicals and external pollutants that can exacerbate central neurological inflammation including neuronal degeneration [67].

Another important family of microbially generated neurotoxic chemicals are lipopolysaccharides (LPS), which are essential parts of the outer protective layer of Gram-negative bacteria (such as *Escherichia coli* and Enterobacteriaceae). In a healthy gut, LPS remains compartmentalized within the intestinal lumen [68]. However, in conditions of gut dysbiosis and barrier dysfunction, LPS translocates into systemic circulation, where it binds to TLR4-CD14-MD2 complexes on immune cells, triggering a cascade of NF-κB-mediated pro-inflammatory cytokine release. Importantly, LPS has been detected in postmortem AD brains, where it colocalizes with amyloid-beta plaques and contributes to chronic microglial activation, oxidative stress, and neuronal apoptosis [69]. LPS's function as a peripheral instigator of central neurodegenerative is further supported by experimental infusion of the hormone into animal models, which replicates important clinical characteristics of AD, such as tau hyperphosphorylation and cognitive impairments [70].

The tryptophan metabolism pathway, another critical axis of gut-brain metabolic interaction, is profoundly influenced by gut microbiota. Under physiological conditions, tryptophan is metabolized into serotonin via the serotonin pathway or into kynurenine and its downstream metabolites via the kynurenine pathway. Gut microbial enzymes like tryptophanase and indoleamine 2,3-dioxygenase (IDO) orchestrate this metabolic routing [71]. In dysbiotic states and chronic inflammation, the kynurenine pathway becomes dominant, leading to excessive generation of neurotoxic metabolites such as quinolinic acid, 3-hydroxykynurenine, and picolinic acid, which act as NMDA receptor agonists, exacerbate excitotoxicity, and induce reactive oxygen species (ROS) generation in neurons [72]. Simultaneously, decreased serotonergic signaling—due to reduced microbial conversion of tryptophan into

5-hydroxytryptophan—impairs neurogenesis, mood regulation, and cognitive resilience [73]. The depressed symptoms and cognitive deterioration commonly seen in AD patients have been closely associated with this physiological remodeling [74].

Bile acids, though traditionally associated with lipid digestion, have emerged as critical signaling molecules in gut-brain communication [75]. Deoxycholic acid (DCA) and lithocholic acid (LCA), which are secondary bile acids produced by microbes from primary bile acids produced in the liver, can interact alongside host receptors including Takeda G-protein receptor 5 (TGR5) and farnesoid X receptor (FXR) [76]. Both plasma and cerebral fluid have been found to have changed bile acid profiles in AD, with higher concentrations of cytotoxic bile acids that are secondary and a corresponding drop in neurologically protective bile acid substitutes [77]. These changes may impair membrane integrity, disturb lipid rafts, and affect cholesterol homeostasis—processes essential for the processing of amyloid precursor protein (APP) and the aggregation of Aβ. Furthermore, it has been shown that dysregulated bile acid signaling via FXR/TGR5 modulates insulin resistance and neuroinflammatory processes, all of which are closely linked to the development of AD [78].

11.6 Inflammation, Gut Barrier Dysfunction, and Neurodegeneration

An increasing amount of data highlights the critical roles that systemic inflammation and gut barrier integrity play in the etiology of Alzheimer's disease (AD) [79]. As a two-way channel, the gut-brain axis is greatly affected by breaches in the blood-brain barrier (BBB) along with intestinal epithelium barrier (also known as "leaky gut") [80]. Together, these compromised interfaces allow the systemic translocation of microbial products, pro-inflammatory cytokines, and neurotoxic molecules, thereby initiating and sustaining a chronic inflammatory state that accelerates neurodegeneration (Fig. 11.2).

Under normal circumstances, the intestinal barrier—which is made up of epithelial cells joined by proteins at the tight junction, including occludin, claudins, as well as zonula occludens-1—acts as a vital barrier against microbial translocation and luminal antigens [81]. This barrier is commonly weakened in AD, which results in increased intestinal permeability, or "leaky gut." Mucosal layer disturbance brought on by dysbiosis and a reduction in SCFA synthesis, especially butyrate, decrease goblet cell mucus secretion and impair tight junction expression [82]. As a result, microbe-associated molecular patterns (MAMPs) such as flagellin, peptidoglycans, and lipopolysaccharides (LPS) enter the bloodstream and trigger systemic immune activation [83].

Once in systemic circulation, these microbial components interact with innate immune receptors such as Toll-like receptors (TLRs) and NOD-like receptors (NLRs) on monocytes, dendritic cells, and intestinal macrophages, leading to the release of pro-inflammatory cytokines, notably interleukin-1β (IL-1β), tumor necrosis factor-alpha (TNF-α), and interleukin-6 (IL-6) [84]. These cytokines can act peripherally or cross the compromised BBB, creating a feed-forward loop of

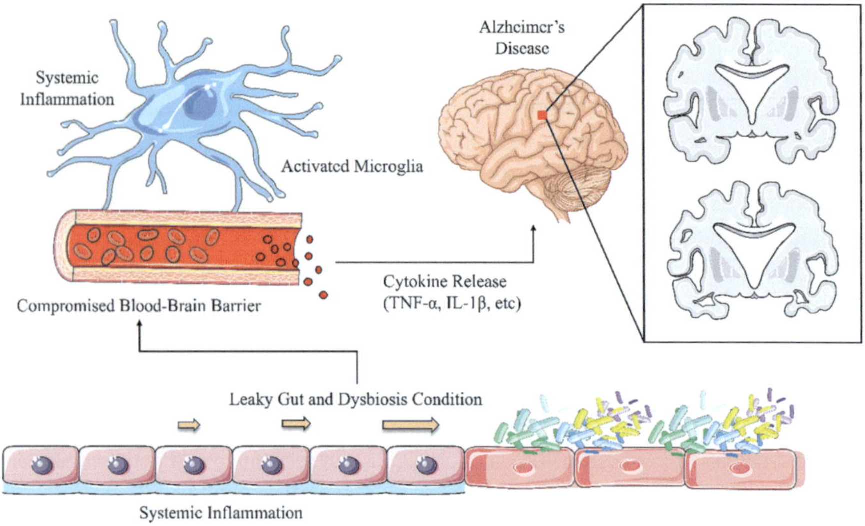

Fig. 11.2 Schematic representation of gut barrier dysfunction and systemic inflammation contributing to blood-brain barrier compromise and neurodegeneration in Alzheimer's disease

neuroinflammation. Notably, elevated systemic levels of IL-1β and TNF-α have been reported in both AD patients and transgenic AD models, with direct correlations to cognitive decline and plaque load [85].

The blood-brain barrier (BBB), a selectively permeable interface between the CNS and the peripheral vasculature, is similarly vulnerable in the context of aging and chronic inflammation [86]. In AD, BBB integrity is increasingly shown to be disrupted, both structurally and functionally. Tight junction proteins between endothelial cells are downregulated, pericyte loss is evident, and astrocytic end foot processes exhibit abnormal morphology [87]. These changes make it easier for immune cells, microbial metabolites, and peripheral poisons to enter the central nervous system, for instance, it has been demonstrated that pro-inflammatory cytokines and LPS that penetrate the blood-brain barrier activate microglial cells, which are the immune surveillance cells that dwell in the brain [88].

Gut-derived immunological signals cause microglia to change morphologically and functionally into an inflammatory mediator (M1) phenotype. This leads to the production of reactive oxygen species (ROS), nitric oxide (NO), and pro-inflammatory cytokines (e.g., IL-1β, TNF-α), which collectively contribute to neuronal injury, synaptic dysfunction, and tau hyperphosphorylation [89]. Additionally, chronic microglial activation impairs the phagocytic clearance of amyloid-beta, exacerbating plaque accumulation and spreading of neurotoxic aggregates [90]. Importantly, microglial activation via TLR4, a receptor also engaged by LPS, has been identified as a key molecular mechanism linking peripheral inflammation with CNS immune responses [91].

Moreover, the synergistic interaction between a leaky gut and a leaky BBB establishes a vicious cycle, wherein gut-derived inflammatory signals perpetuate

CNS inflammation, and neurodegeneration further disrupts neurovascular regulation, fostering continued BBB breakdown [92]. This self-propagating loop ensures the persistence of low-grade systemic inflammation ("inflammaging"), a recognized hallmark of both aging and AD [93].

11.7 Role of Diet, Aging, and Antibiotics in Gut Dysbiosis

The dynamic composition particularly functional ability of the gut microbiome is strongly impacted by environmental along with lifestyle variables, among which nutrition, longevity, and exposure to antibiotics are particularly essential [94]. These modulators not only shape microbial diversity and abundance but also determine the metabolic and immunological tone of the host, thus influencing susceptibility to gut dysbiosis and downstream neurodegenerative outcomes, including Alzheimer's disease (AD) (Fig. 11.3).

Microbial dysbiosis has been closely linked to the Western diet, which is heavy in animal proteins, refined carbohydrates, saturated oils, and highly processed foods [95]. This diet reduces microbial diversity and selectively enriches pathobionts, such as *Escherichia coli*, *Bilophila wadsworthia*, and *Clostridium* spp., while diminishing beneficial taxa including *Bifidobacterium*, *Lactobacillus*, and *Faecalibacterium prausnitzii*. Importantly, the Western diet is typically low in dietary fiber—an essential substrate for the fermentation of short-chain fatty acids (SCFAs), including butyrate, acetate, and propionate, by saccharolytic bacteria.

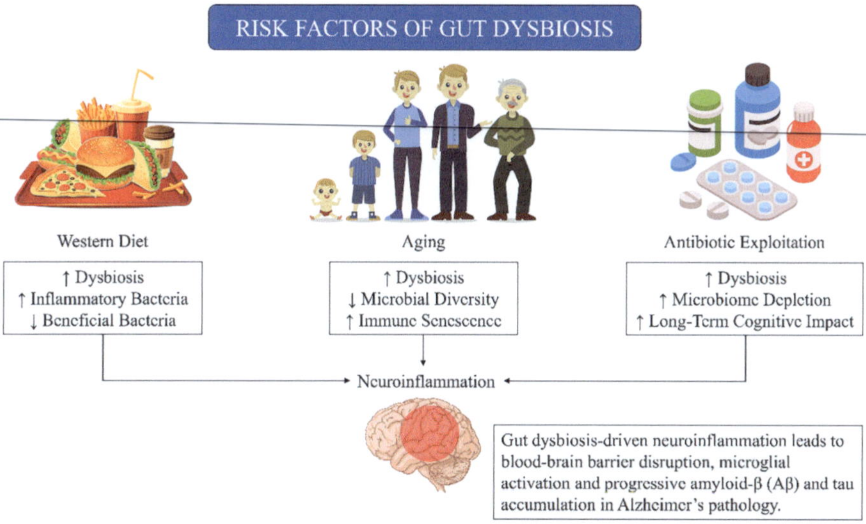

Fig. 11.3 Illustration of the multifactorial influences of Western diet, aging, and antibiotics on gut dysbiosis, highlighting downstream effects on immune senescence, microbiome depletion, and brain dysfunction linked to Alzheimer's disease

These SCFAs are vital for maintaining intestinal epithelial integrity, regulating inflammation, and supporting blood-brain barrier (BBB) function. Another natural cause of the imbalance in gut microbes is aging. Senescence is associated with a decrease in microbial diversity, which is frequently characterized by an increase in opportunistic pathogens and a decrease in compatible anaerobes. Fiber deficiency consequently reduces SCFA production, impairs mucosal immunity, and increases intestinal permeability, all of which contribute to a pro-inflammatory milieu conducive to cognitive decline [96, 97].

Age-related dysbiosis is closely linked with immunosenescence and inflammaging—the chronic, low-grade inflammation characteristic of aging [98]. Reduced mucosal barrier function, lower SCFA levels, and diminished microbial resistance all contribute to elevated translocation of bacteria and chronic inflammation. This increases microglial stimulation and adds to the buildup of tau and amyloid-beta pathology seen in AD [99, 100]. Furthermore, the aging gut becomes more susceptible to environmental insults, and its regenerative capacity diminishes, reinforcing the fragility of the gut-brain axis in elderly populations [101].

Antibiotic exposure dramatically alters the gut microbiome by inducing broad-spectrum reduction of both transmissible and beneficial microorganisms, especially in aging adults and during important windows of development [102]. This microbial depletion disrupts colonization resistance, allowing for the expansion of harmful species and opportunistic infections [103]. Studies in animal models have demonstrated that prolonged or repeated antibiotic use leads to persistent alterations in microbial ecology, reduced SCFA production, and exacerbation of cognitive deficits [104, 105]. In murine models of AD, antibiotic-induced dysbiosis has been shown to worsen amyloid deposition and impair neurogenesis. Moreover, antibiotic use can blunt the immunomodulatory and neuroprotective effects of beneficial bacteria, thereby impairing host resilience against neuroinflammatory insults [106].

Together, these factors—Western dietary patterns, age-associated microbial drift, and antibiotic-induced perturbations—synergistically contribute to the disruption of gut homeostasis, weakening of epithelial and BBB integrity, and activation of systemic and neuroinflammation [107–109]. In order to improve the course of Alzheimer's disease, preventative and therapeutic approaches that target the gut microbiota must take into account and mitigate the effects of these modifiable along with non-modifiable variables [110, 111].

11.8 Experimental and Clinical Evidence

The importance of gut microbiota in regulating the etiology and course of Alzheimer's disease (AD) is highlighted by an expanding corpus of experimental along with clinical data [112]. Preclinical animal models and human cohort studies have shed light on the complex relationships between gut dysbiosis and the hallmarks of AD, including tau hyperphosphorylation, amyloid-beta buildup, neuroinflammation, and cognitive loss [113].

Human clinical research using metagenomics, metabolomics, and 16S rRNA gene sequencing has repeatedly shown that AD patients' gut microbial profiles differ from those of age-matched healthy controls [114]. These changes include an enrichment of inflammatory substances species, like *Escherichia/Shigella* and Proteobacteria, and a notable decrease in microbial diversity along with a depletion of anti-inflammatory taxa including *Bifidobacterium*, *Lactobacillus*, and *Eubacterium rectale* [115]. Many investigations have also documented associations between the makeup of the gut microbiota and cerebrospinal fluid (CSF) markers of AD, such as phosphorylated tau and Aβ42, indicating a concrete connection between peripheral microbial changes and central neuropathology [116].

Preclinical models, particularly murine and zebrafish systems, have been instrumental in elucidating causal relationships. Germ-free transgenic mouse models of AD display exacerbated amyloid and tau pathology, neuroinflammation, and behavioral deficits, all of which can be modulated by the introduction of specific microbial consortia [117]. Zebrafish models, due to their transparency and conserved gut-brain axis, have offered high-throughput insights into microbial-induced neuroinflammatory pathways, blood-brain barrier (BBB) integrity, and neurobehavioral outcomes. These models have revealed that antibiotic-induced dysbiosis, high-fat diet exposure, or genetic mutations can synergistically modulate the trajectory of neurodegeneration [118].

Fecal microbiota transplantation (FMT) studies have offered compelling interventional evidence. In AD mouse models, transplantation of fecal microbiota from healthy donors ameliorated cognitive decline, reduced amyloid deposition, and normalized microglial activation [119]. Conversely, FMT from AD patients accelerated neurodegeneration in recipient animals, thereby providing direct evidence for the transmissibility of dysbiotic phenotypes. These findings hold translational promise but require cautious clinical validation [120].

In parallel, advances in microbiome profiling tools—including shotgun metagenomics, metabolomics, and transcriptomics—have enabled comprehensive characterization of gut microbial communities and their functional outputs [121, 122]. These instruments are being used to find biomarkers generated from the microbiome for the early diagnosis of AD, for example, some metabolite profiles and microbial signatures (such as SCFAs, indoles, and LPS) are being researched as possible noninvasive diagnostic methods that might provide information about the preclinical phases of neurodegeneration [123, 124]. Precision medicine approaches to the treatment of AD may be made possible by the combination of fluid-based diagnostics, neuroimaging, and microbiome-based biomarkers [125, 126].

11.9 Therapeutic Approaches Targeting Gut Microbiota

The growing understanding of the role of the microbiota in the gut in the pathophysiology of Alzheimer's disease (AD) has increased interest in treatment approaches that target the microbiome [127]. Modulating the gut ecosystem offers a promising adjunct to conventional neurodegenerative disease treatments by addressing systemic inflammation, metabolic dysfunction, and neuroimmune cross talk at the gut-brain interface [127].

The fundamental treatments for re-establishing microbial balance are probiotics and prebiotics. Preclinical models have shown that certain probiotic strains, including *Lactobacillus rhamnosus*, *Lactobacillus plantarum*, and *Bifidobacterium longum*, exhibit neuroprotective qualities via reducing oxidative stress, controlling inflammatory cytokines, and altering neurotransmitter pathways [127]. These strains have been shown to enhance hippocampal neurogenesis, reduce amyloid-beta accumulation, and improve cognitive performance. Prebiotics, such as inulin, fructooligosaccharides (FOS), and galactooligosaccharides (GOS), promote the growth of beneficial gut bacteria and increase the production of short-chain fatty acids (SCFAs), which exert anti-inflammatory and neuroprotective effects [127].

Probiotics and prebiotics work in concert to improve microbial survival as well as function in the digestive tract. This is known as a symbiotic [128]. Postbiotics, which are defined as metabolic byproducts and nonviable microbial products (such as butyrate, lactate, and indole derivatives), have demonstrated effectiveness in reducing systemic inflammation and gut barrier function, making them a desirable substitute for live bacterial administration in patients where risks may arise [129].

The gut microbiota is significantly affected by dietary interventions. Rich in fiber, polyphenols, and unsaturated fats, the Mediterranean diet has been linked to a lower risk of dementia and more microbial diversity [130]. It fosters the proliferation of SCFA-producing bacteria and reduces pro-inflammatory taxa. The ketogenic diet, though more controversial, may modulate gut microbiota in ways that increase ketone body availability to neurons and reduce neuroinflammation, potentially benefiting patients in early AD stages [131].

Microbiota-derived drug candidates represent a frontier in drug development. Molecules like urolithin A, derived from polyphenol metabolism by gut microbes, have shown mitochondrial and anti-inflammatory benefits in neurodegenerative models. Similarly, bile acid derivatives and tryptophan catabolites are being explored as modulators of microglial activation and synaptic plasticity [132].

In animal models of AD, fecal microbiota transplantation (FMT) has demonstrated promise by restoring microbial balance, lowering neuroinflammation, and improving cognitive outcomes by transplanting of healthy donor microbiota [133, 134]. However, clinical translation faces challenges including donor variability, risk of pathogen transmission, regulatory hurdles, and patient acceptance. Standardization of protocols, donor screening, and long-term safety assessments remain key obstacles before FMT can be adopted widely for neurological indications [135, 136].

11.10 Challenges and Future Directions

Despite the growing body of evidence implicating gut microbiota in Alzheimer's disease (AD) pathogenesis, several critical challenges hinder the translational potential of microbiome-based interventions. A major limitation lies in the heterogeneity and scarcity of large-scale, longitudinal human studies [137]. Most current findings are derived from small cohort sizes with limited ethnic and dietary diversity, short follow-up periods, and cross-sectional designs that fail to establish causality between gut dysbiosis and AD progression [138].

The notable inter-individual variation in microbiome composition, which is impacted by a number of variables including genetics, nutrition, age, environment, drugs, and multiple illnesses, is another important obstacle. It is challenging to identify universal microbial fingerprints linked to AD because of this diversity, which also makes it challenging to identify consistent diagnostics or treatment targets across groups. Furthermore, interpreting data and creating successful therapies are made more difficult by the microbiome's changing nature throughout time [139].

Standardization of microbiome-based diagnostics remains elusive. Differences in sampling techniques, DNA extraction protocols, sequencing platforms, and bioinformatic pipelines contribute to inconsistent and sometimes conflicting findings across studies. Establishing standardized guidelines for microbiome profiling is imperative to improve reproducibility, facilitate data sharing, and develop clinically viable diagnostic tools [140].

The combination of multi-omics techniques, including transcriptomics, epigenomics, metabolomics, and metagenomics, is crucial to overcoming these constraints. These platforms facilitate the identification of metabolite-mediated signaling pathways pertinent to neurodegeneration, reveal functional microbial changes, and offer a systems-level knowledge of host-microbiome interactions. Such integrative analyses can help pinpoint causative microbial drivers and their mechanistic roles in AD pathophysiology [141].

Looking ahead, the creation of customized microbiome-based therapeutics is the future of microbiome research in AD. There is potential for improving treatment results and reducing side effects by customizing therapies according to a person's microbial profile, family history, eating habits, and immunological condition [142]. This precision medicine framework, however, will require robust validation in diverse clinical settings and the establishment of regulatory standards for microbiome-based therapeutics [143]. Ultimately, advancing our understanding of the gut-brain axis through rigorous, interdisciplinary research will be pivotal in harnessing the microbiome as a targetable axis in Alzheimer's disease management [144].

11.11 Conclusion

Alzheimer's disease (AD), a complex neurodegenerative condition, has traditionally been explained through the accumulation of amyloid-beta plaques and tau tangles. However, recent research highlights the significant influence of the gut microbiota in both the onset and the progression of AD. The gut-brain axis—a dynamic, two-way communication system involving neural, hormonal, immune, and metabolic pathways—serves as the conduit through which gut microbes can affect brain function. Findings from both clinical and preclinical studies reveal that gut dysbiosis, marked by decreased microbial diversity and alterations in specific bacterial populations (such as lower levels of *Bifidobacterium* and higher levels of

Escherichia/Shigella), is associated with heightened neuroinflammation, compromised blood-brain barrier integrity, and the buildup of amyloid-beta and hyperphosphorylated tau. These microbial imbalances also impact critical neuroactive compounds—including short-chain fatty acids, lipopolysaccharides, and tryptophan metabolites—thereby intensifying neurodegeneration and accelerating cognitive impairment. Gut dysbiosis is increasingly recognized not just as a consequence but as a modifiable risk factor in Alzheimer's disease (AD). This insight paves the way for therapeutic approaches focused on restoring microbial balance. Dietary interventions—such as the Mediterranean or ketogenic diets—along with probiotics, prebiotics, synbiotics, and fecal microbiota transplantation (FMT), are under investigation for their potential neuroprotective benefits. Additionally, advancements in multi-omics technologies are enhancing our understanding of the complex interactions between the host and gut microbiota, supporting the development of personalized microbiome-based treatments. Despite ongoing challenges like individual variability and the absence of standardized clinical protocols, the gut microbiome is emerging as both a diagnostic marker and a treatment target in AD. Ultimately, therapies aimed at modulating the microbiome may serve as valuable complements to traditional treatments, offering a fresh approach to preventing and managing Alzheimer's disease.

Acknowledgments The authors greatly acknowledge and convey our sincere gratitude to REVA University and Parul University for providing an opportunity and platform for research.

Declarations

Conflict of Interest The authors declare that there is no potential conflict of interest.

Competing Interests The author(s) has no competing interests to declare that are relevant to the content of this manuscript.

Ethics Approval The author(s) declares that this manuscript does not include primary research with human participants.

The author(s) declares that this manuscript does not include primary studies involving animals.

References

1. Askenazi M, Kavanagh T, Pires G, Ueberheide B, Wisniewski T, Drummond E. Compilation of reported protein changes in the brain in Alzheimer's disease. Nat Commun. 2023;14(1):4466.
2. Dubois B, Villain N, Frisoni GB, Rabinovici GD, Sabbagh M, Cappa S, et al. Clinical diagnosis of Alzheimer's disease: recommendations of the International Working Group. Lancet Neurol. 2021;20(6):484–96.
3. Ebrahimi A, Luo S, Disease Neuroimaging Initiative ftAs. Convolutional neural networks for Alzheimer's disease detection on MRI images. J Med Imaging. 2021;8(2):024503.
4. Jack JCR, Andrews JS, Beach TG, Buracchio T, Dunn B, Graf A, et al. Revised criteria for diagnosis and staging of Alzheimer's disease: Alzheimer's Association Workgroup. Alzheimers Dement. 2024;20(8):5143–69.

5. Jia J, Ning Y, Chen M, Wang S, Yang H, Li F, et al. Biomarker changes during 20 years preceding Alzheimer's disease. N Engl J Med. 2024;390(8):712–22.
6. Ju Y, Tam KY. Pathological mechanisms and therapeutic strategies for Alzheimer's disease. Neural Regen Res. 2022;17(3):543–9.
7. Mathys H, Boix CA, Akay LA, Xia Z, Davila-Velderrain J, Ng AP, et al. Single-cell multiregion dissection of Alzheimer's disease. Nature. 2024;632(8026):858–68.
8. Porsteinsson AP, Isaacson RS, Knox S, Sabbagh MN, Rubino I. Diagnosis of early Alzheimer's disease: clinical practice in 2021. J Prev Alzheimers Dis. 2021;8(3):371–86.
9. Rostagno AA. Pathogenesis of Alzheimer's disease. MDPI; 2022. p. 107.
10. Scheltens P, De Strooper B, Kivipelto M, Holstege H, Chételat G, Teunissen CE, et al. Alzheimer's disease. Lancet. 2021;397(10284):1577–90.
11. Sperling RA, Donohue MC, Raman R, Rafii MS, Johnson K, Masters CL, et al. Trial of solanezumab in preclinical Alzheimer's disease. N Engl J Med. 2023;389(12):1096–107.
12. Twarowski B, Herbet M. Inflammatory processes in alzheimer's disease—pathomechanism, diagnosis and treatment: a review. Int J Mol Sci. 2023;24(7):6518.
13. Walsh S, Merrick R, Milne R, Brayne C. Aducanumab for Alzheimer's disease? British Medical Journal Publishing Group; 2021.
14. Zhang XX, Tian Y, Wang ZT, Ma YH, Tan L, Yu J-T. The epidemiology of Alzheimer's disease modifiable risk factors and prevention. J Prev Alzheimers Dis. 2021;8(3):313–21.
15. Agans RT. Modeling effects of diet on human gut microbiota. Wright State University; 2016.
16. Do TH, Dao TK, Nguyen HD, Truong NH. Understanding the role of free-living bacteria in the gut of the lower termite Coptotermes gestroi based on metagenomic DNA analysis. Insects. 2023;14(11):832.
17. Jaago M, Timmusk US, Timmusk T, Palm K. Drastic effects on the microbiome of a young rower engaged in high-endurance exercise after a month usage of a dietary fiber supplement. Front Nutr. 2021;8:654008.
18. Aggarwal V, Sunder S, Verma SR. Disease-associated dysbiosis and potential therapeutic role of Akkermansia muciniphila, a mucus degrading bacteria of gut microbiome. Folia Microbiol. 2022;67(6):811–24.
19. Baldelli V, Scaldaferri F, Putignani L, Del Chierico F. The role of Enterobacteriaceae in gut microbiota dysbiosis in inflammatory bowel diseases. Microorganisms. 2021;9(4):697.
20. Fu X, Liu Z, Zhu C, Mou H, Kong Q. Nondigestible carbohydrates, butyrate, and butyrate-producing bacteria. Crit Rev Food Sci Nutr. 2019;59(sup1):S130–S52.
21. Linares DM, Fitzgerald G, Hill C, Stanton C, Ross P. Production of vitamins, exopolysaccharides and bacteriocins by probiotic bacteria. In: Probiotic dairy products. Wiley; 2017. p. 359–88.
22. Takeuchi T, Nakanishi Y, Ohno H. Microbial metabolites and gut immunology. Annu Rev Immunol. 2024;42(1):153–78.
23. Jaagura M. The effect of diet and dietary fiber on the human gut microbiota in vitro and in vivo Tallinna Tehnikaülikooli Kirjastus; 2021.
24. Kmezik C, Franzén CJ. Exploring and exploiting plant biomass degradation by Bacteroidetes. Chalmers Tekniska Hogskola (Sweden); 2021.
25. Kober O. The role of mucus in the cross talk between gut bacteria and the host. University of East Anglia; 2013.
26. McKee LS, La Rosa SL, Westereng B, Eijsink VG, Pope PB, Larsbrink J. Polysaccharide degradation by the Bacteroidetes: mechanisms and nomenclature. Environ Microbiol Rep. 2021;13(5):559–81.
27. Overbeeke ANJ. Bacterial degradation of polysaccharides in the gut muco-sal eco-system; 2022.
28. Perez-Munoz ME. Characterization of the gut microbiota and identification of colitogenic bacterial species in core 1 O-glycans deficient mice. The University of Nebraska-Lincoln; 2013.
29. Raimondi S, Musmeci E, Candeliere F, Amaretti A, Rossi M. Identification of mucin degraders of the human gut microbiota. Sci Rep. 2021;11(1):11094.

30. Santos JWD. The Effect of non-starch polysaccharides degrading enzymes on gut health of pigs. Universitat Autònoma de Barcelona; 2017.
31. Tedros Y. How do gut bacteria boost our immune system? University of East Anglia; 2021.
32. Aburto MR, Cryan JF. Gastrointestinal and brain barriers: unlocking gates of communication across the microbiota–gut–brain axis. Nat Rev Gastroenterol Hepatol. 2024;21(4):222–47.
33. Agirman G, Yu KB, Hsiao EY. Signaling inflammation across the gut-brain axis. Science. 2021;374(6571):1087–92.
34. Ahmed H, Leyrolle Q, Koistinen V, Kärkkäinen O, Layé S, Delzenne N, et al. Microbiota-derived metabolites as drivers of gut–brain communication. Gut Microbes. 2022;14(1):2102878.
35. Ge L, Liu S, Li S, Yang J, Hu G, Xu C, et al. Psychological stress in inflammatory bowel disease: psychoneuroimmunological insights into bidirectional gut–brain communications. Front Immunol. 2022;13:1016578.
36. Gershon MD, Margolis KG. The gut, its microbiome, and the brain: connections and communications. J Clin Invest. 2021;131(18):e143768.
37. Hattori N, Yamashiro Y. The gut-brain axis. Ann Nutr Metab. 2021;77(Suppl. 2):1–3.
38. Jiang M, Kang L, Wang Y-L, Zhou B, Li H-Y, Yan Q, et al. Mechanisms of microbiota-gut-brain axis communication in anxiety disorders. Front Neurosci. 2024;18:1501134.
39. Kasarello K, Cudnoch-Jedrzejewska A, Czarzasta K. Communication of gut microbiota and brain via immune and neuroendocrine signaling. Front Microbiol. 2023;14:1118529.
40. Lu S, Zhao Q, Guan Y, Sun Z, Li W, Guo S, et al. The communication mechanism of the gut-brain axis and its effect on central nervous system diseases: a systematic review. Biomed Pharmacother. 2024;178:117207.
41. Mayer EA, Nance K, Chen S. The gut–brain axis. Annu Rev Med. 2022;73(1):439–53.
42. Morais LH, Schreiber IV HL, Mazmanian SK. The gut microbiota–brain axis in behaviour and brain disorders. Nat Rev Microbiol. 2021;19(4):241–55.
43. Ortega MA, Álvarez-Mon MA, García-Montero C, Fraile-Martínez Ó, Monserrat J, Martinez-Rozas L, et al. Microbiota–gut–brain axis mechanisms in the complex network of bipolar disorders: potential clinical implications and translational opportunities. Mol Psychiatry. 2023;28(7):2645–73.
44. Ortiz-Samur NS, Vijaya AK, Burokas A, Mela V. Exploring the role of microglial cells in the gut–brain axis communication: a systematic review. J Neurochem. 2025;169(7):e70154.
45. Ortlek BE, Akan OB. Modeling and analysis of SCFA-driven vagus nerve signaling in the gut-brain axis via molecular communication. IEEE Trans Mol Biol Multi-Scale Commun. 2025;11:588.
46. Pirolli NH, Bentley WE, Jay SM. Bacterial extracellular vesicles and the gut-microbiota brain axis: emerging roles in communication and potential as therapeutics. Adv Biol. 2021;5(7):2000540.
47. Chen Q, Shi J, Yu G, Xie H, Yu S, Xu J, et al. Gut microbiota dysbiosis in patients with Alzheimer's disease and correlation with multiple cognitive domains. Front Aging Neurosci. 2024;16:1478557.
48. Cho J, Park YJ, Gonzales-Portillo B, Saft M, Cozene B, Sadanandan N, et al. Gut dysbiosis in stroke and its implications on Alzheimer's disease-like cognitive dysfunction. CNS Neurosci Ther. 2021;27(5):505–14.
49. Janeiro MH, Ramírez MJ, Solas M. Dysbiosis and Alzheimer's disease: cause or treatment opportunity? Cell Mol Neurobiol. 2022;42(2):377–87.
50. Jemimah S, Chabib CMM, Hadjileontiadis L, AlShehhi A. Gut microbiome dysbiosis in Alzheimer's disease and mild cognitive impairment: a systematic review and meta-analysis. PLoS One. 2023;18(5):e0285346.
51. Kozhakhmetov S, Kaiyrlykyzy A, Jarmukhanov Z, Vinogradova E, Zholdasbekova G, Alzhanova D, et al. Inflammatory manifestations associated with gut dysbiosis in Alzheimer's disease. Int J Alzheimers Dis. 2024;2024(1):9741811.

52. Lazar E, Sherzai A, Adeghate J, Sherzai D. Gut dysbiosis, insulin resistance and Alzheimer's disease: review of a novel approach to neurodegeneration. Front Biosci-Sch. 2021;13(1):17–29.
53. Ling Z, Zhu M, Yan X, Cheng Y, Shao L, Liu X, et al. Structural and functional dysbiosis of fecal microbiota in Chinese patients with Alzheimer's disease. Front Cell Dev Biol. 2021;8:634069.
54. Navalón-Monllor V, Soriano-Romaní L, Silva M, de Las Hazas M-CL, Hernando-Quintana N, Diéguez TS, et al. Microbiota dysbiosis caused by dietetic patterns as a promoter of Alzheimer's disease through metabolic syndrome mechanisms. Food Funct. 2023;14(16):7317–34.
55. Qian X-H, Liu X-L, Chen G, Chen S-D, Tang H-D. Injection of amyloid-β to lateral ventricle induces gut microbiota dysbiosis in association with inhibition of cholinergic anti-inflammatory pathways in Alzheimer's disease. J Neuroinflammation. 2022;19(1):236.
56. Qu C, Xu Q-Q, Yang W, Zhong M, Yuan Q, Xian Y-F, et al. Gut dysbiosis aggravates cognitive deficits, amyloid pathology and lipid metabolism dysregulation in a transgenic mouse model of Alzheimer's disease. J Pharm Anal. 2023;13(12):1526–47.
57. Roe K. An alternative explanation for Alzheimer's disease and Parkinson's disease initiation from specific antibiotics, gut microbiota dysbiosis and neurotoxins. Neurochem Res. 2022;47(3):517–30.
58. Shabbir U, Arshad MS, Sameen A, Oh D-H. Crosstalk between gut and brain in Alzheimer's disease: the role of gut microbiota modulation strategies. Nutrients. 2021;13(2):690.
59. Thakkar A, Vora A, Kaur G, Akhtar J. Dysbiosis and Alzheimer's disease: role of probiotics, prebiotics and synbiotics. Naunyn Schmiedeberg's Arch Pharmacol. 2023;396(11):2911–23.
60. Sowmiya S, Dhivya LS, Harikrishnan N, Ankul SS. Exploring the potential of probiotics in Alzheimer's disease and gut dysbiosis. IBRO Neurosci Rep. 2024;17:441–55.
61. Choi H, Mook-Jung I. Functional effects of gut microbiota-derived metabolites in Alzheimer's disease. Curr Opin Neurobiol. 2023;81:102730.
62. Cui J, Xiao S, Cao Y, Zhang Y, Yang J, Zheng L, et al. Organophosphate insecticide malathion induces Alzheimer's disease-like cognitive impairment in mice: evidence of the microbiota-gut-brain axis. Environ Sci Technol. 2024;58(50):21966–77.
63. Jo J-K, Lee G, Nguyen CD, Park S-E, Kim E-J, Kim H-W, et al. Effects of donepezil treatment on brain metabolites, gut microbiota, and gut metabolites in an amyloid beta-induced cognitive impairment mouse pilot model. Molecules. 2022;27(19):6591.
64. Jung YH, Chae CW, Han HJ. The potential role of gut microbiota-derived metabolites as regulators of metabolic syndrome-associated mitochondrial and endolysosomal dysfunction in Alzheimer's disease. Exp Mol Med. 2024;56(8):1691–702.
65. Missiego-Beltrán J, Beltrán-Velasco AI. The role of microbial metabolites in the progression of neurodegenerative diseases—therapeutic approaches: a comprehensive review. Int J Mol Sci. 2024;25(18):10041.
66. Nie Y, Yang J, Zhou L, Yang Z, Liang J, Liu Y, et al. Marine fungal metabolite butyrolactone I prevents cognitive deficits by relieving inflammation and intestinal microbiota imbalance on aluminum trichloride-injured zebrafish. J Neuroinflammation. 2022;19(1):39.
67. Ntranos A, Park H-J, Wentling M, Tolstikov V, Amatruda M, Inbar B, et al. Bacterial neurotoxic metabolites in multiple sclerosis cerebrospinal fluid and plasma. Brain. 2022;145(2):569–83.
68. Paul S, Dey M, Roy B, Dhara B, Mitra AK, editors. Potentiality of marine microbial metabolites in the remedy of Alzheimer's disease: a comprehensive review. Springer; 2024.
69. Salminen A. Activation of aryl hydrocarbon receptor (AhR) in Alzheimer's disease: role of tryptophan metabolites generated by gut host-microbiota. J Mol Med. 2023;101(3):201–22.
70. Sun P, Zhu H, Li X, Shi W, Guo Y, Du X, et al. Comparative metagenomics and metabolomes reveals abnormal metabolism activity is associated with gut microbiota in Alzheimer's disease mice. Int J Mol Sci. 2022;23(19):11560.
71. Tinkov AA, Martins AC, Avila DS, Gritsenko VA, Skalny AV, Santamaria A, et al. Gut microbiota as a potential player in Mn-induced neurotoxicity. Biomolecules. 2021;11(9):1292.

72. Wang Y. The role of the gut microbiota and microbial metabolites in the pathogenesis of Alzheimer's disease. CNS Neurol Disord Drug Targets. 2023;22(4):577–98.
73. Zhao Y, Jaber V, Lukiw WJ. Gastrointestinal tract microbiome-derived pro-inflammatory neurotoxins in Alzheimer's disease. J Aging Sci. 2021;9(Suppl 5):002.
74. Zhao Y, Jaber VR, Pogue AI, Sharfman NM, Taylor C, Lukiw WJ. Lipopolysaccharides (LPSs) as potent neurotoxic glycolipids in Alzheimer's disease (AD). Int J Mol Sci. 2022;23(20):12671.
75. Johnson SL, Park HY, Vattem DA, Grammas P, Ma H, Seeram NP. Equol, a blood–brain barrier permeable gut microbial metabolite of dietary isoflavone daidzein, exhibits neuroprotective effects against neurotoxins induced toxicity in human neuroblastoma SH-SY5Y cells and Caenorhabditis elegans. Plant Foods Hum Nutr. 2020;75(4):512–7.
76. Vogt NM, Romano KA, Darst BF, Engelman CD, Johnson SC, Carlsson CM, et al. The gut microbiota-derived metabolite trimethylamine N-oxide is elevated in Alzheimer's disease. Alzheimers Res Ther. 2018;10(1):124.
77. Yu T, Xing Y, Gao Q, Wang D, Chen H, Wang H, et al. Ginkgo biloba extract drives gut flora and microbial metabolism variation in a mouse model of Alzheimer's disease. Pharmaceutics. 2023;15(12):2746.
78. Yuan T, Ma H, Liu W, Niesen DB, Shah N, Crews R, et al. Pomegranate's neuroprotective effects against Alzheimer's disease are mediated by urolithins, its ellagitannin-gut microbial derived metabolites. ACS Chem Neurosci. 2016;7(1):26–33.
79. Beltran-Velasco AI, Clemente-Suárez VJ. Impact of peripheral inflammation on blood–brain Barrier dysfunction and its role in neurodegenerative diseases. Int J Mol Sci. 2025;26(6):2440.
80. Goyal D, Ali SA, Singh RK. Emerging role of gut microbiota in modulation of neuroinflammation and neurodegeneration with emphasis on Alzheimer's disease. Prog Neuro-Psychopharmacol Biol Psychiatry. 2021;106:110112.
81. Jackson A, Engen PA, Forsyth CB, Shaikh M, Naqib A, Wilber S, et al. Intestinal barrier dysfunction in the absence of systemic inflammation fails to exacerbate motor dysfunction and brain pathology in a mouse model of Parkinson's disease. Front Neurol. 2022;13:882628.
82. Kearns R. Gut–brain axis and neuroinflammation: the role of gut permeability and the kynurenine pathway in neurological disorders. Cell Mol Neurobiol. 2024;44(1):64.
83. Kim JS, Chen M-H, Wang HE, Lu C-L, Wang Y-P, Zhang B. Inflammatory bowel disease and neurodegenerative diseases. Gut Liver. 2023;17(4):495.
84. Knox EG, Aburto MR, Clarke G, Cryan JF, O'Driscoll CM. The blood-brain barrier in aging and neurodegeneration. Mol Psychiatry. 2022;27(6):2659–73.
85. Lukiw WJ, Arceneaux L, Li W, Bond T, Zhao Y. Gastrointestinal (GI)-tract microbiome derived neurotoxins and their potential contribution to inflammatory neurodegeneration in Alzheimer's disease (AD). J Alzheimers Dis Parkinsonism. 2021;11(6):525.
86. Mou Y, Du Y, Zhou L, Yue J, Hu X, Liu Y, et al. Gut microbiota interact with the brain through systemic chronic inflammation: implications on neuroinflammation, neurodegeneration, and aging. Front Immunol. 2022;13:796288.
87. Padhi P, Worth C, Zenitsky G, Jin H, Sambamurti K, Anantharam V, et al. Mechanistic insights into gut microbiome dysbiosis-mediated neuroimmune dysregulation and protein misfolding and clearance in the pathogenesis of chronic neurodegenerative disorders. Front Neurosci. 2022;16:836605.
88. Pellegrini C, Fornai M, D'Antongiovanni V, Antonioli L, Bernardini N, Derkinderen P. The intestinal barrier in disorders of the central nervous system. Lancet Gastroenterol Hepatol. 2023;8(1):66–80.
89. Shemtov SJ, Emani R, Bielska O, Covarrubias AJ, Verdin E, Andersen JK, et al. The intestinal immune system and gut barrier function in obesity and ageing. FEBS J. 2023;290(17):4163–86.
90. Singh A, Dawson TM, Kulkarni S. Neurodegenerative disorders and gut-brain interactions. J Clin Invest. 2021;131(13):e143775.
91. Solanki R, Karande A, Ranganathan P. Emerging role of gut microbiota dysbiosis in neuroinflammation and neurodegeneration. Front Neurol. 2023;14:1149618.

92. Untersmayr E, Brandt A, Koidl L, Bergheim I. The intestinal barrier dysfunction as driving factor of inflammaging. Nutrients. 2022;14(5):949.
93. Wei W, Wang S, Xu C, Zhou X, Lian X, He L, et al. Gut microbiota, pathogenic proteins and neurodegenerative diseases. Front Microbiol. 2022;13:959856.
94. Bosco N, Noti M. The aging gut microbiome and its impact on host immunity. Genes Immun. 2021;22(5):289–303.
95. Daily JW, Park S. Sarcopenia is a cause and consequence of metabolic dysregulation in aging humans: effects of gut dysbiosis, glucose dysregulation, diet and lifestyle. Cells. 2022;11(3):338.
96. Kiran NS, Chatterjee A, Yashaswini C, Deshmukh R, Alsaidan OA, Bhattacharya S, et al. The gastrointestinal mycobiome in inflammation and cancer: unraveling fungal dysbiosis, pathogenesis, and therapeutic potential. Med Oncol. 2025;42(6):195.
97. Kiran NS, Rajagopal S. Neuroinflammation—the role of heteroreceptor complexes. Explor Neuroprot Ther. 2025;5:1004105.
98. Kiran NS, Vaishnavi G, Singh S, Yashaswini C, Parihar A, Pal S, et al. Biomaterials comprising implantable and dermal drug delivery targeting brain in management of Alzheimer's disease: a review. Regen Eng Transl Med. 2024:1–24.
99. Kiran NS, Yashaswini C, Chatterjee A. Noxious ramifications of cosmetic pollutants on gastrointestinal microbiome: a pathway to neurological disorders. Life Sci. 2024;336:122311.
100. Kiran NS, Yashaswini C, Chatterjee A. Zebrafish: a trending model for gut-brain axis investigation. Aquat Toxicol. 2024;270:106902.
101. Kumar A, Pramanik J, Batta K, Bamal P, Gaur M, Rustagi S, et al. Impact of metallic nanoparticles on gut microbiota modulation in colorectal cancer: a review. Cancer Innov. 2024;3(6):e150.
102. Kiran NS, Yashaswini C, Maheshwari R, Bhattacharya S, Prajapati BG. Advances in precision medicine approaches for colorectal cancer: from molecular profiling to targeted therapies. ACS Pharmacol Transl Sci. 2024;7(4):967–90.
103. Kumar A, Pramanik J, Goyal N, Chauhan D, Sivamaruthi BS, Prajapati BG, et al. Gut microbiota in anxiety and depression: unveiling the relationships and management options. Pharmaceuticals. 2023;16(4):565.
104. Smith KM, Francisco SG, Zhu Y, LeRoith T, Davis ML, Crott JW, et al. Dietary prevention of antibiotic-induced dysbiosis and mortality upon aging in mice. FASEB J. 2024;38(23):e70241.
105. Zhang Q, Cheng L, Wang J, Hao M, Che H. Antibiotic-induced gut microbiota dysbiosis damages the intestinal barrier, increasing food allergy in adult mice. Nutrients. 2021;13(10):3315.
106. Kumar A, Sivamaruthi BS, Dey S, Kumar Y, Malviya R, Prajapati BG, et al. Probiotics as modulators of gut-brain axis for cognitive development. Front Pharmacol. 2024;15:1348297.
107. Li R, Roy R. Gut microbiota and its role in anti-aging phenomenon: evidence-based review. Appl Biochem Biotechnol. 2023;195(11):6809–23.
108. Ling Z, Liu X, Cheng Y, Yan X, Wu S. Gut microbiota and aging. Crit Rev Food Sci Nutr. 2022;62(13):3509–34.
109. Patel P, Butani K, Kumar A, Singh S, Prajapati BG. Effects of fermented food consumption on non-communicable diseases. Foods. 2023;12(4):687.
110. Prajapati BGG. Advances in probiotics for health and nutrition. BoD–Books on Demand; 2024.
111. Raghani NR, Chorawala MR, Prajapati BG. Gut fungal, friends or foes? Exploring the role of mycobiome in pancreatic cancer. IP Int J Comprehensive Adv Pharmacol. 2023;8(3):134–7.
112. Bello-Medina PC, Corona-Cervantes K, Zavala Torres NG, González A, Pérez-Morales M, González-Franco DA, et al. Chronic-antibiotics induced gut microbiota dysbiosis rescues memory impairment and reduces β-amyloid aggregation in a preclinical Alzheimer's disease model. Int J Mol Sci. 2022;23(15):8209.
113. Bello-Medina PC, Hernández-Quiroz F, Pérez-Morales M, González-Franco DA, Cruz-Pauseno G, García-Mena J, et al. Spatial memory and gut microbiota alterations are already present in early adulthood in a pre-clinical transgenic model of Alzheimer's disease. Front Neurosci. 2021;15:595583.

114. Bonfili L, Cuccioloni M, Gong C, Cecarini V, Spina M, Zheng Y, et al. Gut microbiota modulation in Alzheimer's disease: focus on lipid metabolism. Clin Nutr. 2022;41(3):698–708.
115. Chandra S, Sisodia SS, Vassar RJ. The gut microbiome in Alzheimer's disease: what we know and what remains to be explored. Mol Neurodegener. 2023;18(1):9.
116. D'Argenio V, Veneruso I, Gong C, Cecarini V, Bonfili L, Eleuteri AM. Gut microbiome and mycobiome alterations in an in vivo model of Alzheimer's disease. Genes. 2022;13(9):1564.
117. Dhami M, Raj K, Singh S. Relevance of gut microbiota to Alzheimer's disease (AD): potential effects of probiotic in management of AD. Aging and Health Research. 2023;3(1):100128.
118. Ferreiro AL, Choi J, Ryou J, Newcomer EP, Thompson R, Bollinger RM, et al. Gut microbiome composition may be an indicator of preclinical Alzheimer's disease. Sci Transl Med. 2023;15(700):eabo2984.
119. Guilherme MS, Nguyen VTT, Reinhardt C, Endres K. Impact of gut microbiome manipulation in 5xFAD mice on Alzheimer's disease-like pathology. Microorganisms. 2021;9(4):815.
120. Hashim HM, Makpol S. A review of the preclinical and clinical studies on the role of the gut microbiome in aging and neurodegenerative diseases and its modulation. Front Cell Neurosci. 2022;16:1007166.
121. Heston MB, Hanslik KL, Zarbock KR, Harding SJ, Davenport-Sis NJ, Kerby RL, et al. Gut inflammation associated with age and Alzheimer's disease pathology: a human cohort study. Sci Rep. 2023;13(1):18924.
122. Jung JH, Kim G, Byun MS, Lee JH, Yi D, Park H, et al. Gut microbiome alterations in preclinical Alzheimer's disease. PLoS One. 2022;17(11):e0278276.
123. Maitre Y, Mahalli R, Micheneau P, Delpierre A, Amador G, Denis F. Evidence and therapeutic perspectives in the relationship between the oral microbiome and Alzheimer's disease: a systematic review. Int J Environ Res Public Health. 2021;18(21):11157.
124. Mulak A. Bile acids as key modulators of the brain-gut-microbiota axis in Alzheimer's disease. J Alzheimer's Dis. 2021;84(2):461–77.
125. Nguyen NM, Cho J, Lee C. Gut microbiota and Alzheimer's disease: how to study and apply their relationship. Int J Mol Sci. 2023;24(4):4047.
126. Zhang T, Gao G, Kwok L-Y, Sun Z. Gut microbiome-targeted therapies for Alzheimer's disease. Gut Microbes. 2023;15(2):2271613.
127. Bonfili L, Cecarini V, Gogoi O, Gong C, Cuccioloni M, Angeletti M, et al. Microbiota modulation as preventative and therapeutic approach in Alzheimer's disease. FEBS J. 2021;288(9):2836–55.
128. Fernandes MR, Aggarwal P, Costa RGF, Cole AM, Trinchieri G. Targeting the gut microbiota for cancer therapy. Nat Rev Cancer. 2022;22(12):703–22.
129. Liu L, Zhang J, Cheng Y, Zhu M, Xiao Z, Ruan G, et al. Gut microbiota: a new target for T2DM prevention and treatment. Front Endocrinol. 2022;13:958218.
130. Ruo SW, Alkayyali T, Win M, Tara A, Joseph C, Kannan A, et al. Role of gut microbiota dysbiosis in breast cancer and novel approaches in prevention, diagnosis, and treatment. Cureus. 2021;13(8):e17472.
131. Schupack DA, Mars RAT, Voelker DH, Abeykoon JP, Kashyap PC. The promise of the gut microbiome as part of individualized treatment strategies. Nat Rev Gastroenterol Hepatol. 2022;19(1):7–25.
132. Shahini A, Shahini A. Role of interleukin-6-mediated inflammation in the pathogenesis of inflammatory bowel disease: focus on the available therapeutic approaches and gut microbiome. J Cell Commun Signaling. 2023;17(1):55–74.
133. Swer NM, Venkidesh BS, Murali TS, Mumbrekar KD. Gut microbiota-derived metabolites and their importance in neurological disorders. Mol Biol Rep. 2023;50(2):1663–75.
134. Wang J, Chen Y, Zhong H, Chen F, Regenstein J, Hu X, et al. The gut microbiota as a target to control hyperuricemia pathogenesis: potential mechanisms and therapeutic strategies. Crit Rev Food Sci Nutr. 2022;62(14):3979–89.
135. Yang R, Chen Z, Cai J. Fecal microbiota transplantation: emerging applications in autoimmune diseases. J Autoimmun. 2023;141:103038.

136. Zhu M, Liu X, Ye Y, Yan X, Cheng Y, Zhao L, et al. Gut microbiota: a novel therapeutic target for Parkinson's disease. Front Immunol. 2022;13:937555.
137. Dittami SM, Arboleda E, Auguet J-C, Bigalke A, Briand E, Cárdenas P, et al. A community perspective on the concept of marine holobionts: current status, challenges, and future directions. PeerJ. 2021;9:e10911.
138. Hashimoto K. Emerging role of the host microbiome in neuropsychiatric disorders: overview and future directions. Mol Psychiatry. 2023;28(9):3625–37.
139. Karl JP, Margolis LM, Fallowfield JL, Child RB, Martin NM, McClung JP. Military nutrition research: contemporary issues, state of the science and future directions. Eur J Sport Sci. 2022;22(1):87–98.
140. Li Z, Ren Z, Zhao L, Chen L, Yu Y, Wang D, et al. Unique roles in health promotion of dietary flavonoids through gut microbiota regulation: current understanding and future perspectives. Food Chem. 2023;399:133959.
141. Routy B, Jackson T, Mählmann L, Baumgartner CK, Blaser M, Byrd A, et al. Melanoma and microbiota: current understanding and future directions. Cancer Cell. 2024;42(1):16–34.
142. Sanna S, Kurilshikov A, van der Graaf A, Fu J, Zhernakova A. Challenges and future directions for studying effects of host genetics on the gut microbiome. Nat Genet. 2022;54(2):100–6.
143. Sharpton SR, Schnabl B, Knight R, Loomba R. Current concepts, opportunities, and challenges of gut microbiome-based personalized medicine in nonalcoholic fatty liver disease. Cell Metab. 2021;33(1):21–32.
144. Sood A, Kaur K, Mehrotra A, Tuli HS. Future directions in gut microbiota research: prospects and challenges. In: Gut microbiota; 2025. p. 221–44.

Gut Microbiome, Machine Learning, and Artificial Intelligence Way Forward with Precision Medicine

12

Riya Patel, Pragya Dubey, Shivani Patel, Nisarg Patel, and Bhupendra G. Prajapati

12.1 Introduction

The gut microbiota is essential to human health maintenance because it affects several physiological functions, such as metabolism, immunological control, and brain function [129]. According to a recent research, abnormalities in the makeup of the gut microbiota can have a role in the development and course of a number of illnesses, including metabolic diseases like diabetes and obesity as well as neurodegenerative diseases like Alzheimer's and Parkinson's [16]. Certain types of cell death, such as ferroptosis and pyroptosis, are strongly associated with these disorders, which are frequently characterized by dysregulated immunological responses and persistent inflammation. Probiotics have been shown to lessen these effects by regulating immunological responses, but pathogenic bacteria in the gut can cause these cell death pathways by releasing toxins [124]. In addition, the gut microbiota may serve as a biomarker for a number of diseases, including inflammatory bowel disease (IBD) [165]. At the moment, traditional therapies that target pro-inflammatory cytokines and pathways in dysbiosis linked to IBD are ineffective. There are now more treatments that specifically target the dysbiotic microbiota to

R. Patel · P. Dubey · S. Patel · N. Patel
School of Pharmacy, Indrashil University, Rajpur, Kadi, Gujarat, India
e-mail: nisarg.patel@indrashiluniversity.edu.in

B. G. Prajapati (✉)
Department of Pharmaceutics, Parul Institute of Pharmacy, Faculty of Pharmacy, Parul University, Waghodia, Vadodara, Gujarat, India
e-mail: bhupendra.prajapati40731@paruluniversity.ac.in

M. H. Rahman, Muhasina K M (eds.), *The Microbiome's Impact on the Brain*,
https://doi.org/10.1007/978-981-95-7494-0_12

achieve positive results [175]. Traditional medicine had a "one size fits all" approach, which meant that all patients with a certain condition were treated with the same medication [81]. However, this strategy has a number of issues, such as the fact that only a small portion of these people would benefit from the medication, a sizable portion would not, and a considerable portion would experience negative side effects [118]. The phrase "precision medicine" describes adapting a treatment to a subgroup of individuals who have different risk factors for developing a certain illness or reactions to a given medication [125]. Precision medicine is not particularly a novel idea. It was there even when Sir William Osler stated, "Knowing what kind of disease a patient has is far more important than knowing what kind of disease a patient has" [110]. Effective treatment is based on the development of a solid therapeutic relationship between the therapist and the client. However building trust can be challenging, particularly if the client has been betrayed or has suffered trauma in the past [1]. The stigma associated with mental health and therapy persists despite growing knowledge, which may discourage people from seeking or completely embracing therapy [76]. The diagnosis and therapy of immune-mediated diseases have advanced significantly during the past 10 years. With the development of technology, artificial intelligence (AI) has brought this fantasy to life, promising to improve diagnosis and treatment [7]. Image identification for cancer, heart illness, and neurological problems is just one example of how AI algorithms, such machine learning and deep learning, may identify patterns and connections in patient data. By assessing a patient's reaction to a certain drug and choosing the best course of action depending on the patient's genetic and clinical characteristics, AI may also assist in the treatment of diseases. AI may also assist in drug research and discovery, target identification, clinical trial design, and predicting the safety and effectiveness of novel medications [145]. This review consolidates recent findings on the impact of gut microbiota in these immune-mediated and inflammation-associated conditions. It also identifies gaps in current research and explores the potential of advanced AI-based technologies.

12.2 Composition and Functional Role of the Human Gut Microbiome

The term "microbiota" describes the whole population of microorganisms that live in a specific area. This population includes not only bacteria but also fungus, viruses, archaea, and protozoa [95]. As a complex entity, the diversity of gut bacteria can manifest itself in various ways. First, the taxonomy of bacteria: domain (archaea and bacteria), phylum, class, order, family, genus, species, and subspecies (strains) can be used to discuss the diversity of bacteria [82]. An isolated bacterium was first assigned to a taxonomic category according to its phenotypic. The 16S rDNA sequence is currently the most widely used taxonomic technique for bacterial classification, including in the human gut. Functions and genes can also be used to explain the diversity of gut microorganisms [69]. Diversity based on culturing methods: In contrast to bacteria found in soil and seawater, gut bacteria typically do not exist in the viable but not culturable (VBNC) form. A culturability of 87%27 was discovered through quantitative examination of fecal samples utilizing an anaerobic culture-dependent method (plating technique) and a culture-independent method (flow cytometry) [66]. This is not surprising because the gut does not contain the stressful conditions that cause the VBNC state, such as oxidative stress, low temperature, high osmolarity, UV radiation, and deprivation of vital nutrients [144]. Diversity within unity As was previously mentioned, there is bacterial diversity in the human gut among individuals, between individuals, and among other groups [48]. However, the entire microbiome is a component of a single unit that we refer to as the holobiont with its hologenome, meaning that it is a component of three distinct units: functional, genetic, and evolutionary [84]. Three ways exist for the holobiont and its microbiome to be united: the microorganisms inside the microbiome, the human host and its microbiome, and the human holobiont population as a whole in relation to its microbiome [5].

12.2.1 Functional Contributions

The term "microbiota" describes the whole population of microorganisms that live in a specific area. This population includes not only bacteria but also fungus, viruses, archaea, and protozoa [123]. In a healthy person, the gut microbiota contributes significant metabolic, immunological, and gut protective activities while maintaining a symbiotic interaction with the gut mucosa [28]. With its large metabolic capacity and significant functional plasticity, the gut microbiota is an organ unto itself. It obtains its nutrients from the host's diet and sheds epithelial cells [75]. Through fermentation by species like *Bacteroides*, *Roseburia*, *Bifidobacterium*, and *Faecalibacterium*, the gut microbiota uses dietary carbohydrates, particularly those that are undigested in the proximal gut, to produce short-chain fatty acids (SCFAs) like butyrate, propionate, and acetate [27]. Through the G-protein–coupled receptor 41 (Gpr41) receptor and peptide YY hormone signaling, these SCFAs act as energy sources and mediate host energy balance. A major fermenter of carbohydrates, *Bacteroides thetaiotaomicron*, has more hydrolase genes than the human genome, with around 260. D-Lactate and other harmful byproducts are likewise avoided by butyrate. *Bifidobacterium*, *Lactobacillus*, and *Oxalobacter formigenes* also aid in reducing oxalate levels, which lowers the incidence of kidney stones [73]. By upregulating the expression of a colipase that pancreatic lipase needs for lipid digestion, *Bacteroides thetaiotaomicron* has been shown to increase the efficiency of lipid hydrolysis [37]. Another important metabolic role of the gut bacteria is the synthesis of vitamin K and several vitamin B components. It has been demonstrated that members of the genus *Bacteroides* produce conjugated linoleic acid, which has been found to have immunomodulatory, antidiabetic, antiatherogenic, antiobesogenic, and hypolipidemic effects. In the human colon, the gut microbiota, particularly *Bacteroides intestinalis*, possesses the ability to deconjugate and dehydrate the primary bile acids, transforming them into the secondary bile acids deoxycholic and lithocholic acids [78].

12.2.2 Gut-Organ Axes

Gut-brain axis (*GBA*), which connects the brain's emotional and cognitive regions with peripheral intestinal functions, is made up of two-way communication between the central and enteric nerve systems. Neuroimmune and neuroendocrine signals, as well as afferent spinal and efferent vagal nerve routes, facilitate communication between the gut and the brain. Intestinal microorganisms have neurotransmitter-receptor sites that facilitate effective brain-gut communication [18]. By regulating the synthesis of neurotransmitters, such as the conversion of tryptophan to serotonin, gut microorganisms affect how the brain functions [99]. Gut motility and microbial balance are disturbed by stress-induced cortisol release. Anxiety is associated with lower levels of bacteria that produce SCFA and higher levels of pro-inflammatory cytokines, underscoring the importance of the gut-brain axis (GBA) for neurological and emotional well-being [172].

Gut-liver axis is one of the most crucial connections between the GM and the extraintestinal organs [143]. A close physical, functional, and reciprocal relationship between the gastrointestinal tract and its microbiota and the liver via the biliary tract, portal vein, and systemic circulation is known as the "gut-liver axis" [169]. By releasing bile acids and antimicrobial compounds into the biliary tract and systemic circulation, the liver interacts with the gut [22]. Simultaneously, the host and microbiota convey their products to the liver through the portal vein, influencing liver function, as a result of their metabolism of endogenous (bile acids, amino acids) and exogenous (dietary and environmental exposure) substrates [46]. The portal vein is the direct venous outflow from the gut that transports blood to the liver, where it filters the blood to eliminate gastrointestinal waste products and redistributes and accumulates carbohydrates, lipids, and amino acids. Members of the GM, which include the genera *Bacteroides*, *Clostridium*, *Eubacterium*, *Lactobacillus*, and *Escherichia*, then help transform the primary bile acids into secondary bile acids [47].

Gut-immune axis: In order to give the host a number of vital advantages, the human gut microbiota plays a critical role in many facets of human health, particularly the immune system [140]. According to recent research, the gut microbiota's early development is essential for immune system health and the prevention of autoimmune diseases [23]. The development of the immune system, which encompasses both innate and adaptive immune responses, depends on the intestinal microbiome [36]. The physical barrier of the epithelia, specialized cells, and circulating chemicals are all part of innate immunity, which quickly recognizes and eliminates a wide range of foreign antigens. According to recent research, the microbiota in the gastrointestinal tract affects the migration and function of neutrophils as well as the differentiation of T-cell populations into regulatory T cells or different types of T helper cells (Th), such as Th1, Th2, and Th17. Th17 cells are a subpopulation of TCD4+ cells that influence inflammation and immunological homeostasis by secreting several cytokines [176]. One of the main benefits of intestinal antigen-presenting cells (APCs), which co-evolved with the gut microbiota, is their capacity to protect the body from infection while maintaining immunological tolerance to the normal gut microbiota [21]. The generation of APCs is significantly regulated by the gut microbiome. Numerous investigations have shown that gut microbiota also has a role in a variety of intestinal and extraintestinal autoimmune disorders [19].

12.2.3 Dysbiosis and Associated Disorders

When a microbial ecology experiences a compositional and/or functional imbalance that surpasses its capacity for resistance and resilience, it is referred to as dysbiosis [14]. Dysbiosis and GI tract-related disorders like Inflammatory bowel disease- The two most common types of IBD, Crohn's disease (CD) and ulcerative colitis (UC), are characterized by persistent, recurrent inflammation of the anal mucosa [163]. Dysbiosis has been strongly implicated in systemic metabolic disorders. Obese mice and humans have been shown to exhibit a higher relative abundance of Firmicutes and reduced levels of Bacteroidetes [97]. Dysbiosis and other GI tract disorders: Apart from type 2 diabetes (T2D), metabolic disorders, obesity, and irritable bowel syndrome (IBS), the intestinal microbiota has also been linked to a number of other (chronic) GI-related illnesses and conditions, including celiac disease, colorectal cancer (CRC), and IBS [112]. Dysbiosis and CNS-related disorders: Numerous research have examined the potential for the intestinal microbiota to affect behavior and cognitive function by direct reprogramming of the hypothalamic-pituitary-adrenal axis, a common route that is triggered by psychological stressors and activated in response to infection. Enteric infections are known to induce cognitive impairment, anxiety, and sadness [20].

12.3 Microbiome Profiling Technologies and Data Generation

12.3.1 Sequencing Platforms

DNA extraction and sequencing of metagenomes and 16S rRNA gene amplicons: Total microbial DNA must be isolated from a sample in order to perform DNA extraction and sequencing of metagenomes and 16S rRNA genes. To identify and categorize bacteria according to conserved areas, the 16S rRNA gene is amplified using polymerase chain reaction [166]. For functional insights, metagenomic sequencing examines the complete genetic composition. Microbial diversity and potential activity are revealed by this method [141]. 16S rRNA gene amplicon processing and analysis: Raw sequencing reads are trimmed and quality-filtered as part of the 16S rRNA gene amplicon processing procedure. After that, the sequences are grouped into either amplicon sequence variations (ASVs) or operational taxonomic units (OTUs) [132]. To achieve taxonomic classification, reference databases are used. This technique aids in profiling the variety and composition of microbial communities [41]. Metagenome data processing and analysis: Quality control and the elimination of contaminants and low-quality reads are the first steps in processing metagenome data [67]. Genes, pathways, and functional components are subsequently identified by assembling and annotating the cleaned sequences [24]. The microbial makeup is ascertained using taxonomic profiling. This research sheds light on the dynamics of ecosystems and the roles of microbes [161]. Statistical analysis: In microbiome studies, statistical analysis includes assessing relationships between microbial characteristics and metadata, diversity measures (alpha and beta diversity), and differential abundance [34]. Data visualization and hypothesis testing are done with tools like R, Python, and QIIME. Multivariate techniques, such as principal component analysis (PCA) and principal coordinate analysis (PCoA), aid in the interpretation of intricate patterns found in microbial communities [51]. Analyzing primer coverage: Assessing how effectively chosen primers amplify target microbial sequences across various taxa is known as primer coverage analysis [49]. This guarantees thorough and precise identification of microbial communities in sequencing investigations [49].

12.3.2 Omics Approaches

Metatranscriptomics: Metatranscriptomics provides insights into the active functional profile of a microbial community by concentrating on the genes that are expressed by the entire population [130]. By capturing the entire mRNA, the metatranscriptome offers a snapshot of the gene expression in a sample at a certain time and under particular circumstances [86]. Metabolomics: The study of all the tiny molecules, or metabolites, in a biological sample is known as metabolomics [17]. Omics techniques can be used in metabolomics research to give a system-level understanding of metabolism and to facilitate high-throughput metabolite analysis. Omics techniques can be used in metabolomics research to enable high-throughput study of metabolites and to offer a system-level understanding of metabolism [62]. Biomarker discovery: Potential biomarkers for a number of illnesses, including diabetes, cardiovascular disease, and cancer, have been found using metabolomics [154]. Pharmacological discovery: Metabolomics has been utilized to investigate the processes of pharmacological action and determine the targets of medications. For instance, metabolomics was utilized in a study to determine a possible anticancer drug's target. Another study examined the mechanism of action of an antituberculosis medication using metabolomics [106]. Environmental monitoring: Metabolomics has been used to track pollution levels in the environment and look into the different ways that environmental pollutants affect the microbiome [15].

Agriculture and food sciences: Metabolomics has been utilized in agriculture and food sciences to investigate plant metabolism and pinpoint the metabolites that give plants their desired flavors and nutritional qualities [108]. Proteomics: With new advances in methodology and technology, the fascinating area of proteomics is developing quickly. It is a crucial tool for comprehending how proteins function in biological systems and for creating novel diagnostic and therapeutic instruments [120]. Quantitative proteomics: One effective method for assessing variations in protein abundance across samples is quantitative proteomics [135]. Structural proteomics: The study of protein structure and function at the molecular level is known as structural proteomics. It entails figuring out the three-dimensional structure of proteins using methods like cryo-EM (cryo-electron microscopy), nuclear magnetic resonance spectroscopy, and X-ray crystallography [9]. Posttranslational modification proteomics: The study of protein modifications that take place after translation, such as phosphorylation, acetylation, and glycosylation, is known as posttranslational modification proteomics. It entails the detection and enrichment of altered peptides through the use of methods like mass spectrometry and tests based on antibodies [58]. Functional proteomics: The study of protein connections and function in relation to cellular networks and pathways is known as functional proteomics [107]. To obtain a thorough grasp of cellular function, proteomics data must be integrated with other omics data, including transcriptomics and metabolomics [52].

12.3.3 Bioinformatics Pipelines and Tools

More and more bioinformatics analyses are being offered as services that users may access via a web interface with a robust backend that performs the analysis. Research institutes like European Molecular Biology Laboratory-European Bioinformatics Institute, commercial businesses like Illumina, and research initiatives like Galaxy may offer general services [91]. META-Pipe 1.0 Metagenomics Pipeline: For metagenomics samples, our META-pipe workflow offers pre-processing, assembly, taxonomic categorization, and functional analysis [26]. Genome Analysis Toolkit (GATK) Variant Calling Reference Pipeline: Scalable genomics data analysis workflows frequently refer to the GATK best practices pipeline for germline single nucleotide polymorphism and indel identification in whole-genome and whole-exome sequences [56]. This pipeline offers callset refining, variant calling, and pre-processing. ADAM Variant Calling Pipeline: ADAM is a genomics pipeline that is built on top of the Apache Spark big data processing engine Avro data serialization system and Parquet columnar storage system to improve the performance and reduce the cost of variant calling [57]. GESALL Variant Calling Pipeline: A genomic analysis platform for unaltered analysis tools that utilize the POSIX file system interface is called GESALL. Implemented as MapReduce applications, GESALL pipelines use resources allotted by YARN [134]. Toil: TCGA RNA-Seq Reference Pipeline—Toil is a workflow program used in high-performance computing or cloud environments to execute scientific workflows on a large scale. It is intended for use with large-scale analytic pipelines, such The Cancer Genome Atlas (TCGA) which calculates gene- and isoform-level expression values from RNA-seq data using the best practices [101].

12.3.4 Data Complexity and Interpretation Needs

Technologies for microbiome profiling include metagenomics, metatranscriptomics, and 16S rRNA sequencing lean to light on the makeup and roles of microorganism. High-throughput complex data generated by these techniques require classy computational tools for precise interpretation [174]. Eco-friendly variability, sequencing depth, and microbial community variety all contribute to data complexity. Understanding the function of microbes in health and disease is improved by integrating multi-omics data [72]. Application in environmental monitoring, diagnostics, and tailored medicine all depend on accurate in the data [77].

12.4 Analytical Challenges in Microbiome Data Science

Enormous effects on health stem from the rapidly expanding field of the human microbiome. Although researchers can study microbial populations using high-throughput microbiome sequencing, they encounter methodological hurdles. Specifically, researchers face the lack of targeted pre-processing methods to improve data quality, which prevents effective bias minimization before downstream analysis [177].

12.4.1 Data Sparsity and Dimensionality

The variance-oriented compositional PCA biplot provides a statistically sound substitute for conventional β-diversity methodologies in microbiome research. It simultaneously represents inter-OTU variability and inter-sample dissimilarities, thereby facilitating more nuanced data interpretation. In contrast to PCoA, it consistently maintains performance under data subsetting and reduces sensitivity to sparsity and binary abundance structures. Because it does not require a phylogenetic tree, it becomes more reproducible and simpler to update with new taxa. This straightforward approach enhances the utility of exploratory data analysis [61]. The biplot delivers valuable qualitative insights into data quality and group relationships. While challenges such as those tackled by methods like zero-inflated modeling and dimensionality reduction techniques, including PCA, t-distributed stochastic neighbor embedding (t-SNE), and uniform manifold approximation and projection (UMAP), are often addressed, issues related to interpretability and overall effectiveness can also be posed by them [98].

12.4.2 Inter-individual and Temporal Variability

Attention is being gained by the temporal dynamics of host-associated microbial populations (the microbiota) due to their importance to health. The stability of human microbiota is typically maintained for months, if not years. However, it is indicated by research on mice and humans that prevalent characteristics of the modern Western lifestyle, including antibiotics and high-fat diets, can lead to permanent changes in commensal microbial ecosystems [40]. Dietary habits can vary greatly, making diet an important confounding factor in microbiome research. One option to compensate for this is for all food to be provided during the trial, although this is expensive and logistically difficult. Controlled feeding trials are not involved in most microbiome studies. An additional strategy is for household controls who eat comparable diets to be included. Similar environmental circumstances are also subjected to them, which may affect microbiome makeup [103]. Advanced statistical tools like mixed-effect models and longitudinal clustering are used, but high-quality metadata and frequent sampling are required [168]. Batch effects are experienced when biological signals are mixed up by changes in technology during sample collection, sequencing, DNA extraction, or data processing [148]. Systematic changes between batches may be apparent rather than real biological variation. Similarly, results might be altered by contamination from chemicals, lab equipment, or the environment, particularly in low-biomass samples like skin or air microbiomes [50]. Incorrect associations might be created or actual ones obscured by these artifacts. To address these challenges, tight experimental techniques, the inclusion of negative controls, and computational methods such as surrogate variable analysis or the use of tools like Decontam to identify and mitigate pollutants are required [42]. Microbiome data is found to be unsuitable for traditional statistical procedures, which are frequently built for Gaussian-distributed and independent data. The compositional, sparse, and over-dispersed nature of the data is not explained by the traditional procedures. Deceptive results may be yielded when relative abundance data is subjected to PCA without sufficient transformation [111]. It has been noted that zero-inflated distributions cannot be adequately handled by differential abundance testing with t-tests or analysis of variance and that customized approaches such as DESeq2, ALDEx2, or ANCOM must be used instead. These data aspects must be taken into account by integrative and machine learning algorithms to enable valid biological interpretation [127].

12.5 Machine Learning in Microbiome Research

Substantial advances in microbiome research have been made by ML through allowing complex patterns and predictive features to be extracted from high-dimensional, noisy, and heterogeneous datasets. Given the rapid expansion of next-generation sequencing (NGS) microbiome datasets, powerful tools for microbial community classification, clustering, feature selection, and temporal modeling are provided by ML approaches. Significant ML approaches used in microbiome research are examined in this section, with an emphasis on their utility and limitations.

12.5.1 Supervised and Unsupervised Learning Techniques

Microbiome investigations can be divided into two types of machine learning approaches: supervised and unsupervised. The process of training models with labeled data with the purpose of predicting or classifying outcomes based on microbiological properties is known as supervised learning. Common approaches are composed of support vector machines (SVM), random forests (RF), gradient boosting machine, and deep neural networks [83]. Disease classification and biomarker development are particularly supported by these models. Latent patterns in unlabeled data are detected by unsupervised learning, allowing for exploratory analysis like clustering, dimensionality reduction, and community structure inference. K-means clustering, hierarchical clustering, and PCA are included as common techniques [159]. The display of complicated microbiological datasets is enabled by more advanced methods, such as t-SNE and UMAP.

12.5.2 Feature Selection and Dimensionality Reduction

Frequently, hundreds of ASVs or OTUs are contained in data that are sparse and high-dimensional. The speed of computation is enhanced, and model understanding is improved; the selection of features is considered essential. Relevant taxa or genes are commonly identified using methods including shared data gain, least absolute shrinkage and selection operator, and recursive feature elimination [164]. A lower-dimensional space is used to project high-dimensional data for analysis using dimensionality reduction techniques, which can be either linear or nonlinear. Underlying gradients in ecological structuring and community composition are frequently revealed by these techniques, which are associated with host phenotypes or environmental gradients [104].

12.5.3 Disease Classification Models

Predictive models that have been used in microbial signatures to differentiate between unhealthy and healthy states have been made possible largely due to machine learning. Illnesses such as Type 2 Diabetes, obesity, colorectal cancer, and inflammatory bowel disease for which disease-specific classifiers have been developed [121]. The popularity of RF and SVM for correlated and noisy features is made to their robustness. The interpretability of disease models is progressively improved by integrating local interpretable model-agnostic explanations or Shapley Additive exPlanations, which pinpoint important microbial taxa influencing categorization choice [63]. However, caution is called for by batch effects, demographic heterogeneity, and variables that could impact model generalizability.

12.5.4 Clustering Enterotypes and Microbiome Signatures

The microbial community known as "enterotypes," which are defined by the predominance of particular bacterial families like *Bacteroides*, *Prevotella*, or *Ruminococcus*, was identified by unsupervised clustering. The biological significance and stability of enterotypes are still up for debate, and a framework for comprehending host-microbiome interactions and inter-individual variability is offered by these stratifications. Based on composition, microbial communities are grouped into discrete enterotypes or metacommunities by Dirichlet multinomial mixture (DMM) models. These models, which may be implemented using the Laplace approximation in the microbe DMM program, can evaluate treatment effects and assist with categorization tasks. Four clusters, two dominated by *Bacteroides* and two with more variable compositions, were found when the gut microbiota data from obese and lean twins were compared. Fractional membership across a collection of subjects in linear discriminant analysis can be done by documents, which is a generalization of DMM modeling. The value of this extra-flexibility in microbiome research is to be illustrated by one objective of the case study in Sect. 12.4, as the primary drawback of DMM modeling is tackled by it [142]. Dynamic unigram model and non-negative matrix factorization are gaining popularity in community-level patterns in metagenomics data.

12.5.5 Dynamic and Time-Series Modeling

The microbiome is dependent upon a number of variables, like illness, antibiotics, nutrition, and circadian rhythms. Time-series modeling techniques can be used to observe microbial succession and stability over time. These changes have been modeled by autoregressive models, dynamic Bayesian networks, and hidden Markov models [60]. The prediction of the beginning of disease or the effectiveness of treatment may be made using long-term memory models of recurrent neural networks, which are new tools for modeling sequential microbiome data [29]. To investigate microbiome biomarkers for CRC, results on two additional cohorts were verified by this study, and a meta-analysis of 969 fecal metagenomes from seven cohorts (including five public and two new datasets) was performed. It was discovered that oral-derived species contributed to the continuously greater microbial richness observed in CRC patients. Additionally, it was found that functional pathways (such as fermentation and gluconeogenesis) were abundant in CRC, whereas healthy controls were associated with starch degradation pathways. Choline metabolism genes were found to be overrepresented in CRC, and good accuracy (average AUC = 0.84) was demonstrated by predictive microbiome models trained on pooled datasets, indicating that reliable, repeatable biomarkers for therapeutic use were identified [153].

12.6 Artificial Intelligence and Precision Medicine

12.6.1 AI Versus ML

Artificial intelligence is the science and engineering of creating intelligent computers that replicate human cognitive processes like learning and problem-solving by following algorithms or a set of rules [155]. Artificial intelligence systems function in a deliberate, intelligent, and adaptable way because they can either foresee problems or handle them as they arise [89]. For instance, AI systems could convert a patient's complete medical history into a single number that indicates a likely diagnosis [87]. AI's strength lies in its capacity to learn and identify patterns and relationships from large multidimensional and multimodal datasets. AI is a collection of subfields (including machine learning and deep learning) that, either separately or in combination, provide application intelligence [128]. It is not a single, all-pervasive, universal technology. The study of techniques that enable computer programs to automatically get better with experience is known as machine learning (ML) [13]. ML itself can be divided into three categories: "supervised," "unsupervised," and "reinforcement learning". Research is also being conducted in a number of subfields, such as "semi-supervised," "self-supervised," and "multi-instance" [89]. Using tagged X-ray images of known cancers to identify tumors in new images is an example of how supervised learning makes use of labeled data or annotated information [128]. In order to find a common cause, unsupervised learning classifies groups of patients with similar symptoms in an effort to extract information from data without labeling [128]. Artificial intelligence is the development of computer systems that are inherently able to perform tasks that call for human intelligence. These programs simulate how people learn, reason, and make decisions [150]. ML, a subfield of AI, is the study of statistical models and algorithms. They enable computers to learn from data and progressively enhance their performance without the need for explicit programming [65]. Software development has changed as a result of the combination of AI and ML, spurring innovation and progress. Like many other industries, software development has seen a significant change as a result of AI and ML technology [152]. Developers can unlock new possibilities and the best conventional conditions by utilizing AI and ML techniques [167].

12.6.2 Deep Learning Applications in Healthcare

Deep learning has already made an impact on the medical field. In order to help clinicians manage patient data and outcomes, Google has devoted a great deal of work to investigating the potential of deep learning models to generate prediction [45]. The blog post, "Deep learning for Electronic Health Records," continued by outlining how deep learning may be applied to lessen administrative burdens while enhancing understanding of patient needs and care. Because it may reduce the administrative burden while enabling medical professionals to concentrate on what they do best—provide treatment—this is the finest application of deep learning in healthcare [147]. The National Health Service (NHS) in the UK is dedicated to being a leader in healthcare using AI, ML, and deep learning. The benefits that this technology offers have been acknowledged by the healthcare provider [31]. The ability to improve patient care through the use of deep learning toolkits and intelligence analysis is undoubtedly tempting to the NHS, which is struggling with chronic talent shortages, Brexit, and cost cutting [8]. By investing in deep learning technologies, the company may be able to streamline patient care while avoiding some of the legacy issues that have affected efficiency. Additionally, it might offer much-needed assistance to the medical staff. The fact that AI, ML, and deep learning would be a part of the government institution's future was solidified in August 2019 when Boris Johnson invested £250 million in deep learning healthcare efforts for the NHS [119]. Although there are concerns about the NHS's possible use of AI, a recent study published in *The Lancet Digital Health* journal greatly enhanced its legitimacy. According to the paper, "deep learning models' performance is comparable to that of health-care professionals." A remarkable statement that did come with some caveats, but ultimately emphasized how deep learning in healthcare, could benefit patients and health systems in clinical practice [170]. Medical imaging solutions, chatbots that can recognize patterns in patient symptoms, deep learning algorithms that can detect particular cancer types, and imaging solutions that use deep learning to detect rare diseases or particular pathologies are already examples of deep learning applications in the healthcare industry. Deep learning has proved essential in giving doctors the knowledge they need to spot problems early and provide far more individualized and pertinent service to patients [3].

12.6.3 AI-Powered Prognostic Tools

Predicting a disease or condition's future course and results after diagnosis is known as prognosis [39]. AI has a revolutionary impact in this field. It can find patterns and correlations that human clinicians might not see right away by using big datasets and sophisticated algorithms. This involves examining past patient outcome data, response rates to different therapies, and trends in the course of diseases [74]. In order to provide a more comprehensive and accurate prognosis, AI systems can also take into account a variety of factors, such as lifestyle choices and genetic information [6]. AI-driven prediction is especially helpful for degenerative and chronic illnesses like cancer, heart disease, and neurological conditions [44]. It aids in comprehending the anticipated course of various illnesses throughout time, allowing medical professionals to foresee potential issues and modify treatment plans appropriately. In addition to lowering long-term healthcare expenses, this proactive strategy to controlling chronic diseases can greatly enhance patient quality of life [158]. AI in prognosis helps with personalized medicine as well. Treatments can be better adapted to each patient's unique needs by knowing how their condition progresses, increasing efficacy and reducing side effects. This personalization is especially important in conditions with high variability in outcomes among patients [80].

12.6.4 Integration with EHRs and Real-World Data

The integration of gut microbiome research with electronic health records (EHRs) and real-world data (RWD) shows a major breakthrough in the generation of precision medicine [113]. EHRs have the capability to have historical health data, and longitudinal patient data, diagnostic codes, medications, laboratory generated data, and clinical notes are abundant sources of interpreting health trends. On the other hand, RWD can be extracted from devices which are wearable, patient registries, claim insurance, and mobile health platforms that captures behavior and outcomes on everyday basis, making it highly preferable for microbiome studies [113]. A combination of these data sources, researchers can make predictive models by using ML and AI that helps make personalized diagnostics, prognostics, and various treatment techniques [90]. For example, correlation of particular microbiome composition and treatment response of treatment can be validated for historical EHR data and can enable clinical researchers to classify patients based on microbiome profile. This is auspicious conditions such as IBD, metabolic disorders, and cancers where microbiomes plays an important role [30]. Ethical scaffolding and securing data sharing process will be necessary to fully tackle the power of EHR-microbiome integration. With vigorous AI-driven analytics and harmonized data, convergency of EHR, RWD, and microbiome holds a massive potential to reshape amplified care and upgrade clinical decision-making in real-world needs [162].

12.6.5 AI in Drug Response Prediction

Due to a number of important aspects, AI offers a convincing approach for drug response prediction. Conventional methods of assessing drug response frequently depend on small datasets and basic regression models, which may ignore the intricate interactions between variables affecting treatment results [59]. Additionally, by concentrating on a limited number of variables, these approaches may overlook important details about the unique traits of each patient and how they respond to treatment. However, there is a chance to use enormous volumes of data, including genomics data, EHRs, and real-world patient data, thanks to the development of AI, especially ML algorithms [114]. AI makes it possible to conduct a more thorough analysis by taking into account several variables and confounders at once. It has the potential to improve our comprehension of drug response mechanisms by looking at data holistically and spotting complex patterns across many information sources [113]. All special features of machine learning has been given in Fig. 12.1.

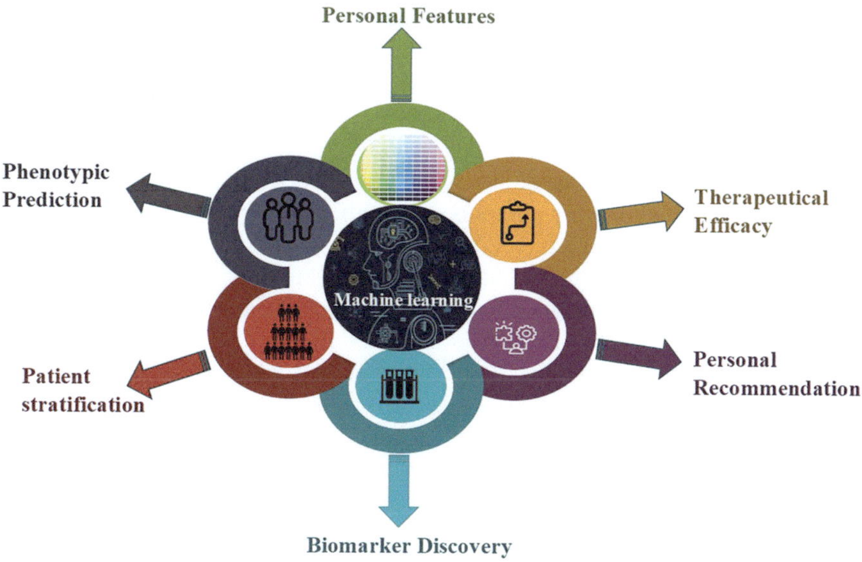

Fig. 12.1 Special features of machine learning

12.7 Synergizing AI with Gut Microbiome for Personalized Medicine

12.7.1 Identifying Microbial Biomarkers Via AI

In gut microbiome studies, computational platforms like QIIME, Mothur, and DADA2 are frequently utilized to process and analyze sequencing data [100]. Certain genetic markers, such as the 16S rRNA gene, which offer differing levels of taxonomic precision and phylogenetic information about the microbial population, are targeted and sequenced by these marker-gene surveys [105]. XGBoost: A collection of decision trees trained via an iterative gradient boosting procedure is used by the XGBoost method. In this procedure, important topics in decision trees are addressed at every stage as the trees progress. Sparsity-aware split finding is used by XGBoost to address the problem of missing values [133]. Random forest: The RF algorithm uses bootstrapping to create an ensemble of decision trees by repeatedly dataset [156]. Support vector machine: The SVM minimizes a loss function called hinge loss and adds a penalty term to determine the best boundary between two or more classes in the data space [116].

12.7.2 Personalized Nutrition and Therapeutic Response

Microbiome as a determinant of human therapeutics: A key element affecting human treatments is the human microbiome, which is made up of trillions of bacteria that live in and on our bodies [93]. Microbiome and disease risk: Disease risk and microbiome composition are closely related. A number of illnesses, such as inflammatory ailments, metabolic abnormalities, and even some types of cancer, have been linked to dysbiosis or the imbalance of microbial communities [53]. Microbiome as a diagnostic biomarker: The potential of microbial signatures as diagnostic biomarkers has been revealed by developments in microbiome research. A person's microbiome's particular makeup can act as a unique fingerprint that indicates their current state of health [122]. Microbiome and personalized treatment: Therapeutic approaches for a variety of diseases are changing as a result of the microbiome's role in personalized treatment [79]. A paradigm shift toward adjusting therapies to a person's unique microbial profile has been spurred by the complex relationships between the microbiome and human health [55]. Microbiome dysbiosis: The disruption of the normal equilibrium of the microbial communities that live inside and on the human body, known as microbiome dysbiosis, results from a complex interaction of several factors [33]. Microbiome's impact on nutrient metabolism: The body's ability to consume and absorb food components is influenced by the microbiome, which has a major impact on nutritional metabolism [88]. Microbes facilitate the synthesis of certain vitamins and short-chain fatty acids, the fermentation of fiber, and the breakdown of complex carbohydrates [157].

12.7.3 Microbiome-Modulating Interventions

Prebiotics are indigestible food ingredients that promote the growth and/or activity of beneficial bacteria un the gut [102]. Examples of these include fiber, inulin, and oligosaccharides. They serve as substrates for particular microorganisms, increasing their metabolic activity and ability to proliferate [96]. Probiotics: They are live bacteria that improve health when consumed in sufficient amounts. They can be obtained through pills or fermented foods like kimchi or yogurt. They either alter the activity patterns of already-existing populations within the gut microbiome or directly introduce beneficial species into it [54]. Symbiotics are a combination of prebiotics and probiotics which have synergistic effects on the gut microbiota. Fecal microbiota transplantation (FMT): FMT is the process of giving fecal material from a healthy donor to a person whose microbiota has been disturbed or has become unhealthy due to illness [35]. Although FMT is primarily used to treat recurrent *Clostridium difficile* infections, it also investigates other dysbiosis-related illnesses such as IBD and metabolic disorders. FMT improves health outcomes by restoring the variety of bacteria in a person's intestinal [136].

12.7.4 Clinical Decision Support Tools and Software Platforms

The phenomenon of the clinical decision support (CDS) tools and software platforms has gained a fundamental place in the process of contemporary healthcare since it provides real-time support to health professionals by helping them make effective clinical decisions. They apply data collected on patients, clinicians, practice guides, and medical knowledge databases to give advice in the form of reminders, warnings, diagnostic decisions, and recommendations [25]. Such systems as the IBM Watson Health, Epic Systems, and Cerner Millennium use the power of artificial intelligence and machine learning algorithms to analyze extensive bodies of data and forecast the results, making it possible to increase the accuracy of diagnosis and minimize the occurrence of errors. EHRs are quite frequently equipped with CDS systems that can be integrated with clinical workflows easily. They are also used as alerts of drug interaction, dose calculation, prediction of disease risks, and personalized medicine as well. These systems increase efficiency, ensure optimal results of patients, and facilitate evidence-based practice. The data standardization, interoperability, and the issue of clinician fatigue caused by too many alerts and patient privacy should continue to be the source of more research and system optimization [115].

12.7.5 Ethical, Regulatory, and Societal Implications

Data protection, vulnerable groups, and informed consent are examples of ethical issues. The goal, dangers, and rewards must all be completely understood by participants. The individuals affected should get extra attention. Strong database security is necessary to protect the privacy and information of participants. Strict regulatory standards, efficacy and safety, and scientific integrity are among the regulatory challenges [64]. To adhere to regulatory requirements, there are regulatory agencies in India such as Central Drugs Standard Control Organization, Department of Biotechnology, and Ministry of Environment, Forest and Climate Change. It is their responsibility to guarantee that research satisfies the requirements for efficacy, safety, and ethics. When choosing a medication delivery method for brain-gut repair, these considerations are crucial. Important privacy and security ethical questions are brought up when AI is combined with surveillance technologies in the healthcare industry. Sensitive health information should be protected from unauthorized access by taking some crucial precautions: ensuring patients are aware of how data is used, shared, and stored; stopping biased AI systems from causing healthcare disparities; and creating an AI paradigm that guarantees accountability and transparency [126]. Gaining patient trust, upholding individual rights, and promoting social trust all depend on doing these sorts of actions. We can guarantee that AI in healthcare adheres to ethical standards by taking a few actions into consideration [139]. Regulations are necessary for the incorporation of AI in brain-gut axis treatments. Regulations pertaining to data protection ought to come first. Patient data security and privacy are guaranteed under General Data Protection Regulation

(EU) and Health Insurance Portability and Accountability Act (USA). The creation of ethical AI should be accomplished by encouraging openness, equity, and the eradication of prejudice. The FDA and other regulatory bodies should make sure that everything satisfies the requirements. It will guarantee the safety and effectiveness of AI-driven treatments [139].

12.7.6 Federated Learning and Interoperability

Federated learning is a decentralized machine learning technique in which several clients work together to train a common model without exchanging data directly [146]. This approach is especially helpful in situations where data privacy is crucial or if centralized data storage is impractical because of limitations like bandwidth or data ownership laws [92]. Traditional machine learning involves gathering data in one place and using it to train a model. However in FL, only model updates—like weights or gradients—are sent to a central server; all other data stays on the client devices [146]. This protects privacy and lowers the risks of data breaches by guaranteeing that private information never leaves the device or organization. The utility of FL and interoperability is gaining traction in microbiome research, more specifically in terms of ML and precision medicine [2]. It gains more importance in fields like healthcare, where patient privacy is more important, ownership of data, and regulatory compliance. Interoperability plays tribute by guaranteeing seamless communications and integrative heterogeneous systems, such as electronic health records (EHRs), genomic databases, and microbiome repositories [151]. The standards like fast healthcare interoperability (HL7 FHIR) and observational medical outcomes partnership (OMOP) facilitate framework for data exchange and enable FL models to learn from wide spectrum of real-world, multimodal data sources [38].

12.7.7 Explainable AI for Clinical Use

Explainable AI (XAI) investigates if ML models can be made more comprehensible to humans and seeks to improve their efficacy while also making it possible for non-experts to use them in a variety of contexts [12]. Researchers have tried numerous times, but precise definitions of terms like interpretability and explainability are still hard to come by. Resolving the AI "black box" conundrum has led to a rise in transparency and XAI appeal. XAI techniques could help us better understand the output of an ML model. Furthermore, explainability refers to the explanation as a trustworthy and understandable way of communicating with decision-makers [109]. Furthermore, AI-based healthcare systems can meet the growing need for top-notch medical care. Most of these AI-based products, however, are only prototypes and never reach the market [4]. The models relied on differentiating qualities to predict outcomes and failed to meet performance goals and critical risky assessments. When put into test or production, a number of projects, including Google Health, have all failed in some fashion [138]. In order to categorize the explainability of deep learning, Misztal-Radecka and Indurkhya established a taxonomy that encompasses understanding deep network representation, processing, and system construction. Finally, creating clear or understandable models and encouraging interdisciplinary collaboration among medical professionals are essential for information exchange [117]. A three-group taxonomy of neural network interpretability techniques was developed by the authors. However the main emphasis was on DL. In order to shed light on the connections between the inputs and outputs of the model, the first category includes techniques that simulate data processing [85].

12.7.8 Real-Time Monitoring Via Smart Biosensors

Biomarkers, or indications of disease, can be found using a variety of biological materials. A biomarker is defined as "a defined characteristic that is measured as an indicator of normal biological processes, pathogenic processes, or responses to an exposure or intervention," according to the Food and Drug Administration (FDA) and the National Institutes of Health [11]. Susceptibility, prognostic, diagnostic, prediction, monitoring, reaction, and safety are the seven different categories into which biomarkers fall, according to the US FDA. Understanding these categories makes it feasible to implement standard procedures that support a biomarker with its unique context and purpose, improving the effectiveness of creating precise diagnostic and therapeutic approaches [71]. Clinical experts are always searching for new biomarkers, and recently, they have focused on digital, unusual signals. In order to create an intermediate biomarker, these digital biomarkers usually combine biological, neurological, socioeconomic, and environmental data. The application of ML and DL techniques to the search for biomarkers has grown in popularity recently [68]. Biomarkers are extracted from a range of biological data using both supervised and unsupervised learning techniques. Supervised learning utilizes methods such as RF, SVM, and logistic regression, along with DL techniques like CNNs and RNNs. On the other hand, unsupervised learning employs PCA, ICA, and clustering algorithms to uncover novel biomarkers. Researching disorders that require the combined analysis of data from multiple sources will benefit more from the integrative digital biomarker [43]. This method will help with early identification and more accurate symptom forecasting by making sure no subtle patient signals are missed and accounting for the interaction between various signals. The development of biomarkers, particularly in cancer, has been revolutionized by advances in NGS and our understanding of the human genome [131]. Traditional biomarkers usually depend on markers that could be found by imaging techniques or that circulated in blood, plasma, and serum. Deeper investigation has been made possible by the development of genomics, which has shown specific genetic differences that have a major impact on disease pathophysiology, especially in cancers where genetic mutations usually predominate [32]. AI-based technologies show notable gains in oncologic clinical trials, reducing patient screening time by 25% and increasing the number of potentially eligible patients by 50%. The examination of the biofluid should be carried out on the subject as well as confirmed using conventional laboratory techniques to validate consistent results in order to confirm the effectiveness of this unique sensor [171]. The sensor needs to have a wide linear response range, a low detection threshold, and high sensitivity in order to detect the pertinent physiological concentration ranges. Without requiring sample preparation, the sensor should be biocompatible and able to identify the target molecule in the intricate biofluid environment. Fixed receptors that bind selectively to the desired targets, such as enzymes, peptides, aptamers, or molecularly imprinted polymers, are used to achieve this precision [10, 11]. Obtaining molecular data from participants in an effective manner is crucial for the successful discovery of biomarkers. Physical activity-induced signal interference and the effects of pH, temperature, conductivity, and fluid renewal on sensor performance are major barriers to sensors collecting reliable data [11]. Selecting the proper equipment for fluid sampling and putting into practice suitable filtration and calibration techniques are crucial. For biomarker validation, cross-sectional and longitudinal research are essential, producing large datasets [71].

12.7.9 Integrated Multi-omics for Deep Personalization

Technologies that validate a patient's basic biology, such as DNA, RNA, or protein analysis, are crucial for precise disease diagnosis and are a major component of contemporary advances in personalized medicine [94]. The synergies between precision medicine and DNA analysis are anticipated to grow as technology develops further, opening up new possibilities for managing, treating, and preventing disease. For example, genome sequencing and other customized methods can identify DNA changes that affect conditions like cancer and cystic fibrosis [149]. By studying the aggregate characterization of biological molecules, omics studies shed light on the dynamics, structure, and function of an organism. Multi-omics in precision medicine combines information from transcriptomics, proteomics, metabolomics, genomes, and other fields to reveal how individual genetic variations affect the onset, course, and response to therapy of disease [70]. By customizing treatments to each patient's own molecular composition, this holistic approach supports customized healthcare by deciphering intricate biological processes and disease mechanisms. The application of multi-omics approaches to precision medicine is currently the focus of several international networks and projects, including the Human Proteome Project, the Precision Medicine Initiative Cohort Program, TCGA, and the Pan-Cancer Analysis of Whole Genomes [160]. For instance, by combining multi-omics data to better understand disease causes and improve therapeutic precision, the European Joint Programme on Rare Diseases, a Europe-wide project, seeks to speed up the diagnosis and treatment of rare diseases [137]. A 2018 study on pancreatic ductal adenocarcinoma improved patient classification for tailored therapy by combining genomes, transcriptomics, and proteomics data to find unique molecular subgroups that show different responses to treatment. Likewise, in 2020, a study used a multi-omics technique to analyze cerebrospinal fluid samples from Alzheimer's patients, combining proteomics and metabolomics data to find new

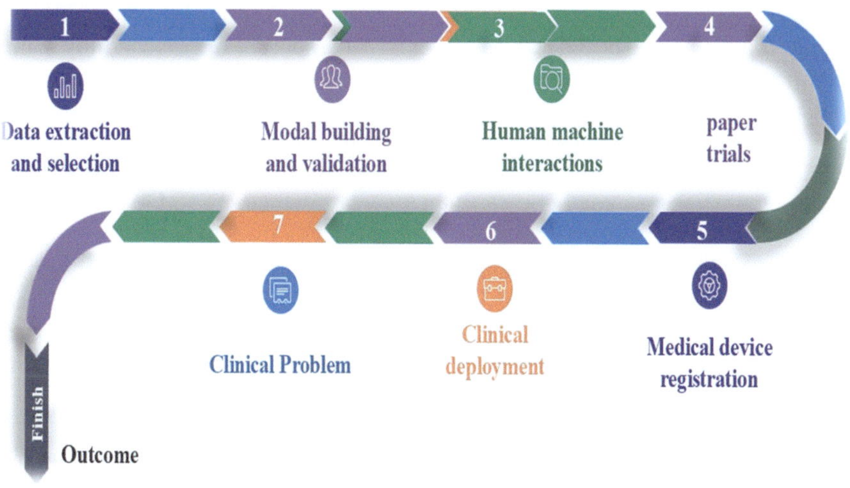

Fig. 12.2 Future outlook and research roadmap

biomarkers linked to the illness and expanding our knowledge of its molecular underpinnings. These instances show how multi-omics techniques have revolutionized our knowledge of intricate biological systems and illnesses, enhancing diagnostic and treatment approaches in a variety of domains. The entire process has been explained to roadmap (Fig. 12.2).

12.8 Conclusion

In summary, using AI to microbiology offers a promising way to improve our comprehension of microbial systems and their intricate behaviors. We have examined the many uses of AI in this article, ranging from drug discovery to predictive modeling, and we have emphasized how revolutionary it may be for microbiological research. However in addition to its enormous potential, artificial intelligence in microbiology also presents formidable obstacles. To guarantee the appropriate and efficient use of AI tools in this field, issues including data quality, the interpretability of AI models, and ethical considerations must be properly addressed. The future of microbiology will surely be shaped in the upcoming years by developments in AI technology as well as a better understanding of microbial ecosystems. We may steer toward a more profound and impactful understanding of the microbial world by embracing innovation while being aware of the limitations and difficulties inherent in AI. By analyzing enormous datasets to find patterns linked to health and illness, directing individualized nutritional and probiotic therapies, and forecasting treatment outcomes, AI has the potential to completely transform our knowledge and management of gut microbiota. Advanced AI-driven tools could help discover new treatments for a variety of illnesses, from digestive disorders to other health problems, by providing insights into intricate microbial ecosystems. Although AI-enhanced microbial technologies have significantly improved healthcare, a number of issues must be resolved to fully realize their potential. The requirement for extremely accurate models is implied by the fact that even a small percentage of false positives or false negatives in the medical field might result in incorrect diagnoses or wrong treatment recommendations. The historical fight against new microorganisms has brought attention to the links between environmental interactions and public health issues. Combining AI and microbiological methods has an enormous potential for comprehending these intricate dynamic systems. This strategy, which promotes collaboration between the public and private sectors, emphasizes a comprehensive view that transcends personal health and acknowledges the interdependence of environmental, microbial, and human systems. This cooperative strategy, powered by AI and microbiological technologies, has the potential to solve global health issues holistically [173].

Acknowledgments The authors are very thankful to the School of Pharmacy, Indrashil University, for providing the necessary library facilities to get the information about PGB, etc. Dr. Prajapati extends his sincere appreciation to the Faculty of Pharmacy, Silpakorn University, Thailand, for their generous support that enabled the completion of this work.

Funding
Not applicable.

Declarations

Declaration of Generative AI and AI-Assisted Technologies in the Writing Process During the preparation of this work, the authors used QuillBot and Grammarly service in order to improve scientific writing and the English language. After using this tool/service, the authors reviewed and edited the content as needed and take full responsibility for the content of the publication.

Competing Interests The authors have no competing interests to declare that are relevant to the content of this manuscript.

Ethics Approval The authors declare that this manuscript does not include primary research with human participants.

The authors declare that this manuscript does not include primary studies involving animals.

References

1. 10 Common challenges in therapy and how to overcome them | JiYu Psychologist Digest. n.d. Retrieved 29 July 2025, from https://jiyushe.com/psychologist/10-common-challenges-in-therapy-and-how-to-overcome-them.html.
2. Abbas SR, Abbas Z, Zahir A, Lee SW. Federated learning in smart healthcare: a comprehensive review on privacy, security, and predictive analytics with IoT integration. Healthcare. 2024;12(24):2587. https://doi.org/10.3390/HEALTHCARE12242587.
3. Abdel-Jaber H, Devassy D, Al Salam A, Hidaytallah L, El-Amir M. A review of deep learning algorithms and their applications in healthcare. Algorithms. 2022;15(2):71. https://doi.org/10.3390/A15020071.

4. Adadi A, Berrada M. Peeking inside the black-box: a survey on explainable artificial intelligence (XAI). IEEE Access. 2018;6:52138–60. https://doi.org/10.1109/ACCESS.2018.2870052.
5. Afzaal M, Saeed F, Shah YA, Hussain M, Rabail R, Socol CT, Hassoun A, Pateiro M, Lorenzo JM, Rusu AV, Aadil RM. Human gut microbiota in health and disease: unveiling the relationship. Front Microbiol. 2022;13:1–14. https://doi.org/10.3389/fmicb.2022.999001.
6. Al Kuwaiti A, Nazer K, Al-Reedy A, Al-Shehri S, Al-Muhanna A, Subbarayalu AV, Al Muhanna D, Al-Muhanna FA. A review of the role of artificial intelligence in healthcare. J Pers Med. 2023;13(6). https://doi.org/10.3390/JPM13060951.
7. Alum EU, Ugwu OPC. Artificial intelligence in personalized medicine: transforming diagnosis and treatment. Discover Appl Sci. 2025;7(3):1–5. https://doi.org/10.1007/S42452-025-06625-X/METRICS.
8. Applications of deep learning in healthcare – GeeksforGeeks. n.d. Retrieved 28 July 2025, from https://www.geeksforgeeks.org/deep-learning/applications-of-deep-learning-in-healthcare/.
9. Asano Y, Hiramoto T, Nishino R, Aiba Y, Kimura T, Yoshihara K, Koga Y, Sudo N. Critical role of gut microbiota in the production of biologically active, free catecholamines in the gut lumen of mice. Am J Physiol Gastrointest Liver Physiol. 2012;303(11). https://doi.org/10.1152/AJPGI.00341.2012.
10. Au R, Kolachalama VB, Paschalidis IC. Redefining and validating digital biomarkers as fluid, dynamic multi-dimensional digital signal patterns. Front Digit Health. 2022;3. https://doi.org/10.3389/FDGTH.2021.751629.
11. Babrak LM, Menetski J, Rebhan M, Nisato G, Zinggeler M, Brasier N, Baerenfaller K, Brenzikofer T, Baltzer L, Vogler C, Gschwind L, Schneider C, Streiff F, Groenen PMA, Miho E. Traditional and digital biomarkers: two worlds apart? Digit Biomark. 2019;3(2):92–102. https://doi.org/10.1159/000502000.
12. Bach S, Binder A, Montavon G, Klauschen F, Müller KR, Samek W. On pixel-wise explanations for non-linear classifier decisions by layer-wise relevance propagation. PLoS One. 2015;10(7). https://doi.org/10.1371/JOURNAL.PONE.0130140.
13. Bajwa J, Munir U, Nori A, Williams B. Artificial intelligence in healthcare: transforming the practice of medicine. Future Healthc J. 2021;8(2):e188. https://doi.org/10.7861/FHJ.2021-0095.
14. Barajon I, Serrao G, Arnaboldi F, Opizzi E, Ripamonti G, Balsari A, Rumio C. Toll-like receptors 3, 4, and 7 are expressed in the enteric nervous system and dorsal root ganglia. J Histochem Cytochem. 2009;57(11):1013–23. https://doi.org/10.1369/JHC.2009.953539.
15. Barrett E, Ross RP, O'Toole PW, Fitzgerald GF, Stanton C. γ-Aminobutyric acid production by culturable bacteria from the human intestine. J Appl Microbiol. 2012;113(2):411–7. https://doi.org/10.1111/J.1365-2672.2012.05344.X.
16. Bauer PV, Hamr SC, Duca FA. Regulation of energy balance by a gut-brain axis and involvement of the gut microbiota. Cell Mol Life Sci. 2016;73(4):737–55. https://doi.org/10.1007/S00018-015-2083-Z.
17. Beilharz JE, Kaakoush NO, Maniam J, Morris MJ. Cafeteria diet and probiotic therapy: cross talk among memory, neuroplasticity, serotonin receptors and gut microbiota in the rat. Mol Psychiatry. 2018;23(2):351–61. https://doi.org/10.1038/MP.2017.38.
18. Bellono NW, Bayrer JR, Leitch DB, Castro J, Zhang C, O'Donnell TA, Brierley SM, Ingraham HA, Julius D. Enterochromaffin cells are gut chemosensors that couple to sensory neural pathways. Cell. 2017;170(1):185–198.e16. https://doi.org/10.1016/j.cell.2017.05.034.
19. Benakis C, Brea D, Caballero S, Faraco G, Moore J, Murphy M, Sita G, Racchumi G, Ling L, Pamer EG, Iadecola C, Anrather J. Commensal microbiota affects ischemic stroke outcome by regulating intestinal γδ T cells. Nat Med. 2016;22(5):516–23. https://doi.org/10.1038/NM.4068.
20. Bercik P, Park AJ, Sinclair D, Khoshdel A, Lu J, Huang X, Deng Y, Blennerhassett PA, Fahnestock M, Moine D, Berger B, Huizinga JD, Kunze W, Mclean PG, Bergonzelli GE, Collins SM, Verdu EF. The anxiolytic effect of Bifidobacterium longum NCC3001 involves vagal pathways for gut-brain communication. Neurogastroenterol Motil. 2011;23(12):1132–9. https://doi.org/10.1111/J.1365-2982.2011.01796.X.

21. Bercik P, Collins SM, Verdu EF. Microbes and the gut-brain axis. Neurogastroenterol Motil. 2012;24(5):405–13. https://doi.org/10.1111/J.1365-2982.2012.01906.X.
22. Berg G, Rybakova D, Fischer D, Cernava T, Vergès MCC, Charles T, Chen X, Cocolin L, Eversole K, Corral GH, Kazou M, Kinkel L, Lange L, Lima N, Loy A, Macklin JA, Maguin E, Mauchline T, McClure R, et al. Microbiome definition re-visited: old concepts and new challenges. Microbiome. 2020;8(1). https://doi.org/10.1186/S40168-020-00875-0.
23. Bowey E, Adlercreutz H, Rowland I. Metabolism of isoflavones and lignans by the gut microflora: a study in germ-free and human flora associated rats. Food Chem Toxicol. 2003;41(5):631–6. https://doi.org/10.1016/S0278-6915(02)00324-1.
24. Bravo JA, Forsythe P, Chew MV, Escaravage E, Savignac HM, Dinan TG, Bienenstock J, Cryan JF. Ingestion of Lactobacillus strain regulates emotional behavior and central GABA receptor expression in a mouse via the vagus nerve. Proc Natl Acad Sci USA. 2011;108(38):16050–5. https://doi.org/10.1073/PNAS.1102999108.
25. Bubenik GA, Brown GM. Pinealectomy reduces melatonin levels in the serum but not in the gastrointestinal tract of rats. Neurosignals. 1997;6(1):40–4. https://doi.org/10.1159/000109107.
26. Cani PD, Lecourt E, Dewulf EM, Sohet FM, Pachikian BD, Naslain D, De Backer F, Neyrinck AM, Delzenne NM. Gut microbiota fermentation of prebiotics increases satietogenic and incretin gut peptide production with consequences for appetite sensation and glucose response after a meal. Am J Clin Nutr. 2009;90(5):1236–43. https://doi.org/10.3945/AJCN.2009.28095.
27. Carabotti M, Scirocco A, Maselli MA, Severi C. The gut-brain axis: interactions between enteric microbiota, central and enteric nervous systems. Ann Gastroenterol. 2015;28(2):203–9.
28. Carding S, Verbeke K, Vipond DT, Corfe BM, Owen LJ. Dysbiosis of the gut microbiota in disease. Microb Ecol Health Dis. 2015;26(0). https://doi.org/10.3402/mehd.v26.26191.
29. Carrieri AP, Haiminen N, Maudsley-Barton S, Gardiner LJ, Murphy B, Mayes AE, Paterson S, Grimshaw S, Winn M, Shand C, Hadjidoukas P, Rowe WPM, Hawkins S, MacGuire-Flanagan A, Tazzioli J, Kenny JG, Parida L, Hoptroff M, Pyzer-Knapp EO. Explainable AI reveals changes in skin microbiome composition linked to phenotypic differences. Sci Rep. 2021;11(1). https://doi.org/10.1038/S41598-021-83922-6.
30. Casagni M, Llewellyn N, Kokolus M, Chan M, Dingwell R, Chow S, Campbell N, Elrahi C, Piantadosi S, Quina A. Integrated electronic health record tools to access real-world data in oncology research. JAMIA Open. 2024;7(4):ooae144. https://doi.org/10.1093/JAMIAOPEN/OOAE144.
31. Chakraborty C, Bhattacharya M, Pal S, Lee SS. From machine learning to deep learning: advances of the recent data-driven paradigm shift in medicine and healthcare. Curr Res Biotechnol. 2024;7:100164. https://doi.org/10.1016/J.CRBIOT.2023.100164.
32. Chen Y, Guan S, Guan Y, Tang S, Zhou Y, Wang X, Bi H, Huang M. Novel clinical biomarkers for drug-induced liver injury. Drug Metab Dispos. 2022;50(5):671–84. https://doi.org/10.1124/DMD.121.000732.
33. Clarke MB, Hughes DT, Zhu C, Boedeker EC, Sperandio V. The QseC sensor kinase: a bacterial adrenergic receptor. Proc Natl Acad Sci USA. 2006;103(27):10420–5. https://doi.org/10.1073/PNAS.0604343103.
34. Clarke G, Grenham S, Scully P, Fitzgerald P, Moloney RD, Shanahan F, Dinan TG, Cryan JF. The microbiome-gut-brain axis during early life regulates the hippocampal serotonergic system in a sex-dependent manner. Mol Psychiatry. 2013;18(6):666–73. https://doi.org/10.1038/MP.2012.77.
35. Cogan TA, Thomas AO, Rees LEN, Taylor AH, Jepson MA, Williams PH, Ketley J, Humphrey TJ. Norepinephrine increases the pathogenic potential of Campylobacter jejuni. Gut. 2007;56(8):1060–5. https://doi.org/10.1136/GUT.2006.114926.
36. Cowan CSM, Callaghan BL, Richardson R. The effects of a probiotic formulation (Lactobacillus rhamnosus and L. Helveticus) on developmental trajectories of emotional learning in stressed infant rats. Transl Psychiatry. 2016;6(5). https://doi.org/10.1038/TP.2016.94.

37. Crumeyrolle-Arias M, Jaglin M, Bruneau A, Vancassel S, Cardona A, Daugé V, Naudon L, Rabot S. Absence of the gut microbiota enhances anxiety-like behavior and neuroendocrine response to acute stress in rats. Psychoneuroendocrinology. 2014;42:207–17. https://doi.org/10.1016/j.psyneuen.2014.01.014.
38. Data Standardization – OHDSI. n.d. Retrieved 29 July 2025, from https://www.ohdsi.org/data-standardization/.
39. Davenport T, Kalakota R. The potential for artificial intelligence in healthcare. Future Healthc J. 2019;6(2):94–8. https://doi.org/10.7861/futurehosp.6-2-94.
40. David LA, Materna AC, Friedman J, Campos-Baptista MI, Blackburn MC, Perrotta A, Erdman SE, Alm EJ. Host lifestyle affects human microbiota on daily timescales. Genome Biol. 2014;15(7):1–15. https://doi.org/10.1186/GB-2014-15-7-R89/TABLES/1.
41. David LA, Maurice CF, Carmody RN, Gootenberg DB, Button JE, Wolfe BE, Ling AV, Devlin AS, Varma Y, Fischbach MA, Biddinger SB, Dutton RJ, Turnbaugh PJ. Diet rapidly and reproducibly alters the human gut microbiome. Nature. 2014;505(7484):559–63. https://doi.org/10.1038/NATURE12820.
42. Davis NM, Proctor DM, Holmes SP, Relman DA, Callahan BJ. Simple statistical identification and removal of contaminant sequences in marker-gene and metagenomics data. Microbiome. 2018;6(1). https://doi.org/10.1186/S40168-018-0605-2.
43. de Miguel-Perez D, Russo A, Gunasekaran M, Buemi F, Hester L, Fan X, Carter-Cooper BA, Lapidus RG, Peleg A, Arroyo-Hernández M, Cardona AF, Naing A, Hirsch FR, Mack PC, Kaushal S, Serrano MJ, Adamo V, Arrieta O, Rolfo C. Baseline extracellular vesicle TGF-β is a predictive biomarker for response to immune checkpoint inhibitors and survival in non–small cell lung cancer. Cancer. 2023;129(4):521–30. https://doi.org/10.1002/CNCR.34576.
44. De Panfilis L, Peruselli C, Tanzi S, Botrugno C. AI-based clinical decision-making systems in palliative medicine: ethical challenges. BMJ Support Palliat Care. 2023;13(2):183–9. https://doi.org/10.1136/BMJSPCARE-2021-002948.
45. Deep learning in healthcare [Full eGuide]. n.d. Retrieved 28 July 2025, from https://www.aidoc.com/learn/blog/deep-learning-in-healthcare/#deep-learning-applications-in-healthcare%C2%A0.
46. Desai MS, Seekatz AM, Koropatkin NM, Kamada N, Hickey CA, Wolter M, Pudlo NA, Kitamoto S, Terrapon N, Muller A, Young VB, Henrissat B, Wilmes P, Stappenbeck TS, Núñez G, Martens EC. A dietary fiber-deprived gut microbiota degrades the colonic mucus barrier and enhances pathogen susceptibility. Cell. 2016;167(5):1339–1353.e21. https://doi.org/10.1016/j.cell.2016.10.043.
47. Desbonnet L, Garrett L, Clarke G, Kiely B, Cryan JF, Dinan TG. Effects of the probiotic Bifidobacterium infantis in the maternal separation model of depression. Neuroscience. 2010;170(4):1179–88. https://doi.org/10.1016/j.neuroscience.2010.08.005.
48. Di Carlo P, Serra N, Alduina R, Guarino R, Craxì A, Giammanco A, Fasciana T, Cascio A, Sergi CM. A systematic review on omics data (metagenomics, metatranscriptomics, and metabolomics) in the role of microbiome in gallbladder disease. Front Physiol. 2022;13:1–14. https://doi.org/10.3389/fphys.2022.888233.
49. Duca FA, Swartz TD, Sakar Y, Covasa M. Increased oral detection, but decreased intestinal signaling for fats in mice lacking gut microbiota. PLoS One. 2012;7(6). https://doi.org/10.1371/JOURNAL.PONE.0039748.
50. Eisenhofer R, Minich JJ, Marotz C, Cooper A, Knight R, Weyrich LS. Contamination in low microbial biomass microbiome studies: issues and recommendations. Trends Microbiol. 2019;27(2):105–17. https://doi.org/10.1016/J.TIM.2018.11.003.
51. Engevik AC. The age of angiogenesis: a novel role of NGF in gastric repair. Cell Mol Gastroenterol Hepatol. 2018;6(2):227–8. https://doi.org/10.1016/j.jcmgh.2018.06.001.
52. Erny D, De Angelis ALH, Jaitin D, Wieghofer P, Staszewski O, David E, Keren-Shaul H, Mahlakoiv T, Jakobshagen K, Buch T, Schwierzeck V, Utermöhlen O, Chun E, Garrett WS, Mccoy KD, Diefenbach A, Staeheli P, Stecher B, Amit I, Prinz M. Host microbiota constantly control maturation and function of microglia in the CNS. Nat Neurosci. 2015;18(7):965–77. https://doi.org/10.1038/NN.4030.

53. Everard A, Lazarevic V, Derrien M, Girard M, Muccioli GM, Neyrinck AM, Possemiers S, Van Holle A, François P, De Vos WM, Delzenne NM, Schrenzel J, Cani PD. Responses of gut microbiota and glucose and lipid metabolism to prebiotics in genetic obese and diet-induced leptin-resistant mice. Diabetes. 2011;60(11):2775–86. https://doi.org/10.2337/DB11-0227.
54. Falony G, Joossens M, Vieira-Silva S, Wang J, Darzi Y, Faust K, Kurilshikov A, Bonder MJ, Valles-Colomer M, Vandeputte D, Tito RY, Chaffron S, Rymenans L, Verspecht C, De Sutter L, Lima-Mendez G, D'hoe K, Jonckheere K, Homola D, et al. Population-level analysis of gut microbiome variation. Science. 2016;352(6285):560–4. https://doi.org/10.1126/SCIENCE.AAD3503.
55. Fernandez-Real JM, Serino M, Blasco G, Puig J, Daunis-I-Estadella J, Ricart W, Burcelin R, Fernández-Aranda F, Portero-Otin M. Gut microbiota interacts with brain microstructure and function. J Clin Endocrinol Metab. 2015;100(12):4505–13. https://doi.org/10.1210/JC.2015-3076.
56. Fung TC, Olson CA, Hsiao EY. Interactions between the microbiota, immune and nervous systems in health and disease. Nat Neurosci. 2017;20(2):145–55. https://doi.org/10.1038/NN.4476.
57. Furet JP, Kong LC, Tap J, Poitou C, Basdevant A, Bouillot JL, Mariat D, Corthier G, Doré J, Henegar C, Rizkalla S, Clément K. Differential adaptation of human gut microbiota to bariatric surgery-induced weight loss: links with metabolic and low-grade inflammation markers. Diabetes. 2010;59(12):3049–57. https://doi.org/10.2337/DB10-0253.
58. Furness JB, Rivera LR, Cho HJ, Bravo DM, Callaghan B. The gut as a sensory organ. Nat Rev Gastroenterol Hepatol. 2013;10(12):729–40. https://doi.org/10.1038/NRGASTRO.2013.180.
59. Galicia-Garcia U, Benito-Vicente A, Jebari S, Larrea-Sebal A, Siddiqi H, Uribe KB, Ostolaza H, Martín C. Pathophysiology of type 2 diabetes mellitus. Int J Mol Sci. 2020;21(17):1–34. https://doi.org/10.3390/IJMS21176275.
60. Gibbons SM, Duvallet C, Alm EJ. Correcting for batch effects in case-control microbiome studies. 2017. https://doi.org/10.1101/165910.
61. Gloor GB, Macklaim JM, Pawlowsky-Glahn V, Egozcue JJ. Microbiome datasets are compositional: and this is not optional. Front Microbiol. 2017;8. https://doi.org/10.3389/FMICB.2017.02224.
62. Grill MF, Maganti RK. Neurotoxic effects associated with antibiotic use: management considerations. Br J Clin Pharmacol. 2011;72(3):381–93. https://doi.org/10.1111/J.1365-2125.2011.03991.X.
63. Guyon I. Advances in neural information processing systems 30: annual conference on neural information processing systems 2017, Long Beach, CA, USA, 4–9 Dec 2017.
64. Heerman WJ, White RO, Barkin SL. Advancing informed consent for vulnerable populations. Pediatrics. 2015;135(3):e562. https://doi.org/10.1542/PEDS.2014-3041.
65. Howard J. Artificial intelligence: implications for the future of work. Am J Ind Med. 2019;62(11):917–26. https://doi.org/10.1002/AJIM.23037.
66. Hrncir T. Gut microbiota dysbiosis: triggers, consequences, diagnostic and therapeutic options. Microorganisms. 2022;10(3). https://doi.org/10.3390/microorganisms10030578.
67. Hughes DT, Sperandio V. Inter-kingdom signalling: communication between bacteria and their hosts. Nat Rev Microbiol. 2008;6(2):111–20. https://doi.org/10.1038/NRMICRO1836.
68. Hunter B, Hindocha S, Lee RW. The role of artificial intelligence in early cancer diagnosis. Cancer. 2022;14(6). https://doi.org/10.3390/CANCERS14061524.
69. Huttenhower C, Gevers D, Knight R, Abubucker S, Badger JH, Chinwalla AT, Creasy HH, Earl AM, Fitzgerald MG, Fulton RS, Giglio MG, Hallsworth-Pepin K, Lobos EA, Madupu R, Magrini V, Martin JC, Mitreva M, Muzny DM, Sodergren EJ, et al. Structure, function and diversity of the healthy human microbiome. Nature. 2012;486(7402):207–14. https://doi.org/10.1038/nature11234.
70. Jacob M, Lopata AL, Dasouki M, Abdel Rahman AM. Metabolomics toward personalized medicine. Mass Spectrom Rev. 2019;38(3):221–38. https://doi.org/10.1002/MAS.21548.
71. Janjusevic M, Fluca AL, Ferro F, Gagno G, D'alessandra Y, Beltrami AP, Sinagra G, Aleksova A. Traditional and emerging biomarkers in asymptomatic left ventricular dysfunction—

promising non-coding RNAs and exosomes as biomarkers in early phases of cardiac damage. Int J Mol Sci. 2021;22(9). https://doi.org/10.3390/IJMS22094937.

72. Jeffery IB, O'Toole PW, Öhman L, Claesson MJ, Deane J, Quigley EMM, Simrén M. An irritable bowel syndrome subtype defined by species-specific alterations in faecal microbiota. Gut. 2012;61(7):997–1006. https://doi.org/10.1136/GUTJNL-2011-301501.
73. Johnson KVA. Gut microbiome composition and diversity are related to human personality traits. Hum Microbiome J. 2020;15:100069. https://doi.org/10.1016/j.humic.2019.100069.
74. Johnson KB, Wei WQ, Weeraratne D, Frisse ME, Misulis K, Rhee K, Zhao J, Snowdon JL. Precision medicine, AI, and the future of personalized health care. Clin Transl Sci. 2021;14(1):86–93. https://doi.org/10.1111/CTS.12884.
75. Jovel J, Patterson J, Wang W, Hotte N, O'Keefe S, Mitchel T, Perry T, Kao D, Mason AL, Madsen KL, Wong GKS. Characterization of the gut microbiome using 16S or shotgun metagenomics. Front Microbiol. 2016;7:1–17. https://doi.org/10.3389/fmicb.2016.00459.
76. Kane PB, Bittlinger M, Kimmelman J. Individualized therapy trials: navigating patient care, research goals and ethics. Nat Med. 2021;27(10):1679–86. https://doi.org/10.1038/S41591-021-01519-Y.
77. Keita ÅV, Söderholm JD. The intestinal barrier and its regulation by neuroimmune factors. Neurogastroenterol Motil. 2010;22(7):718–33. https://doi.org/10.1111/J.1365-2982.2010.01498.X.
78. Kelly JR, Borre Y, O'Brien C, Patterson E, El Aidy S, Deane J, Kennedy PJ, Beers S, Scott K, Moloney G, Hoban AE, Scott L, Fitzgerald P, Ross P, Stanton C, Clarke G, Cryan JF, Dinan TG. Transferring the blues: depression-associated gut microbiota induces neurobehavioural changes in the rat. J Psychiatr Res. 2016;82:109–18. https://doi.org/10.1016/j.jpsychires.2016.07.019.
79. Kerckhoffs APM, Ben-Amor K, Samsom M, Van Der Rest ME, De Vogel J, Knol J, Akkermans LMA. Molecular analysis of faecal and duodenal samples reveals significantly higher prevalence and numbers of Pseudomonas aeruginosa in irritable bowel syndrome. J Med Microbiol. 2011;60(2):236–45. https://doi.org/10.1099/JMM.0.022848-0.
80. Khalifa M, Albadawy M. Artificial intelligence for clinical prediction: exploring key domains and essential functions. Comput Methods Programs Biomed Update. 2024;5:100148. https://doi.org/10.1016/J.CMPBUP.2024.100148.
81. Khalil M, Di Ciaula A, Mahdi L, Jaber N, Di Palo DM, Graziani A, Baffy G, Portincasa P. Unraveling the role of the human gut microbiome in health and diseases. Microorganisms. 2024;12(11):2333. https://doi.org/10.3390/MICROORGANISMS12112333.
82. Knight R, Vrbanac A, Taylor BC, Aksenov A, Callewaert C, Gonzalez A, Kosciolek T, Mccall L, Mcdonald D, Alexey V, Morton JT, Navas J, Quinn RA, Sanders JG, Swafford AD. Best practices for analyzing microbiomes. Nat Rev Microbiol. 2018;16(7):410–22.
83. Knights D, Costello EK, Knight R. Supervised classification of human microbiota. FEMS Microbiol Rev. 2011;35(2):343–59. https://doi.org/10.1111/J.1574-6976.2010.00251.X.
84. Kumar B, Lorusso E, Fosso B, Pesole G. A comprehensive overview of microbiome data in the light of machine learning applications: categorization, accessibility, and future directions. Front Microbiol. 2024;15. https://doi.org/10.3389/fmicb.2024.1343572.
85. Kumeda B, Zhang F, Zhou F, Hussain S, Almasri A, Assefa M. Classification of road traffic accident data using machine learning algorithms. In: 2019 IEEE 11th international conference on communication software and networks, ICCSN 2019; 2019. p. 682–7. https://doi.org/10.1109/ICCSN.2019.8905362.
86. Kunze WA, Mao YK, Wang B, Huizinga JD, Ma X, Forsythe P, Bienenstock J. Lactobacillus reuteri enhances excitability of colonic AH neurons by inhibiting calcium-dependent potassium channel opening. J Cell Mol Med. 2009;13(8 B):2261–70. https://doi.org/10.1111/J.1582-4934.2009.00686.X.
87. Lasko TA, Denny JC, Levy MA. Computational phenotype discovery using unsupervised feature learning over noisy, sparse, and irregular clinical data. PLoS One. 2013;8(6). https://doi.org/10.1371/JOURNAL.PONE.0066341.

88. Lauffer A, Vanuytsel T, Vanormelingen C, Vanheel H, Salim Rasoel S, Tóth J, Tack J, Fornari F, Farré R. Subacute stress and chronic stress interact to decrease intestinal barrier function in rats. Stress. 2016;19(2):225–34. https://doi.org/10.3109/10253890.2016.1154527.
89. Lecun Y, Bengio Y, Hinton G. Deep learning. Nature. 2015;521(7553):436–44. https://doi.org/10.1038/NATURE14539.
90. Lehne M, Sass J, Essenwanger A, Schepers J, Thun S. Why digital medicine depends on interoperability. NPJ Digit Med. 2019;2(1). https://doi.org/10.1038/S41746-019-0158-1.
91. Li JV, Ashrafian H, Bueter M, Kinross J, Sands C, Le Roux CW, Bloom SR, Darzi A, Athanasiou T, Marchesi JR, Nicholson JK, Holmes E. Metabolic surgery profoundly influences gut microbial – host metabolic cross-talk. Gut. 2011;60(9):1214–23. https://doi.org/10.1136/GUT.2010.234708.
92. Li Y, Guo Z, Yang N, Chen H, Yuan D, Ding W. Threats and defenses in the federated learning life cycle: a comprehensive survey and challenges. IEEE Trans Neural Netw Learn Syst. 2025. https://doi.org/10.1109/TNNLS.2025.3563537.
93. Lin HV, Frassetto A, Kowalik EJ, Nawrocki AR, Lu MM, Kosinski JR, Hubert JA, Szeto D, Yao X, Forrest G, Marsh DJ. Butyrate and propionate protect against diet-induced obesity and regulate gut hormones via free fatty acid receptor 3-independent mechanisms. PLoS One. 2012;7(4). https://doi.org/10.1371/JOURNAL.PONE.0035240.
94. Liu L. Editorial: Therapeutic and diagnosis target discovery based on metabolomics. Front Pharmacol. 2022;13. https://doi.org/10.3389/FPHAR.2022.893905.
95. Loh JS, Mak WQ, Tan LKS, Ng CX, Chan HH, Yeow SH, Foo JB, Ong YS, How CW, Khaw KY. Microbiota–gut–brain axis and its therapeutic applications in neurodegenerative diseases. Signal Transduct Target Ther. 2024;9(1):29–39. https://doi.org/10.1038/s41392-024-01743-1.
96. Lyte M. The role of microbial endocrinology in infectious disease. J Endocrinol. 1993;137(3):343–5. https://doi.org/10.1677/JOE.0.1370343.
97. Mallett AK, Bearne CA, Rowland IR, Farthing MJG, Cole CB, Fuller R. The use of rats associated with a human faecal flora as a model for studying the effects of diet on the human gut microflora. J Appl Bacteriol. 1987;63(1):39–45. https://doi.org/10.1111/J.1365-2672.1987.TB02415.X.
98. Mandal S, Van Treuren W, White RA, Eggesbø M, Knight R, Peddada SD. Analysis of composition of microbiomes: a novel method for studying microbial composition. Microb Ecol Health Dis. 2015;26(0). https://doi.org/10.3402/MEHD.V26.27663.
99. Marchix J, Goddard G, Helmrath MA. Host-gut microbiota crosstalk in intestinal adaptation. Cell Mol Gastroenterol Hepatol. 2018;6(2):149–62. https://doi.org/10.1016/j.jcmgh.2018.01.024.
100. Mayer EA, Padua D, Tillisch K. Altered brain-gut axis in autism: comorbidity or causative mechanisms? BioEssays. 2014;36(10):933–9. https://doi.org/10.1002/BIES.201400075.
101. Mayer EA, Labus JS, Tillisch K, Cole SW, Baldi P. Towards a systems view of IBS. Nat Rev Gastroenterol Hepatol. 2015;12(10):592–605. https://doi.org/10.1038/NRGASTRO.2015.121.
102. Mayer EA, Nance K, Chen S. The gut-brain axis. Annu Rev Med. 2022;73:439–53. https://doi.org/10.1146/ANNUREV-MED-042320-014032.
103. McGuinness AJ, Stinson LF, Snelson M, Loughman A, Stringer A, Hannan AJ, Cowan CSM, Jama HA, Caparros-Martin JA, West ML, Wardill HR. From hype to hope: considerations in conducting robust microbiome science. Brain Behav Immun. 2024;115:120–30. https://doi.org/10.1016/J.BBI.2023.09.022.
104. McMurdie PJ, Holmes S. Waste not, want not: why rarefying microbiome data is inadmissible. PLoS Comput Biol. 2014;10(4). https://doi.org/10.1371/JOURNAL.PCBI.1003531.
105. Mertsalmi TH, Aho VTE, Pereira PAB, Paulin L, Pekkonen E, Auvinen P, Scheperjans F. More than constipation – bowel symptoms in Parkinson's disease and their connection to gut microbiota. Eur J Neurol. 2017;24(11):1375–83. https://doi.org/10.1111/ENE.13398.
106. Messaoudi M, Violle N, Bisson JF, Desor D, Javelot H, Rougeot C. Beneficial psychological effects of a probiotic formulation (Lactobacillus helveticus R0052 and Bifidobacterium

longum R0175) in healthy human volunteers. Gut Microbes. 2011;2(4). https://doi.org/10.4161/GMIC.2.4.16108.
107. Minuk GY. Gamma-aminobutyric acid (GABA) production by eight common bacterial pathogens. Scand J Infect Dis. 1986;18(5):465–7. https://doi.org/10.3109/00365548609032366.
108. Morton GJ, Kaiyala KJ, Foster-Schubert KE, Cummings DE, Schwartz MW. Carbohydrate feeding dissociates the postprandial FGF19 response from circulating bile acid levels in humans. J Clin Endocrinol Metab. 2014;99(2). https://doi.org/10.1210/JC.2013-3129.
109. Nahavandi S, Alizadehsani R, Nahavandi D, Mohamed S, Mohajer N, Rokonuzzaman M, Hossain I. A comprehensive review on autonomous navigation. ACM Comput Surv. 2025;57(9):1–67. https://doi.org/10.1145/3727642.
110. Naithani N, Sinha S, Misra P, Vasudevan B, Sahu R. Precision medicine: concept and tools. Med J Armed Forces India. 2021;77(3):249. https://doi.org/10.1016/J.MJAFI.2021.06.021.
111. Navarro-Lopez C, Linares-Mustaros S, Mulet-Forteza C. "The statistical analysis of compositional data" by John Aitchison (1986): a bibliometric overview. SAGE Open. 2022;12(2). https://doi.org/10.1177/21582440221093366.
112. Neufeld KM, Kang N, Bienenstock J, Foster JA. Reduced anxiety-like behavior and central neurochemical change in germ-free mice. Neurogastroenterol Motil. 2011;23(3). https://doi.org/10.1111/J.1365-2982.2010.01620.X.
113. Nordo AH, Levaux HP, Becnel LB, Galvez J, Rao P, Stem K, Prakash E, Kush RD. Use of EHRs data for clinical research: historical progress and current applications. Learn Health Syst. 2019;3(1). https://doi.org/10.1002/LRH2.10076.
114. Obermeyer Z, Emanuel EJ. Predicting the future — big data, machine learning, and clinical medicine. N Engl J Med. 2016;375(13):1216–9. https://doi.org/10.1056/NEJMP1606181.
115. Ochoa-Repáraz J, Mielcarz DW, Ditrio LE, Burroughs AR, Foureau DM, Haque-Begum S, Kasper LH. Role of gut commensal microflora in the development of experimental autoimmune encephalomyelitis. J Immunol. 2009;183(10):6041–50. https://doi.org/10.4049/JIMMUNOL.0900747.
116. Ochoa-Repáraz J, Mielcarz DW, Ditrio LE, Burroughs AR, Begum-Haque S, Dasgupta S, Kasper DL, Kasper LH. Central nervous system demyelinating disease protection by the human commensal Bacteroides fragilis depends on polysaccharide A expression. J Immunol. 2010;185(7):4101–8. https://doi.org/10.4049/JIMMUNOL.1001443.
117. Oestreich MA, Wyler F, Frauchiger BS, Latzin P, Ramsey KA. Breath detection algorithms affect multiple-breath washout outcomes in pre-school and school age children. PLoS One. 2022;17(10):e0275866. https://doi.org/10.1371/JOURNAL.PONE.0275866.
118. Overview of precision medicine – IN. n.d. Retrieved 29 July 2025, from https://www.thermofisher.com/in/en/home/clinical/precision-medicine/precision-medicine-learning-center/precision-medicine-resource-library/precision-medicine-articles/overview-precision--medicine.html.
119. Ozcan T, Toprak AN, Aruk İ, Sahin O, Ozcan I. Applications of deep learning techniques in healthcare systems: a review. J Clin Pract Res. 2024;46(6):527–36. https://doi.org/10.14744/cpr.2024.25381.
120. Özogul F. Effects of specific lactic acid bacteria species on biogenic amine production by foodborne pathogen. Int J Food Sci Technol. 2011;46(3):478–84. https://doi.org/10.1111/J.1365-2621.2010.02511.X.
121. Pasolli E, Truong DT, Malik F, Waldron L, Segata N. Machine learning meta-analysis of large metagenomic datasets: tools and biological insights. PLoS Comput Biol. 2016;12(7). https://doi.org/10.1371/JOURNAL.PCBI.1004977.
122. Pedram P, Wadden D, Amini P, Gulliver W, Randell E, Cahill F, Vasdev S, Goodridge A, Carter JC, Zhai G, Ji Y, Sun G. Food addiction: its prevalence and significant association with obesity in the general population. PLoS One. 2013;8(9). https://doi.org/10.1371/JOURNAL.PONE.0074832.
123. Peterson D, Bonham KS, Rowland S, Pattanayak CW, Klepac-Ceraj V, Deoni SCL, D'Sa V, Bruchhage M, Volpe A, Beauchemin J, Wallace C, Rogers J, Cano R, Fernandes J, Walsh E, Rhodes B, Huentelman M, Lewis C, De Both MD, et al. Comparative analysis of 16S

rRNA gene and metagenome sequencing in pediatric gut microbiomes. Front Microbiol. 2021;12:1–13. https://doi.org/10.3389/fmicb.2021.670336.
124. Petrut SM, Bragaru AM, Munteanu AE, Moldovan AD, Moldovan CA, Rusu E. Gut over mind: exploring the powerful gut–brain axis. Nutrients. 2025;17(5):842. https://doi.org/10.3390/NU17050842.
125. Precision Medicine | FDA. n.d. Retrieved 29 July 2025, from https://www.fda.gov/medical-devices/in-vitro-diagnostics/precision-medicine.
126. Prevention and control of intestinal parasitic infections: report of a WHO Expert Committee. 1987. p. 86.
127. Quinn TP, Erb I, Richardson MF, Crowley TM. Understanding sequencing data as compositions: an outlook and review. Bioinformatics. 2018;34(16):2870–8. https://doi.org/10.1093/BIOINFORMATICS/BTY175.
128. Quinn TP, Senadeera M, Jacobs S, Coghlan S, Le V. Trust and medical AI: the challenges we face and the expertise needed to overcome them. J Am Med Inform Assoc. 2021;28(4):890–4. https://doi.org/10.1093/JAMIA/OCAA268.
129. Ramadan YN, Alqifari SF, Alshehri K, Alhowiti A, Mirghani H, Alrasheed T, Aljohani F, Alghamdi A, Hetta HF. Microbiome gut-brain-axis: impact on brain development and mental health. Mol Neurobiol. 2025;62(8):10813–33. https://doi.org/10.1007/S12035-025-04846-0/FIGURES/4.
130. Rhee SH, Pothoulakis C, Mayer EA. Principles and clinical implications of the brain-gut-enteric microbiota axis. Nat Rev Gastroenterol Hepatol. 2009;6(5):306–14. https://doi.org/10.1038/NRGASTRO.2009.35.
131. Robin J, Harrison JE, Kaufman LD, Rudzicz F, Simpson W, Yancheva M. Evaluation of speech-based digital biomarkers: review and recommendations. Digit Biomark. 2020;4(3):99–108. https://doi.org/10.1159/000510820.
132. Rousseaux C, Thuru X, Gelot A, Barnich N, Neut C, Dubuquoy L, Dubuquoy C, Merour E, Geboes K, Chamaillard M, Ouwehand A, Leyer G, Carcano D, Colombel JF, Ardid D, Desreumaux P. Lactobacillus acidophilus modulates intestinal pain and induces opioid and cannabinoid receptors. Nat Med. 2007;13(1):35–7. https://doi.org/10.1038/NM1521.
133. Ruddick JP, Evans AK, Nutt DJ, Lightman SL, Rook GAW, Lowry CA. Tryptophan metabolism in the central nervous system: medical implications. Expert Rev Mol Med. 2006;8(20):1–27. https://doi.org/10.1017/S1462399406000068.
134. Russell WR, Gratz SW, Duncan SH, Holtrop G, Ince J, Scobbie L, Duncan G, Johnstone AM, Lobley GE, Wallace RJ, Duthie GG, Flint HJ. High-protein, reduced-carbohydrate weight-loss diets promote metabolite profiles likely to be detrimental to colonic health. Am J Clin Nutr. 2011;93(5):1062–72. https://doi.org/10.3945/AJCN.110.002188.
135. Ryan KK, Kohli R, Gutierrez-Aguilar R, Gaitonde SG, Woods SC, Seeley RJ. Fibroblast growth factor-19 action in the brain reduces food intake and body weight and improves glucose tolerance in male rats. Endocrinology. 2013;154(1):9–15. https://doi.org/10.1210/EN.2012-1891.
136. Saad RJ, Rao SSC, Koch KL, Kuo B, Parkman HP, McCallum RW, Sitrin MD, Wilding GE, Semler JR, Chey WD. Do stool form and frequency correlate with whole-gut and colonic transit results from a multicenter study in constipated individuals and healthy controls. Am J Gastroenterol. 2010;105(2):403–11. https://doi.org/10.1038/AJG.2009.612.
137. Sadee W, Wang D, Hartmann K, Toland AE. Pharmacogenomics: driving personalized medicine. Pharmacol Rev. 2023;75(4):789–814. https://doi.org/10.1124/PHARMREV.122.000810.
138. Sadeghi Z, Alizadehsani R, Cifci MA, Kausar S, Rehman R, Mahanta P, Bora PK, Almasri A, Alkhawaldeh RS, Hussain S, Alatas B, Shoeibi A, Moosaei H, Hladík M, Nahavandi S, Pardalos PM. A review of explainable artificial intelligence in healthcare. Comput Electr Eng. 2024;118:109370. https://doi.org/10.1016/J.COMPELECENG.2024.109370.
139. Salles-Loustau G, Le T, Najafizadeh L, Zonouz S, Javanmard M. Cytocoded passwords: BioMEMS based barcoding of biological samples for user authentication in microflu-

idic diagnostic devices. Biomed Microdevices. 2018;20(3). https://doi.org/10.1007/S10544-018-0306-4.
140. Sampson TR, Debelius JW, Thron T, Janssen S, Shastri GG, Ilhan ZE, Challis C, Schretter CE, Rocha S, Gradinaru V, Chesselet MF, Keshavarzian A, Shannon KM, Krajmalnik-Brown R, Wittung-Stafshede P, Knight R, Mazmanian SK. Gut microbiota regulate motor deficits and neuroinflammation in a model of Parkinson's disease. Cell. 2016;167(6):1469–1480.e12. https://doi.org/10.1016/j.cell.2016.11.018.
141. Samuel BS, Shaito A, Motoike T, Rey FE, Backhed F, Manchester JK, Hammer RE, Williams SC, Crowley J, Yanagisawa M, Gordon JI. Effects of the gut microbiota on host adiposity are modulated by the short-chain fatty-acid binding G protein-coupled receptor, Gpr41. Proc Natl Acad Sci USA. 2008;105(43):16767–72. https://doi.org/10.1073/PNAS.0808567105.
142. Sankaran K, Holmes SP. Latent variable modeling for the microbiome. Biostatistics. 2019;20(4):599–614. https://doi.org/10.1093/BIOSTATISTICS/KXY018.
143. Savignac HM, Kiely B, Dinan TG, Cryan JF. Bifidobacteria exert strain-specific effects on stress-related behavior and physiology in BALB/c mice. Neurogastroenterol Motil. 2014;26(11):1615–27. https://doi.org/10.1111/NMO.12427.
144. Saxami G, Kerezoudi EN, Eliopoulos C, Arapoglou D, Kyriacou A. The gut–organ axis within the human body: gut dysbiosis and the role of prebiotics. Life. 2023;13(10):1–41. https://doi.org/10.3390/life13102023.
145. Sebastiani M, Vacchi C, Manfredi A, Cassone G. Personalized medicine and machine learning: a roadmap for the future. J Clin Med. 2022;11(14):4110. https://doi.org/10.3390/JCM11144110.
146. Sen J, Waghela H, Rakshit S. Privacy in federated learning. 2024. https://doi.org/10.5772/intechopen.1003421.
147. Shamshirband S, Fathi M, Dehzangi A, Chronopoulos AT, Alinejad-Rokny H. A review on deep learning approaches in healthcare systems: taxonomies, challenges, and open issues. J Biomed Inform. 2021;113:103627. https://doi.org/10.1016/J.JBI.2020.103627.
148. Sinha R, Abu-Ali G, Vogtmann E, Fodor AA, Ren B, Amir A, Schwager E, Crabtree J, Ma S, Abnet CC, Knight R, White O, Huttenhower C. Assessment of variation in microbial community amplicon sequencing by the Microbiome Quality Control (MBQC) project consortium. Nat Biotechnol. 2017;35(11):1077–86. https://doi.org/10.1038/NBT.3981.
149. Smith M, Staley C, Kooby D, Styblo T, Wood W, Yang L. Multiplexed fluorescence imaging of tumor biomarkers in gene expression and protein levels for personalized and predictive medicine. Curr Mol Med. 2009;9(8):1017–23. https://doi.org/10.2174/156652409789712765.
150. Soori M, Arezoo B, Dastres R. Artificial intelligence, machine learning and deep learning in advanced robotics, a review. Cogn Robot. 2023;3:54–70. https://doi.org/10.1016/J.COGR.2023.04.001.
151. The basics of OMOP – data standardization in healthcare – Nashville Biosciences. n.d. Retrieved 29 July 2025, from https://nashbio.com/blog/healthcare-data/the-basics-of-omop-data-standardization-in-healthcare/.
152. The role of AI and machine learning in modern software development – GeeksforGeeks. n.d. Retrieved 25 July 2025, from https://www.geeksforgeeks.org/blogs/role-of-ai-and-machine-learning-in-modern-software-development/.
153. Thomas AM, Manghi P, Asnicar F, Pasolli E, Armanini F, Zolfo M, Beghini F, Manara S, Karcher N, Pozzi C, Gandini S, Serrano D, Tarallo S, Francavilla A, Gallo G, Trompetto M, Ferrero G, Mizutani S, Shiroma H, et al. Metagenomic analysis of colorectal cancer datasets identifies cross-cohort microbial diagnostic signatures and a link with choline degradation. Nat Med. 2019;25(4):667–78. https://doi.org/10.1038/S41591-019-0405-7.
154. Tolhurst G, Heffron H, Lam YS, Parker HE, Habib AM, Diakogiannaki E, Cameron J, Grosse J, Reimann F, Gribble FM. Short-chain fatty acids stimulate glucagon-like peptide-1 secretion via the G-protein-coupled receptor FFAR2. Diabetes. 2012;61(2):364–71. https://doi.org/10.2337/DB11-1019.
155. Topol EJ. High-performance medicine: the convergence of human and artificial intelligence. Nat Med. 2019;25(1):44–56. https://doi.org/10.1038/S41591-018-0300-7.

156. Topping DL, Clifton PM. Short-chain fatty acids and human colonic function: roles of resistant starch and nonstarch polysaccharides. Physiol Rev. 2001;81(3):1031–64. https://doi.org/10.1152/PHYSREV.2001.81.3.1031.
157. Turnbaugh PJ, Ley RE, Mahowald MA, Magrini V, Mardis ER, Gordon JI. An obesity-associated gut microbiome with increased capacity for energy harvest. Nature. 2006;444(7122):1027–31. https://doi.org/10.1038/NATURE05414.
158. Van Calster B, Wynants L, Timmerman D, Steyerberg EW, Collins GS. Predictive analytics in health care: how can we know it works? J Am Med Inform Assoc. 2019;26(12):1651–4. https://doi.org/10.1093/JAMIA/OCZ130.
159. Vázquez-Baeza Y, Pirrung M, Gonzalez A, Knight R. EMPeror: a tool for visualizing high-throughput microbial community data. GigaScience. 2013;2(1). https://doi.org/10.1186/2047-217X-2-16.
160. Vellekoop H, Huygens S, Versteegh M, Szilberhorn L, Zelei T, Nagy B, Koleva-Kolarova R, Tsiachristas A, Wordsworth S, Rutten-van Mölken M. Guidance for the harmonisation and improvement of economic evaluations of personalised medicine. PharmacoEconomics. 2021;39(7):771–88. https://doi.org/10.1007/S40273-021-01010-Z.
161. Vijay-Kumar M, Aitken JD, Carvalho FA, Cullender TC, Mwangi S, Srinivasan S, Sitaraman SV, Knight R, Ley RE, Gewirtz AT. Metabolie syndrome and altered gut microbiota in mice lacking toll-like receptor 5. Science. 2010;328(5975):228–31. https://doi.org/10.1126/SCIENCE.1179721.
162. Vrontis D, Christofi M, Pereira V, Tarba S, Makrides A, Trichina E. Artificial intelligence, robotics, advanced technologies and human resource management: a systematic review. Int J Hum Resour Manag. 2022;33(6):1237–66. https://doi.org/10.1080/09585192.2020.1871398.
163. Vuong HE, Hsiao EY. Emerging roles for the gut microbiome in autism spectrum disorder. Biol Psychiatry. 2017;81(5):411–23. https://doi.org/10.1016/j.biopsych.2016.08.024.
164. Wang J, Jia H. Metagenome-wide association studies: fine-mining the microbiome. Nat Rev Microbiol. 2016;14(8):508–22. https://doi.org/10.1038/NRMICRO.2016.83.
165. Wang X, Peng J, Cai P, Xia Y, Yi C, Shang A, Akanyibah FA, Mao F. The emerging role of the gut microbiota and its application in inflammatory bowel disease. Biomed Pharmacother. 2024;179. https://doi.org/10.1016/j.biopha.2024.117302.
166. Wikoff WR, Anfora AT, Liu J, Schultz PG, Lesley SA, Peters EC, Siuzdak G. Metabolomics analysis reveals large effects of gut microflora on mammalian blood metabolites. Proc Natl Acad Sci USA. 2009;106(10):3698–703. https://doi.org/10.1073/PNAS.0812874106.
167. Woschank M, Rauch E, Zsifkovits H. A review of further directions for artificial intelligence, machine learning, and deep learning in smart logistics. Sustainability (Switzerland). 2020;12(9). https://doi.org/10.3390/SU12093760.
168. Xia Y, Sun J. Hypothesis testing and statistical analysis of microbiome. Genes Dis. 2017;4(3):138–48. https://doi.org/10.1016/J.GENDIS.2017.06.001.
169. Xuelian M, Mao YK, Wang B, Huizinga JD, Bienenstock J, Kunze W. Lactobacillus reuteri ingestion prevents hyperexcitability of colonic DRG neurons induced by noxious stimuli. Am J Physiol Gastrointest Liver Physiol. 2009;296(4). https://doi.org/10.1152/AJPGI.90511.2008.
170. Yang S, Zhu F, Ling X, Liu Q, Zhao P. Intelligent health care: applications of deep learning in computational medicine. Front Genet. 2021;12:607471. https://doi.org/10.3389/FGENE.2021.607471.
171. Yang J, Tu X, Zhao H, Lin X, Xie Y, Li S, Wang M. A comparative study of levels of serum biomarkers in patients with different stages of asthma. Chin Gen Pract. 2022;25(14):1700–6. https://doi.org/10.12114/J.ISSN.1007-9572.2022.0031.
172. Yissachar N, Zhou Y, Ung L, Lai NY, Mohan JF, Ehrlicher A, Weitz DA, Kasper DL, Chiu IM, Mathis D, Benoist C. An intestinal organ culture system uncovers a role for the nervous system in microbe-immune crosstalk. Cell. 2017;168(6):1135–1148.e12. https://doi.org/10.1016/j.cell.2017.02.009.
173. Yu T, Chae M, Wang Z, Ryu G, Kim GB, Lee SY. Microbial technologies enhanced by artificial intelligence for healthcare applications. Microb Biotechnol. 2025;18(3):e70131. https://doi.org/10.1111/1751-7915.70131.

174. Zhang H, DiBaise JK, Zuccolo A, Kudrna D, Braidotti M, Yu Y, Parameswaran P, Crowell MD, Wing R, Rittmann BE, Krajmalnik-Brown R. Human gut microbiota in obesity and after gastric bypass. Proc Natl Acad Sci USA. 2009;106(7):2365–70. https://doi.org/10.1073/PNAS.0812600106.
175. Zhang Y, Wang H, Sang Y, Liu M, Wang Q, Yang H, Li X. Gut microbiota in health and disease: advances and future prospects. MedComm. 2024;5(12):e70012. https://doi.org/10.1002/MCO2.70012.
176. Zheng P, Zeng B, Zhou C, Liu M, Fang Z, Xu X, Zeng L, Chen J, Fan S, Du X, Zhang X, Yang D, Yang Y, Meng H, Li W, Melgiri ND, Licinio J, Wei H, Xie P. Gut microbiome remodeling induces depressive-like behaviors through a pathway mediated by the host's metabolism. Mol Psychiatry. 2016;21(6):786–96. https://doi.org/10.1038/MP.2016.44.
177. Zhou R, Ng SK, Sung JJY, Goh WWB, Wong SH. Data pre-processing for analyzing microbiome data – a mini review. Comput Struct Biotechnol J. 2023;21:4804–15. https://doi.org/10.1016/J.CSBJ.2023.10.001.

Altering Gut Microbiome Interactions: Clinical Trial Results and Implications

13

Vara Prasad Saka, Vidyasrilekha Sanapalli, Shrestha Palit, Ashwini Deshpande, Md Abul Hassan, Md Ekhtiar Uddin, and Bharat Kumar Reddy Sanapalli

13.1 Introduction

13.1.1 Overview of the Gut Microbiome's Role in Human Health and Disease

The gut microbiome comprises a diverse and complex community of trillions of microorganisms, including bacteria, viruses, fungi, and archaea, that inhabit the gastrointestinal (GI) tract. The members of the microbiome contribute to various host functions, including the regulation of metabolism, immunity, and overall health, as well as the risk of disease [1]. The gut microbiota regulates local

V. P. Saka
Department of Pharmacology, Drug Standardization, Dr. Anjali Chatterji, Regional Research Institute for Homeopathy, Under Central Council for Research in Homeopathy, Ministry of AYUSH, Kolkata, West Bengal, India

V. Sanapalli
Department of Pharmaceutical Chemistry, School of Pharmacy & Technology Management, SVKM's Narsee Monjee Institute of Management Studies (NMIMS) Deemed to be University, Jadcherla, Hyderabad, India

S. Palit
School of Pharmacy & Technology Management, SVKM's Narsee Monjee Institute of Management Studies (NMIMS) Deemed to be University, Jadcherla, Hyderabad, India

A. Deshpande
Department of Pharmaceutics, School of Pharmacy & Technology Management, SVKM's Narsee Monjee Institute of Management Studies (NMIMS) Deemed to be University, Jadcherla, Hyderabad, India

M. A. Hassan
Obihiro University of Agriculture and Veterinary Medicine, Obihiro, Japan

M. H. Rahman, Muhasina K M (eds.), *The Microbiome's Impact on the Brain*,
https://doi.org/10.1007/978-981-95-7494-0_13

functions of the intestine and cellular functions of the host, contributing to systemic functions such as energy metabolism and immune homeostasis, which highlights the strong relationship between the host-microbiome system and its essential role in health [2].

In recent decades, the understanding of gut microbiota-host interactions has increased dramatically, in part due to the introduction of rigorous technologies and methods, such as high-throughput sequencing, metagenomics, metabolomics, and metatranscriptomics, which have enabled the precise characterization of microbiome composition and function [3]. Research has gradually shifted from descriptive, association-based studies of the microbiome to investigations that probe underlying mechanisms. Earlier work primarily relied on taxonomic surveys to identify correlations between microbial composition and disease. In contrast, more recent studies emphasize how microbial metabolism, host immune responses, and changes in gene expression contribute to the microbiome's impact on physiology [4].

The gut microbiome plays a complex role in human health and presents a potential therapeutic avenue for various conditions, including metabolic and autoimmune diseases, neurodegenerative disorders, and cancer [5]. The rationale for investigating the gut microbiome in this manner lies in its distinctive ability to change in response to external factors, such as diet, antibiotics, probiotics, and fecal microbiome transplantation, offering a more precise approach to disease modulation [6]. The application of gut microbiome-modulating strategies as a therapeutic approach is an emerging focus in clinical research aimed at reducing disease severity, preventing treatment failure, and restoring health. Future research will depend significantly on methodologically sound clinical trials.

13.1.2 Significance of Clinical Trials in Microbiome Modulation

There is a substantial amount of preclinical evidence supporting the role of the microbiome in the disease process; however, a significant gap exists in translating experimental results into practical application. The controlled clinical trial is the most critical effort in realizing this goal, as it enables the evaluation of the safety, efficacy, and mechanistic validity of microbiome-modulating options in humans [7]. Clinical trials in the area of the human microbiome are particularly important, as establishing causality from correlations alone is not

M. E. Uddin
The School of Medicine, Medical Sciences and Nutrition, University of Aberdeen, Aberdeen, UK

B. K. R. Sanapalli (✉)
Department of Pharmacology, School of Pharmacy & Technology Management, SVKM's Narsee Monjee Institute of Management Studies (NMIMS) Deemed to be University, Jadcherla, Hyderabad, India

possible, and a substantial amount of clinical data is needed to decipher the complexity of the human microbiome, which is both interindividual and intricate in nature [5]. It is essential to establish clear assignments of causality and demonstrate a distinct clinical benefit in our microbiome-directed therapies. Unfortunately, the most significant limitation and challenge is the heterogeneity of patient populations, diversity of microbial communities, complexity of microbial modulation kinetics, and the abundance of metric options applicable to monitoring changes in the microbiome [4]. Additionally, the host, environmental factors (such as diet and medications), and the host's immune system work together as a complex tapestry, making it difficult to ascribe a clinical benefit solely to modulation of the microbiome [8]. Clinical trials serve more than just as proof of concept, providing informative insights into host-microbiome interactions with therapeutic interventions.

Both clinical trial designs and the results of trials involving interventions demonstrate the strengths and weaknesses of the field. Several clinical trials are examining dietary changes, probiotics, prebiotics, synbiotics, and caloric microbiota transplants, which demonstrate a notable amount of variability—a crucial reminder of the complexity of microbiota-host interactions, not to mention therapeutic specificity [9]. Missing from many clinical trials is the fact that the intervention will be dependent, in large part, on the baseline microbiome, the context of the disease, and the host-specific factors, which is an essential reminder that clinical trials are intended to generate knowledge to better clinical approaches to inform specific interventions rather than blanket one-size-fits-all recommendations [10].

13.1.3 Scope and Objectives of the Review

This review systematically examines the clinical trial data available for interventions designed to alter aspects of gut microbiome interactions. We consider a wide range of interventions, including various dietary interventions, probiotics, and fecal microbiota transplantation (FMT), as supported by the clinical literature. We ask whether the interventions are supported theoretically and clinically across different categories of disease, including metabolic syndromes, autoimmune and inflammatory diseases, degenerative diseases of brain health, and cancer.

The second aim is to critically examine clinical research into clinical and methodological gaps, variable treatment responses, and the lack of standardized treatment protocols. By examining what we know and considering clinical outcomes, we present a comprehensive perspective on clinical evidence that can inform future research, trial implementation, and the organization of a pathway for the clinical application of gut microbiome-targeted interventions in health and disease.

13.2 Types of Interventions Targeting Gut Microbiome

13.2.1 Dietary Modulation

Diet is one of the most powerful and flexible ways to influence the gut microbiome, with dietary components capable of rapidly altering the composition of the microbial community and the activity of its members. Research shows that relatively small changes in diet, even for a short term (48–72 h), have been linked to significant shifts in the abundances of microbial taxa and the profiles of various metabolites; these experiments point to the extraordinary degree to which the microbiome is sensitive to dietary changes and nutritional inputs [11]. Such plasticity presents actual prospects for applying diet therapeutically, for instance, in disease prevention and management.

Clinical studies have begun to investigate the impacts of diverse dietary interventions on gut microbiome composition in chronic disease states. For example, in chronic kidney disease (CKD), a dietary restriction of protein with the addition of prebiotics has been studied both in the lab and the clinic, and changes in gut microbiota toward a more favorable composition, along with decreases in uremic toxins and reductions in systemic inflammation, have been demonstrated [12]. Similarly, after bariatric surgery, alterations to both dietary patterns and gut anatomy have been correlated with changes in the diversity and function of the microbiome, and many of these functional changes have been associated with improvements in metabolic outcomes [13].

Dietary patterns, such as the Mediterranean, Western, and keto diets, also interact with the gut microbiome in different ways to produce variable health outcomes [14]. The Mediterranean diet promotes greater fungal and microbial diversity, consisting of fiber-rich foods rather than diets that favor high-fat, processed foods, which can lead to dysbiosis in the gut microbiota. This is associated with an increase in proinflammatory bacteria, correlated with metabolic dysfunction. Clinical and preclinical research on the keto diet has demonstrated neuroprotection from dietary patterns that alter the gut microbiota, although studies are limited [15]. In conclusion, research suggests that the interactions between diet and the gut microbiome may hold a promise for dietary modification and interventions when developed based on a specific microbiome profile associated with the management of chronic conditions.

13.2.2 Probiotics and Synbiotics

Probiotics have been widely studied in clinical trials as a form of supplementation (i.e., live microorganisms intended to confer health benefits to the host) for various inflammatory and dermatological diseases. Specifically for atopic dermatitis (AD), probiotic strains have been shown to reduce disease severity in children. For example, *Lacticaseibacillus rhamnosus* GG has been shown to reduce severity in pediatric populations with AD and increase quality-of-life measures [16].

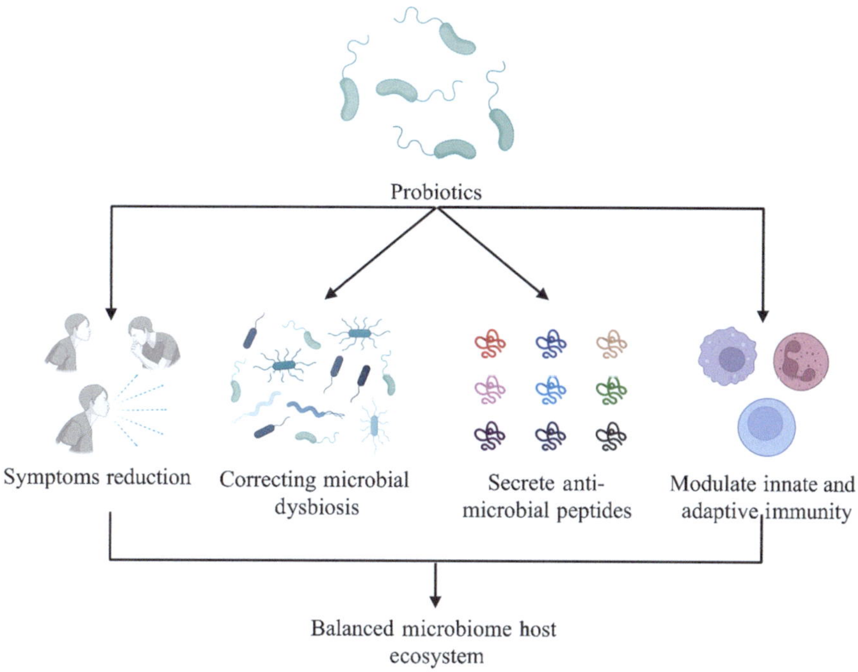

Fig. 13.1 Probiotics role in maintaining gut microbiome ecosystem

Probiotics are believed to function by modulating the host immune response, enhancing the gut barrier, and preventing pathogen colonization, ultimately leading to a reduction in symptoms.

Mechanistically, probiotics produce multiple effects that likely contribute to this symptom reduction, including modulation of innate and adaptive immunity, secretion of antimicrobial peptides, and correction of microbial dysbiosis to create a balanced microbiome-host ecosystem [17] (Fig. 13.1). The immunomodulatory properties of probiotics may also make them suitable for treating autoimmune diseases, allergies, and systemic inflammatory diseases [18]. However, clinical studies report significant variability in strains of probiotics, dosages, and duration, which result in variable measurements of efficacy [7]. There are also host-specific factors, including the initial baseline microbiota and genetic factors, that may predict variation in treatment response, highlighting the importance of personalized approaches [10].

13.2.3 Fecal Microbiota Transplantation

FMT, or fecal microbiota transplantation, is an emerging treatment modality that involves transferring gut microbiota from a healthy donor to a recipient to restore microbiota balance and function. FMT has been shown to have a known efficacy for recurrent *Clostridium difficile* infections, with cure rates of 80% or higher in clinical trials [4]. The literature examining FMT use in metabolic syndrome is mixed regarding insulin sensitivity and other aspects of gut microbiome composition [9].

Several barriers exist to optimizing a clinical protocol with FMT, including donor screening, FMT preparation approaches, dosing regimens, and the route of administration, all of which can impact the success of engraftment and clinical effect [3]. Additionally, long-term safety, clinical indications, and purpose remain largely unanswered questions being investigated for FMT. Current evidence from clinical trials does produce preliminary evidence indicating some potential for FMT in autoimmune disease and possibly neurodegenerative diseases such as inflammatory bowel disease (IBD) and Parkinson's disease (PD), through regulating systemic immune response and neuroinflammation [19]. However, it is clear from both the autoimmune and neurodegenerative examples that larger, well-controlled trials are needed to validate these findings and establish a mechanism of action in FMT [20].

13.3 Clinical Trial Outcomes in Metabolic Diseases

13.3.1 Obesity and Type 2 Diabetes

Alterations to gut microbial community structure and diversity are increasingly associated with metabolic syndrome and obesity. Certain microbial signatures have been linked to insulin resistance and an inflammatory burden [21]. Dysbiosis can present as an overall decrease in beneficial microbes, including *Akkermansia muciniphila*, and an increase in opportunistic pathogens. Clinical trials designed to modify the gut microbiome using dietary interventions, including prebiotics and probiotics, have shown moderate improvements in whole-physiology metabolic health, as indicated by changes in fasting glucose and body mass index [9].

However, while fecal microbiota transplantation studies in humans with metabolic dysregulation have demonstrated inconsistent results, some have concluded with transient metabolic benefits, but have not described sustained amelioration [9]. The variability in results may arise from donor microbiota composition, recipient baseline characteristics, and the study design. The multi-causal pathways associated with both obesity and diabetes support combining microbiome-targeted therapies with lifestyle interventions and pharmacological approaches in a practical continuum of options for improving long-term metabolic health outcomes.

13.3.2 Chronic Kidney Disease

Gut microbiota-derived uremic toxins (i.e., indoxyl sulfate and p-cresyl sulfate) promote systemic inflammation and contribute to CKD progression, emphasizing the salient role of the microbiome in renal pathophysiology [12]. Prior clinical experiments utilizing low-protein diets with prebiotics, such as inulin, have successfully altered gut microbiota profiles, reduced circulating inflammatory markers, and decreased oxidative stress in patients with CKD [12]; therefore, these findings offer support for the notion that manipulation of gut microbiota may prevent detrimental metabolic and inflammatory processes associated with renal injury.

In addition to dietary strategies, developing therapies that target the microbiome aim to reduce the accumulation of toxic metabolites and improve gut barrier function to limit the translocation of proinflammatory compounds [22]. Although preliminary data provide some evidence of therapeutic effects, further randomized controlled trials are needed to confirm efficacy and establish protocols for clinical application.

13.3.3 Nonalcoholic Fatty Liver Disease

The pathogenesis of nonalcoholic fatty liver disease (NAFLD) encompasses complex interactions whereby changes in gut microbes trigger hepatic steatosis and inflammation through mechanisms such as increased intestinal permeability and production of microbial metabolites [23]; observed changes in susceptibility to NAFLD have been associated with changes in bacteria of the gut, as well as increased levels of endotoxin in the circulation [23] (Fig. 13.2). Clinical studies have also demonstrated a correlation between specific shifts in bacterial taxa and the severity of NAFLD, supporting a dysbiotic contribution of the gut microbiota to the disease.

Interventional studies have demonstrated that gut microbiota and NAFLD-related metabolic diseases can be altered (toward more favorable outcomes) with diet modification (reduction of sugars and fats, increase in fiber) to reduce inappropriate exposure to energy substrates and boost overall health [23]. While it is clear that there is a gap in the microbiome that has not yet been evaluated in clinical studies for diseases like NAFLD, gut-targeted therapies such as probiotics or fecal microbiome transplant have been relatively unexplored in the context of NAFLD. This represents a significant gap in translational research. Research that is more microbiome-focused is greatly needed to determine therapeutic interventions that can be combined to support lifestyle interventions and pharmacotherapy.

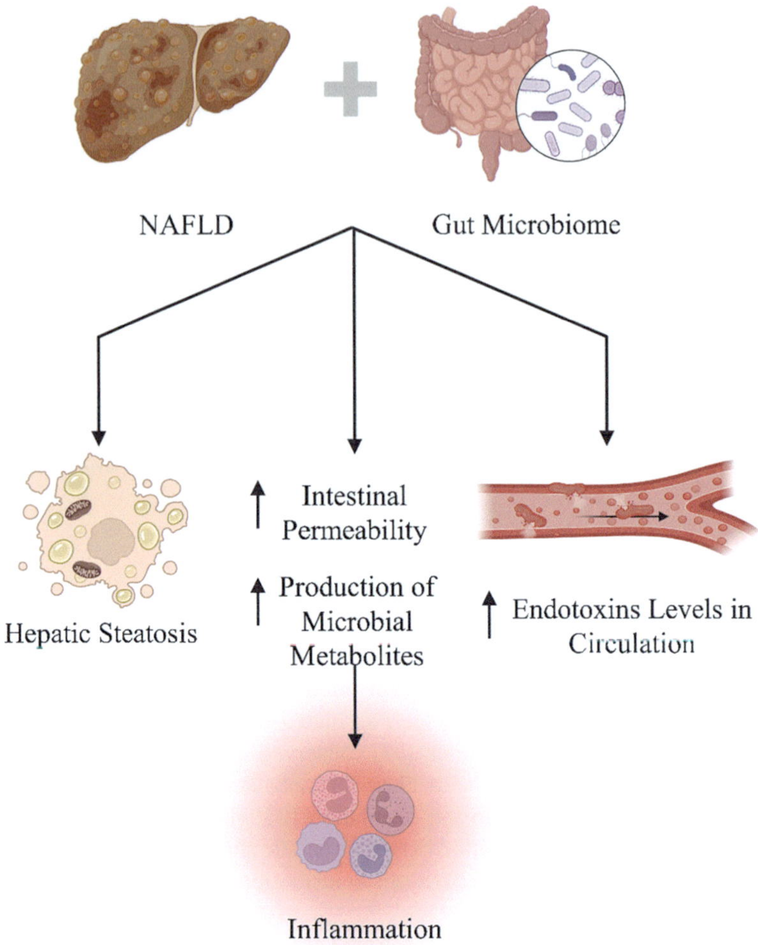

Fig. 13.2 Connection between NAFLD and gut microbiome in causing inflammation

13.4 Microbiome Modulation in Autoimmune and Inflammatory Disorders

13.4.1 Rheumatoid Arthritis and Systemic Autoimmune Diseases

There is an increasing evidence that identifies gut microbiota dysbiosis as a potential mechanism of pathogenesis and progression of rheumatoid arthritis and other systemic autoimmune diseases [24]. Microbial community structure changes occur before the appearance of clinically apparent disease, suggesting a role for the microbiome in the loss of immune tolerance and the induction of systemic inflammation.

Clinical approaches under investigation range from antibiotic therapy to eliminate specific pathogenic species, probiotic supplementation to restore microbial balance, dietary strategies to alter the metabolite profile, and fecal microbiota transplantation, which functions to reset the gut ecosystem [24]. Immunomodulatory effects are exerted through microbiome interactions by regulating T-cell differentiation, representing a potential mechanism through cytokine production, and maintaining gut barrier integrity. The inability to maintain gut barrier function may have implications for the loss of systemic tolerance, leading to autoimmunity [17]. While these clinical trials may offer compelling signals, study heterogeneity and small sample sizes warrant more scientifically robust studies.

13.4.2 Inflammatory Bowel Diseases

Microbiome-centered therapies have specific implications in treating IBD, due to the gut-based pathophysiology of the disease. There is clinical evidence for approaches aimed at resolving microbial dysbiosis, thereby restoring homeostasis in the patient and reducing inflammation [20]. Studies of probiotics, prebiotics, dietary changes, and fecal microbiota transplant all inherently have some variability, in part due to the subtype of the disease, patient heterogeneity, and treatment protocols used [25].

While IBD interventions are promising for the future, there are challenges related to standardizing treatment protocols, adherence to duration, dosing, and responders to treatment within patient subgroups. There could be a greater use of mechanistic biomarkers and multi-omics to inform studies, which would better target interventions and provide more specificity in the treatment of patients [7].

13.4.3 Dermatological Conditions: Atopic Dermatitis and Skin Microbiome

The gut-skin axis represents a global microbial component extending up to skin diseases such as atopic dermatitis, in which the dysbiosis of gut microbiome(s) and skin microbiome(s) contributes to the pathophysiology [26]. Clinical trials of probiotics in children with AD have shown improvement in AD severity and quality of life, along with immune modulation and microbiome changes [16].

Systemic and topical interventions that utilize the microbiome are emerging as potential adjunctive therapies in dermatology to restore microbial balance and enhance host immune defense mechanisms [27]. Further investigation is needed into the efficacy of strain, routes of delivery, and long-term tolerability.

13.5 Neurodegenerative Diseases and Gut Microbiome Interactions

13.5.1 Parkinson's Disease

The concept of alterations in the gut microbiome contributing to the etiology of PD is gaining wider acceptance, and clinical evidence suggests an association between gut dysbiosis and neuroinflammatory mechanisms that lead to the degeneration of dopaminergic neurons [14]. Initial clinical trials using probiotics, prebiotics, and food-based interventions to alter gut microbiomes have shown a promise in reducing both motor and non-motor symptoms of PD by decreasing inflammation and oxidative stress at the gut-brain axis, suggesting the potential to improve specific symptoms associated with PD. These initial clinical trials have encountered various obstacles, including heterogeneous patient populations, differing protocols and interventions, and neurologic outcomes that are more challenging to quantify and formally measure. These discrepancies significantly limit conclusions [19]. Future trials should incorporate mechanistic endpoints and standardized neuroimaging metrics to enhance the delineation of how the microbiome contributes to the physiological and behavioral markers of PD over time.

13.5.2 Alzheimer's Disease

Recently conducted observational studies have indicated distinct alterations in gut microbiota in individuals diagnosed with Alzheimer's disease (AD) compared to healthy controls, including both differences in microbial diversity and changes in specific bacterial groups associated with amyloid deposition [28]. The clinical applications exploring the microbiome, although modest, are likely to be achieved through probiotics and dietary changes, which aim to influence neuroinflammation and cognitive decline. This is because the gut microbiome can produce microbial metabolites that may regulate neuroinflammatory pathways and the blood-brain barrier [19].

Robust randomized controlled trials are necessary to evaluate these approaches and assess their potential for translating laboratory research into clinical benefits in AD treatment.

13.5.3 Psychiatric Disorders: Anxiety, Impulsivity, and Cognitive Function

Gut microbiome-derived metabolites, such as indoles, have been associated with clinically relevant symptomatology, including anxiety and depression. Evidence from functional neuroimaging demonstrates associations between metabolite quantities and the behavior and function of brain areas involved in emotional processing [29]. Clinical trials with probiotics and synbiotics have suggested potential additive effects on impulsivity, mood, and cognition through the possible modulation of neurotransmitter systems and neuroimmune signaling [18].

These therapeutic decisions offer a novel and exciting approach to treating mental health disorders, mainly when used adjunctively with standard pharmacotherapy. However, there is a need to replicate efficacy and mechanisms involved, using larger and multisite studies that utilize organized measures of effect and outcomes [30].

13.6 Cancer Immunotherapy and Microbiome Modulation

13.6.1 Gut Microbiome Influence on Cancer Immunotherapy Outcomes

The gut microbiome has a significant impact on host antitumor immune responses and is crucial in predicting clinical responses to immune checkpoint inhibitors across a wide range of cancers [31]. Clinical studies are manipulating the microbiome by using dietary interventions, probiotics, or fecal transplantation to enhance the effects of immunotherapy. These interventions aim to facilitate the development of beneficial microbial communities and metabolites that promote immune activation [4].

Disentangling the complex interactions between the microbiome and cancer creates limitations in understanding causative relationships, optimal delivery mechanisms, and successful patient selection strategies. Current investigations are underway to generate microbial signatures predictive of treatment response, as well as microbiome-targeted adjuncts to immunotherapy.

13.6.2 Microbiome Modulation in Gastrointestinal Cancers

Dysbiosis has been implicated in the onset, development, and dissemination of GI cancers through effects on the tumor microenvironment, immunoediting, and treatment outcomes [32]. Microbiome-targeted strategies may help improve the standard of care in oncology, either by optimizing responses to cancer treatments or reducing the side effects of therapy.

However, we lack clinical trials evaluating the direct influence of microbiota manipulation on GI cancer; nonetheless, clinical trials that incorporate targeted microbiome profiling or modulation strategies may lead to significant advances in personalized therapy in cancer care.

13.6.3 Dietary Interventions Improving Cancer Treatment Responses

Preclinical and clinical data suggest that dietary patterns can alter the microbiome, leading to improvements in the outcomes of cancer patients treated with immunotherapy [15]. Personalized nutrition plans enhance current treatment strategies by utilizing diet-microbiome interactions to optimize immune function. Additional research is needed to develop evidence-based dietary guidelines tailored to individual patient microbiome profiles, thereby supporting cancer therapy.

13.7 Mechanistic Insights from Microbial Metabolites and Host Interactions

13.7.1 Short-Chain Fatty Acids and Immune Regulation

Bacterial fermentation of dietary fibers produces various short-chain fatty acids (SCFAs), such as acetate, propionate, and butyrate. SCFAs are significant regulators of gut-brain communication and systemic immunity [33]. Clinical trial data indicate a correlation between SCFA abundance and favorable metabolic health, reduced inflammation, and enhanced gut barrier function [13].

SCFAs influence microbial targets associated with G protein-coupled receptor-mediated signaling pathways, such as GPR41 and GPR43, thereby modifying immune activation and metabolic pathways [4].

13.7.2 Microbial Metabolites Influencing Neuroinflammation and Cognition

Microbiota-derived metabolites, particularly those derived from tryptophan metabolism, such as indoles, have been linked to neuroinflammatory processes and psychiatric symptoms. Evidence indicates that indole levels correlate with functional magnetic resonance imaging signatures in anxiety disorders [29]. Furthermore, preclinical and clinical studies suggest that these metabolites may modulate neurodegenerative mechanisms, indicating potential for therapeutic targeting [14]. Clinical trials are currently examining various strategies involving metabolites as an intervention. These studies are promising; however, it remains essential to correlate dosing, timing, and the potential for specific patient subsets to benefit from the intervention.

13.7.3 Microbiome-Host Immune Crosstalk Mechanisms

Microbial components activate pattern-recognition receptors, particularly Toll-like and nucleotide-binding and oligomerization domain-like receptors, which trigger innate and adaptive immune responses that maintain homeostasis and offer protection against disease [17]. The mechanisms by which the microbiome influences T-cell differentiation, particularly in terms of the balance between regulatory T cells (Tregs) and proinflammatory Th17 cells, may provide targets for modulating autoimmune diseases and inflammation [4].

Clinical experiences with immune modulation through microbiome-targeted therapy indicate that modifying the microbiome may serve as a viable therapeutic target to alter outcomes, enhance function, and/or address various diseases.

13.8 Clinical Trial Methodologies and Challenges

13.8.1 Design Considerations for Microbiome-Targeted Interventions

The host microbiome systems present a complex challenge for trial engagement, necessitating consideration of participant variability, implementation variability, and significant endpoints [7]. Over time, clinical trials have increasingly incorporated multi-omics and deep phenotyping design elements to address nuanced distinctions in treatment. Such studies typically employ a significantly more complex methodology and often utilize more sophisticated analytical techniques [2].

Utilizing valid clinical and microbiome outcomes, including endpoints, is essential to ensure the interpretability of data with translational implications.

13.8.2 Limitations and Variability in Responses

Variations in participants' microbiomes, dietary history, medical and medication history, and other lifestyle characteristics all influence participants' responses to treatment [4]. The interaction of these variables indicates that antibiotics and environmental factors intensified the treatment effects. The future study should account for accommodations to mitigate extraneous effects, thereby ensuring clear reporting of findings [34].

Clinical outcomes are influenced by both placebo and nocebo effects, which can be variable and contribute to the overall results. This underscores the importance of maintaining blinding and controlling participant exposure and study environments that may affect participant awareness [10].

13.9 Therapeutic Implications and Clinical Applications

13.9.1 Personalized Nutrition and Precision Microbiome Modulation

There may be ways to improve the effectiveness and adherence to treatment for illnesses through dietary interventions and personalized nutrition therapies that modify the microbiota according to a person's genetics, phenotypic traits, and microbiota [35]. Simultaneously, precision nutrition aims to enhance the management of chronic diseases by modulating host-microbiome relationships through personalized dietary recommendations.

13.9.2 Microbiome Engineering and Synthetic Biology Approaches

We can now create synthetic consortia and designer probiotics with desired therapeutic properties, such as immune modulation and the production of specific metabolites, thanks to advancements in microbial engineering [36]. It should be noted that numerous questions regarding safety, colonization, and regulatory frameworks remain unanswered, despite the potential clinical applications of these technologies for restoring healthy microbiomes [2].

13.9.3 Adjunctive Microbiome-Targeted Therapies in Mainstream Clinical Practice

The use of probiotics, prebiotics, and FMT in standard of care protocols is supported by a growing body of clinical evidence, especially when treating autoimmune, oncologic, and metabolic disorders [9, 20]. These are seen as helpful supplementary treatments to primary therapies, merely increasing the range of patient care options to support and maximize clinical results. To support clinical practice, education, and ongoing training on microbiome, therapeutics remain essential for healthcare professionals.

13.10 Future Directions and Research Gaps

To demonstrate original findings and generalize interventions, large randomized controlled trials with appropriate dosing and outcome measures are necessary. Meanwhile, longitudinal and multi-omics data will help clarify microbiome-host interactions, thereby refining therapies. Reliable biomarkers of microbiota and metabolites are crucial for benchmarking clinical endpoints and patient stratification. Emerging technologies, such as clustered regularly interspaced short palindromic repeats-based microbial engineering, synthetic biology, and artificial intelligence-driven modeling, will enhance mechanistic and predictive studies. Translational progress requires collaborative approaches across microbiology, immunology, systems biology, and clinical sciences, alongside consideration of ethical, regulatory, and societal implications to ensure the equitable acceptance of microbiome-based interventions. Current evidence supports dietary changes, probiotics, synbiotics, and fecal microbiota transplants as the most studied and effective strategies, with microbial metabolite-immune system interactions providing a strong rationale for targeted therapeutics. Future work should focus on clinical efficacy, standardized measurement approaches, and the integration of emerging technologies to advance microbiome-based personalized medicine.

Declarations

Competing Interests The authors have no competing interests to declare that are relevant to the content of this manuscript.

Ethics Approval The authors declare that this manuscript does not include primary research with human participants.

The authors declare that this manuscript does not include primary studies involving animals.

References

1. De Vos WM, et al. Gut microbiome and health: mechanistic insights. Gut. 2022;71(5):1020–32.
2. Aggarwal N, et al. Microbiome and human health: current understanding, engineering, and enabling technologies. Chem Rev. 2022;123(1):31–72.
3. Derrien M, van Hylckama Vlieg JE. Fate, activity, and impact of ingested bacteria within the human gut microbiota. Trends Microbiol. 2015;23(6):354–66.
4. Fischbach MA. Microbiome: focus on causation and mechanism. Cell. 2018;174(4):785–90.
5. Zhang Z, et al. Demystifying the manipulation of host immunity, metabolism, and extraintestinal tumors by the gut microbiome. Signal Transduct Target Ther. 2019;4(1):41.
6. Mutalub YB, et al. Gut microbiota modulation as a novel therapeutic strategy in cardiometabolic diseases. Foods. 2022;11(17):2575.
7. Quigley EM, Gajula P. Recent advances in modulating the microbiome. F1000Res. 2020;9:F1000 Faculty. Rev-46.
8. Kustrimovic N, et al. Microbiome and prostate cancer: a novel target for prevention and treatment. Int J Mol Sci. 2023;24(2):1511.
9. Horvath A, et al. Human gut microbiome: therapeutic opportunities for metabolic syndrome—hype or hope? Endocrinol Diabetes Metab. 2024;7(1):e436.
10. Ford AC, et al. American College of Gastroenterology monograph on management of irritable bowel syndrome. Off J Am Coll Gastroenterol ACG. 2018;113:1–18.
11. Singh RK, et al. Influence of diet on the gut microbiome and implications for human health. J Transl Med. 2017;15(1):73.
12. Lai S, et al. Effect of low-protein diet and inulin on microbiota and clinical parameters in patients with chronic kidney disease. Nutrients. 2019;11(12):3006.
13. Guo Y, et al. Modulation of the gut microbiome: a systematic review of the effect of bariatric surgery. Eur J Endocrinol. 2018;178(1):43–56.
14. Salim S, et al. Gut microbiome and Parkinson's disease: perspective on pathogenesis and treatment. J Adv Res. 2023;50:83–105.
15. Somodi C, et al. Gut microbiome changes and cancer immunotherapy outcomes associated with dietary interventions: a systematic review of preclinical and clinical evidence. J Transl Med. 2025;23(1):756.
16. Carucci L, et al. Therapeutic effects elicited by the probiotic Lacticaseibacillus rhamnosus GG in children with atopic dermatitis. The results of the ProPAD trial. Pediatr Allergy Immunol. 2022;33(8):e13836.
17. Zheng D, Liwinski T, Elinav E. Interaction between microbiota and immunity in health and disease. Cell Res. 2020;30(6):492–506.
18. Arteaga-Henríquez G, et al. Treating impulsivity with probiotics in adults (PROBIA): study protocol of a multicenter, double-blind, randomized, placebo-controlled trial. Trials. 2020;21(1):161.
19. Loh JS, et al. Microbiota–gut–brain axis and its therapeutic applications in neurodegenerative diseases. Signal Transduct Target Ther. 2024;9(1):37.

20. Hu KA, Gubatan J. Gut microbiome–based therapeutics in inflammatory bowel disease. Clin Transl Discov. 2023;3(2):e182.
21. Boulangé CL, et al. Impact of the gut microbiota on inflammation, obesity, and metabolic disease. Genome Med. 2016;8(1):42.
22. Tuttle KR, et al. Molecular mechanisms and therapeutic targets for diabetic kidney disease. Kidney Int. 2022;102(2):248–60.
23. Tilg H, Adolph TE, Moschen AR. Multiple parallel hits hypothesis in nonalcoholic fatty liver disease: revisited after a decade. Hepatology. 2021;73(2):833–42.
24. Lin L, et al. Gut microbiota in preclinical rheumatoid arthritis: from pathogenesis to preventing progression. J Autoimmun. 2023;141:103001.
25. Durack J, Lynch SV. The gut microbiome: relationships with disease and opportunities for therapy. J Exp Med. 2019;216(1):20.
26. Paller AS, et al. The microbiome in patients with atopic dermatitis. J Allergy Clin Immunol. 2019;143(1):26–35.
27. De Pessemier B, et al. Gut–skin axis: current knowledge of the interrelationship between microbial dysbiosis and skin conditions. Microorganisms. 2021;9(2):353.
28. Vogt NM, et al. Gut microbiome alterations in Alzheimer's disease. Sci Rep. 2017;7(1):13537.
29. Brydges CR, et al. Indoxyl sulfate, a gut microbiome-derived uremic toxin, is associated with psychic anxiety and its functional magnetic resonance imaging-based neurologic signature. Sci Rep. 2021;11(1):21011.
30. Bell L, et al. A randomized, placebo-controlled trial investigating the acute and chronic benefits of American ginseng (Cereboost®) on mood and cognition in healthy young adults, including in vitro investigation of gut microbiota changes as a possible mechanism of action. Eur J Nutr. 2022;61(1):413–28.
31. Li X, et al. Gut microbiome in modulating immune checkpoint inhibitors. EBioMedicine. 2022;82:104163.
32. Smet A, et al. The role of microbiota in gastrointestinal cancer and cancer treatment: chance or curse? Cell Mol Gastroenterol Hepatol. 2022;13(3):857–74.
33. Silva YP, Bernardi A, Frozza RL. The role of short-chain fatty acids from gut microbiota in gut-brain communication. Front Endocrinol. 2020;11:508738.
34. Thatcher M. Facing the new challenge. In: Women and social policy: a reader. Springer; 1985. p. 213–7.
35. Antwi J. Precision nutrition to improve risk factors of obesity and type 2 diabetes. Curr Nutr Rep. 2023;12(4):679–94.
36. Han K, et al. Engineering strategies to modulate the gut microbiome and immune system. J Immunol. 2024;212(2):208–15.

The Microbiome-Brain Axis in Neurodegenerative Disease: Emerging Insights and Therapeutic Horizons

14

Priya Prasannakumar, M. Sruthi, I. Ameena, and A. Robert Antony

14.1 Introduction

Neurodegenerative diseases (NDDs) represent a complex group of progressive disorders featuring irreversible loss of neuronal structure and function, finally impairing cognition, motor control, and behavior. These conditions are especially prevalent among older populations and pose a significant global health burden due to their complex and multifactorial etiologies [1]. The progressive nature of these diseases means that once neuronal loss begins, it typically continues in a relentless fashion, leading to the gradual deterioration of critical brain functions and creating clinical manifestations that are as varied as they are devastating.

The demographic shift toward an aging global population has transformed neurodegenerative diseases from a relatively rare condition to being a major public health concern. Alzheimer's disease (AD) alone affects over 50 million people worldwide, with prevalence rates doubling approximately every 5 years after age 65. Parkinson's disease (PD) impacts more than ten million people globally, whereas amyotrophic lateral sclerosis (ALS) has an annual incidence of 1–2 per 100,000 people. Huntington's disease, which is genetically inherited and a rare disease, typically manifests between ages 30 and 50 with relentless progression. Overall, these conditions create an enormous burden on healthcare systems and the society, with economic impacts which include direct medical costs as well as indirect burdens in the form of lost productivity, caregiver burden, and provision for specialized long-term care facilities [2].

Each major neurodegenerative disease presents unique clinical signatures while sharing key pathological features. Alzheimer's disease manifests as memory loss, marked disorientation, and slow cognitive decline, featuring extracellular

P. Prasannakumar (✉) · M. Sruthi · I. Ameena · A. Robert Antony
Government Arts and Science College, Kozhinjampara, Palakkad, Kerala, India

M. H. Rahman, Muhasina K M (eds.), *The Microbiome's Impact on the Brain*, https://doi.org/10.1007/978-981-95-7494-0_14

amyloid-beta (Aβ) plaques and intracellular neurofibrillary tangles composed of hyperphosphorylated tau protein in the hippocampus and cerebral cortex regions. Parkinson's disease mainly presents as a movement disorder with tremors, bradykinesia, rigidity, and postural instability, driven by degeneration of dopaminergic neurons in the substantia nigra and accumulation of alpha-synuclein into Lewy bodies. ALS involves progressive degeneration of upper and lower motor neurons, resulting in weakness of muscles and their atrophy, with ultimate respiratory failure, characterized by cytoplasmic aggregates of TAR DNA-binding protein 43 (TDP-43) protein that interferes with RNA processing. Huntington's disease, caused by cytosine, adenine, and guanine trinucleotide repeat expansion in the Huntingtin (HTT) gene, presents with chorea, psychiatric symptoms, and cognitive decline as the mutant huntingtin protein disrupts transcription and mitochondrial function, particularly in the striatum [3].

Despite their distinct clinical manifestations, these diseases share similar molecular mechanisms that both complicate and inform our understanding of neurodegeneration. Protein misfolding and aggregation represent recurrent hallmarks across conditions. They show involvement of amyloid plaques and tau tangles in Alzheimer's disease, alpha-synuclein in Parkinson's disease, or TDP-43 in frontotemporal dementia and ALS. Mitochondrial dysfunction and oxidative stress exacerbate neuronal damage, while chronic neuroinflammation hastens disease progression. Additionally, defects in autophagy and proteostasis stop the clearance of toxic proteins, and genetic mutations or epigenetic alterations predispose individuals to specific conditions [4]. These mechanisms frequently display region-specific neuronal vulnerability, ultimately shaping each disease's unique phenotype.

Neurodegenerative diseases thus arise from a complex marriage comprising genetic predisposition, environmental toxins (pollutants, heavy metals, pesticides), and gut microbiota dysbiosis [5]. This can trigger neuroinflammation and protein aggregation through compromised blood-brain and intestinal barriers. Despite intensive research and investment, no curative treatments exist—current therapies like cholinesterase inhibitors and dopamine replacement only provide symptomatic relief while the diseases progress. The therapeutic challenge stems from the brain's unique barriers, non-regenerating neurons, and complex neural networks that resist intervention. Advanced biomarkers and emerging therapies (gene therapy, immunotherapy, stem cells) offer hope, but the growing societal burden of caring for aging populations with these diseases raises pertinent questions about the quality of life and healthcare resource allocation. Early detection through improved biomarkers may eventually enable more effective interventions before irreversible neural damage occurs [6].

14.2 The Human Gut Microbiome and the Gut-Brain Axis

The human gut microbiome (GMB) represents a highly complex ecosystem within the human body, comprising trillions of microorganisms including bacteria, archaea, fungi, viruses, and protists. It is largely populated by bacterial phyla such as *Firmicutes*, *Bacteroidetes*, *Actinobacteria*, and *Proteobacteria* [7]. This microbiota exhibits remarkable diversity with over 1000 different bacterial species identified to date. The composition varies significantly between individuals as it is influenced by factors including genetics, diet, age, antibiotic use, and environmental exposures. Functionally, the gut microbiome serves as a metabolic organ, contributing to digestion, vitamin synthesis (particularly B vitamins and vitamin K), immune system development, pathogen resistance, and maintenance of intestinal barrier integrity. These microorganisms have a collective encoding power of 150 times more than the human genome. They consequently provide several metabolic capabilities that humans lack, such as the breakdown of complex carbohydrates and producing bioactive compounds [8].

The gut-brain axis represents an advanced bidirectional communication network linking the enteric nervous system of the gastrointestinal tract with the central nervous system (CNS). This communication occurs through many-pronged interconnected pathways, with the vagus nerve serving as the primary neural highway. The vagus nerve, the longest cranial nerve, provides direct parasympathetic connection between the brainstem and the gut, allowing for rapid transmission of signals in both directions. Afferent vagal fibers detect changes in gut chemistry, mechanical stimuli, and microbial metabolites, transmitting this information to the nucleus tractus solitarius in the brainstem, which then projects to various brain regions including the hypothalamus, amygdala, and prefrontal cortex. Simultaneously, efferent vagal pathways modulate gut motility, secretion, and immune responses, creating a dynamic feedback loop that influences both gastrointestinal function and central nervous system activity [9].

Immune modulation represents another important mechanism of gut-brain communication, mediated through both local and systemic inflammatory responses. The gut-associated lymphoid tissue, which contains approximately 70% of the body's immune cells, serves as a crucial interface between the microbiome and the host immune system. Microbial components such as lipopolysaccharides (LPS), peptidoglycans, and flagellin are recognized by pattern recognition receptors on immune cells, triggering cascades that influence neuroinflammation and brain function. Pro-inflammatory cytokines including interleukin-1β, tumor necrosis factor-α, and interleukin-6 can cross the blood-brain barrier (BBB) or signal through vagal pathways to activate microglia and astrocytes in the brain [10]. Conversely, anti-inflammatory signals from beneficial microbes can promote regulatory T-cell responses and produce anti-inflammatory cytokines that support neuroprotection and cognitive function.

Short-chain fatty acids (SCFAs), the most abundant of microbial products, are recognized as acetate, propionate, and butyrate. They represent key metabolic messengers in gut-brain communication. These compounds are produced through bacterial fermentation of dietary fiber and resistant starch by beneficial microbes such as *Bifidobacterium*, *Lactobacillus*, and *Faecalibacterium prausnitzii*. SCFAs exert varied effects on brain function through different mechanisms: they can cross the blood-brain barrier directly or through monocarboxylate transporters and serve as energy substrates for brain cells, modulate microglial activation, influence neurotransmitter synthesis, and strengthen intestinal barrier function to prevent harmful bacterial translocation [11]. Butyrate, in particular, acts as a histone deacetylase inhibitor, influencing gene expression related to neuroplasticity, inflammation, and

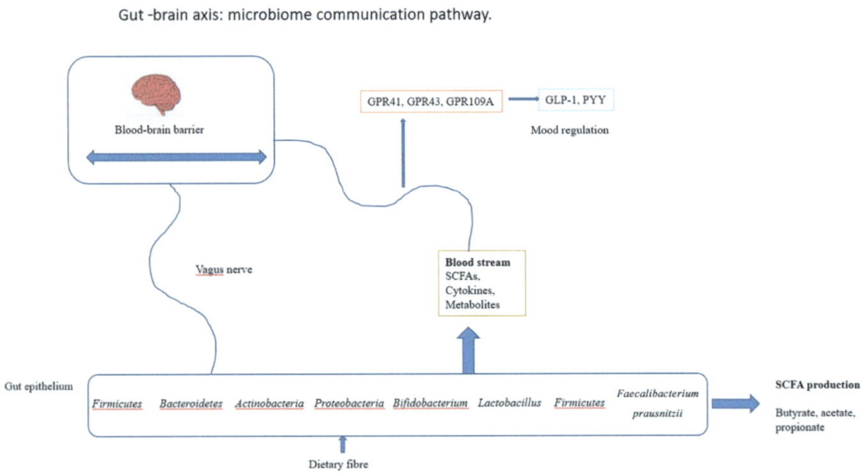

Fig. 14.1 The microbiome communication pathway in the gut-brain axis

stress responses. Additionally, SCFAs activate G-protein-coupled receptors on enteroendocrine cells, stimulating the release of hormones like glucagon-like peptide-1 and peptide YY, which can influence satiety, mood, and cognitive function through both local and central mechanisms [12].

The BBB plays a pivotal role in mediating gut-brain interactions, functioning not only as a protective barrier but also as a selective gateway for microbial signals. This highly specialized structure consists of a selectively permeable barrier comprising of tightly joined endothelial cells, pericytes, astrocytes, and basement membrane that normally prevents harmful substances from entering the brain while allowing essential nutrients and signaling molecules to pass through [13]. The gut microbiome can influence BBB permeability through multiple mechanisms: bacterial lipopolysaccharides can increase barrier permeability through Toll-like receptor activation, while beneficial microbial metabolites like SCFAs can strengthen barrier integrity by upregulating tight junction proteins [14]. Certain microbial metabolites, including tryptophan derivatives and gamma-aminobutyric acid, can cross the BBB through specific transporters, directly influencing neurotransmitter systems. Additionally, the BBB contains specialized transport systems for cytokines and hormones, allowing immune and endocrine signals originating from gut-microbe interactions to reach the brain and modulate neuronal function, mood, and behavior (Fig. 14.1).

14.3 Microbiome Alterations in Specific Neurodegenerative Disorders

14.3.1 Alzheimer Disease

AD patients show distinctive gut microbiome alterations characterized by decrease in microbial diversity and change in distinct bacterial communities compared to

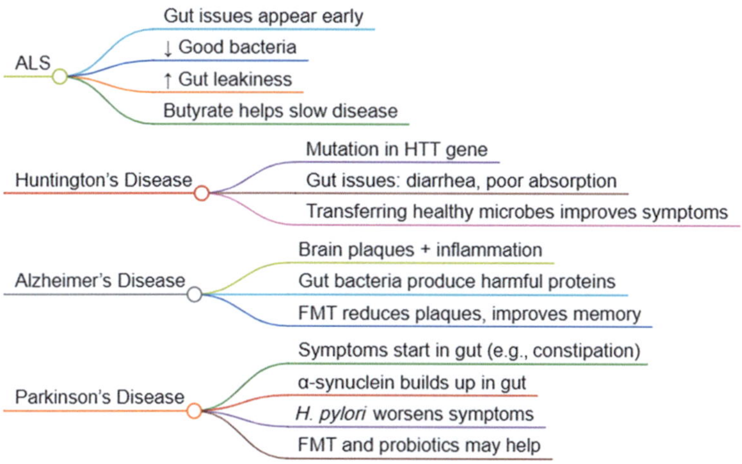

Fig. 14.2 Gut microbiota and product alterations implicated in neurodegenerative disorders

age-matched controls [15]. Specific dysbiosis patterns include decreased *Firmicutes* and increased *Bacteroidetes* at the phylum level with significant reductions in beneficial bacteria such as *Bifidobacterium* and increases in potentially pathogenic species. This gut microbiota dysbiosis is closely linked to the onset and progression of AD, affecting neuroinflammation and gastrointestinal permeability. The altered microbiome composition correlates with cognitive decline severity, suggesting that specific bacterial taxa may serve as biomarkers for disease progression [16]. Importantly, emerging evidence suggests that disruption of the normal gut microbiome in early life leads to a cascade of molecular events within brain circuits suggesting that microbiome alterations are a precursor to classical AD pathology (Fig. 14.2).

The mechanistic relationship between gut dysbiosis and AD pathogenesis involves multiple interconnected pathways, with microbial amyloids playing a particularly significant role. Recent research has revealed that many bacterial species produce functional amyloid proteins that serve various physiological functions, including biofilm formation and cell adhesion [17]. However, these microbial amyloids may contribute to neurodegenerative disease progression through cross-seeding mechanisms. These amyloids can act as seeds for the host amyloid protein aggregation, including Aβ in Alzheimer's disease and α-synuclein in Parkinson's disease. This cross-seeding effect may accelerate the formation of pathological protein aggregates and contribute to neurodegeneration [18]. Exposure to microbial amyloids may prime the innate immune system to have enhanced and potentially pathogenic responses to neuronal amyloids. This priming effect could contribute to the chronic neuroinflammation observed in certain bacterial species that produce functional amyloids that structurally resemble human amyloid-β peptides, potentially serving as seeds for cerebral amyloid aggregation through molecular mimicry. These bacterial amyloids can cross the compromised blood-brain barrier and interact with endogenous amyloid-β, accelerating plaque formation and neurodegeneration. Additionally, the dysbiotic microbiome produces increased levels of lipopolysaccharides and other inflammatory mediators that trigger systemic inflammation and activate brain microglia. Altered microbial diversity can regulate bowel-brain axis involvement in the AD through changes in metabolite production including reduced short-chain fatty acid synthesis and increased production of neurotoxic compounds such as trimethylamine N-oxide and phenolic compounds that damage neurons and promote tau hyperphosphorylation [19].

14.3.2 Parkinson's Disease

PD shows evidence of gut microbiota alterations, with gastrointestinal issues usually occurring before motor symptoms [20]. The presence of gastrointestinal symptoms (constipation, dysphagia, and hypersalivation), altered gut microbiota, and leaky gut have been observed in PD patients few years prior to the clinical onset of the disease. The characteristic dysbiosis pattern in PD includes significant reductions in beneficial bacteria such as *Prevotellaceae*, *Lactobacillaceae*, and *Bifidobacteriaceae*. There is also a marked increase in potentially harmful species including *Enterobacteriaceae* and *Verrucomicrobiaceae*. This dysbiosis might result in a pro-inflammatory status which could be linked to the recurrent gastrointestinal symptoms affecting PD patients [21]. The altered microbiome composition has been shown to reflect on disease severity and motor function decline, with bacterial signatures which may serve as early diagnostic markers for PD risk assessment. The gut-brain connection in PD is exceptionally striking due to the potential role of α-synuclein pathology originating in the enteric nervous system. Emerging studies reveal that disruptions in the gut microbiome may drive disease progression through mechanisms such as increased intestinal permeability, inflammation, and neurotransmitter imbalances [22]. The dysbiotic microbiome produces increased levels of curli fibroids and other bacterial proteins that act as α-synuclein aggregation triggers, promoting the formation of Lewy bodies in enteric neurons. These pathological protein aggregates then propagate from the gut to the brain via the vagus nerve, following a prion-like spreading pattern that eventually reaches the substantia nigra. Additionally, the altered microbiome affects dopamine metabolism through tyrosine hydroxylase activity changes and heightened production of neurotoxic metabolites that cross the blood-brain barrier and directly damage dopaminergic neurons [23].

14.3.3 Amyotrophic Lateral Sclerosis

Research on ALS and gut microbiome alterations has shown alterations in fecal community diversity, with recent-onset ALS patients showing distinct microbiome patterns compared to healthy individuals. Alterations to microbiome communities within the gut flora have been consistently observed in ALS patients, with these changes often correlated with disease progression and patient outcome [24]. The dysbiosis pattern in ALS includes increased abundance of pro-inflammatory bacteria such as *Ruminococcus torques* and *Streptococcus*, alongside decreased levels of beneficial bacteria that produce anti-inflammatory metabolites. ALS patients show altered short-chain fatty acid production patterns, with specific changes in butyrate-producing bacteria that may affect host metabolism and neuroinflammation [25]. The gut microbiota alterations show differential effects depending on the ALS model, with microbiome depletion worsening disease in superoxide dismutase 1 models while ameliorating disease in C9orf72 models suggesting that microbiome-host interactions may vary based on underlying genetic mutations [26]. The mechanistic hypotheses linking gut microbiome to ALS pathogenesis focus on immune dysregulation and metabolic dysfunction. The altered gut microbiota contributes to disease by modulating immunologic, metabolic, and neuronal signaling pathways, with the gut microbiota influencing neuroinflammation through its effects on immune system activation. The dysbiotic microbiome produces higher levels of lipopolysaccharides and other inflammatory mediators that activate peripheral immune cells and promote the infiltration of activated microglia and macrophages into the central nervous system. Additionally, dysbiosis is correlated with intestinal inflammation and changes in intestinal integrity in ALS [27]. Furthermore, emerging evidence suggests that specific bacterial metabolites may directly affect neuromuscular junction integrity and motor neuron survival, although the precise molecular mechanisms remain under investigation.

14.3.4 Cross-Disease Patterns and Therapeutic Implications

Across these neurodegenerative disorders, several common microbiome patterns emerge, including reduced microbial diversity, decreased abundance of beneficial short-chain fatty acid-producing bacteria, increased intestinal permeability, and elevated systemic inflammation markers. These shared features suggest that gut dysbiosis may represent a common pathway contributing to neurodegeneration, albeit with disease-specific variations in bacterial taxa and metabolic profiles [28]. The temporal relationship between gut symptoms and neurological manifestations varies among disorders, with gastrointestinal dysfunction often preceding motor symptoms in PD by years, while appearing more concurrently with neurological symptoms in AD and ALS. The correlation between microbiome alterations and disease progression across these disorders suggests that gut-brain axis modulation may have a significant therapeutic potential opening new avenues for early intervention strategies that target microbiomes before irreversible neurological damage occurs. Understanding these disease-specific microbiome signatures may ultimately lead to personalized therapeutic approaches that restore beneficial microbial communities and interrupt the pathological cascades linking gut dysbiosis to neurodegeneration [29].

14.4 Mechanistic Insights

The gut microbiota is of paramount importance in regulating CNS activity through the production of microbial metabolites such as SCFAs and tryptophan derivatives. SCFAs, primarily acetate, propionate, and butyrate, influence the integrity of the blood-brain barrier, promote anti-inflammatory pathways, and support neurogenesis. Meanwhile, tryptophan metabolites, including indole derivatives, kynurenine, and serotonin precursors, exert neuromodulatory effects by interacting with host neurotransmitter systems and modulating immune signaling across the gut-brain axis [30]. Beyond direct metabolite signaling, the gut microbiota influences neuroinflammation through immune priming mechanisms. Dysbiosis can lead to increased permeability of the intestinal barrier, allowing microbial products such as LPS to enter systemic circulation. This starts inflammatory responses both peripherally and centrally, activating microglia and astrocytes in the brain. Chronic exposure to such microbial stimuli is implicated in neurodegenerative conditions such as Parkinson's disease and multiple sclerosis (MS) [31].

Molecular mimicry is another pathogenic mechanism whereby microbial antigens share structural similarities with host proteins, leading to cross-reactive immune responses that can damage CNS tissues [32]. It happens when certain bacterial proteins contain amino acid sequences or structural motifs that closely resemble those found in neural proteins. When the immune system mounts a response against these microbial antigens, the resulting antibodies or T cells may cross-react with host neural tissues, leading to autoimmune diseases. Molecular mimicry has been implicated in several neurological conditions like multiple sclerosis: Certain gut bacteria express proteins that share homology with myelin basic protein, potentially triggering autoimmune demyelination [33]. *Campylobacter jejuni* lipopolysaccharides mimic ganglioside structures in peripheral nerves, leading to cross-reactive antibody production. Molecular mimicry between *Streptococcus pyogenes* antigens and hypocretin receptor proteins may contribute to autoimmune destruction of hypocretin-producing neurons [34]. Microbial-derived neurotoxins such as β-methylamino-L-alanine, produced by cyanobacteria, and ammonia from protein-fermenting bacteria can directly impair neuronal function and promote excitotoxicity. These compounds have been associated with motor neuron degeneration and cognitive impairment in both experimental and clinical contexts. The identification of such toxins has catalyzed interest in the microbiome as both a reservoir and regulator of neuroactive compounds [35]. Understanding the complex interplay between microbial factors and the CNS opens avenues for therapeutic intervention. Probiotic and prebiotic strategies, microbial metabolite modulation, and targeting immune pathways offer promising approaches to mitigate neuropathology. Future research should emphasize strain-specific effects, host-genetic interactions, and longitudinal tracking to fully elucidate microbiome-derived neuroregulatory networks [36].

14.5 Preclinical and Clinical Evidence

14.5.1 Animal Models Demonstrating Causative Links

Experimental studies using animal models have demonstrated the role of GMB in neurodegenerative diseases such as PD, AD, and ALS. Pathogen-free mice that lack intestinal microbes show less neuroinflammation and better cognitive function compared to conventionally raised rodents. This study highlights that dysbiotic gut microbiota contributes to neuroinflammation and neuronal dysfunction. The gut-brain axis has a supporting role in neurodegenerative diseases [37]. Amyloid beta oligomer and plaque formation, tau aggregation, and neuroinflammation are the primary causes of AD. Researchers further found that the GMB influences the above features and AD progression in humans. This microbiota can modulate Aβ clearance rather than production [38]. This can be studied with the help of various animal models. Their age, sex, and disease model, housing, diet, and genetics can result in different results. AD mouse models like 5XFAD, APP/PS1, and APPPS1–21 showed changes in GMB compared to healthy mice, similar to those seen in AD patients. Overall gut microbial diversity decreases as the *Bacteroidetes* and *Proteobacteria* increase with the progression of disease [39]. In 5XFAD mice, an increased number of *Firmicutes* and a lower level of *Bacteroidetes* were found, but in APPPS1–21 mice, the case is vice versa. APP/PS1 mice had higher levels of *Escherichia-Shigella*, *Desulfovibrio*, *Akkermansia*, and *Blautia*, and a drop in beneficial bacteria like *Prevotella* was reported. Antibiotic treatment consistently reduces the Aβ deposition in males, which confirms the role of gut microbes in neurodegeneration. Imbalance in gut microbiota may trigger PD. The dysbiotic gut microbiome in PD can disrupt the balance of a healthy microbiome. Th17-mediated immune homeostasis is also affected. This initiates a series of events that extend to the brain [40]. It also disrupts the integrity of the intestinal membrane. They can increase the production of a metabolite that influences the aggregation of a protein, linked to PD pathology in the brain, called α-synuclein [41]. The reactive astrocytes associated with prion disease were found to secrete IL-6, which reduces the trans-endothelial electrical resistance in healthy endothelial cells.

The co-cultivation of healthy endothelial and infected astrocytes demonstrated this [42]. The gut microbiome axis communication is a bidirectional process. The vagus nerve passes the information between the gut microbiome and the brain. Vagotomy of the animal showed no neurotoxicity in them [41]. Research involving animals supports the theory that Parkinson's disease might begin in the gut. It has been shown that harmful α-synuclein can develop in the gut and then move to the brain through the vagus nerve. These studies illustrate that inflammation in the gut, changes in the microbiome, and problems with the intestinal barrier can initiate the aggregation of α-synuclein and its spread to the brain, which contributes to the degeneration of dopaminergic neurons [43].

14.5.2 Observational and Interventional Human Studies

In a study, researchers analyzed gut microbiomes from 420 people, including PD patients, individuals with early PD signs, constipated controls, and healthy controls. They found that people with PD or early signs had fewer beneficial anaerobic bacteria. Also, their gut microbial metabolic pattern gets affected, especially the utilization of carbohydrates. A microbiome-based test could moderately distinguish PD cases (AUC ~0.76). These results suggest that gut microbiome changes may serve as early biomarkers for PD [44]. In another study analyzing stool samples from PD patients and control groups across two US cohorts, shotgun metagenomic sequencing revealed a consistent shift in microbial composition, mainly in *Intestinimonas butyriciproducens*, and disruption in lipid and carbohydrate metabolism. The same result was obtained in Germany and China, where they confirmed that consistent changes in gut microbes may serve as biomarkers and therapeutic targets in PD [45]. There is a difference between the gut microbiome in the body-first and brain-first types of PD. Comparing the results from PD patients and healthy individuals revealed that body-first PD patients had a different microbial load, characterized by higher levels of *E. coli* and *Akkermansia muciniphila*, as well as fewer short-chain fatty acid-producing beneficial bacteria. Hence, microbial genes have a role in inflammation and neurodegeneration. These results may help plan specific treatment based on the subtype of disease [43]. *Actinobacteria* sp. has been reported to affect PD by changing the specific cytokine levels. This finding suggests that GMB has a causal role in PD by altering inflammatory cytokines [46]. An interventional trial was conducted in PD patients to study the effect of fecal microbiota transplantation (FMT). The treatment group was supplied with FMT and the control group with a placebo. The result showed that FMT increased the gut microbial diversity and is safe to consume orally without any side effects, which helps manage PD symptom [47]. A new study reveals that the Parkinson's drug entacapone significantly disrupts the gut microbiome by causing iron deficiency, highlighting the overlooked effects of human-targeted drugs on gut health [48].

14.5.3 Challenges in Reproducibility and Heterogeneity of Findings

Research involving both animals and humans underscores significant challenges in the reproducibility and variability of studies examining the influence of gut microbiome on neurodegenerative diseases. In rodent models, factors such as the source of the animals, facility conditions, diet, bedding, cage effects, and husbandry practices contribute to considerable variations both within and between studies in terms of gut microbial composition and disease phenotypes. These elements hinder experimental reproducibility and applicability. Hence, new methods such as co-housing, littermate controls, gnotobiotic microbes, and strict laboratory protocols are employed to ensure consistent results in these studies [49]. However, even studies deemed "identical" often produce differing outcomes due to hidden microbiome-driven modifiers of phenotype. Importantly, meta-analyses of dietary interventions in mice have shown that certain microbial changes, such as those induced by a high-fat diet, are consistently reproducible across different laboratories [50]. However, issues like contamination can still affect reproducibility and introduce variability in experimental results (e.g., *Lactococcus*), and methodological discrepancies continue to complicate interpretation. In human studies, inconsistency often stems from methodological heterogeneity—such as the use of 16S rRNA versus shotgun metagenomics, differences in sample collection and bioinformatics pipelines, and the statistical analysis of microbiome data—which can yield divergent microbial signatures even in similar cohorts [51]. Diverse patient populations also contribute to variability, with cohort differences in disease stage, sex, geography, diet, medications, comorbidities, and lifestyle factors significantly influencing gut microbiota composition and confounding attempts to identify consistent neurodegenerative disease signatures [52]. Prescription medications, especially non-antibiotic drugs like proton pump inhibitors, metformin, antipsychotics, and entacapone, are known to reshape the microbiome by altering microbial diversity and metabolic pathways, yet these effects are often inadequately controlled or reported in neurodegenerative disease studies [53]. In interventional trials such as FMT in Parkinson's disease, donor-to-donor microbial variation, non-standardized delivery protocols (e.g., frequency, route), and limited follow-up durations contribute to mixed clinical outcomes and poor reproducibility. Many studies rely primarily on symptom scores or diversity indices, without robust functional biomarkers such as SCFA levels or neuroinflammatory markers, further limiting mechanistic insights and comparability across trials. To tackle the issues of reproducibility and variability in gut microbiome studies concerning neurodegenerative diseases, the field needs to implement stringent best practices. This involves employing standardized procedures for collecting, storing, sequencing, and analyzing samples to reduce methodological differences between studies [54]. In animal research, methods like co-housing, using littermate controls, and breeding F2 generations can help decrease microbiome-induced phenotypic differences and enhance experimental reliability [55]. Human research should include larger, multi-center cohorts with consistent clinical metadata, facilitating cross-cohort comparisons and minimizing confounding factors related to demographics, diet, and lifestyle. Employing multi-omics approaches—such as metagenomics, metabolomics, immunomics, and neuroimaging—provides a more thorough understanding of the gut-brain connection and its impact on disease progression. For instance, recent longitudinal studies have effectively associated gut microbiota profiles with fecal metabolites, brain imaging, and cognitive performance in Alzheimer's disease groups.

Programs like the Alzheimer's Disease Neuroimaging Initiative demonstrate how standardized, multi-site collaborations can aid in robust biomarker discovery and translational insights. Additionally, consistently including functional biomarkers like SCFAs, inflammatory mediators, and neuroimaging markers is crucial for mechanistic clarity and reproducibility in both observational and interventional studies. Altogether, these strategies are essential for advancing microbiome-based neurodegeneration research into reproducible and clinically applicable findings [56].

14.6 Therapeutic and Diagnostic Potential

14.6.1 Probiotics, Prebiotics, and Dietary Interventions

The therapeutic promise of modulating the gut microbiome in NDDs has gained significant traction, with probiotics, prebiotics, and dietary strategies emerging as key tools in personalized interventions. These approaches aim to restore microbial balance, reduce neuroinflammation, and enhance neuroprotective signaling through the gut-brain axis. Probiotics are live microorganisms that, when administered in

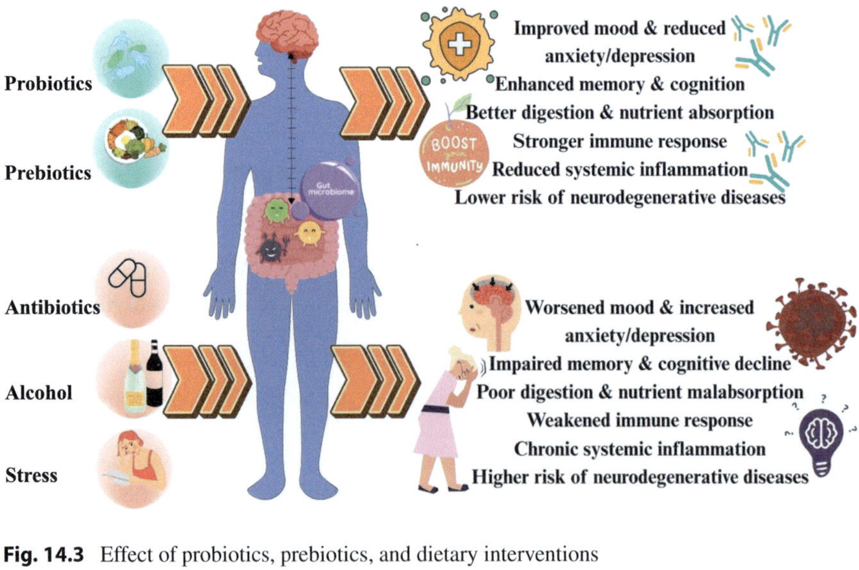

Fig. 14.3 Effect of probiotics, prebiotics, and dietary interventions

adequate amounts, confer health benefits to the host [57]. In the context of NDDs, strains such as *Lactobacillus*, *Bifidobacterium*, *Clostridium butyricum*, and *Faecalibacterium prausnitzii* have shown promise in modulating neuroimmune responses, enhancing gut barrier integrity, and producing neuroactive metabolites like SCFAs. These metabolites influence microglial activation, reduce oxidative stress, and regulate neurotransmitter synthesis, thereby mitigating neurodegenerative pathology (Fig. 14.3). Clinical and preclinical studies have demonstrated that probiotic supplementation can improve cognitive function, reduce amyloid-beta deposition, and alleviate motor symptoms in models of Alzheimer's and Parkinson's disease [58]. Prebiotics are non-digestible dietary fibers that selectively stimulate the growth and activity of beneficial gut bacteria. Compounds such as inulin, fructooligosaccharides, and galactooligosaccharides enhance the abundance of SCFA-producing microbes, which in turn support anti-inflammatory and neuroprotective pathways [59]. By promoting microbial diversity and resilience, prebiotics help counteract dysbiosis—a common feature in NDDs—and may delay disease progression. Emerging evidence suggests that tailored prebiotic formulations, based on individual microbiome profiles, could optimize therapeutic outcomes in a personalized medicine framework.

Diet plays a pivotal role in shaping the gut microbiota and its metabolic output. Diets rich in whole grains, legumes, fruits, vegetables, and fermented foods foster microbial diversity and stability, while Western-style diets high in saturated fats and refined sugars promote dysbiosis and inflammation. The Mediterranean diet has been associated with a reduced risk of cognitive decline and improved gut-brain axis function [60]. Nutritional strategies that emphasize polyphenols, omega-3 fatty acids, and dietary fiber can enhance microbial production of neuroprotective compounds and support brain health.

14.6.2 Fecal Microbiota Transplantation

FMT, which involves transferring gut microbiota from a healthy donor to a recipient, is gaining momentum as a therapeutic strategy for NDDs. Originally developed for managing recurrent *Clostridioides difficile* infections, FMT is now under investigation for its ability to rectify dysbiosis implicated in the progression of AD, PD, MS, and ALS [61]. By introducing a balanced microbial consortium, FMT aims to re-establish homeostasis within the gut environment and positively influence the complex neuroimmune signaling that underpins these conditions. Mechanistically, FMT acts on the gut-brain axis, a dynamic communication pathway integrating neural, endocrine, immune, and metabolic signals between the gastrointestinal system and the central nervous system. In NDDs, microbial imbalance often triggers increased intestinal permeability and systemic inflammation, disrupting neurochemical homeostasis. FMT offers a comprehensive reset of the gut ecosystem, enhancing SCFA production and regulating neuroactive metabolites, such as tryptophan derivatives [62]. Studies in transgenic animal models of AD reveal that FMT from healthy donors can significantly reduce amyloid-beta plaque deposition, improve synaptic connectivity, and bolster cognitive function.

Clinical studies further support FMT's potential, demonstrating behavioral improvement and neuroinflammatory modulation in preclinical models of AD and PD. Particularly notable are findings that donor age and microbial diversity impact outcomes, with younger donor profiles offering enhanced neuroprotection. In human trials, FMT has shown benefits in alleviating gastrointestinal symptoms and mood disturbances in ALS patients, although its effects on motor function remain inconclusive. Challenges such as standardized protocols, long-term microbiome stability, and personalized donor-recipient matching must be addressed to optimize therapeutic efficacy. Emerging innovations—including tailored microbiota consortia and encapsulated delivery systems—are paving the way for safer, more targeted interventions in neurodegenerative disease care [63].

14.6.3 Microbiome-Based Biomarkers for Early Detection and Disease Monitoring

The gut microbiome has emerged as a promising source of biomarkers for the early detection and monitoring of NDDs, offering a non-invasive and dynamic window into CNS pathology [64]. Unlike traditional biomarkers that often rely on imaging or cerebrospinal fluid analysis, microbiome-derived indicators—such as microbial taxa, metabolites, and extracellular vesicles—can be accessed through stool, blood, or saliva, making them highly amenable to routine clinical use. These biomarkers reflect systemic changes in inflammation, metabolism, and neuroimmune signaling, which are intricately linked to the gut-brain axis.

Recent studies have identified disease-specific microbial signatures in conditions such as AD, PD, and ALS. For instance, reduced abundance of *Prevotella* and *Bifidobacterium*, alongside increased levels of *Enterobacteriaceae* and *Desulfovibrio*, have been associated with PD progression. Similarly, alterations in SCFA profiles, bile acid metabolism, and tryptophan catabolites have been correlated with cognitive decline and neuroinflammation in AD. These microbial shifts often precede clinical symptoms, suggesting their utility as predictive biomarkers [65]. Moreover, circulating microbial metabolites and microbe-derived extracellular vesicles have been detected in blood and cerebrospinal fluid, enabling liquid biopsy approaches for longitudinal disease monitoring.

Advances in multi-omics technologies—such as metagenomics, metabolomics, and transcriptomics—combined with artificial intelligence-driven analytics, have enabled the development of robust predictive models based on microbiome data [66]. These models can stratify patients by disease risk, track therapeutic response, and guide personalized interventions. However, challenges remain in standardizing sample collection, controlling for confounding variables like diet and medication, and validating biomarkers across diverse populations. Future efforts should focus on harmonized protocols, longitudinal cohort studies, and integration of microbiome analytics into clinical decision-making frameworks to fully harness their diagnostic and prognostic potential.

14.7 Standardization Challenges in Microbiome Research

Despite the growing promise of microbiome-based interventions in NDD therapies, the field faces significant hurdles in achieving methodological standardization. Variability in sample collection, sequencing platforms, bioinformatics pipelines, and data interpretation has led to inconsistent findings across studies, limiting reproducibility and clinical translation. These discrepancies are particularly problematic when attempting to identify disease-specific microbial signatures or validate microbiome-derived biomarkers for early detection and monitoring.

One major challenge lies in the lack of harmonized protocols for microbiome sampling and processing [67]. Stool samples, commonly used for gut microbiome analysis, are highly sensitive to collection methods, storage conditions, and DNA extraction techniques. Even minor deviations can alter microbial composition and abundance profiles, complicating cross-study comparisons. Furthermore, differences in sequencing depth, primer selection for 16S rRNA gene amplification, and taxonomic classification algorithms introduce additional layers of variability. Shotgun metagenomics, while more comprehensive, is also subject to biases stemming from library preparation and reference database limitations. Standardization is further complicated by host-specific factors such as age, diet, medication, and comorbidities, which can confound microbiome data and obscure disease-related patterns. In neurodegenerative research, where patient cohorts often vary in disease stage and symptomatology, controlling for these variables is essential but challenging. Moreover, the dynamic nature of the gut microbiome—its responsiveness to environmental and physiological changes—necessitates longitudinal sampling and robust statistical frameworks to distinguish transient fluctuations from disease-relevant shifts [68].

To address these issues, international efforts are underway to develop consensus guidelines and reference standards. Initiatives such as the Human Microbiome Project and the NEuroBioStand consortium aim to establish validated protocols, certified reference materials, and traceable measurement systems for microbiome research in clinical contexts [69]. Incorporating metrological principles—such as uncertainty quantification, assay calibration, and inter-laboratory reproducibility—will be critical for integrating microbiome analytics into precision medicine. Ultimately, overcoming standardization challenges will pave the way for reliable, scalable, and clinically actionable microbiome-based diagnostics and therapeutics in neurodegenerative disease care.

14.7.1 Need for Longitudinal and Multi-Omics Studies

The gut microbiome is a dynamic ecosystem that fluctuates in response to host physiology, environmental exposures, and disease progression. In the context of NDDs, where pathogenesis unfolds over years or decades, cross-sectional microbiome studies offer only a limited snapshot of microbial states. To unravel causal relationships and temporal patterns, *longitudinal studies* are essential. These designs enable tracking of microbial trajectories, host-microbiome interactions, and therapeutic responses over time, providing insights into disease onset, progression, and remission phases [70].

Moreover, the complexity of host-microbiome interactions in NDDs demands a *multi-omics approach* that integrates metagenomics, metatranscriptomics, metaproteomics, metabolomics, and host-derived omics such as transcriptomics and epigenomics [71]. This holistic strategy allows researchers to move beyond taxonomic profiling and explore functional dynamics, microbial gene expression, metabolite production, and immune modulation. For example, combining gut microbial metabolomics with host neuroinflammatory markers can illuminate pathways linking microbial dysbiosis to neurodegeneration.

Despite their promise, longitudinal multi-omics studies face several challenges. These include high costs, data heterogeneity, missing time points, and the need for robust statistical models that account for compositionality, sparsity, and temporal correlation. Additionally, standardized protocols for sample collection, processing, and data integration are still evolving. Reference-independent approaches, such as de novo assembly and unsupervised clustering, are gaining traction for identifying novel microbial taxa and functions that may be missed by reference-based pipelines.

Recent initiatives, such as the Human Microbiome Project's integrative multi-omics arm and longitudinal cohorts like the iHMP and NEuroBioStand, are paving the way for more rigorous and reproducible studies [72]. These efforts underscore the importance of *temporal resolution and functional depth* in understanding the gut-brain axis and its role in personalized NDD therapies. Ultimately, the convergence of longitudinal sampling and multi-omics integration will be pivotal in identifying predictive biomarkers, stratifying patients, and tailoring microbiome-based interventions for neurodegenerative conditions.

14.7.2 Ethical Considerations in Microbiome-Targeted Therapies

As microbiome-targeted therapies gain traction in the personalized treatment of NDDs, ethical scrutiny becomes increasingly vital. These interventions—ranging from probiotics and prebiotics to FMT and live biotherapeutic products—raise complex questions around safety, consent, privacy, equity, and regulatory oversight. Unlike conventional pharmaceuticals, microbiome therapies involve living organisms with dynamic interactions, making their long-term effects difficult to predict and necessitating rigorous ethical frameworks [73].

One of the foremost concerns is *informed consent*, particularly in therapies involving donor-derived materials such as FMT [74]. Patients must be made aware of the potential risks, including transmission of antimicrobial resistance genes, unknown pathogens, and unpredictable host-microbiome interactions. Moreover, the concept of "ownership" of microbiome data and biological samples introduces novel challenges in biobanking and data governance. Given that microbiome profiles can reveal sensitive information about lifestyle, ancestry, and disease susceptibility, robust privacy protections and transparent data-sharing policies are essential.

Equity and access also pose ethical dilemmas. Personalized microbiome therapies may be prohibitively expensive or limited to populations with access to advanced sequencing technologies and clinical infrastructure [75]. This raises concerns about healthcare disparities and the potential for stratified medicine to widen the gap between resource-rich and underserved communities. Additionally, the uniqueness of each individual's microbiome complicates the development of universal standards, potentially excluding certain groups from clinical trials or therapeutic eligibility.

From a regulatory standpoint, microbiome-based interventions often fall into gray zones between food, drug, and biologic classifications. Ethical oversight must therefore extend beyond clinical trial design to include post-marketing surveillance, adverse event reporting, and long-term monitoring of microbiome dynamics.

Finally, the *principles of beneficence, non-maleficence, respect, and justice*—cornerstones of biomedical ethics—must be reinterpreted in light of microbiome science [75]. There is a significant need for interdisciplinary collaboration among microbiologists, ethicists, clinicians, and patient advocates to guide the responsible development and deployment of microbiome-targeted therapies in neurodegenerative disease care.

14.8 Conclusion

The emerging understanding of the gut-brain axis has fundamentally transformed our approach to neurodegenerative diseases, revealing the gut microbiome as a critical modulator of neurological health and disease progression. This comprehensive examination of the bidirectional communication between the gastrointestinal tract and central nervous system demonstrates that microbial dysbiosis is not merely a consequence of neurodegeneration but may serve as both a contributing factor and a therapeutic target in conditions such as Alzheimer's disease, Parkinson's disease, and amyotrophic lateral sclerosis.

The mechanistic insights gleaned from both preclinical and clinical studies underscore the multifaceted role of gut microbiota in influencing neuroinflammation, protein aggregation, and neurotransmitter balance through diverse pathways including vagal signaling, immune modulation, and metabolite production. The identification of disease-specific microbial signatures and the demonstration of therapeutic efficacy through interventions such as probiotics, prebiotics, dietary modifications, and fecal microbiota transplantation represent significant advances toward personalized medicine approaches in neurodegenerative disease management.

However, the field faces substantial challenges that must be addressed to fully realize the therapeutic potential of microbiome-based interventions. Issues of reproducibility, methodological standardization, and inter-study variability continue to hinder progress, while the complexity of host-microbiome interactions demands more sophisticated longitudinal and multi-omics approaches. The development of robust biomarkers for early detection and disease monitoring, though promising, requires further validation across diverse populations and clinical settings.

As we advance toward clinical implementation, ethical considerations surrounding informed consent, data privacy, therapeutic equity, and regulatory oversight must be carefully navigated. The unique nature of microbiome therapies, involving living organisms with dynamic and potentially unpredictable interactions, necessitates new frameworks for safety assessment and long-term monitoring. As microbiome profiling becomes more precise, future approaches are expected to combine host genetics, lifestyle factors, and microbiome information to create personalized interventions. To effectively translate research into practical clinical applications, it will be essential to develop strong translational frameworks and scalable models for delivering therapies and ensure that underrepresented groups are included in clinical trials.

The future of microbiome research in neurodegenerative diseases lies in the integration of standardized methodologies, comprehensive multi-omics platforms, and personalized therapeutic strategies. By addressing current limitations through collaborative international efforts, rigorous scientific standards, and ethical frameworks, the gut microbiome holds transformative promise as both a diagnostic tool and therapeutic frontier in the fight against neurodegenerative diseases. This emerging field represents a paradigm shift toward understanding neurodegeneration as a system-level disorder, where the health of the gut ecosystem is inextricably linked to the fate of the aging brain.

Declarations

Competing Interests The authors have no competing interests to declare that are relevant to the content of this manuscript.

Ethics Approval The authors declare that this manuscript does not include primary research with human participants.

The authors declare that this manuscript does not include primary studies involving animals.

References

1. Wang S, Jiang Y, Yang A, Meng F, Zhang J. The expanding burden of neurodegenerative diseases: an unmet medical and social need. Aging Dis. 2024. https://doi.org/10.14336/AD.2024.1071.
2. Bloomingdale P, Karelina T, Ramakrishnan V, Bakshi S, Véronneau-Veilleux F, Moye M, Sekiguchi K, Meno-Tetang G, Mohan A, Maithreye R, Thomas VA. Hallmarks of neurodegen-

erative disease: a systems pharmacology perspective. CPT Pharmacometrics Syst Pharmacol. 2022;11(11):1399–429.
3. Gao W, Jing S, He C, Saberi H, Sharma HS, Han F, Chen L. Advancements in neurodegenerative diseases: pathogenesis and novel neurorestorative interventions. J Neurorestoratology. 2025;13(2):100176.
4. Imam F, Saloner R, Vogel JW, Krish V, Abdel-Azim G, Ali M, An L, Anastasi F, Bennett D, Pichet Binette A, Boxer AL. The global neurodegeneration proteomics consortium: biomarker and drug target discovery for common neurodegenerative diseases and aging. Nat Med. 2025;15:1–1.
5. Cannon JR, Greenamyre JT. The role of environmental exposures in neurodegeneration and neurodegenerative diseases. Toxicol Sci. 2011;124(2):225–50.
6. Chen Y, Wang D, Liao Y, Zuo G, Lu X. The complex relationship between the microbiota-gut-brain axis and neurodegenerative diseases and future perspectives. Discov Med. 2024;1(1):139.
7. Yadav H, Bhardwaj R, Bakshi A, Gupta S, Maurya SK. Unveiling the role of gut-brain axis in regulating neurodegenerative diseases: a comprehensive review. Life Sci. 2023;330:122022.
8. Cavalcante DP, dos Santos Nunes AÍ, da Silva ER, de Carvalho GA, Chiareli RA, Oliveira-Lima OC, Ortiz-Leoncini G, Ulrich H, Gomez RS, Pinto MC. GlyT1 inhibition promotes neuroprotection in the middle cerebral artery occlusion model through the activation of GluN2A-containing NMDAR. Exp Neurol. 2025;383:115006.
9. Wang Y, Duan C, Du X, Zhu Y, Wang L, Hu J, Sun Y. Vagus nerve and gut-brain communication. Neuroscientist. 2025;31(3):262–78.
10. Rauf A, Badoni H, Abu-Izneid T, Olatunde A, Rahman MM, Painuli S, Semwal P, Wilairatana P, Mubarak MS. Neuroinflammatory markers: key indicators in the pathology of neurodegenerative diseases. Molecules. 2022;27(10):3194.
11. Silva YP, Bernardi A, Frozza RL. The role of short-chain fatty acids from gut microbiota in gut-brain communication. Front Endocrinol. 2020;11:508738.
12. Sahu N, Upadhyay P, Mishra SK. Role of short-chain fatty acids from gut microbiota in neuroendocrine pathogenesis management. In: Gut microbiome in neurological health and disorders. Singapore: Springer Nature Singapore; 2022. p. 139–51.
13. Parker A, Fonseca S, Carding SR. Gut microbes and metabolites as modulators of blood-brain barrier integrity and brain health. Gut Microbes. 2020;11(2):135–57.
14. Zhu S, Jiang Y, Xu K, Cui M, Ye W, Zhao G, Jin L, Chen X. The progress of gut microbiome research related to brain disorders. J Neuroinflammation. 2020;17(1):25.
15. Seo D, Holtzman DM. Current understanding of the Alzheimer's disease-associated microbiome and therapeutic strategies. Exp Mol Med. 2024;56:86–94. https://doi.org/10.1038/s12276-023-01146-2.
16. Vogt NM, Kerby RL, Dill-McFarland KA, et al. Gut microbiome alterations in Alzheimer's disease. Sci Rep. 2017;7:13537. https://doi.org/10.1038/s41598-017-13601-y.
17. Chapman MR, Robinson LS, Pinkner JS, et al. Role of Escherichia coli curli operons in directing amyloid fiber formation. Science. 2002;295(5556):851–5.
18. Friedland RP. Mechanisms of molecular mimicry involving the microbiota in neurodegeneration. J Alzheimer's Dis. 2015;45(2):349–62.
19. Zou X, Zou G, Zou X, Wang K, Chen Z. Gut microbiota and its metabolites in Alzheimer's disease: from pathogenesis to treatment. PeerJ. 2024;12:e17061.
20. Zhang X, Tang B, Guo J. Parkinson's disease and gut microbiota: from clinical to mechanistic and therapeutic studies. Transl Neurodegener. 2023;12(1):59.
21. Romano S, Savva GM, Bedarf JR, Charles IG, Hildebrand F, Narbad A. Meta-analysis of the Parkinson's disease gut microbiome suggests alterations linked to intestinal inflammation. NPJ Parkinsons Dis. 2021;7(1):27.
22. Du Y, Li Y, Xu X, Li R, Zhang M, Cui Y, Zhang L, Wei Z, Wang S, Tuo H. Probiotics for constipation and gut microbiota in Parkinson's disease. Parkinsonism Relat Disord. 2022;103:92–7.

23. Philip Mani A, Balasubramanian B, Mali LA, Joseph KS, Meyyazhagan A, Pappuswamy M, Joseph BV. The role of the gut microbiota in neurodegenerative diseases. Microbiol Res. 2024;15(2):489–507.
24. Noor Eddin A, Alfuwais M, Noor Eddin R, Alkattan K, Yaqinuddin A. Gut-modulating agents and amyotrophic lateral sclerosis: current evidence and future perspectives. Nutrients. 2024;16(5):590.
25. Fontdevila L, Povedano M, Domínguez R, Boada J, Serrano JC, Pamplona R, Ayala V, Portero-Otín M. Examining the complex interplay between gut microbiota abundance and short-chain fatty acid production in amyotrophic lateral sclerosis patients shortly after onset of disease. Sci Rep. 2024;14(1):23497.
26. Kaul M, Mukherjee D, Weiner HL, Cox LM. Gut microbiota immune cross-talk in amyotrophic lateral sclerosis. Neurotherapeutics. 2024;21(6):e00469.
27. Sun J, Zhang Y. Microbiome and micronutrient in ALS: from novel mechanisms to new treatments. Neurotherapeutics. 2024;21(6):e00441.
28. Lloyd-Price J, Mahurkar A, Rahnavard G, et al. Strains, functions and dynamics in the expanded human microbiome project. Nature. 2017;550(7674):61–6.
29. Sun J, Zhang Y. Microbiome and micronutrient in ALS: from novel mechanisms to new treatments. Neurotherapeutics. 2024;21(6):e00441. ISSN 1878-7479.
30. Roager HM, Licht TR. Microbial tryptophan catabolites in health and disease. Nat Commun. 2018;9(1):3294.
31. Bicknell B, Liebert A, Borody T, Herkes G, McLachlan C, Kiat H. Neurodegenerative and neurodevelopmental diseases and the gut-brain axis: the potential of therapeutic targeting of the microbiome. Int J Mol Sci. 2023;24(11):9577.
32. Oldstone MB. Molecular mimicry: its evolution from concept to mechanism as a cause of autoimmune disease. Monoclon Antib Immunodiagn Immunother. 2014;33(3):158–65.
33. Sospedra M, Martin R. Immunology of multiple sclerosis. Annu Rev Immunol. 2005;23:683–747.
34. de Vos WM, Tilg H, Van Hul M, Cani PD. Gut microbiome and health: mechanistic insights. Gut. 2022;71(5):1020–32.
35. Zeng MY, Inohara N, Nuñez G. Mechanisms of inflammation-driven bacterial dysbiosis in the gut. Mucosal Immunol. 2017;10(1):18–26.
36. Cao Y, Fanning S, Proos S, et al. Age-related alterations in the gut microbiome and host immunity impact vaccine efficacy. NPJ Vaccines. 2019;4:47.
37. Jain A, Madkan S, Patil P. The role of gut microbiota in neurodegenerative diseases: current insights and therapeutic implications. Cureus. 2023;15(10):e47861.
38. Schaible P, Henschel J, Erny D. How the gut microbiota impacts neurodegenerative diseases by modulating CNS immune cells. J Neuroinflammation. 2025;22(1):60.
39. Chandra S, Popovic J, Singhal NK, Watkins EA, Dodiya HB, Weigle IQ, Salvo MA, Ramakrishnan A, Chen Z, Watson T, Shetti A. The gut microbiome controls reactive astrocytosis during aβ amyloidosis via propionate-mediated regulation of IL-17. J Clin Invest. 2025;135(13):e180826.
40. Munoz-Pinto MF, Candeias E, Melo-Marques I, Esteves AR, Maranha A, Magalhães JD, Carneiro DR, Sant'Anna M, Pereira-Santos AR, Abreu AE, Nunes-Costa D. Gut-first Parkinson's disease is encoded by gut dysbiome. Mol Neurodegener. 2024;19(1):78.
41. Islam MM, Mahbub NU, Hong ST, Chung HJ. Gut bacteria: an etiological agent in human pathological conditions. Front Cell Infect Microbiol. 2024;14:1291148.
42. Kushwaha M, Chaudhary S, Singh AK, Makharia GK, Kumar A. Bacterial profiling of colorectal cancer biopsies: a culture-based study in Indian patients. Front Cell Infect Microbiol. 2025;15:1535477.
43. Park J, Lee YS, Lee DK, Hong J, Park SJ, Lee BW, Park SM, Lee HJ. Effects of a balanced diet and probiotics on blood biomarkers and gut microbiota in the elderly: a community-based intervention study. Nutrients. 2025;17(11):1933.
44. Palacios N, Gordon S, Wang T, Burk R, Qi Q, Huttenhower C, Gonzalez HM, Knight R, De Carli C, Daviglus M, Lamar M. Gut microbiome multi-omics and cognitive function in the

Hispanic community health study/study of Latinos-investigation of neurocognitive aging. J Alzheimers Dis. 2025;108(1):84–97. medRxiv 2024 2024–05.
45. Boktor JC, Adame MD, Rose DR, Schumann CM, Murray KD, Bauman MD, Careaga M, Mazmanian SK, Ashwood P, Needham BD. Global metabolic profiles in a non-human primate model of maternal immune activation: implications for neurodevelopmental disorders. Mol Psychiatry. 2022;27(12):4959–73.
46. Caiyun M, Hebao W, Wenhao Y, Changqing L, Changqing L, Xiaojiang Z. Dissecting causal links between gut microbiota, inflammatory cytokines, and Parkinson's disease: a mendelian randomization study. Brain Behav. 2024;14(11):e70169.
47. Cheng CK, Ye L, Zuo Y, Wang Y, Wang L, Li F, Chen S, Huang Y. Aged gut microbiome induces metabolic impairment and hallmarks of vascular and intestinal aging in young mice. Antioxidants. 2024;13(10):1250.
48. El Boukhari R, Matin M, Bouissane L, Ławiński M, Lushchak O, Singla RK, Mickael ME, Mayneris-Perxachs J, Grafakou ME, Xu S, Liu B. Enhancing human gut health: global innovations in dysbiosis management. Imeta. 2025;4(3):e70028.
49. McCoy KD, Geuking MB, Ronchi F. Gut microbiome standardization in control and experimental mice. Curr Protoc Immunol. 2017;117(1):23.
50. Bisanz JE, Upadhyay V, Turnbaugh JA, Ly K, Turnbaugh PJ. Meta-analysis reveals reproducible gut microbiome alterations in response to a high-fat diet. Cell Host Microbe. 2019;26(2):265–72.
51. Ebadpour N, Abavisani M, Sahebkar A. Microbiome-driven precision medicine: advancing drug development with pharmacomicrobiomics. J Drug Target. 2025;20:1–6.
52. Nagata N, Nishijima S, Miyoshi-Akiyama T, Kojima Y, Kimura M, Aoki R, Ohsugi M, Ueki K, Miki K, Iwata E, Hayakawa K. Population-level metagenomics uncovers distinct effects of multiple medications on the human gut microbiome. Gastroenterology. 2022;163(4):1038–52.
53. Vich Vila A, Collij V, Sanna S, Sinha T, Imhann F, Bourgonje AR, Mujagic Z, Jonkers DM, Masclee AA, Fu J, Kurilshikov A. Impact of commonly used drugs on the composition and metabolic function of the gut microbiota. Nat Commun. 2020;11(1):362.
54. Caruso R, Ono M, Bunker ME, Núñez G, Inohara N. Dynamic and asymmetric changes of the microbial communities after cohousing in laboratory mice. Cell Rep. 2019;27(11):3401–12.
55. Robertson SJ, Lemire P, Maughan H, Goethel A, Turpin W, Bedrani L, Guttman DS, Croitoru K, Girardin SE, Philpott DJ. Comparison of co-housing and littermate methods for microbiota standardization in mouse models. Cell Rep. 2019;27(6):1910–9.
56. Bezirtzoglou E, Plaza-Diaz J, Song J, Xie G, Stavropoulou E. Rodent model organisms: therapeutic treatments and drugs interaction with the gut microbiome. Front Microbiol. 2025;16:1581166.
57. Sánchez B, Delgado S, Blanco-Míguez A, Lourenço A, Gueimonde M, Margolles A. Probiotics, gut microbiota, and their influence on host health and disease. Mol Nutr Food Res. 2017;61(1):1600240.
58. Kumar MR, Azizi NF, Yeap SK, Abdullah JO, Khalid M, Omar AR, Osman MA, Leow AT, Mortadza SA, Alitheen NB. Clinical and preclinical studies of fermented foods and their effects on Alzheimer's disease. Antioxidants. 2022;11(5):883.
59. Kolida S, Tuohy K, Gibson GR. Prebiotic effects of inulin and oligofructose. Br J Nutr. 2002;87(S2):S193–7.
60. Rinninella E, Tohumcu E, Raoul P, Fiorani M, Cintoni M, Mele MC, Cammarota G, Gasbarrini A, Ianiro G. The role of diet in shaping human gut microbiota. Best Pract Res Clin Gastroenterol. 2023;62:101828.
61. D'Haens GR, Jobin C. Fecal microbial transplantation for diseases beyond recurrent clostridium difficile infection. Gastroenterology. 2019;157(3):624–36.
62. Su SH, Wu YF, Lin Q, Zhang L, Wang DP, Hai J. Fecal microbiota transplantation and replenishment of short-chain fatty acids protect against chronic cerebral hypoperfusion-induced colonic dysfunction by regulating gut microbiota, differentiation of Th17 cells, and mitochondrial energy metabolism. J Neuroinflammation. 2022;19(1):313.

63. Mkilima T. Engineering artificial microbial consortia for personalized gut microbiome modulation and disease treatment. Ann N Y Acad Sci. 2025;1548:29.
64. Nair AT, Ramachandran V, Joghee NM, Antony S, Ramalingam G. Gut microbiota dysfunction as reliable non-invasive early diagnostic biomarkers in the pathophysiology of Parkinson's disease: a critical review. J Neurogastroenterol Motil. 2018;24(1):30.
65. Yang J, Liang J, Hu N, He N, Liu B, Liu G, Qin Y. The gut microbiota modulates neuroinflammation in Alzheimer's disease: elucidating crucial factors and mechanistic underpinnings. CNS Neurosci Ther. 2024;30(10):e70091.
66. Ali H. Artificial intelligence in multi-omics data integration: advancing precision medicine, biomarker discovery and genomic-driven disease interventions. Int J Sci Res Arch. 2023;8(1):1012–30.
67. Amos GC, Logan A, Anwar S, Fritzsche M, Mate R, Bleazard T, Rijpkema S. Developing standards for the microbiome field. Microbiome. 2020;8(1):98.
68. LaGrow TJ. Exploration of spatiotemporal dynamics in neurodegenerative functional brain networks. [PhD dissertation]. Atlanta (GA): Georgia Institute of Technology. 2025.
69. Giangrande C, Vaneeckhoutte H, Delatour V, Vinh J, Verdier Y, Vanmechelen E, Koertvelyessy P, Blennow K, Zetterberg H, Gobom J. NEuroBioStand project: a European metrological research framework for standardising p-tau measurements in plasma. Alzheimers Dement. 2024;20:e092203.
70. Sherwani MK, Ruuskanen MO, Feldner-Busztin D, Firbas PN, Boza G, Móréh Á, Borman T, Erawijantari PP, Scheuring I, Gopalakrishnan S, Lahti L. Multi-omics time-series analysis in microbiome research: a systematic review. bioRxiv. 2025;26(5):bbaf502. 2025–07.
71. Ferrocino I, Rantsiou K, McClure R, Kostic T, De Souza RS, Lange L, FitzGerald J, Kriaa A, Cotter P, Maguin E, Schelkle B. The need for an integrated multi-OMICs approach in microbiome science in the food system. Compr Rev Food Sci Food Saf. 2023;22(2):1082–103.
72. Villette R, Ortís Sunyer J, Novikova PV, Aho VT, Petrov VA, Hickl O, Busi SB, De Rudder C, Kunath B, Heintz-Buschart A, Trezzi JP. Integrated multi-omics highlights alterations of gut microbiome functions in prodromal and idiopathic Parkinson's disease. bioRxiv. 2024;17:200. 2024–12.
73. Newman KL, Allegretti JR. Emerging noninfectious indications for live biotherapeutic products in gastroenterology. Am J Gastroenterol. 2024;119(1S):S30–5.
74. Merrick B, Allen L, Zain NM, Forbes B, Shawcross DL, Goldenberg SD. Regulation, risk and safety of faecal microbiota transplant. Infect Prev Pract. 2020;2(3):100069.
75. Mukherjee R, Chang CM. Exploring the ethical, legal, and societal implications of gut microbiome research in metabolic syndrome. In: Gut health and metabolic syndrome. Singapore: Springer; 2025. p. 195–211.

Gut Microbiome in Chronic Disorder Management

15

Sunil Kumar Patnaik, Amarjith Thiyyar Kandy, Essakimuthukumar Mariappan, Preeya Negi, Shrikant Shivkumar Nilewar, Nagaraju Bandaru, and Akey Krishna Swaroop

15.1 Introduction

The human body contains billions of bacteria referred to as microbiota, and its collective genome is known as the microbiome [1–3]. Microbes flourish in several regions of the human body, which involves the epidermis, respiratory system, and gastrointestinal (GI) tract (GIT), the latter being the most widely inhabited area. Numerous exogenous and endogenous variables, including nutrition, metabolism, age, geography, stress, and antibiotic treatment, influence the microbial community by disturbing the balance between commensals and pathogens. The gut microbiota plays a vital role in sustaining immunological function, metabolic equilibrium, vitamin synthesis, pathogen defence, promoting angiogenesis, and preserving the intestinal barrier [4]. Numerous parts of the human body, such as the skin, respiratory system, and gastrointestinal tract, are home to microbes, the latter being the most densely populated area. The gut microbiota is essential for regulating immunity, metabolism, and vitamin production while also defending against pathogens, stimulating blood vessel growth, and maintaining the intestinal barrier [1, 2, 5].

S. K. Patnaik · S. S. Nilewar · N. Bandaru
School of Pharmaceutical Sciences, Sandip University, Nashik, Maharashtra, India

A. T. Kandy · E. Mariappan
Department of Pharmaceutical Chemistry, JSS College of Pharmacy, JSS Academy of Higher Education & Research, Ooty, Tamil Nadu, India

P. Negi
Department of Pharmaceutical Sciences and Technology, Institute of Chemical Technology, Mumbai, Maharashtra, India

A. K. Swaroop (✉)
Department of Pharmaceutical Chemistry, SVKM's College of Pharmacy, Shirpur, Maharashtra, India

M. H. Rahman, Muhasina K M (eds.), *The Microbiome's Impact on the Brain*,
https://doi.org/10.1007/978-981-95-7494-0_15

Recently, microbiome-based therapeutics such as prebiotics, probiotics, and synbiotics are advocated for mental and metabolic health [6].

15.1.1 The Human Microbiome: An Overview

The collective genomes of trillions of microorganisms, including bacteria, viruses, fungi, and archaea, that live in various body parts—the gut being the most densely inhabited area—are referred to as the human microbiome [7]. These microbial communities exhibit remarkable diversity and complexity, influenced by host genetics, diet, environment, and age. The gut microbiota alone may consist of over 1000 species-level taxa, performing essential functions in digestion, immune modulation, vitamin synthesis, and protection against pathogens [8].

The composition of microbiota is not fixed; instead, it demonstrates adaptability during the course of life. A normal microbiome promotes health, but alterations known as dysbiosis are associated with several ailments, including inflammatory bowel disease, metabolic problems, allergies, and neuropsychiatric disorders [9]. Notably, emerging evidence links microbiome imbalances with systemic inflammation and altered neural signalling, suggesting relevance to neurodegenerative diseases (NDs) as well.

15.1.2 The Brain-Microbiome Axis: Historical Perspective

The complex collection of bacteria, viruses, fungi, and other microorganisms found in the human brain is known as the brain microbiome. Generally speaking, the brain region is thought to be sterile and germ-free. On the other hand, studies have shown that the brain has a vibrant and sophisticated microbial ecosystem [10, 11]. The brain-gut axis, a bilateral interaction channel between the gut microbiota and the brain, highlights the connections and effects of bacteria on brain function and health. The gut microbiota's immune compounds, neurotransmitters, and microbial metabolites can affect behaviour and brain function through a variety of signalling pathways [4, 12]. The brain can regulate the gut microbiota via neurological, endocrine, and immunological pathways [10]. Focusing on the microbiota and its regulation by the brain, modifications can be used for the treatment of neurological illnesses.

The gut-brain axis (GBA) refers to the connection between the gut bacteria and the brain [4, 13]. The notion of a gut-brain connection originates in early medicine, with anecdotal links between digestive health and mood found in Hippocratic writings. Nonetheless, the scientific examination of this connection started in earnest throughout the twentieth century, when research observed stress-induced alterations in gut motility and microbial composition [14, 15]. These findings suggested bidirectional connection between the central nervous system (CNS) and intestines, facilitated by neuronal, endocrine, immunological, and metabolic pathways. The GBA's direct anatomical and physiological linkages may involve the vagus nerve, hormonal signalling, the metabolism of particular chemicals like tryptophan (Trp), and the immune system, all of which correspond to essential routes for investigating the GBA microbiota [16, 17].

For the last 20 years, scientific exploration research has matured into the formal concept of the brain-gut-microbiome axis. Landmark studies have demonstrated that gut microbes can influence behaviour, stress response, and even neurodevelopment in animal models. For example, germ-free mice exhibit altered anxiety-like behaviour and neurochemical profiles, which can be partially reversed by microbial colonization [18]. Furthermore, faecal microbiota transplantation (FMT) experiments have shown that transferring microbiota from affected individuals (e.g., those with Parkinson's disease (PD)) into animal models can replicate disease-like phenotypes [19].

These discoveries have expanded the scope of neuroscience, leading to cross-disciplinary investigations of how microbial metabolites, immune signalling, and vagal nerve pathways influence brain health. The field has shifted from a one-way understanding of gut effects on the brain to a dynamic multidirectional network, emphasizing reciprocal regulation and feedback loops.

Researchers use high-throughput sequencing technologies, including metagenomics, metatranscriptomics, and 16S rRNA gene sequencing, to investigate the microbiome. These tools enable taxonomic profiling and functional characterization of microbial communities at unprecedented resolution [20]. Advancements in bioinformatics and systems biology now allow integration of microbiome data with host physiology, laying the groundwork for microbiome-based diagnostics and therapeutics. Bacterial deoxyribonucleic acid (DNA) has been found in the brain tissue by metagenomic analysis and high-throughput sequencing of DNA. The in situ tagging of bacterial peptidoglycan and the amplification of 16S ribosomal ribonucleic acid genes confirmed this [10].

The involvement of the microbiome in neurodegenerative illness is covered in the sections that follow:

15.1.2.1 Alzheimer's Disease

Table 15.1 summarizes microbial species detected in AD patients, their localization, and possible pathogenic mechanisms.

15.1.2.2 Parkinson's Disease

Patients with Parkinson's disease have been found to have microorganisms in their central nervous systems. *Streptococcus* is recurrently observed in the main cortex, whereas *Pseudomonas* in the medulla and *C. acnes* in different regions of CNS. Fungi such as *Botrytis*, *Candida*, *Fusarium*, and *Malassezia* were also observed in PD patients [25].

15.1.2.3 Amyotrophic Lateral Sclerosis

Several bacteria such as *C. acnes*, *Corynebacterium*, *Fusobacterium nucleatum*, *Lawsonella clevelandensis*, *Streptococcus thermophiles*, *Actinomycetales*, *Burkholderiales*, *Rhizobiales*, and *Xanthomonadales* were observed in ALS central nervous tissue [26].

Table 15.1 Reported bacterial and fungal associations with Alzheimer's disease and their potential effects

Bacteria	Site	Effect in AD	References
Chlamydia pneumoniae	DNA in astrocytes, microglia, and neurons adjacent to amyloid plaques and neurofibrillary tangles	Yet to be reported	[10]
Borrelia burgdorferi	Co-localizes with amyloid and tau aggregates	Triggers astrocytes and neurons to release pro-inflammatory cytokines including interleukins-6, 8, 10, interferon-γ, and tumour necrosis factor-α and modulates oxidative stress	[21, 22]
Helicobacter pylori	Anti-*H. pylori*-specific IgG levels in cerebrospinal fluid	Complement component 3 (C3) receptor signalling pathway-mediated OMV-induced interactions between astrocytes, glial cells, and neurons	[23]
Alternaria *Botrytis* *Candida* *Malassezia*	Frontal cortices	Yet to be reported	[24]

A significant amount of the fungal antigens of *C. albicans*, *Trichoderma viride*, *Cryptococcus magnus*, *Malassezia*, *Fusarium*, and *Botrytis* was reported in cerebrospinal fluid of ALS [27].

Both the bacterial and the fungal infiltrations are expected to be playing a role in the neuroinflammation, and the studies are yet be confirmed [26].

15.1.2.4 Multiple Sclerosis

Proteobacteria was observed in a white matter and is associated with to increased inflammatory markers [28]. Non-*albicans Candida* can reach the CNS [29].

15.1.2.5 Huntington's Disease

A broad range of microbes, such as *Escherichia coli*, *Cutibacterium*, and *Pseudomonas*. Patients with Huntington's disease (HD) have *Acinetobacter* and *Burkholderia* detected in their frontal cortices and striata. It was observed that fungi such as *Davidiella*, *Malassezia*, *Rhodotorula*, *Ramularia*, and *Candida* have contributed to the HD [30].

15.1.3 Rationale for a Paradigm Shift in Neurodegenerative Research

With the collective reported evidence of the microbiome in different regions of the brain, now the researchers have focused on targeting these organisms which use antibiotics to contribute to the neurodegenerative condition. Permeability to the blood-brain barrier (BBB), however, is crucial to the treatment. Drugs like metronidazole, fluroquinolones, and fluconazole are capable of crossing BBB and be a building bridge in the progression of research for targeting microbiome-based neurodegenerative disease therapy. Neurodegenerative illnesses are now treated with symptomatic measures. Studies on neurodegenerative diseases, such as Alzheimer's disease (AD), PD, and amyotrophic lateral sclerosis (ALS), typically concentrate on hereditary predisposition, protein aggregation, and neuronal pathology. While these approaches have yielded valuable insights, they have not translated into curative therapies, and the mechanisms underlying disease onset and progression remain incompletely understood [31]. Moreover, the narrow CNS-centric perspective often overlooks systemic contributors to neurodegeneration. An increasing amount of data endorses a paradigm shift that incorporates the gut microbiota as a crucial regulator of neurodegenerative mechanisms. Dysbiosis has been linked with increased intestinal permeability, systemic inflammation, and the production of neuroactive and neurotoxic metabolites all of which may influence the onset and progression of CNS disorders [16]. This paradigm reframes neurodegenerative diseases as multisystem disorders, where gut-brain interactions play a central role. This chapter aims to set the stage for exploring this evolving framework. First, it introduces the basic biology of the human microbiome and its systemic roles. It then traces the conceptual development of the brain-microbiome axis, highlighting pivotal discoveries. Finally, it argues for the necessity of integrative, microbiome-inclusive approaches to better understand and treat neurodegenerative diseases.

15.2 The Gut-Brain Axis: Mechanisms and Communication Pathways

A highly complex bidirectional communication network exists between the digestive system and the CNS, coordinating interactions that affect behaviour, emotion, and cognition alongside digestion. The bidirectional link between the brain and the stomach is regulated by this complex network, which crosses several overlapping axes of neuronal, immunological, endocrine, and metabolic signalling. Recent advances in microbiome research have brought attention to the impact of gut flora and metabolites on brain function, offering fresh insights into the pathophysiology of neuropsychiatric and neurodegenerative disorders. Determining these pathways of communication will be essential to developing therapeutic approaches that act upon the gut-brain pathway to promote mental health and overall well-being [32].

15.2.1 Neural Pathways: Vagus Nerve and Enteric Nervous System

The vagus nerve, a significant neural highway that connects the brain and the gut, is the most noticeable of the neural pathways. It originates in the brainstem and travels through the thorax and abdomen to innervate vital organs and supplies the majority of the gastrointestinal tract. This is around 80% afferent fibres (up to the brain from the gut) and 20% efferent fibres (down from the brain with motor signals to the gut). The vagal nerve is critical in the regulation of gastric motility, secretion, and inflammation. New studies have emphasized its involvement in transmitting the effects of probiotics and gut-derived metabolites on emotional and cognitive functions, and consequently vagal modulation in the therapy of depression and anxiety could be considered conflicting its use [33]. The enteric nervous system (ENS), sometimes referred to as the "second brain", assists. The enteric nervous system makes up the ENS with a giant network of about 100 million neurons distributed throughout the walls of the GI tract, primarily consisting of the myenteric and submucosal plexuses. This intrinsic network controls basic gut motor and secretory functions such as peristalsis, fluid secretion, and local blood flow in large measure independently of the CNS. The ENS communicates bidirectional with the CNS to provide information on the current status of the gut. The bidirectional conversation between the ENS and CNS allows the relevance of stress, feeding, and microbiota changes, illustrating the profound neuronal integration behind the gut-brain axis [34]. This interaction between the vagus nerve and the ENS serves as a functional scaffold for gut-brain communication. For example, gut hormones and microbial metabolites can activate vagal afferents, influencing neural circuits in the brain that control mood and emotional behaviour. By contrast, psychological stress can modify both vagal tone and ENS function, which in turn alters gut motility and permeability. The neural communication is therefore a fast, yet finely controlled, channel connecting gut physiology to higher brain functions [35].

Table 15.2 outlines key gut-brain communication mechanisms, including neural, immune, metabolic, and endocrine pathways, with their roles and effects.

Table 15.2 Major communication pathways of the gut-brain axis and their functional relevance

Pathway	Components	Function/role	Examples and effects	References
Neural pathways	Vagus nerve, enteric nervous system	Rapid bidirectional signalling; regulates motility, secretion, mood, and stress response	Vagal stimulation → reduced anxiety; ENS reflexes → peristalsis and digestion	[36]
Immune system interactions	Cytokines, GALT, microglia	Transmits peripheral immune signals; modulates neuroinflammation	LPS → systemic inflammation; elevated IL-6 in depression; microglial activation	[37]
Microbial metabolites and NTs	SCFAs, serotonin, GABA, tryptophan metabolites	Modulate neurotransmission, epigenetics, and inflammation	Butyrate → HDAC inhibition and neuroprotection; gut-derived serotonin affects mood	[38]
Endocrine signalling	Cortisol, gut hormones (e.g., ghrelin, CCK)	Stress and feeding-related modulation of brain activity	Ghrelin → stimulates appetite; cortisol → stress-related GI changes	[39]

15.2.2 Immune System Interactions

The immune system is essential for gut-brain connection and the advancement of neural processes. The largest immunological organ in the body is the gut-associated lymphoid tissue, which consists of mesenteric lymph nodes, isolated lymphoid follicles, and Peyer's patches. This mechanism consistently collects antigens from the gut lumen, coordinating tolerance to beneficial microorganisms and protection against infections. Significantly, immune mediators generated in the stomach can directly and indirectly affect brain function [40].

One of the basic techniques uses cytokines, which are small signalling proteins, TNF-α, interleukins, and interferons, few example. Under certain circumstances, these substances can pass through the BBB or communicate through the circumventricular organs, where the BBB is intrinsically permeable. Elevated pro-inflammatory cytokines have been linked to alterations in neurotransmitter metabolism, neuroendocrine function, and synaptic plasticity, potentially contributing to mood disorders, cognitive decline, and neurodegeneration. The idea that "neuroinflammation" is a common pathogenic pathway is supported by the observation that depression and Alzheimer's disease are associated with increased peripheral inflammation. The gut-brain axis's immunological component is also impacted by the microbiome. Beneficial bacteria may promote the production of the anti-inflammatory cytokine IL-10 and the proliferation of Treg cells. On the other hand, dysbiosis describes an imbalance of microbial populations that may weaken the intestinal barrier and allow bacterial substances like lipopolysaccharides (LPS) to enter the bloodstream. By activating Toll-like receptors on immune cells, which might affect CNS function, LPS causes systemic inflammation. An excellent example of how changes in gut microorganisms might result in immune-mediated signals to the brain is this inflammatory cascade of events [41]. Moreover, microglia, the immune cells of the central nervous system, respond to circulating immunological signals. Gut microbiome may influence microglial growth and function, hence impacting neuroinflammatory responses and synaptic plasticity. Such discoveries illustrate a complex conversation in which immune signalling acts as a molecular switchboard between gut wellness and brain health with immense implications for diseases as diverse as multiple sclerosis (MS), autism, and Parkinson's disease.

15.2.3 Microbial Metabolites and Neurotransmitter Modulation

The third and emerging path of communication includes the role of microbial metabolites on host neurotransmitter systems. The gut microbiota produces short-chain fatty acids (SCFAs) like acetate, propionate, and butyrate by fermenting food fibres and other substrates. This SCFA can act not only as a by-product of energy in colonocytes but also as a regulator of systemic immunity, a BBB integrity, and a neuronal activity. For example, butyrate has anti-inflammatory properties, upregulates neurotropic factor expression, and modulates histone deacetylase activity, thus potentially acting on gene expression in the CNS. In addition to SCFA production, gut bacteria directly influence the synthesis and breakdown of key neurotransmitters and neuromodulators. Specific species of *Lactobacillus* and *Bifidobacterium* can synthesize gamma-aminobutyric acid (GABA), the brain's primary inhibitory neurotransmitter. Similarly, other microbes influence the availability of serotonin (5-HT) by modulating tryptophan metabolism along the kynurenine pathway (KP). Considering that around 90% of the body's serotonin is synthesized in the gastrointestinal tract, changes in microbial composition can profoundly influence systemic serotonin levels, thereby impacting mood, appetite, and sleep [42]. Microbes also produce metabolites such as indoles from tryptophan, which may influence neuroinflammatory processes and behaviour by acting on CNS receptors and crossing the blood-brain barrier. Furthermore, bacterial peptidoglycans and cell wall components can modulate neuroimmune responses, illustrating how microbial-derived molecules impact CNS health beyond classical neurotransmitter pathways. Another important class of microbial metabolites includes bile acids and their derivatives. These substances have the ability to function as signalling agents via the Takeda G-protein-coupled receptor 5 and the farnesoid X receptor (FXR), influencing metabolic regulation, inflammation, and even cognitive processes. The gut microbiota thus emerges as a biochemical factory that shapes the host's neurochemical landscape, linking diet, microbial composition, and mental health in an intricate web of interactions [43]. Figure 15.1 illustrates how gut microbiota metabolites modulate

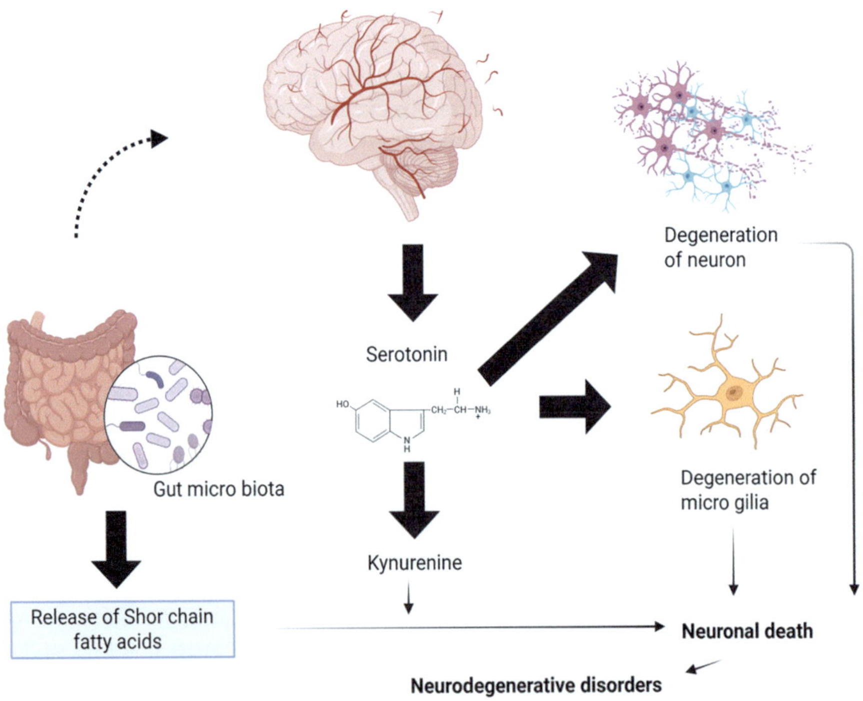

Fig. 15.1 Gut microbiota-derived metabolites influencing serotonin and kynurenine pathways in neurodegeneration

serotonin and kynurenine pathways, contributing to neuronal degeneration and neurodegenerative disorders.

It is clear from studying the intricate web of immunological, neurological, and metabolic communication pathways that make up the gut-brain axis that the health and makeup of the gut microbiota play a critical role in preserving the homeostasis of the central nervous system [44]. Through mechanisms such as vagal nerve signalling, modulation of neuroinflammatory pathways, and production of key metabolites like short-chain fatty acids, kynurenine, and serotonin, the gut microbiome dynamically influences neuronal health, synaptic plasticity, and even behaviour. Disruptions in this delicate balance whether due to environmental factors, aging, or chronic disease can initiate or exacerbate maladaptive immune responses, neuroinflammation, and oxidative stress within the brain. Increasing evidence now points to these microbiome alterations as significant contributors to the aetiology of neurodegenerative diseases, such as multiple sclerosis, Parkinson's disease, and Alzheimer's disease. Understanding how specific shifts in microbial diversity and function intersect with neural degeneration offers a promising frontier for uncovering novel therapeutic targets and diagnostic biomarkers in neurodegenerative disease research [45].

15.3 Microbiome Alterations in Neurodegenerative Diseases

The human body harbours roughly 100 trillion bacteria, predominantly located in the GIT. A microbial imbalance known as dysbiosis can result in increased intestinal permeability, systemic inflammation, and neuroinflammation [46]. The GBA is a reciprocal communication network that connects the brain and gastrointestinal system [47]. Dietary modification, probiotics, or FMT methods are employed to preserve microbial equilibrium [48]. Growing research suggests that imbalances in gut microbiota, known as gut dysbiosis, are closely associated with the onset and progression of various neurological disorders, including Alzheimer's, Parkinson's, multiple sclerosis, and others. These findings highlight the significance of the gut-brain axis in neurodegenerative disease mechanisms [46, 48].

15.3.1 Alzheimer's Disease and Dysbiosis

Alzheimer's disease is the most prevalent cause, making for 60–70% of all cases. Abnormal intracellular neurofibrillary tangle deposits made up of hyperphosphorylated protein and extracellular beta-amyloid (Aβ) plaques are indicators of AD, which causes diminished memory, behavioural problems, and a slow decline in cognitive function [49]. Significant focus has been given to the function of the gut microbiota and neuroimmune interactions, notably involving microglia, in the pathogenesis of AD [50].

Microglia are essential for sustaining brain homeostasis by detecting and eliminating damaged cells, protein aggregates (such as Aβ, tau, and alpha-synuclein (α-syn)), and pathogens. Microglia are the primary responders to Aβ plaques among glial cells. However, with age and under neurodegenerative stress, their phagocytic ability becomes dysregulated. This results in reduced clearance of deleterious proteins and unintended phagocytosis of stressed yet viable neurones, hence leading to neuronal death and cognitive dysfunction. Patients treated with sodium oligomannate, a prebiotic made from brown algae, showed a 2-point improvement in cognitive abilities on the 12-item Alzheimer's Disease Assessment Scale—Cognitive Subscale, according to a systematic multicentric study conducted in China on patients with Alzheimer's disease who were not taking medication [50]. In a study conducted on mice, probiotic treatment with *Lactobacillus* and *Bifidobacterium* improved Mini-Mental State Examination (MMSE) scores, glucose metabolism, and lipid profiles in AD patients. In another study, administration of *Lactobacillus fermentum*, *Lactobacillus helveticus*, or other probiotics restored spatial memory and decreased neuroinflammation in mice subjected to stress or antibiotic-induced dysbiosis. Notably, the elimination of *Helicobacter pylori*, a pathogen linked to amyloid and tau dysregulation, led to cognitive benefits in human AD patients [47].

Recent meta-analyses and systematic reviews have revealed consistent microbial signatures in AD patients, including increased abundance of pro-inflammatory *Proteobacteria* and a decrease in beneficial microbes such as *Anaerobacterium*, *Clostridiaceae*, *Eubacterium*, *Firmicutes*, *Lachnospiraceae*, *Odoribacter*, *Papillibacter*, and *Ruminococcus*. The decrease in these bacteria is linked with lower cognitive scores (e.g., MMSE, MoCA), increased gut permeability, and higher brain amyloid load [51]. FMT from AD patients into germ-free mice restored typical AD pathologies and pro-inflammatory microglial states, showing a causative connection between dysbiotic microbiota and AD progression [48]. Beyond Aβ, gut microbiota also influences tau pathology of AD. When P301S tau transgenic mice expressing human APOE4 were treated with antibiotics during early life, they demonstrated less brain shrinkage, less glial activation, and lower tau load. However, these benefits were reversed following FMT from conventionally bred APOE4 mice, showing the microbiota's direct impact on tauopathy [48].

AD patients with moderate cognitive impairment (aMCI) and dementia reveal different gut microbial patterns compared to healthy controls. Elevated levels of pro-inflammatory *Enterobacteriaceae*, including *E. coli*, were reported in both aMCI and AD groups. These bacteria create LPS and amyloid curli fibres, which may contribute to early disease progression by stimulating innate immune responses in the brain [51].

The idea of a "youth-like" microbiota is gaining popularity. Studies have revealed that FMT from young mice reverses age-associated microglial gene expression patterns, including downregulation of pro-inflammatory markers (C1qb, Dap12, Fcgr2b, Gpr84, Tlr13, and Trem2), and restoration of cognitive function as compared to FMT from older mice [48]. Dysbiosis, characterized by an overgrowth of pro-inflammatory bacteria and loss of SCFA makers, leads to systemic inflammation, neuroimmune dysregulation, and aggravation of amyloid and tau pathology. Microglia, functioning as key mediators between the peripheral microbiota and the CNS, are becoming recognized as both regulators and amplifiers of neurodegenerative processes. The importance of the gut-brain axis in the aetiology of Alzheimer's disease is consistently shown by the cumulative data from numerous mouse models.

15.3.2 Parkinson's Disease: Microbial Signatures and Pathology

PD is a progressive disorder of the nervous system that primarily results from the degeneration of nerve cells responsible for producing dopamine in a brain region called the substantia nigra. It is estimated to impact approximately 8.5 million individuals worldwide [49, 52]. Tremors, bradykinesia, stiffness, and postural instability are distinct motor symptoms that are caused by this neuronal loss. PD non-motor symptoms (NMS) can include gastrointestinal problems, cognitive decline, and emotional instability [46, 49]. One of the clinical characteristics of Parkinson's disease is the accumulation of Lewy bodies, which are aberrant clusters of the protein α-syn within neural cells [49].

Research has consistently proven GIT disturbances like constipation, malnutrition, and *H. pylori* infection as indicators of PD [47]. This correlates with observations of α-syn accumulation in the ENS. Animal studies indicate that truncal vagotomy inhibits the movement of α-syn from the GIT to the brain [47].

Elevated levels of LPS-producing bacteria such as *Bacteroides*, *Oscillospira*, *Lactobacillus*, *Bifidobacterium*, and *Akkermansia* have been documented in PD patients, alongside decreasing levels of butyrate-producing, anti-inflammatory genera such as *Blautia*, *Roseburia*, *Faecalibacterium*, *Prevotellaceae*, and *Coprococcus* [47, 52]. *Bilophila wadsworthia* and *Akkermansia muciniphila* form hydrogen sulphide, which disrupts the intestinal mucus barrier, enhances toxin absorption, and exacerbates both GI and motor symptoms. *Desulfovibrio* bacteria, also enhanced in PD, create LPS and magnetite nanoparticles that may pass the blood-brain barrier and exacerbate α-syn aggregation [51]. Constipation combined with a decline in *Prevotellaceae* can be a predictive biomarker for PD with over 90% specificity [51, 52]. The Unified Parkinson's Disease Rating Scale III indicates a positive correlation between the relative abundance of *Ruminococcaceae* and the severity of Parkinson's disease. These microbes proliferate as the disease progresses, providing a possible biomarker for Parkinson's disease staging [51]. α-Syn overexpressing (ASO) mice show more severe Parkinson's disease abnormalities compared to their germ-free and antibiotic-treated counterparts, indicating that the gut microbiome affects the motor impairments and microglial activation in ASO mice. In healthy mice, FMT from 1-methyl-4-phenyl-1,2,3,6-tetrahydropyridine (MPTP)-treated animals showed neurotransmitter depletion and motor deficits. In contrast, faecal microbiota transplantation from healthy mice improved gut dysbiosis and Parkinson's disease in MPTP-induced animals by addressing gut inflammation, glial activation, neurotransmitter abnormalities, and motor impairment. Only mice grown traditionally, not germ-free, showed signs of gastrointestinal dysfunction and motor complaints following prolonged exposure to rotenone. These studies shed light on the role of the gut-brain axis in the pathogenesis of Parkinson's disease

[48]. Several studies demonstrate that gut bacteria can alter the metabolism and efficacy of drugs like levodopa (L-dopa). *Dorea* and *Phascolarctobacterium* adversely correlate with L-dopa equivalent doses, while *Clostridium* cluster XI species like *Peptoniphilus* and *Finegoldia* may degrade L-dopa, lowering its therapeutic impact [51]. Changes in gut microbiota composition, particularly increases in pro-inflammatory, neurotoxic, and L-dopa-metabolizing bacteria, are substantially correlated with the symptoms and progression of Parkinson's disease. With new biomarkers and mechanistic clarity, the microbiome offers not only a lens into PD aetiology but a roadmap for future therapies.

15.3.3 Other Disorders: ALS, Huntington's Disease, and Multiple Sclerosis

15.3.3.1 Amyotrophic Lateral Sclerosis

Amyotrophic lateral sclerosis, also referred to as Lou Gehrig's disease, is a gradually worsening neurodegenerative disorder marked by the degeneration of motor neurons in the brainstem, motor cortex, and spinal cord. This neuronal loss results in progressive muscle weakness, loss of movement, and, ultimately, failure of respiratory function. ALS manifests in two basic forms: spontaneous (with no known aetiology) and familial (related to genetic alterations like SOD1 and C9orf72) [47, 49]. It presently affects over 200,000 people globally, with roughly 5000 new cases recorded each year. The disease proceeds through four clinical stages, beginning with muscular twitching and speech difficulties, rising to restricted movement and severe breathing complications, and ultimately terminating in total paralysis and end-of-life care [52]. Too far, barely ten research have addressed gut microbiota in individuals with ALS. Decreased levels of E-cadherin and zonula occludens (ZO)-1 result in intestinal wall injury. In the SOD1G93A mouse model of ALS, there was a reduction in butyrate-producing bacteria *Peptostreptococcus* and *Butyrivibrio fibrisolvens*, alongside an increase in *B. fibrisolvens*. Butyrate supplementation (2%) in drinking water retarded the progression of ALS. Similarly, ALS patient samples indicate lesser butyrate-producing bacteria like *Anaerostipes*, *Lachnospira*, and *Oscillibacter* and an increased presence of *Dorea*, a genus associated with ethanol production from glucose metabolism, causing toxicity. Elevated circulating LPS and neurotoxins like β-methylamino-L-alanine can potentially disrupt the BBB and induce neuroinflammation in ALS [47]. Studies reveal increased *Bacteroidetes* and *Escherichia*, while decreased *Firmicutes*, specifically a lower *Firmicutes/Bacteroidetes* ratio, which may serve as a dysbiosis marker. Butyrate-producing bacteria such as *Eubacterium rectale* and *Roseburia intestinalis* are in lower abundance, suggesting a potential mechanism by which SCFA shortage may promote systemic inflammation via NF-κB activation and poor T-cell control [51]. Some researchers reported no significant differences in the relative abundance of the faecal microbiota between ALS patients and control subjects, while certain researchers observed a change in abundance of *Ruminococcaceae*, *Faecalibacterium*, *Prevotella*, and *Clostridium* [51, 52].

The role of the gut microbiota in genetically related ALS has also been examined in C9orf72-mutant mice. Gut microbiota composition strongly influenced illness severity, with FMT from mice with anti-inflammatory microbiota boosting survival and lowering neuroinflammation. Similarly, the beneficial bacterium *Akkermansia muciniphila* was reported to increase motor function and lifespan in SOD1G93A mice. However, the role of antibiotics remains contentious. While some studies claim favourable effects by lowering inflammation and SOD1 aggregation, others suggest that long-term antibiotic use worsens illness by generating dysbiosis and increasing motor neuron death [48].

15.3.3.2 Huntington's Disease

Approximately 2.7 occurrences of HD, an autosomal dominant neurological disorder, occur worldwide for every 100,000 people. Typically emerging between the ages of 35 and 44, HD is defined by three key clinical features: cognitive decline, mental problems, and motor dysfunction. One of the most severe effects of motor dysfunction is dysphagia, which leads to weight loss and aspiration and can eventually be deadly [47]. HD is caused by an expansion of the cytosine-adenine-guanine repeat in the huntingtin (HTT) gene by causing the mutant huntingtin protein to misfold and aggregate in neurons, astrocytes, and microglia. Apart from the typical neurological symptoms, gastrointestinal problems including diarrhoea, nutritional deficits, and unexpected weight loss are also important. Emerging studies indicate a potential link between gut microbiota and the onset of Huntington's disease. Changes in the composition and diversity of gut microorganisms observed in both human patients and animal models suggest that the GBA may play a role in the disease's progression [48].

Sex-based variations in gut microbiota composition have been reported in HD mice models. Male HD mice demonstrated increased levels of *Bacteroidales* and *Lactobacillales*, while female HD mice revealed higher *Bacteroidales*, *Burkholderiales*, *Coriobacteriales*, and *Erysipelotrichales*. Both male and female HD mice reportedly had lower *Clostridiales* [47]. This sexual dimorphism extended to research on FMT, where R6/1 male mice demonstrated increased resistance to FMT engraftment and no cognitive enhancement, unlike R6/1 female counterparts [48]. Microbiota-depleted mice revealed a loss in mature oligodendrocytes and myelin proteins in the prefrontal cortex, affecting white matter plasticity and callosal myelination. This shows that gut microbiota may be crucial for maintaining central nervous system integrity and that its reduction can exacerbate HD symptoms [47]. In a study, R6/1 and R6/2 HD mice models reported to have higher abundance of *Bacteroidetes* and lower abundance of *Firmicutes* than the wild-type mice, which may reflect inflammation and increased permeability. However, research on HD patients showed reduced abundance of *Akkermansiaceae*, *Firmicutes*, and *Lachnospiraceae* bacteria typically involved in maintaining gut barrier integrity and anti-inflammatory activity through butyrate production [51]. Collectively, the rising role of the gut microbiome in HD pathogenesis potentially afford options for early detection and microbiota-targeted treatments.

15.3.3.3 Multiple Sclerosis

MS is a long-term autoimmune neurological condition that primarily affects young adults, with an estimated global prevalence of 1.8 million individuals. The disease is classified into four clinical forms: relapsing-remitting MS (RRMS), progressive relapsing MS (PRMS), primary progressive MS (PPMS), and secondary progressive MS (SPMS). RRMS is the predominant kind, comprising 85% of patients, and is distinguished by complete recovery following relapses. SPMS is marked by progressive deterioration following relapses, and PPMS is defined by a continuous decline devoid of identifiable relapses, whereas PRMS features slow decline accompanied by intermittent relapses. The pathophysiology of MS entails chronic inflammation inside the CNS, although its aetiology remains incompletely elucidated. The elevation of *Bacteroides* and *Akkermansia* is documented in multiple sclerosis and may affect immunological function and inflammation. The reduction of *Faecalibacterium*, *Prevotella*, *Ruminococcus*, and *Escherichia* indicates a disturbance in the equilibrium of normal gut microbiota [52]. *Collinsella* has been proven to increase gut permeability and promote pro-inflammatory responses, ultimately aggravating MS symptoms. Similarly, *Prevotella* is implicated in various inflammatory illnesses, such as rheumatoid arthritis, by activating pro-inflammatory cytokines. A decrease in *Eggerthella* is a possible biomarker for autoimmune diseases like MS and rheumatoid arthritis. These findings further underline the role of gut microbial composition in MS aetiology. Overall, the gut microbiome appears to play a role in progression and MS development. Targeting certain microbial populations may give promising treatment methods, boosting immunological control, lowering inflammation, and even influencing the course of the disease [51].

15.3.3.4 Creutzfeldt-Jakob Disease

Creutzfeldt-Jakob disease (CJD) is a rare neurological disorder caused by the abnormal folding of prion proteins [53]. CJD patients had elevated abundance of *Actinomycetota* and *Fusobacteriota* phyla, specifically *Bifidobacteriaceae* and *Fusobacteriaceae*. *Actinomycetia* is known to produce SCFAs, which are toxic in excess, just like low butyrate concentrations decrease gut permeability, but high concentrations can trigger apoptosis in gut epithelial cells, demonstrating involvement of SCFAs in NDs like CJD [51]. Altered gut microbiota not only contributes to disease progression but may also act as a significant source of diagnostic and prognostic indicators [51]. A promising therapeutic approach for managing NDs is to target the gut microbiota. The stability of Gut Microbiota (GM) depends on microbial diversity, host-microbe interactions, and microbe-microbe interactions. Therefore, the ultimate aim of microbiome modulation is to preserve or restore this equilibrium depending on an individual's microbial composition. To achieve this, there is a rising need for the standardization and industrial-scale development of tools such as engineered probiotics, gut-derived metabolite mimics, and synthetic microbiota for FMT [54].

15.4 Novel Insights: Microbial Metabolites and Neuroinflammation

15.4.1 Short-Chain Fatty Acids and Brain Health

SCFAs are saturated aliphatic organic acids that contain fewer than six carbon atoms these primarily consist of butyrate (C4), propionate (C3), and acetate (C2). Their principal by-products arise from the anaerobic fermentation of dietary fibres and resistant starches by intestinal bacteria in the large intestine. The processes behind their formation are complex, encompassing diverse microbial populations, several metabolic pathways, and advanced cross-feeding interactions among bacteria. These molecules are important metabolites produced by gut microbiota and play a vital role in maintaining overall health, with a notable impact on the nervous system [55, 56].

15.4.1.1 Production of SCFA

15.4.1.1.1 Substrate Availability and Initial Hydrolysis

The principal substrates for short-chain fatty acid production are complex carbohydrates, including dietary fibres, resistant starches, and certain oligosaccharides that evade digestion in the upper gastrointestinal tract. Hydrolysis: Several carbohydrate-active enzymes (CAZymes), including glycoside hydrolases and polysaccharide lyases, are released by specialized gut bacteria in the colon. These enzymes hydrolyse complex polysaccharides to produce oligosaccharides and monosaccharides [57]. Following hydrolysis, the resultant monosaccharides (e.g., glucose, xylose, arabinose) are taken up by bacteria and channelled into core metabolic pathways, primarily glycolysis and the pentose phosphate pathway, producing pyruvate as a critical intermediary [58].

15.4.1.1.2 Acetate Production

Bacteroides, *Prevotella*, and *Bifidobacterium* are among the most prominent acetate makers. Most gut bacteria make acetate via the acetyl-CoA pathway. Acetyl-CoA synthetase or the phosphotransacetylase-acetate kinase pathways are used to convert pyruvate to acetyl-CoA, which is further transformed into acetate. Acetate serves as a substrate for other bacteria, particularly those that produce butyrate.

15.4.1.1.3 Propionic Acid Production

Three main bacterial processes create propionate: *Bacteroides* species frequently use the succinate pathway, which changes pyruvate into succinate and then propionate. Lactate serves as an intermediary in the acrylate pathway, which is present in some *Clostridium* species. Propanediol pathway produces propionate as a bridge by using deoxy sugars (such fucose and rhamnose) as substrates [59].

Table 15.3 Major SCFAs, their biosynthetic pathways, and key microbial producers

SCFA	Main pathways	Key microbial producers
Acetate	Acetyl-CoA pathway, Wood-Ljungdahl	*Bacteroides*, *Bifidobacterium*, *Prevotella*
Propionate	Succinate, acrylate, propanediol	*Bacteroides*, *Veillonella*, *Propionibacterium*
Butyrate	Butyryl-CoA/acetate CoA-transferase	*Faecalibacterium*, *Roseburia*, *Eubacterium*

15.4.1.1.4 Butyrate Production

The most prevalent bacteria in the butyrate synthesis pathway are *Faecalibacterium prausnitzii*, *Roseburia* spp., *Eubacterium rectale*, and other *Firmicutes* phyla. The butyryl-CoA/acetate CoA-transferase pathway is the most widely used pathway. Pyruvate is first metabolized to acetyl-CoA, then to butyryl-CoA, and lastly to butyrate, with acetate serving as a co-substrate [60].

The generation of SCFAs in the human gut is a complex, multi-step process involving various microbial communities, substrate-specific enzyme reactions, and extensive metabolic networks. Substrate availability, microbiota type, and environmental circumstances all influence the amount and profile of SCFAs generated. These metabolites, in turn, play important roles in gut health and overall physiology. At the physiological level, they help protect neurons by lowering inflammation, maintaining barrier integrity, and improving synaptic function. However, excessive or imbalanced SCFA levels, particularly propionate, can be neurotoxic, emphasizing the need of a healthy gut flora and diet for neurological health [61]. Table 15.3 summarizes the primary SCFAs, their metabolic pathways, and predominant gut microbial producers.

15.4.1.2 Neuroprotective Effects of SCFAs

By controlling microglia activation and blocking pro-inflammatory cytokines like IL-1β and IL-6, SCFAs reduce neuroinflammation [61]. Support for the blood-brain barrier: They encourage the synthesis of tight junction proteins that support the integrity of the blood-brain barrier, including ZO-1, occludin. Neuroprotection in disease models: Animal studies indicate that SCFAs can slow neuronal degeneration, minimize behavioural deficits, and improve outcomes in models of sepsis-associated encephalopathy and stroke [62]. Also, SCFAs boost the expression of synaptic plasticity-related genes (e.g., synaptophysin, TrkB), which may aid in healing from brain injury. Microglial modulation: SCFAs can improve the removal of neurotoxic chemicals by reprogramming microglia from a pro-inflammatory (M1) to an anti-inflammatory (M2) state [63].

15.4.1.3 Neurotoxic Effects of SCFAs

While SCFAs are typically beneficial, some situations or amounts may result in neurotoxic effects. Elevated SCFA levels: High systemic SCFA concentrations, as observed in gut dysbiosis or pathological conditions, can affect neuronal lipid metabolism, mitochondrial function, and ATP generation, resulting in increased neuronal death [64]. Behavioural impairments: Animal studies have connected high SCFA exposure to memory, learning, locomotor activity, and anxiety-related behaviours [65]. Among SCFAs, propionate has been explicitly identified as having possible neurotoxic effects, whereas butyrate and acetate are more commonly neuroprotective [66]. Cellular stress: High amounts of SCFAs can cause mitochondrial fragmentation, disrupt cell cycle progression, and increase neuronal cell death. At the physiological level, they help protect neurons by lowering inflammation, maintaining barrier integrity, and improving synaptic function. However, excessive or imbalanced SCFA levels, particularly propionate, can be neurotoxic, emphasizing the need of a healthy gut flora and diet for neurological health [67].

15.4.1.4 SCFAs and Cognitive Function

Several research in AD mouse models have shown that SCFA supplementation, particularly sodium butyrate, can improve cognitive impairment. Mechanisms include increased memory consolidation genes (e.g., BDNF, CREB, SNAP25), improved synaptic plasticity, and decreased neuroinflammation [68]. SCFA treatment in animal models of SAE decreased behavioural deficits, neuronal degeneration, and pro-inflammatory cytokine levels (IL-1β, IL-6). SCFAs also increased the expression of tight junction proteins, promoting blood-brain barrier integrity and reducing neuroinflammatory signalling. SCFA treatment in rat models of chronic cerebral hypoperfusion increased synaptic plasticity, avoided neuronal and synaptic loss, and reduced cognitive impairments. These benefits were associated with mitochondrial metabolic reprogramming and decreased microglial activation. Table 15.2 shows the effects of SCFA relationship with cognitive functions.

Table 15.4 summarizes experimental and clinical evidence linking SCFAs to cognitive outcomes, underlying mechanisms, and neuroprotective or neurotoxic effects. Current evidence, primarily from animal studies, supports a dual role for SCFAs in cognitive function. At physiological levels, SCFAs especially butyrate exerts neuroprotective effects by enhancing synaptic plasticity, reducing neuroinflammation, and supporting the blood-brain barrier. However, excessive SCFA

Table 15.4 Effects of SCFAs on cognitive functions in various models

Study model	Key findings	Mechanisms involved	References
AD mouse models	Improved cognition, synaptic plasticity	HDAC inhibition, BDNF/CREB upregulation	[68]
SAE mouse models	Reduced cognitive impairment, neuroinflammation	Cytokine suppression, BBB support	[67]
Chronic hypoperfusion rats	Ameliorated cognitive deficits, synaptic loss	Mitochondrial reprogramming, microglia	[69]
Human cohort (AD biomarkers)	Altered SCFA profiles linked to cognitive decline	Not fully elucidated	[70]
High SCFA exposure (animals)	Neurotoxicity, behavioural deficits	Mitochondrial dysfunction, apoptosis	[65]

levels or imbalances may contribute to neurotoxicity and cognitive impairment. Human data are emerging, but further large-scale clinical studies are needed to clarify the therapeutic potential and safety of modulating SCFAs for cognitive health.

15.4.2 Tryptophan Metabolism and Neuroactive Compounds

Tryptophan metabolism is a tightly controlled physiological process that produces a wide range of neuroactive chemicals with significant impacts on brain function and behaviour. Tryptophan, an important amino acid received from the diet, functions not only as a building block for protein synthesis but also as a precursor for several critical metabolic pathways within the central nervous system and peripheral organs [71]. The two principal metabolic routes for tryptophan are kynurenine pathway and serotonin synthesis [72].

15.4.2.1 Kynurenine Pathway

In mammals, the KP serves as the main route for breaking down the essential amino acid Trp, accounting for over 90–95% of its dietary metabolism. Beyond regulating tryptophan levels, this pathway produces diverse bioactive compounds that are essential for immune regulation, neural activity, and maintaining redox equilibrium [73]. Kynurenine is a key metabolite that can be further processed in two primary ways: the neuroprotective branch and the neurotoxic branch. Kynurenine aminotransferase (KAT) is responsible for the production of kynurenic acid (KynA) in the neuroprotective pathway. KynA exerts neuroprotective effects by inhibiting N-methyl-D-aspartate (NMDA) receptors and α7-nicotinic acetylcholine receptors [74]. The neurotoxic branch produces quinolinic acid (QA), 3-hydroxykynurenine (3-HK), and 3-hydroxyanthranilic acid (3HAA). QA is a strong NMDA receptor agonist with neurotoxic and pro-inflammatory effects. The process eventually leads to the synthesis of NAD+ [74]. 3-HK also contributes to oxidative damage. The balance between these branches is tightly regulated and is influenced by inflammatory mediators, which can shift metabolism towards the neurotoxic arm during immune activation or neuroinflammation [75]. The kynurenine pathway is a key component in tryptophan metabolism, integrating immunological, metabolic, and neurological signals. Its metabolites have various and sometimes contradictory effects, and their dysregulation is increasingly recognized as a factor to the aetiology of multiple disorders.

15.4.2.2 Serotonin Pathway

5-HT production requires a much smaller amount of tryptophan (approximately 1–2%) [76]. Serotonin is a neurotransmitter principally produced in the brainstem's raphe nuclei and the gastrointestinal tract. Its synthesis includes two major stages: hydroxylation and decarboxylation. Tryptophan, a crucial amino acid, is hydroxylated by the enzyme tryptophan hydroxylase to yield 5-hydroxytryptophan. Aromatic L-amino acid decarboxylase decarboxylates 5-hydroxytryptophan to produce serotonin [71]. Serotonin controls mood, cognition, anxiety, learning, memory, sleep,

Table 15.5 Alterations in serotonin synthesis and signalling across neurological disorders

Disorder	Serotonin synthesis/ signalling change	Clinical implications
Depression	↓ Synthesis, ↑ reuptake	Mood disturbance, SSRI efficacy [78]
Schizophrenia	Receptor/signalling abnormalities	Cognitive, mood, and psychotic symptoms [79]
Autism spectrum disorder	Altered synthesis, SERT mutations	Neurodevelopmental and behavioural symptoms [80]
Alzheimer's disease	Loss of neurons, ↓ synthesis	Cognitive decline, mood/behavioural changes [81]
Migraine	↓ Synthesis	Headache, pain modulation [82]
Parkinson's disease	Altered synthesis/signalling	NMS, tremor [82]

and pain perception. As indicated in Table 15.3, variations in its synthesis, release, or signalling have a significant effect on brain function and are linked to several neurological and neuropsychiatric disorders [77]. Table 15.5 summarizes changes in serotonin pathways in various neurological disorders and their clinical implications.

Tryptophan metabolism, especially through the kynurenine route, produces a narrow range of neuroactive metabolites directly connected to neurodegenerative disorders. A healthy brain requires a balance of neuroprotective and neurotoxic metabolites [83]. Under normal conditions, metabolites such as KynA provide neuroprotection by antagonizing excitatory glutamate receptors, thereby preventing excitotoxicity. However, during chronic inflammation or immunological activation, the route switches towards increased neurotoxic chemical synthesis, including quinolinic acid (QUIN) and 3-hydroxykynurenine (3-HK). QUIN is a strong NMDA receptor agonist that causes excitotoxicity, oxidative stress, and neuronal death, whereas 3-HK produces reactive oxygen species, which contribute to neuronal damage as shown in Fig. 15.1 [84]. This imbalance is shown in Alzheimer's, Parkinson's, and Huntington's illnesses, when neurotoxic metabolites are raised and neuroprotective metabolites are decreased. Furthermore, increased flux through the kynurenine pathway can reduce tryptophan availability for serotonin synthesis, aggravating mood and cognitive symptoms common in neurodegenerative illnesses [85]. Thus, the change in tryptophan metabolism towards neurotoxic metabolites is an important element in the development and progression of neurodegenerative disease. Figure 15.2 illustrates the kynurenine pathway of tryptophan metabolism, highlighting major metabolites, enzymes, and their roles in neurological function.

Table 15.6 summarizes major tryptophan metabolites, their neuroactive effects, contribution to neurodegeneration, and associated neurological disorders.

15.4.3 Microbiome-Induced Modulation of Neuroinflammation

The gut microbiota influences neuroinflammation via an intricate network of immunological, metabolic, and neurological mechanisms. The microbiota-gut-brain axis

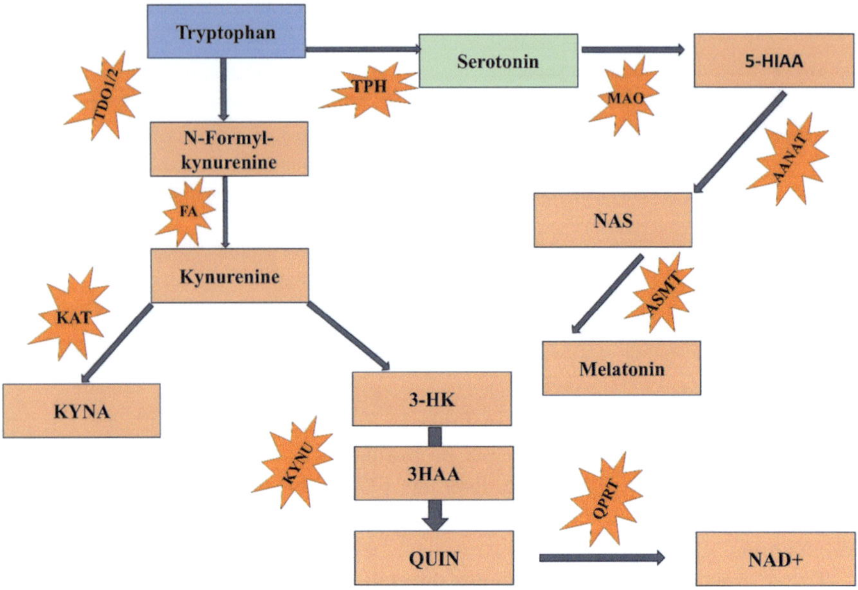

Fig. 15.2 The kynurenine route for tryptophan metabolism. The primary metabolite of the KP is L-kynurenine. The degradation of L-tryptophan by IDO and TDO enzymes is the rate-limiting step in L-kynurenine production. IDO is activated by inflammatory stimuli, whereas TDO is naturally produced in the liver. Kynurenine can be converted into 3-HK, KYNA, or QUIN by different enzymes in the brain, liver, or extrahepatic areas. KAT, QUIN (quinolinic acid), 3-HAA (3-hydroxy anthranilic acid), 3-HK (3-hydroxy kynurenine), NAD (nicotinamide adenine dinucleotide), KynA (kynurenic acid), NAS (N-acetyl serotonin), 5-HIAA (5-hydroxy indoleacetic acid), IDO (indoleamine dioxygenase), TDO (tryptophan dioxygenase), and KMO (kynurenine monooxygenase)

(MGBA), a network of bilateral interactions between the gut and the central nervous system, is largely responsible for this interaction [89].

15.4.3.1 How Microbial Signals Trigger Neuroinflammation

15.4.3.1.1 Molecular Mimicry and Immune Activation

It is possible for microbial compounds to mimic host proteins. Certain peptides from gut bacteria have the ability to activate auto-reactive T lymphocytes. Neuroinflammation may result from these cells' subsequent migration to the central nervous system. This has been observed in animal models of multiple sclerosis, such as experimental autoimmune encephalomyelitis, where exposure to bacterial antigens triggers T-cell activation, leading to inflammation within the central nervous system [90].

15.4.3.1.2 Bacterial Amyloids and Protein Aggregation

Curli, a kind of bacterial amyloid protein, can cause host proteins like α-synuclein and amyloid-beta to aggregate more easily. These aggregates can move from the gut

Table 15.6 Neuroactive tryptophan metabolites, their roles in neurodegeneration, and disease associations

Metabolite	Neuroactive effect	Role in neurodegeneration	Disease associations	References
Kynurenic acid	Neuroprotective	NMDA receptor antagonism, anti-excitotoxic	Reduced in AD, PD, HD	[86]
Quinolinic acid	Neurotoxic	NMDA receptor agonism, excitotoxicity, oxidative stress	Elevated in AD, PD, HD	[84]
3-Hydroxykynurenin	Neurotoxic	ROS generation, oxidative damage	Increased in AD, HD	[87]
Serotonin	Neuroprotective	Mood/cognition, depleted by KP activation	Impaired in depression, dementia	[88]

into the brain, increasing neuroinflammation and neurodegeneration, particularly in conditions like Parkinson's disease [90].

15.4.3.1.3 Microbial Infections and Immune Receptors

Microbial infections can directly engage CNS immune receptors (e.g., TLRs, NODs), activating microglia and astrocytes, and leading to neuroinflammatory responses. Even non-gut microbiomes may influence CNS immune signalling and neuroinflammation [91].

15.4.3.1.4 Dysbiosis and the Blood-Brain Barrier

Gut microbiota dysbiosis can compromise the integrity of the blood-brain barrier, enabling the infiltration of immune cells and microbial components into the central nervous system, which may trigger inflammation [92].

15.4.3.2 How Microbial Signals Dampen Neuroinflammation?

Beneficial gut bacteria degrade dietary fibre to create SCFAs like butyrate and propionate, which can reduce inflammation. These metabolites limit pro-inflammatory cytokine production, boost regulatory immunological responses, and help maintain BBB integrity, hence lowering neuroinflammation [92].

15.4.3.2.1 Tryptophan Metabolism and AHR Activation

Tryptophan is converted into indole derivatives by the gut bacteria, which subsequently stimulate the aryl hydrocarbon receptor (AHR) on microglia and astrocytes. AHR activation reduces central nervous system inflammation and NF-κB signalling in multiple sclerosis models [93].

15.4.3.2.2 Vagus Nerve Stimulation

Certain microbial signals can directly stimulate the vagus nerve, which in turn activates anti-inflammatory pathways in the brain, reducing neuroinflammation [94].

15.4.3.2.3 Promotion of Anti-inflammatory Glial Phenotypes

The microbiome can induce anti-inflammatory astrocyte and microglial phenotypes, further limiting CNS inflammation and supporting neural health [95]. Table 15.7 summarizes microbial metabolites and signals that modulate neuroinflammation through various mechanisms and pathways.

Microbial signals can both provoke and suppress neuroinflammation by interacting with immune cells, producing metabolites, influencing glial cell activity, and modulating barrier integrity and neural pathways. The balance between these opposing effects is crucial for CNS health and is a promising target for therapeutic intervention in neuroinflammatory and neurodegenerative diseases [97].

Table 15.7 Microbial signals influencing neuroinflammation: triggers and dampening mechanisms

Microbial signal/ metabolite	Effect on neuroinflammation	Mechanism/pathway	References
Molecular mimicry (bacterial peptides)	Trigger	Activates auto-reactive T cells	[95]
Bacterial amyloids (curli)	Trigger	Promote protein aggregation, microglial activation	[90]
SCFAs (butyrate, propionate)	Dampen	Suppress cytokines, maintain BBB, modulate microglia	[92]
Indole derivatives (tryptophan metabolites)	Dampen	AHR activation, inhibit NF-κB, suppress astrocyte/microglial inflammation	[96]
Dysbiosis	Trigger	Increases BBB permeability, immune cell trafficking	[97]
Vagus nerve stimulation	Dampen	Activates anti-inflammatory CNS pathways	[95]

15.4.3.3 Therapeutic Implication

Tryptophan is metabolized by intestinal microorganisms into indole derivatives, which in turn trigger the aryl hydrocarbon receptor on microglia and astrocytes. AHR activation reduces central nervous system inflammation and NF-κB signalling in multiple sclerosis models [98]. Preclinical research suggest that FMT can improve gut microbial diversity and reduce neuroinflammation. Its application outside of gastrointestinal problems is still experimental but is being investigated for neurological ailments [99]. Preclinical research suggest that FMT can improve gut microbial diversity and reduce neuroinflammation. Its application outside of gastrointestinal problems is still experimental but is being investigated for neurological ailments [100].

15.5 Therapeutic Strategies Targeting the Microbiome

Research on the microbiome has progressed rapidly over the past two decades, revealing the numerous ways in which these symbiotic microorganisms affect our daily life. Microorganisms have existed on Earth for billions of years and are present in nearly every natural ecosystem. The human microbiome, which includes bacteria, viruses, and fungus, is a varied collection of commensal microorganisms found in the human body. The gastrointestinal tract represents the most vast and diverse microbial ecosystem within the human body. Over 90% of the body's microbiota reside in the GI tract, particularly in the colon. It affects digestion, immunity, metabolism, and notably cognitive function via the gut-brain axis. The gut microbiota interacts with the brain through nerves, hormones, immunological signals, and chemicals, affecting our emotions, cognition, and responses [101].

15.5.1 Probiotics, Prebiotics, and Synbiotics

15.5.1.1 Probiotics

Differences exist in both the composition and activity of the gut microbiota between healthy individuals and those affected by various neurological disorders. Dietary habits are known to influence the makeup of the gut microbiome, thereby impacting the function of the GBA. Among the therapeutic approaches used to address gut microbiota imbalances are probiotics, which help restore the balance of intestinal microflora and potentially enhance neurological health outcomes. According to the World Health Organization, probiotics are defined as live microorganisms that, when consumed in sufficient quantities, provide health benefits to the host [102]. The term "probiotic" was introduced in 1974. Since then, their use as dietary supplements and inclusion in daily nutrition has steadily increased. The majority of probiotics belong to the *Bifidobacterium* genus and lactic acid bacteria, such as *Lactobacillus*. Increasing evidence suggests that the metabolites generated by these probiotics serve as key mediators in the dietary interactions between the human host and gut microbes. Furthermore, certain gut-dwelling bacterial genera including *Bacteroides*, *Clostridium*, *Bifidobacterium*, *Peptostreptococcus*, *Lactobacillus*, and *Ruminococcus* have demonstrated the ability to produce a variety of tryptophan-derived compounds, including indole, 3-methylindole, indolcacetic acid, and tryptamine. Numerous animal studies, particularly in mice, have shown that probiotic therapy can enhance cognitive performance in several neurological conditions such as AD, epilepsy, and autism spectrum disorder (ASD). However, clinical research in humans remains limited regarding the therapeutic efficacy of probiotics for neurological disorders. One study involving children with ASD aged 3–12 years who experienced anxiety and gastrointestinal issues reported that treatment with Visbiome, a probiotic blend containing eight strains, primarily *Lactobacillus* was safe and improved overall well-being in patients retaining *Lactobacillus*. In addition, probiotic interventions have shown encouraging outcomes in human neurodegenerative diseases like Alzheimer's disease. A recent investigation using a genetic mouse model of Parkinson's disease found that prolonged administration of six probiotic strains led to reduced motor dysfunction and provided neuroprotective effects on dopamine-producing neurons [103].

15.5.1.2 Prebiotics

The International Scientific Association for Probiotics and Prebiotics defines a prebiotic as a non-living dietary component that offers health benefits by supporting beneficial microbial communities within the host. Prebiotics can act as an alternative approach to probiotic supplementation for modulating gut microbial ecosystems. This category includes compounds such as soluble and fermentable fibres, human milk oligosaccharides, and nondigestible oligosaccharides, all of which play a significant role in promoting gastrointestinal health [105]. Few research have looked at how prebiotics affect the gut microbiota of both people and animals, despite the possibility that they can increase good bacteria like lactobacilli and bifidobacteria. Studies on the effects of fructo- and galacto-oligosaccharides, either alone or in combination, in male mice have demonstrated that these substances have anxiolytic and antidepressant properties in addition to the ability to lessen the negative consequences of long-term stress. An 8-week course of N-acetylcysteine reduced irritability and repetitive behaviours in infants with ASD, according to a placebo-controlled clinical research. Additionally, children with ASD exhibited behavioural improvements when administered B-GOS (Bimuno), a commercially available prebiotic, alongside a controlled diet. This positive outcome was likely associated with elevated levels of beneficial bacteria such as *Lactobacillus* and *Bifidobacterium*. Recent research also indicates that the prebiotic lactulose may alleviate cognitive impairments in Alzheimer's disease animal models by triggering anti-inflammatory responses and promoting autophagy. These findings suggest that both prebiotics and probiotics hold potential as therapeutic options for neurological disorders. Nevertheless, since a correlation does not establish causation, further studies are required to fully understand the biological mechanisms involved.

15.5.1.3 Synbiotics

Synbiotics refer to a synergistic blend of probiotics and prebiotics, where the prebiotics enhance the population of beneficial microbes in the gut and promote the growth and activity of probiotic strains, thereby increasing their effectiveness and health benefits. For probiotic bacteria to remain viable within the gastrointestinal environment, it is essential that the formulation of synbiotics is appropriately designed. Synbiotics are more beneficial than probiotics or prebiotics taken separately, according to studies. The findings show that in major depressive disorder, a synbiotic formulation containing GOS and a multistrain probiotic with *Lactobacillus helveticus* and *Bifidobacterium longum* reduced depressed symptoms and enhanced tryptophan signalling. A group of Parkinson's disease patients' functional gastrointestinal symptoms improved as a result of treatment using a synbiotic that included multistrain probiotics and a prebiotic, according to the results of a randomized controlled experiment. It has been shown that using *Bifidobacterium infantis* as a synbiotic with oligosaccharides can help people with ASD with their gastrointestinal problems. To investigate the impact of synbiotics on the MGBA, more research is required [104].

Table 15.8 outlines definitions and roles of probiotics, prebiotics, and synbiotics in supporting gut microbial balance.

15.5.2 Faecal Microbiota Transplantation

In order to re-establish a normal gut flora, a procedure known as FMT is performed. This procedure involves implanting faecal matter from a healthy donor into the gastrointestinal tract of a patient. This has emerged as a possible treatment option for microbiota-related illnesses, such as recurrent *Clostridium difficile* infection (rCDI) [105]. Figure 15.3 illustrates the primary therapeutic objectives of FMT, including restoration of microbial diversity, enhanced colonization resistance, immune modulation, and metabolite production.

Table 15.8 Key differences between probiotics, prebiotics, and synbiotics in microbiome modulation

Term	Definition	How it helps the microbiome
Probiotics	Liver beneficial bacteria	Add good microbes to improve balance
Prebiotics	Fiber that feeds good microbes	Nourish and boost the existing good bacteria
Synbiotics	Combo of probiotics + prebiotics	Support both introduced and native microbes

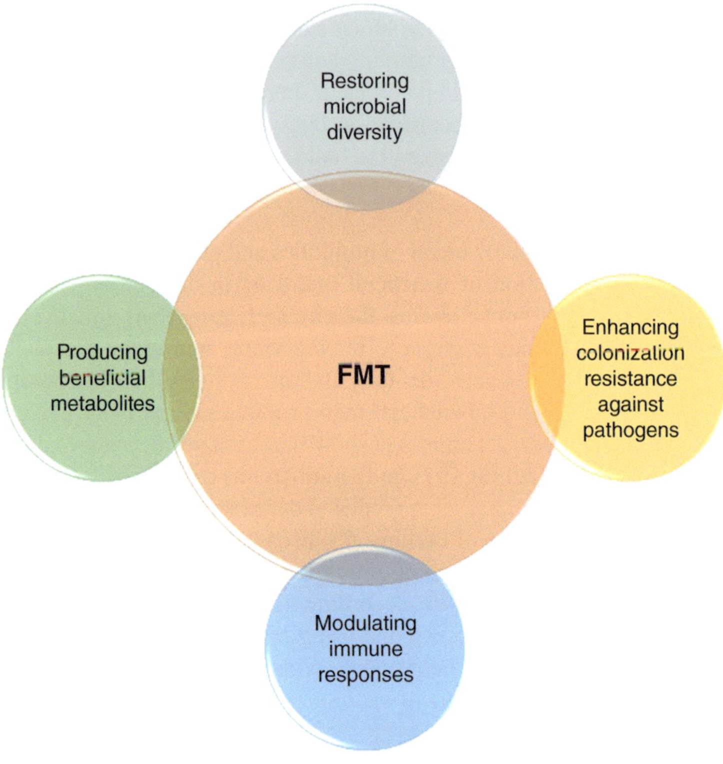

Fig. 15.3 Therapeutic goals of faecal microbiota transplantation

15.5.2.1 FMT Procedures

FMT procedures involve critical steps like donor screening, sample processing, and administration. Donor selection is challenging due to strict screening, with only 4.5–10% qualifying, and international protocols vary due to regional differences. Sample preparation requires anaerobic conditions to maintain bacterial viability and can be fresh, frozen, or encapsulated. FMT is administered via lower (e.g., colonoscopy) or upper (e.g., oral capsules) GI routes, with colonoscopy showing higher success rates (91–95%) in treating CDI. Recipient preparation includes gut microbiota depletion for better engraftment. Standardization efforts aim to improve safety, efficacy, and accessibility of FMT globally. FMT is generally considered safe, especially for recurrent *Clostridioides difficile* infection (CDI). Ongoing monitoring is essential to assess long-term safety and effectiveness across patient groups. Abdominal pain is the most often reported side event attributed to FMT. Additional moderate and self-resolving unpleasant effects encompass diarrhoea, stomach discomfort, and gas. Peri-procedural problems, like faecal regurgitation or vomiting, may arise, underscoring the necessity for vigilant monitoring to identify high-risk patients and select the optimal administration strategy. Serious adverse events (SAEs) associated with FMT are few, exhibiting pooled rates below 1% in certain assessments. Nonetheless, significant adverse effects, such as infections and mortality, have been documented:

- Infection transmission: Severe infections, including fatal cases of drug-resistant *E. coli*, have arisen via FMT despite donor screening, underscoring the necessity for more rigorous screening for resistant pathogens.
- Immunocompromised patients: Severely immunocompromised individuals may exhibit an elevated risk for SAEs, including systemic inflammatory response syndrome or fever, potentially associated with faecal microbiota transplantation. Meticulous selection of this patient cohort for FMT is essential.
- Inflammatory bowel disease (IBD) flares: In individuals with IBD with concurrent CDI, IBD flares have been documented as a potential adverse event following FMT; however certain studies indicate no exacerbation of IBD [106].

15.5.2.2 Present Evidence for FMT in Neurodegenerative Disease Models

Scientists are studying if FMT can help brain diseases by improving the connection between the gut and the brain. Most of the current research is based on animal studies.

Alzheimer's Disease

In studies on mice with Alzheimer's-like conditions, FMT has shown positive signs:

- Enhanced cognitive function and reduced damage: FMT appears to enhance short-term memory in these animals. It also aids in diminishing detrimental protein aggregates (known as amyloid plaques) and tangles in the brain, which are characteristic features of Alzheimer's disease.
- Decreased cerebral oedema: FMT may help mitigate inflammation in the brain, which is believed to contribute to Alzheimer's disease. The donor's age may influence the efficacy of FMT [107].

Parkinson's Disease

FMT is also being looked at for Parkinson's disease, which is linked to an imbalance of gut bacteria:

- Enhanced symptoms: Preliminary human studies on Parkinson's disease indicate that FMT may alleviate both motor dysfunction and non-motor concerns such as constipation and mood fluctuations.
- Enhanced gut health, enhanced cognitive function: In murine models of Parkinson's disease, faecal microbiota transplantation rectified the dysbiosis of gut bacteria. This resulted in enhanced gastrointestinal function and greater mobility. It also decreased inflammation in both the gastrointestinal tract and the brain [108].

15.5.3 Diet, Lifestyle, and Microbiome Modulation

Different diets, like the Mediterranean and ketogenic diets, significantly change the types and amounts of bacteria in your gut, which affects your overall health.

15.5.3.1 Mediterranean Diet

The Mediterranean diet is frequently advantageous for gastrointestinal health. It increases the abundance and variety of beneficial bacteria, including *Lactobacillus* and *Bifidobacterium*, and raises levels of favourable compounds known as SCFAs [109]. This diet can enhance general health, including improved blood sugar regulation and reduced inflammation. Nevertheless, several studies yield inconclusive findings, indicating that further study is necessary to comprehensively comprehend all its consequences [110].

15.5.3.2 Ketogenic Diet

Your gut flora is altered in several ways by the ketogenic diet, which is low in carbohydrates and rich in fat. It may diminish certain advantageous bacteria, such as *Bifidobacterium*, and may lower essential SCFAs. This diet may elevate some detrimental microorganisms. Although it is utilized for illnesses such as epilepsy, its long-term implications for gut health remain under investigation [111].

15.5.3.3 Exercise and the Gut Microbiome

Exercise is becoming recognized as a lifestyle element that positively influences the gut flora. Consistent physical exercise correlates with a more favourable microbial composition and can enhance overall health outcomes [112].

15.5.3.4 Impact on Microbial Diversity and Composition

Physical activity often enhances gut bacteria diversity, an indicator of optimal gut health. Athletes have more variety and distinct microbiota in comparison to sedentary persons. In elderly individuals, physical activity may diminish detrimental microorganisms and enhance advantageous ones. Nonetheless, outcomes in healthy individuals fluctuate, particularly for specific taxa such as *Firmicutes*, *Bacteroidetes*, and *Proteobacteria* probably attributable to variations in research methodologies and exercise regimens [111].

15.5.3.5 Production of Beneficial Metabolites

Physical exercise is associated with increases in beneficial microbial metabolites like SCFAs, such as butyrate, propionate, and acetate. These short-chain fatty acids are essential for gut health because they improve mitochondrial activity and mucin formation, which help regulate the intestinal barrier [113].

15.5.3.6 Exercise Intensity and Type

The type and level of exercise determine how physical activity affects the gut flora. Moderate exercise improves gut health by reducing inflammation and enhancing gut barrier function. Conversely, extended high-intensity exercise may adversely affect the gastrointestinal tract by increasing permeability and inflammation. Various exercise modalities exhibit distinct effects—cardiorespiratory exercise induced transient alterations in the microbiota, but resistance training showed no notable influence [112].

15.5.3.7 Sleep and the Gut Microbiome

Sleep patterns and circadian rhythms exhibit a substantial, bidirectional interaction with the gut microbiome, facilitated by the microbiota-gut-brain axis.

Impact of Sleep Disturbances on Microbiota Sleep disruptions and circadian misalignment can alter gut microbiota composition and metabolism, impacting microbial diversity and metabolite synthesis (e.g., SCFAs, bile acids). These alterations may facilitate chronic inflammation and weight accumulation. Even transient alterations in sleep-wake cycles can modify microbial activities and interactions, albeit consistent microbial abundances [113].

Microbiota's Influence on Sleep The gut microbiota influences sleep by synthesizing bioactive chemicals such as neurotransmitters (e.g., serotonin, melatonin) and modulating metabolism and immunity. A diet abundant in plants facilitates the synthesis of molecules that promote sleep. Probiotics may enhance sleep quality; nevertheless, further consistent data and objective assessments are required [114].

15.5.3.8 Other Lifestyle Factors

In addition to exercise and food, several environmental and lifestyle variables influence the gut microbiota, with considerable consequences for health and illness.

Stress Management By reducing helpful bacteria like *Lactobacillus* and *Bifidobacterium*, stress can have a negative impact on the gut microbiota. Prolonged stress sets off the sympathetic nervous system and the hypothalamic-pituitary-adrenal axis, which results in immunological dysregulation, microbial imbalance, and inflammation. They may result in stress-related illnesses. Although more research is needed, meditation can enhance intestinal health and assist manage stress [115].

Environmental Exposures Environmental exposures such as toxins, metals, and biological agents can disturb the gut microbiome, influencing microbial composition and functionality. These alterations may affect the body's toxin processing and effect general health. Urban air pollution can alter gut microbiota and immunological responses. Preserving microbial equilibrium (eubiosis) may safeguard against these consequences, but dysbiosis can exacerbate chronic illnesses.

15.6 Cutting-Edge Technologies and Future Directions

Microbiome-brain research has changed drastically in the last 5 years. A complete profile of the responses of both microbes and hosts can be created using newer multi-omics methods. Clinical research and diagnosis are made more accurate and precise using artificial intelligence (AI) and predictive modelling. These new technologies have made it possible to create personalized microbiome-based treatments that could change the way we treat neurodegenerative diseases.

15.6.1 Multi-Omics Approaches in Microbiome Research

Multi-omics combines genomics, transcriptomics, proteomics, and metabolomics to make biological snapshots that show many different aspects of life [116–118]. This approach is very useful in microbiome research because different microbial communities can have a big effect on how the nervous system of host works.

Genomics, and especially metagenomics, evaluates the DNA of microbes to Fig. 15.4 out the composition and the function of the microbe [119]. Transcriptomics (also called meta-transcriptomics) evaluates the expression of genes in both microbes and host cells to find out which genes and pathways are active [120]. Proteomics assess the proteins that are made from these genes, which helps understand the functional machinery that controls interactions between microbes and hosts [121]. Small molecules that are indicators and effectors in neuroimmune and neuroinflammatory cascades, including short-chain fatty acids, bile acids, and neurotransmitter derivatives, are measured using metabolomics [122].

Multi-omics studies of human cohorts have found connections between how gut microbes work and how the brain works. A comprehensive study that used 16S sequencing, metagenomics, metabolomics, functional magnetic resonance imaging, and plasma cytokine profiling found that gut-brain signatures linked to memory and hippocampal volume [123]. In studies with rodents, combining microbiome and metabolome data found certain microbial metabolites that aggravate the neuroinflammation after being exposed to air pollution [120]. Researchers found that changes in the gut microbiome and metabolism in older non-human primates were linked to cognitive decline and problems with brain mitochondria [121].

These data layers can be combined using network analysis and systems biology frameworks. Multi-layer weighted gene co-expression network analysis is one of the techniques to detect modules that are linked to aging and neurodegeneration by

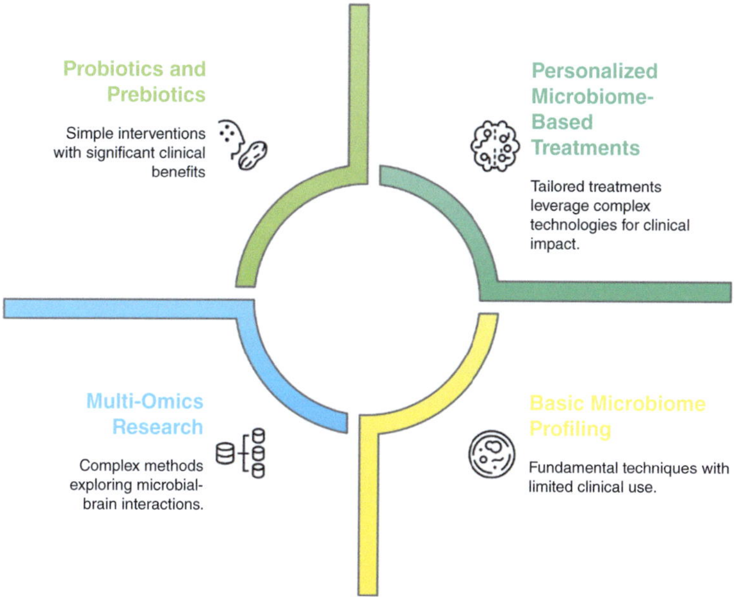

Fig. 15.4 Advancements in microbiome-brain research

looking at how metagenomic taxa are related to brain pathology-associated metabolites and host transcripts [124, 125]. Researchers have also used deep learning models to find omics signatures that can predict Alzheimer's disease in different groups of people [126].

Researchers are now able to map microbe-host interactions at the tissue level in novel ways single-cell and spatially resolved omics. DBiT-seq and Slide-seq combine spatial transcriptomics with proteomic and lipidomic mapping to find out what microbes do in the gut-brain axis at specific sites [127–129]. Even though they are still in the early stages, these technologies hold the promise of making the structural-functional links between host-microbiome interactions in neurodegenerative processes clearer.

15.6.2 Artificial Intelligence and Predictive Modelling

AI and machine learning (ML) are very important for finding useful patterns in complicated, high-dimensional microbiome datasets. Deep learning architectures can classify microbes, make functional inferences, and predict diseases on a remarkable scale [130–132].

Being able to predict AD has been a major achievement. A model that used clinical and cognitive data was able to predict AD with about 72% accuracy up to 7 years before a diagnosis in a UCSF cohort [133]. Another AI framework that looked at speech patterns, imaging, and fluid biomarkers achieved about 78% accuracy right when it came to early MCI-to-AD progression prediction [134].

Advanced AI models are now looking into how to combine microbiome and clinical data. Alzheimer's Disease Analysis Model Generation 1 (ADAM 1) is a retrieval-augmented large language model that uses microbiome profiles, patient histories, and carefully chosen literature to make predictions that are as good as those made by traditional XGBoost in small sample settings [135]. According to broader AI evaluations, more and more people are modelling the course of diseases and the effects of therapy using generative adversarial networks, digital twin modelling, variational autoencoders, recurrent neural networks, and graph neural networks [136].

Using metabolomic, metagenomic, and clinical data, AI-enabled multi-omics integration has found prognostic biomarkers for PD and ALS [137–139]. SHapley Additive exPlanations and Local Interpretable Model-agnostic Explanations are two examples of explainable AI methods that are used to understand how models make decisions and make them more acceptable in clinical settings [140–142]. Privacy-preserving methods, such as federated learning, have also been tested. These methods let AI work in a decentralized way without putting sensitive data at risk [143].

The problems still persist, such as small and unbalanced datasets, AI model transparency is still not complete, data heterogeneity between studies makes standardization harder, and clinical validation is often not thorough enough. Future work must put a lot of emphasis on collaborations between multiple centres, standardized protocols, and data governance that is ethically sound.

15.6.3 Personalized Medicine and Microbiome-Based Interventions

The use of multi-omics and AI in clinical systems has made it possible to test new microbiome-based treatments for neurodegenerative diseases. Probiotics, prebiotics, synbiotics, live biotherapeutic products (LBPs) are all examples of these personalized strategies.

A controlled intervention in PD using prebiotic fibre that was tailored to the baseline microbiome composition led to multiple helpful microbes, higher levels of short-chain fatty acids, lower levels of inflammatory base markers, and lower levels of proteins related to neurodegeneration [144]. A multicentre randomized FMT trial in PD showed that motor function and constipation got better, especially in responders who had certain changes in their microbial communities [145].

Personalized nutrition based on microbiome and metabolomic data for a person with PD has led to long-term improvements in motor severity and quality of life [146]. Precision-engineered LBPs, like an anaerobic microbe made to boost short-chain fatty acid production, are now being tested in early-phase clinical trials [147].

There are still problems with personalized microbiome therapies. For example, it is hard to choose a donor for FMT, the ecological stability of the microbiome after treatment is hard to predict, engineering live microbes raises safety concerns, regulatory frameworks are inconsistent, and there are still ethical issues around consent, ownership, and fair access [148–150].

15.6.4 Future Directions

This multi-omics approaches, AI, and personalized microbiome-targeted treatments are revolutionizing the way we do neurodegenerative research. However, this area must address several challenges and strategic targets that will define the trajectory of the upcoming research efforts for exploiting their full potential. First, it is imperative to note that this has not driven down multi-omics platforms standardization to the same level that of Ilumina. Because different persons prepare samples, gather data, standardize data, and analyse data in different ways, it can be challenging to compare research and replicate results. Establishing standardized protocols, common data ontologies and repositories open to all will be critical in achieving reliable and reproducible results. Second, the new technologies in spatiotemporal and single-cell omics are likely going to provide us with some critical new insights. However new technologies for assessing the expression of genes at the level of single cells have made it possible to do so. SpimuX includes MERFISH, Slide-seq, and spatial metabolomics, which allow scientists to track how genes and metabolites are distributed in tissue networks at the single cell level. Third, interpretable AI is very relevant to the utilization of predictive models in the clinical setting. As AI tools get more complicated, it is increasingly important that they be interpretable so they can get regulatory approval, establish trust with doctors and patients, and make sure no one is put in danger when the machine is turned on.

15.7 Conclusion

The investigation of the gut microbiome's impact on brain health signifies a fundamental transformation in our comprehension of neurodegenerative diseases, transitioning from a predominantly CNS-focused perspective to a comprehensive, multi-omics approach that recognizes the significant interrelation among the gut, immune system, and brain. In the last 20 years, progress in high-throughput sequencing, multi-omics integration, and ML has allowed researchers to find microbial signatures, uncover new metabolites, and elucidate intricate signalling pathways connecting gut and brain homeostasis. Evidence that gut dysbiosis, whether brought on by diet, lifestyle, aging, or environmental factors, can alter metabolic and neurotransmitter pathways, trigger and worsen neuroinflammatory cascades, and ultimately aid in the aetiology and progression of diseases like AD, PD, ALS, HD, MS, and CJD is compiled in this manuscript. The gut-brain axis is recognized as a dynamic, reciprocal communication network that includes immunological (cytokines, microglia), neurological (vagus nerve, enteric nervous system), endocrine, and metabolic pathways. Depending on their balance and context, microbial metabolites, especially SCFAs and compounds derived from tryptophan, can have both neuroprotective and neurotoxic effects. For instance, butyrate and other SCFAs can reduce neuroinflammation, preserve the integrity of the blood-brain barrier, and enhance synaptic plasticity. However, excessive or disproportionate amounts of SCFAs, especially propionate, can cause neuronal stress and cognitive impairments.

Both neuroprotective (kynurenic acid) and neurotoxic (quinolinic acid) metabolites are produced by the kynurenine pathway of tryptophan metabolism; a shift towards the latter is linked to chronic inflammation and neurodegeneration. The microbiome's modulation of neuroinflammation through molecular mimicry, bacterial amyloids, immune receptor activation, and barrier dysfunction further underscores its role as both a trigger and a potential therapeutic target in CNS disorders. The chapter emphasizes a range of microbiome-targeted therapies, such as probiotics, prebiotics, synbiotics, FMT, and individualized dietary strategies. Significant obstacles still exist even though preclinical and some clinical trials show encouraging results, such as improved cognitive function in models of Alzheimer's disease, a decrease in both motor and NMS in Parkinson's disease, and a modification of the severity of the disease in amyotrophic lateral sclerosis and Huntington's disease. The efficacy, safety, and long-term stability of these interventions require rigorous validation in large-scale, randomized controlled trials. Personalized medicine, enabled by multi-omics profiling and AI-driven predictive modelling, holds promise for tailoring therapies to individual microbial and metabolic profiles, but standardization, donor selection for FMT, and regulatory hurdles must be addressed to ensure reproducibility and clinical adoption. Cutting-edge multi-omics platforms (genomics, transcriptomics, proteomics, and metabolomics) and spatial/single-cell technologies are revolutionizing our ability to map host-microbe interactions at unprecedented resolution. Artificial intelligence and ML are transforming data analysis, enabling the identification of predictive biomarkers, disease subtyping, and personalized treatment recommendations. However, the field must prioritize standardized protocols, open data repositories, and interpretable AI models to overcome heterogeneity, enhance reproducibility, and foster clinician and patient trust. Looking ahead, the integration of microbiome science into neurology demands a collaborative, interdisciplinary approach. Key priorities include elucidating causal mechanisms (beyond association), addressing the ecological complexity and stability of engineered microbial communities, and navigating ethical considerations around consent, data ownership, and equitable access to emerging therapies. Environmental and lifestyle factors diet, exercise, sleep, and stress must be considered as modifiable variables that interact with the microbiome to influence brain health across the lifespan. Finally, this chapter consolidates a growing consensus that neurodegenerative diseases are not solely disorders of the brain but manifestations of systemic dysregulation, with the gut microbiome playing a central role. By embracing a holistic, system-level perspective, researchers and clinicians can unlock novel diagnostic tools, therapeutic strategies, and preventive measures. The path forward is marked by both opportunity and challenge: the promise of microbiome-based precision medicine is tempered by the need for robust evidence, ethical vigilance, and global collaboration. As the discipline advances, it may revolutionize our strategy towards neurodegeneration from symptomatic treatment to root cause regulation, heralding a new epoch of brain health grounded in the science of the gut-brain axis.

Declarations

Competing Interests The authors have no competing interests to declare that are relevant to the content of this manuscript.

Ethics Approval The authors declare that this manuscript does not include primary research with human participants.

The authors declare that this manuscript does not include primary studies involving animals.

References

1. Nandwana V, Nandwana NK, Das Y, Saito M, Panda T, Das S, et al. The role of microbiome in brain development and neurodegenerative diseases. Molecules [Internet]. 2022 [cited 2025 Jul 14];27(11):3402. Available from: https://www.mdpi.com/1420-3049/27/11/3402.
2. Nandwana V, Debbarma S. Fecal microbiota transplantation: a microbiome modulation technique for Alzheimer's disease. Cureus [Internet]. 2021 [cited 2025 Jul 14];13(7). Available from: https://www.cureus.com/articles/63028-fecal-microbiota-transplantation-a-microbiome-modulation-technique-for-alzheimers-disease.pdf.
3. Parashar A, Udayabanu M. Gut microbiota: implications in Parkinson's disease. Parkinsonism Relat Disord [Internet]. 2017 [cited 2025 Jul 14];38:1–7. Available from: https://www.sciencedirect.com/science/article/pii/S1353802017300366.
4. Tilocca B, Pieroni L, Soggiu A, Britti D, Bonizzi L, Roncada P, et al. Gut–brain axis and neurodegeneration: state-of-the-art of meta-omics sciences for microbiota characterization. Int J Mol Sci [Internet]. 2020 [cited 2025 Jul 14];21(11):4045. Available from: https://www.mdpi.com/1422-0067/21/11/4045.
5. Cryan JF, O'Riordan KJ, Sandhu K, Peterson V, Dinan TG. The gut microbiome in neurological disorders. Lancet Neurol [Internet]. 2020 [cited 2025 Jul 14];19(2):179–94. Available from: https://www.thelancet.com/article/S1474-4422(19)30356-4/abstract.
6. Moos WH, Faller DV, Harpp DN, Kanara I, Pernokas J, Powers WR, et al. Microbiota and neurological disorders: a gut feeling. Biores Open Access [Internet]. 2016 [cited 2025 Jul 14];5(1):137–45. Available from: https://www.liebertpub.com/doi/10.1089/biores.2016.0010.
7. Lloyd-Price J, Abu-Ali G, Huttenhower C. The healthy human microbiome. Genome Med [Internet]. 2016 [cited 2025 Jul 14];8(1). Available from: http://genomemedicine.biomedcentral.com/articles/10.1186/s13073-016-0307-y.
8. Gilbert JA, Blaser MJ, Caporaso JG, Jansson JK, Lynch SV, Knight R. Current understanding of the human microbiome. Nat Med [Internet]. 2018 [cited 2025 Jul 14];24(4):392–400. Available from: https://www.nature.com/articles/nm.4517.
9. Shreiner AB, Kao JY, Young VB. The gut microbiome in health and in disease. Curr Opin Gastroenterol [Internet]. 2015 [cited 2025 Jul 14];31(1):69–75. Available from: https://journals.lww.com/co-gastroenterology/fulltext/2015/01000/the_gut_microbiome_in_health_and_in_disease.12.aspx.
10. Arabi TZ, Alabdulqader AA, Sabbah BN, Ouban A. Brain-inhabiting bacteria and neurodegenerative diseases: the "brain microbiome" theory. Front Aging Neurosci [Internet]. 2023 [cited 2025 Jul 14];15:1240945. Available from: https://www.frontiersin.org/journals/aging-neuroscience/articles/10.3389/fnagi.2023.1240945/full.
11. Link CD. Is there a brain microbiome? J Exp Neurosci [Internet]. 2021 [cited 2025 Jul 14];16. Available from: https://journals.sagepub.com/doi/10.1177/26331055211018709.
12. Martin CR, Osadchiy V, Kalani A, Mayer EA. The brain-gut-microbiome axis. Cell Mol Gastroenterol Hepatol [Internet]. 2018 [cited 2025 Jul 14];6(2):133–48. Available from: https://www.sciencedirect.com/science/article/pii/S2352345X18300602.

13. Dinan TG, Cryan JF. Gut instincts: microbiota as a key regulator of brain development, ageing and neurodegeneration. J Physiol [Internet]. 2017 [cited 2025 Jul 14];595(2):489–503. Available from: https://physoc.onlinelibrary.wiley.com/doi/10.1113/JP273106.
14. Alcock J, Maley CC, Aktipis CA. Is eating behavior manipulated by the gastrointestinal microbiota? Evolutionary pressures and potential mechanisms. BioEssays [Internet]. 2014 [cited 2025 Jul 14];36(10):940–9. Available from: https://onlinelibrary.wiley.com/doi/10.1002/bies.201400071.
15. Bashir Y, Khan AU. The interplay between the gut-brain axis and the microbiome: A perspective on psychiatric and neurodegenerative disorders. Front Neurosci [Internet]. 2022 [cited 2025 Jul 14];16:1030694. Available from: https://www.frontiersin.org/journals/neuroscience/articles/10.3389/fnins.2022.1030694/full.
16. Quigley EMM. Microbiota-brain-gut axis and neurodegenerative diseases. Curr Neurol Neurosci Rep [Internet]. 2017 [cited 2025 Jul 14];17(12). Available from: http://link.springer.com/10.1007/s11910-017-0802-6.
17. Kowalski K, Mulak A. Brain-gut-microbiota axis in Alzheimer's disease. J Neurogastroenterol Motil [Internet]. 2019 [cited 2025 Jul 14];25(1):48. Available from: https://pmc.ncbi.nlm.nih.gov/articles/PMC6326209/.
18. Heijtz RD, Wang S, Anuar F, Qian Y, Björkholm B, Samuelsson A, et al. Normal gut microbiota modulates brain development and behavior. Proc Natl Acad Sci USA [Internet]. 2011 [cited 2025 Jul 14];108(7):3047–52. Available from: https://pnas.org/doi/full/10.1073/pnas.1010529108.
19. Sampson TR, Debelius JW, Thron T, Janssen S, Shastri GG, Ilhan ZE, et al. Gut microbiota regulate motor deficits and neuroinflammation in a model of Parkinson's disease. Cell [Internet]. 2016 [cited 2025 Jul 14];167(6):1469–80. Available from: https://www.cell.com/cell/fulltext/S0092-8674%2816%2931590-2?mc_cid=f520f5f551&mc_eid=180f8d3834.
20. Knight R, Vrbanac A, Taylor BC, Aksenov A, Callewaert C, Debelius J, et al. Best practices for analysing microbiomes. Nat Rev Microbiol [Internet]. 2018 [cited 2025 Jul 14];16(7):410–22. Available from: https://www.nature.com/articles/s41579-018-0029-9.
21. Senejani AG, Maghsoudlou J, El-Zohiry D, Gaur G, Wawrzeniak K, Caravaglia C, et al. Borrelia burgdorferi co-localizing with amyloid markers in Alzheimer's disease brain tissues. J Alzheimer's Dis [Internet]. 2022 [cited 2025 Jul 14];85(2):889–903. Available from: https://journals.sagepub.com/doi/full/10.3233/JAD-215398.
22. Wawrzeniak K, Gaur G, Sapi E, Senejani AG. Effect of Borrelia burgdorferi outer membrane vesicles on host oxidative stress response. Antibiotics [Internet]. 2020 [cited 2025 Jul 14];9(5):275. Available from: https://www.mdpi.com/2079-6382/9/5/275.
23. Xie J, Cools L, Van Imschoot G, Van Wonterghem E, Pauwels MJ, Vlaeminck I, et al. *Helicobacter pylori*-derived outer membrane vesicles contribute to Alzheimer's disease pathogenesis via C3-C3aR signalling. J Extracellular Vesicles [Internet]. 2023 [cited 2025 Jul 14];12(2). Available from: https://onlinelibrary.wiley.com/doi/10.1002/jev2.12306.
24. Alonso R, Pisa D, Fernández-Fernández AM, Carrasco L. Infection of fungi and bacteria in brain tissue from elderly persons and patients with Alzheimer's disease. Front Aging Neurosci [Internet]. 2018 [cited 2025 Jul 14];10:159. Available from: https://www.frontiersin.org/journals/aging-neuroscience/articles/10.3389/fnagi.2018.00159/full.
25. Pisa D, Alonso R, Carrasco L. Parkinson's disease: a comprehensive analysis of fungi and bacteria in brain tissue. Int J Biol Sci [Internet]. 2020 [cited 2025 Jul 14];16(7):1135. Available from: https://pmc.ncbi.nlm.nih.gov/articles/PMC7053320/.
26. Alonso R, Pisa D, Carrasco L. Searching for bacteria in neural tissue from amyotrophic lateral sclerosis. Front Neurosci [Internet]. 2019 [cited 2025 Jul 14];13:171. Available from: https://www.frontiersin.org/journals/neuroscience/articles/10.3389/fnins.2019.00171/full.
27. Alonso R, Pisa D, Fernández-Fernández AM, Rábano A, Carrasco L. Fungal infection in neural tissue of patients with amyotrophic lateral sclerosis. Neurobiol Dis [Internet]. 2017 [cited 2025 Jul 14];108:249–60. Available from: https://www.sciencedirect.com/science/article/pii/S0969996117302115.

28. Branton WG, Lu JQ, Surette MG, Holt RA, Lind J, Laman JD, et al. Brain microbiota disruption within inflammatory demyelinating lesions in multiple sclerosis. Sci Rep [Internet]. 2016 [cited 2025 Jul 14];6(1):37344. Available from: https://www.nature.com/articles/srep37344.
29. de Fraga-Silva TFC, Munhoz-Alves N, Mimura LAN, de Oliveira LRC, Figueiredo-Godoi LMA, Garcia MT, et al. Systemic infection by non-albicans Candida species affects the development of a murine model of multiple sclerosis. J Fungi [Internet]. 2022 [cited 2025 Jul 14];8(4):386. Available from: https://www.mdpi.com/2309-608X/8/4/386.
30. Alonso R, Pisa D, Carrasco L. Brain microbiota in Huntington's disease patients. Front Microbiol [Internet]. 2019 [cited 2025 Jul 14];10:2622. Available from: https://www.frontiersin.org/articles/10.3389/fmicb.2019.02622/full.
31. Cummings J, Lee G, Ritter A, Sabbagh M, Zhong K. Alzheimer's disease drug development pipeline: 2019. Alzheimers Dement Transl Res Clin Interv [Internet]. 2019 [cited 2025 Jul 14];5:272–93. Available from: https://www.sciencedirect.com/science/article/pii/S2352873719300290.
32. Cryan JF, O'Riordan KJ, Cowan CSM, Sandhu KV, Bastiaanssen TFS, Boehme M, et al. The microbiota-gut-brain axis. Physiol Rev. 2019;99(4):1877–2013.
33. Cong Q, Soteros BM, Wollet M, Kim JH, Sia GM. The endogenous neuronal complement inhibitor SRPX2 protects against complement-mediated synapse elimination during development. Nat Neurosci. 2020;23(9):1067–78.
34. Carabotti M, Scirocco A, Maselli MA, Severi C. The gut-brain axis: interactions between enteric microbiota, central and enteric nervous systems. Ann Gastroenterol. 2015;28(2):203–9.
35. Dalile B, Van Oudenhove L, Vervliet B, Verbeke K. The role of short-chain fatty acids in microbiota-gut-brain communication. Nat Rev Gastroenterol Hepatol. 2019;16(8):461–78.
36. Liu Y, Alookaran JJ, Rhoads JM. Probiotics in autoimmune and inflammatory disorders. Nutrients. 2018;10(10):1537.
37. Ma Q, Xing C, Long W, Wang HY, Liu Q, Wang RF. Impact of microbiota on central nervous system and neurological diseases: the gut-brain axis. J Neuroinflammation. 2019;16(1):53.
38. Sherwin E, Bordenstein SR, Quinn JL, Dinan TG, Cryan JF. Microbiota and the social brain. Science. 2019;366(6465):eaar2016.
39. Rutsch A, Kantsjö JB, Ronchi F. The gut-brain axis: how microbiota and host inflammasome influence brain physiology and pathology. Front Immunol. 2020;11:604179.
40. Roswall J, Olsson LM, Kovatcheva-Datchary P, Nilsson S, Tremaroli V, Simon MC, et al. Developmental trajectory of the healthy human gut microbiota during the first 5 years of life. Cell Host Microbe. 2021;29(5):765–76.e3.
41. Parker A, Fonseca S, Carding SR. Gut microbes and metabolites as modulators of blood-brain barrier integrity and brain health. Gut Microbes. 2020;11(2):135–57.
42. Zheng P, Zeng B, Zhou C, Liu M, Fang Z, Xu X, et al. Gut microbiome remodeling induces depressive-like behaviors through a pathway mediated by the host's metabolism. Mol Psychiatry. 2016;21(6):786–96.
43. Diaz-Arrastia R, Kochanek PM. Pharmacotherapy for traumatic brain injury: the next generation of clinical trials. Neurotherapeutics. 2023;20(6):1428–32.
44. Luczynski P, Tramullas M, Viola M, Shanahan F, Clarke G, O'Mahony S, et al. Microbiota regulates visceral pain in the mouse. Elife. 2017;6:e25887.
45. Skonieczna-Żydecka K, Grochans E, Maciejewska D, Szkup M, Schneider-Matyka D, Jurczak A, et al. Faecal short chain fatty acids profile is changed in polish depressive women. Nutrients. 2018;10(12):1939.
46. López-Villodres JA, Escamilla A, Mercado-Sáenz S, Alba-Tercedor C, Rodriguez-Perez LM, Arranz-Salas I, et al. Microbiome alterations and Alzheimer's disease: modeling strategies with transgenic mice. Biomedicines. 2023;11(7):1846.
47. Khatoon S, Kalam N, Rashid S, Bano G. Effects of gut microbiota on neurodegenerative diseases. Front Aging Neurosci [Internet]. 2023 [cited 2025 Jul 11];15. Available from: https://www.frontiersin.org/journals/aging-neuroscience/articles/10.3389/fnagi.2023.1145241/full.

48. Microbiota–gut–brain axis and its therapeutic applications in neurodegenerative diseases. Signal Transduct Target Ther [Internet]. [cited 2025 Jul 11]. Available from: https://www.nature.com/articles/s41392-024-01743-1.
49. Jain A, Madkan S, Patil P. The role of gut microbiota in neurodegenerative diseases: current insights and therapeutic implications. Cureus. 2023;15(10):e47861.
50. Mincic AM, Antal M, Filip L, Miere D. Modulation of gut microbiome in the treatment of neurodegenerative diseases: a systematic review. Clin Nutr. 2024;43(7):1832–49.
51. Koutsokostas C, Merkouris E, Goulas A, Aidinopoulou K, Sini N, Dimaras T, et al. Gut microbes associated with neurodegenerative disorders: a comprehensive review of the literature. Microorganisms. 2024;12(8):1735.
52. Rob M, Yousef M, Lakshmanan AP, Mahboob A, Terranegra A, Chaari A. Microbial signatures and therapeutic strategies in neurodegenerative diseases. Biomed Pharmacother. 2025;184:117905.
53. Salinas DI, Cruz RA, Rincon Rueda S, Estofan LJ. Creutzfeldt-Jakob disease in South Texas: a case series of three Hispanic patients. Cureus. 2025;17(6):e85699.
54. Ma YY, Li X, Yu JT, Wang YJ. Therapeutics for neurodegenerative diseases by targeting the gut microbiome: from bench to bedside. Transl Neurodegener. 2024;13(1):12.
55. Silva YP, Bernardi A, Frozza RL. The role of short-chain fatty acids from gut microbiota in gut-brain communication. Front Endocrinol [Internet]. 2020 [cited 2025 Jul 11];11. Available from: https://www.frontiersin.org/journals/endocrinology/articles/10.3389/fendo.2020.00025/full.
56. Morrison DJ, Preston T. Formation of short chain fatty acids by the gut microbiota and their impact on human metabolism. Gut Microbes. 2016;7(3):189–200.
57. Gurry T, Nguyen LTT, Yu X, Alm EJ. Functional heterogeneity in the fermentation capabilities of the healthy human gut microbiota. PLoS One. 2021;16(7):e0254004.
58. Deleu S, Machiels K, Raes J, Verbeke K, Vermeire S. Short chain fatty acids and its producing organisms: An overlooked therapy for IBD? eBioMedicine [Internet]. 2021 [cited 2025 Jul 11];66. Available from: https://www.thelancet.com/journals/ebiom/article/PIIS2352-3964(21)00086-4/fulltext.
59. Frolova MS, Suvorova IA, Iablokov SN, Petrov SN, Rodionov DA. Genomic reconstruction of short-chain fatty acid production by the human gut microbiota. Front Mol Biosci [Internet]. 2022 [cited 2025 Jul 11];9. Available from: https://www.frontiersin.org/journals/molecular-biosciences/articles/10.3389/fmolb.2022.949563/full.
60. Macfarlane S, Macfarlane GT. Regulation of short-chain fatty acid production. Proc Nutr Soc. 2003;62(1):67–72.
61. Liu J, Jin Y, Ye Y, Tang Y, Dai S, Li M, et al. The neuroprotective effect of short chain fatty acids against sepsis-associated encephalopathy in mice. Front Immunol. 2021;12:626894.
62. Sadler R, Cramer JV, Heindl S, Kostidis S, Betz D, Zuurbier KR, et al. Short-chain fatty acids improve poststroke recovery via immunological mechanisms. J Neurosci. 2020;40(5):1162–73.
63. Kalyanaraman B, Cheng G, Hardy M. Gut microbiome, short-chain fatty acids, alpha-synuclein, neuroinflammation, and ROS/RNS: relevance to Parkinson's disease and therapeutic implications. Redox Biol. 2024;71:103092.
64. Fillier TA, Shah S, Doody KM, Pham TH, Aubry I, Tremblay ML, et al. Brief exposure of neuronal cells to levels of SCFAs observed in human systemic circulation impair lipid metabolism resulting in apoptosis. Sci Rep. 2022;12:14355.
65. Fillier T. Short-chain fatty acids alter the neurolipidome, neuronal cell function, and neurobehaviour: what are the consequences for overall brain health? [Internet] [Doctoral]. Memorial University of Newfoundland; 2023 [cited 2025 Jul 11]. Available from: https://research.library.mun.ca/16192/.
66. Short chain fatty acid – an overview. ScienceDirect Topics [Internet]. [cited 2025 Jul 13]. Available from: https://www.sciencedirect.com/topics/food-science/short-chain-fatty-acid.
67. Liao H, Li H, Bao H, Jiang L, Du J, Guo Y, et al. Short chain fatty acids protect the cognitive function of sepsis associated encephalopathy mice via GPR43. Front Neurol [Internet].

2022 [cited 2025 Jul 11];13. Available from: https://www.frontiersin.org/journals/neurology/articles/10.3389/fneur.2022.909436/full.
68. Qian XH, Xie RY, Liu XL, Chen SD, Tang HD. Mechanisms of short-chain fatty acids derived from gut microbiota in Alzheimer's disease. Aging Dis. 2022;13(4):1252–66.
69. Su SH, Chen M, Wu YF, Lin Q, Wang DP, Sun J, et al. Fecal microbiota transplantation and short-chain fatty acids protected against cognitive dysfunction in a rat model of chronic cerebral hypoperfusion. CNS Neurosci Ther. 2023;29(S1):98–114.
70. Kuehn JF, Heston MB, Warren MF, Kang JW, Harding S, Davenport-Sis NJ, et al. Relationships between microbiome-derived short chain fatty acids and Alzheimer's disease pathology biomarkers: a human cohort study. Alzheimers Dement. 2025;20(Suppl 8):e094882.
71. Tryptophan metabolic pathways and brain serotonergic activity: a comparative review. Front Endocrinol [Internet]. [cited 2025 Jul 11]. Available from: https://www.frontiersin.org/journals/endocrinology/articles/10.3389/fendo.2019.00158/full.
72. Ruddick JP, Evans AK, Nutt DJ, Lightman SL, Rook GAW, Lowry CA. Tryptophan metabolism in the central nervous system: medical implications. Expert Rev Mol Med. 2006;8(20):1–27.
73. The tryptophan and kynurenine pathway involved in the development of immune-related diseases [Internet]. [cited 2025 Jul 13]. Available from: https://www.mdpi.com/1422-0067/24/6/5742.
74. Muneer A. Kynurenine pathway of tryptophan metabolism in neuropsychiatric disorders: pathophysiologic and therapeutic considerations. Clin Psychopharmacol Neurosci. 2020;18(4):507–26.
75. Campbell BM, Charych E, Lee AW, Möller T. Kynurenines in CNS disease: regulation by inflammatory cytokines. Front Neurosci [Internet]. 2014 [cited 2025 Jul 13];8. Available from: https://www.frontiersin.org/journals/neuroscience/articles/10.3389/fnins.2014.00012/full.
76. Davis I, Liu A. What is the tryptophan kynurenine pathway and why is it important to neurotherapy? Expert Rev Neurother. 2015;15(7):719–21.
77. What does serotonin do? Neurotransmitter function [Internet]. 2025 [cited 2025 Jul 13]. Available from: https://www.simplypsychology.org/what-is-serotonin.html.
78. Moncrieff J, Cooper RE, Stockmann T, Amendola S, Hengartner MP, Horowitz MA. The serotonin theory of depression: a systematic umbrella review of the evidence. Mol Psychiatry 2023;28(8):3243–56.
79. Pourhamzeh M, Moravej FG, Arabi M, Shahriari E, Mehrabi S, Ward R, et al. The roles of serotonin in neuropsychiatric disorders. Cell Mol Neurobiol. 2021;42(6):1671–92.
80. Targeting brain serotonin synthesis: insights into neurodevelopmental disorders with long-term outcomes related to negative emotionality, aggression and antisocial behaviour. Philos Trans R Soc B Biol Sci [Internet]. [cited 2025 Jul 13]. Available from: https://royalsocietypublishing.org/doi/10.1098/rstb.2012.0039.
81. Meltzer CC, Smith G, DeKosky ST, Pollock BG, Mathis CA, Moore RY, et al. Serotonin in aging, late-life depression, and Alzheimer's disease: the emerging role of functional imaging. Neuropsychopharmacology. 1998;18(6):407–30.
82. What is serotonin, and how does it play a role in neurologic disorders? [Internet]. [cited 2025 Jul 13]. Available from: https://www.brainandlife.org/articles/what-is-serotonin-and-how-does-it-play-a-role.
83. Pathak S, Nadar R, Kim S, Liu K, Govindarajulu M, Cook P, et al. The influence of kynurenine metabolites on neurodegenerative pathologies. Int J Mol Sci. 2024;25(2):853.
84. Tan L, Yu JT, Tan L. The kynurenine pathway in neurodegenerative diseases: mechanistic and therapeutic considerations. J Neurol Sci. 2012;323(1–2):1–8.
85. Martins LB, Silveira ALM, Teixeira AL. The involvement of kynurenine pathway in neurodegenerative diseases. Curr Neuropharmacol. 2023;21(2):260–72.
86. Ostapiuk A, Urbanska EM. Kynurenic acid in neurodegenerative disorders—unique neuroprotection or double-edged sword? CNS Neurosci Ther. 2021;28(1):19–35.
87. Staats Pires A, Krishnamurthy S, Sharma S, Chow S, Klistorner S, Guillemin GJ, et al. Dysregulation of the kynurenine pathway in relapsing remitting multiple sclerosis and its

correlations with progressive neurodegeneration. Neurol Neuroimmunol Neuroinflamm. 2025;12(2):e200372.
88. Guillemin GJ, Brew BJ. Implications of the kynurenine pathway and quinolinic acid in Alzheimer's disease. Redox Rep. 2002;7(4):199–206.
89. Bairamian D, Sha S, Rolhion N, Sokol H, Dorothée G, Lemere CA, et al. Microbiota in neuroinflammation and synaptic dysfunction: a focus on Alzheimer's disease. Mol Neurodegener. 2022;17:19.
90. Bostick JW, Schonhoff AM, Mazmanian SK. Gut microbiome-mediated regulation of neuroinflammation. Curr Opin Immunol. 2022;76:102177.
91. Tohidpour A, Morgun AV, Boitsova EB, Malinovskaya NA, Martynova GP, Khilazheva ED, et al. Neuroinflammation and infection: molecular mechanisms associated with dysfunction of neurovascular unit. Front Cell Infect Microbiol [Internet]. 2017 [cited 2025 Jul 13];7. Available from: https://www.frontiersin.org/journals/cellular-and-infection-microbiology/articles/10.3389/fcimb.2017.00276/full.
92. Rutsch A, Kantsjö JB, Ronchi F. The gut-brain axis: how microbiota and host inflammasome influence brain physiology and pathology. Front Immunol [Internet]. 2020 [cited 2025 Jul 13];11. Available from: https://www.frontiersin.org/journals/immunology/articles/10.3389/fimmu.2020.604179/full.
93. Microbiota–gut–brain axis and its therapeutic applications in neurodegenerative diseases. Signal Transduct Target Ther [Internet]. [cited 2025 Jul 13]. Available from: https://www.nature.com/articles/s41392-024-01743-1.
94. Rea K, Dinan TG, Cryan JF. The microbiome: a key regulator of stress and neuroinflammation. Neurobiol Stress. 2016;4:23–33.
95. Microbiome meets microglia in neuroinflammation and neurological disorders [Internet]. [cited 2025 Jul 13]. Available from: https://www.oaepublish.com/articles/2347-8659.2020.13.
96. Farooq RK, Alamoudi W, Alhibshi A, Rehman S, Sharma AR, Abdulla FA. Varied composition and underlying mechanisms of gut microbiome in neuroinflammation. Microorganisms. 2022;10(4):705.
97. Yang. The gut microbiota modulates neuroinflammation in Alzheimer's DISEASE: elucidating crucial factors and mechanistic underpinnings. CNS Neurosci Ther – Wiley Online Library [Internet]. 2024. [cited 2025 Jul 13]. Available from: https://onlinelibrary.wiley.com/doi/10.1111/cns.70091.
98. Murai T, Matsuda S. Therapeutic implications of probiotics in the gut microbe-modulated neuroinflammation and progression of Alzheimer's disease. Life (Basel). 2023;13(7):1466.
99. Cerovic M, Forloni G, Balducci C. Neuroinflammation and the gut microbiota: possible alternative therapeutic targets to counteract Alzheimer's disease? Front Aging Neurosci [Internet]. 2019 [cited 2025 Jul 13];11. Available from: https://www.frontiersin.org/journals/aging-neuroscience/articles/10.3389/fnagi.2019.00284/full.
100. Nohesara S, Mostafavi Abdolmaleky H, Pirani A, Thiagalingam S. Therapeutic horizons: gut microbiome, neuroinflammation, and epigenetics in neuropsychiatric disorders. Cells. 2025;14(13):1027.
101. A comprehensive review on the role of the gut microbiome in human neurological disorders [Internet]. [cited 2025 Jul 18]. Available from: https://journals.asm.org/doi/epub/10.1128/cmr.00338-20.
102. Cunningham M, Azcarate-Peril MA, Barnard A, Benoit V, Grimaldi R, Guyonnet D, et al. Shaping the future of probiotics and prebiotics. Trends Microbiol. 2021;29(8):667–85.
103. Saez-Lara. The role of probiotic lactic acid bacteria and bifidobacteria in the prevention and treatment of inflammatory bowel disease and other related diseases: a systematic review of randomized human clinical trials. BioMed Res Int – Wiley Online Library [Internet]. 2015. [cited 2025 Jul 15]. Available from: https://onlinelibrary.wiley.com/doi/full/10.1155/2015/505878.
104. Zhang XF, Guan XX, Tang YJ, Sun JF, Wang XK, Wang WD, et al. Clinical effects and gut microbiota changes of using probiotics, prebiotics or synbiotics in inflammatory bowel disease: a systematic review and meta-analysis. Eur J Nutr. 2021;60(5):2855–75.

105. Official Journal of the American College of Gastroenterology. ACG [Internet]. [cited 2025 Jul 15]. Available from: https://journals.lww.com/ajg/abstract/2013/04000/fecal_microbiota_transplantation_forclostridium.8.aspx.
106. Systematic review: adverse events of fecal microbiota transplantation. PLoS One [Internet]. [cited 2025 Jul 15]. Available from: https://journals.plos.org/plosone/article?id=10.1371/journal.pone.0161174.
107. Kim N, Jeon SH, Ju IG, Gee MS, Do J, Oh MS, et al. Transplantation of gut microbiota derived from Alzheimer's disease mouse model impairs memory function and neurogenesis in C57BL/6 mice. Brain Behav Immun. 2021;98:357–65.
108. Fecal microbiota transplantation protects rotenone-induced Parkinson's disease mice via suppressing inflammation mediated by the lipopolysaccharide-TLR4 signaling pathway through the microbiota-gut-brain axis. Microbiome [Internet]. [cited 2025 Jul 15]. Available from: https://link.springer.com/article/10.1186/s40168-021-01107-9.
109. Khavandegar A, Heidarzadeh A, Angoorani P, Hasani-Ranjbar S, Ejtahed HS, Larijani B, et al. Adherence to the Mediterranean diet can beneficially affect the gut microbiota composition: a systematic review. BMC Med Genet. 2024;17(1):91.
110. Mediterranean diet and health: food effects on gut microbiota and disease control [Internet]. [cited 2025 Jul 15]. Available from: https://www.mdpi.com/1422-0067/15/7/11678.
111. Rew L, Harris MD, Goldie J. The ketogenic diet: its impact on human gut microbiota and potential consequent health outcomes: a systematic literature review. Gastroenterol Hepatol Bed Bench. 2022;15(4):326–42.
112. Exercise and Sport Sciences Reviews [Internet]. [cited 2025 Jul 15]. Available from: https://journals.lww.com/acsm-essr/fulltext/2017/01000/exercise__the_next_frontier_in_microbiota.4.aspx.
113. Conlon MA, Bird AR. The impact of diet and lifestyle on gut microbiota and human health. Nutrients. 2015;7(1):17–44.
114. Neroni B, Evangelisti M, Radocchia G, Di Nardo G, Pantanella F, Villa MP, et al. Relationship between sleep disorders and gut dysbiosis: what affects what? Sleep Med. 2021;87:1–7.
115. Santangelo A, Corsello A, Spolidoro GCI, Trovato CM, Agostoni C, Orsini A, et al. The influence of ketogenic diet on gut microbiota: potential benefits, risks and indications. Nutrients. 2023;15(17):3680.
116. Wu J, Singleton SS, Bhuiyan U, Krammer L, Mazumder R. Multi-omics approaches to studying gastrointestinal microbiome in the context of precision medicine and machine learning. Front Mol Biosci. 2023;10:1337373.
117. Li Y, Luo Y. Metabolomics of aging and Alzheimer's disease: from single-omics to multi-omics. arXiv e-prints. 2022;arXiv:2212.09870.
118. Nicholson JK, Lindon JC, Holmes E. "Metabonomics": understanding the metabolic responses of living systems to pathophysiological stimuli via multivariate statistical analysis of biological NMR spectroscopic data. Xenobiotica. 1999;29(11):1181–9.
119. Liang X, Fu Y, Cao WT, Wang Z, Zhang K, Jiang Z, et al. Gut microbiome, cognitive function and brain structure: a multi-omics integration analysis. Transl Neurodegener. 2022;11(1):49.
120. Yang Z, Zhang Y, Ran S, Zhang J, Tian F, Shi H, et al. A multi-omics study of neurodamage induced by growth-stage real-time air pollution exposure in mice via the microbiome-gut-brain axis. Toxics. 2025;13(4).
121. Chen X, Liu Y, Pu J, Gui S, Wang D, Zhong X, et al. Multi-omics analysis reveals age-related microbial and metabolite alterations in non-human primates. Microorganisms. 2023;11(10).
122. Warren A, Nyavor Y, Zarabian N, Mahoney A, Frame LA. The microbiota-gut-brain-immune interface in the pathogenesis of neuroinflammatory diseases: a narrative review of the emerging literature. Front Immunol. 2024;15:1365673.
123. Sheng C, Du W, Liang Y, Xu P, Ding Q, Chen X, et al. An integrated neuroimaging-omics approach for the gut-brain communication pathways in Alzheimer's disease. Front Aging Neurosci. 2023;15:1211979.

124. Wang J, Xiang JH, Peng XY, Liu M, Sun LJ, Zhang M, et al. Characteristic alterations of gut microbiota and serum metabolites in patients with chronic tinnitus: a multi-omics analysis. Microbiol Spectr. 2025;13(1):e0187824.
125. Liu Z, Ma A, Mathé E, Merling M, Ma Q, Liu B. Network analyses in microbiome based on high-throughput multi-omics data. Brief Bioinform. 2021;22(2):1639–55.
126. Panizza E. DeepOmicsAE: representing signaling modules in Alzheimer's disease with deep learning analysis of proteomics, metabolomics, and clinical data. JoVE. 2023;(202):e65910.
127. Liu Y, Yang M, Deng Y, Su G, Enninful A, Guo CC, et al. High-spatial-resolution multi-omics sequencing via deterministic barcoding in tissue. Cell. 2020;183(6):1665–81.e18.
128. Rodriques SG, Stickels RR, Goeva A, Martin CA, Murray E, Vanderburg CR, et al. Slide-seq: a scalable technology for measuring genome-wide expression at high spatial resolution. Science. 2019;363(6434):1463–7.
129. Lötstedt B, Stražar M, Xavier R, Regev A, Vickovic S. Spatial host-microbiome sequencing reveals niches in the mouse gut. Nat Biotechnol. 2024;42(9):1394–403.
130. Przymus P, Rykaczewski K, Martín-Segura A, Truu J, Carrillo De Santa Pau E, Kolev M, et al. Deep learning in microbiome analysis: a comprehensive review of neural network models. Front Microbiol. 2024;15:1516667.
131. Hernández Medina R, Kutuzova S, Nielsen KN, Johansen J, Hansen LH, Nielsen M, et al. Machine learning and deep learning applications in microbiome research. ISME Commun. 2022;2(1):98.
132. Lee SJ, Rho M. Multimodal deep learning applied to classify healthy and disease states of human microbiome. Sci Rep. 2022;12(1):824.
133. Tang AS, Rankin KP, Cerono G, Miramontes S, Mills H, Roger J, et al. Leveraging electronic health records and knowledge networks for Alzheimer's disease prediction and sex-specific biological insights. Nat Aging. 2024;4(3):379–95.
134. Amini S, Hao B, Yang J, Karjadi C, Kolachalama VB, Au R, et al. Prediction of Alzheimer's disease progression within 6 years using speech: a novel approach leveraging language models. Alzheimers Dement. 2024;20(8):5262–70.
135. Huang Z, Sekhon VK, Sadeghian R, Vaida ML, Jo C, Ward D, et al. ADAM: an AI reasoning and bioinformatics model for Alzheimer's disease detection and microbiome-clinical data integration [Internet]. 2025. Available from: https://arxiv.org/abs/2501.08324.
136. Hancerliogullari Koksalmis G, Soykan B, Brattain LJ, Huang H. Artificial intelligence for personalized prediction of Alzheimer's disease progression: a survey of methods, data challenges, and future directions. arXiv preprint [Internet]. 2025;arXiv:2504.21189. Available from: https://arxiv.org/abs/2504.21189.
137. Zhang JD, Xue C, Kolachalama VB, Donald WA. Interpretable machine learning on metabolomics data reveals biomarkers for Parkinson's disease. ACS Cent Sci. 2023;9(5):1035–45.
138. Tng TJW, Wong BWY, Sim EHY, Tan EK, Goh WWB, Lim KL. Systematic analysis of multi-omics data reveals component-specific blood-based biomarkers for Parkinson's disease. Transl Med Commun. 2024;9(1):12.
139. Kim SG, Hwang JS, George NP, Jang YE, Kwon M, Lee SS, et al. Integrative metabolome and proteome analysis of cerebrospinal fluid in Parkinson's disease. Int J Mol Sci. 2024;25(21).
140. Alkhanbouli R, Matar Abdulla Almadhaani H, Alhosani F, Simsekler MCE. The role of explainable artificial intelligence in disease prediction: a systematic literature review and future research directions. BMC Med Inform Decis Mak. 2025;25(1):110.
141. Novielli P, Romano D, Magarelli M, Bitonto PD, Diacono D, Chiatante A, et al. Explainable artificial intelligence for microbiome data analysis in colorectal cancer biomarker identification. Front Microbiol [Internet]. 2024;15-2024. Available from: https://www.frontiersin.org/journals/microbiology/articles/10.3389/fmicb.2024.1348974.
142. Vimbi V, Shaffi N, Mahmud M. Interpreting artificial intelligence models: a systematic review on the application of LIME and SHAP in Alzheimer's disease detection. Brain Inform. 2024;11(1):10.
143. Mitrovska A, Safari P, Ritter K, Shariati B, Fischer JK. Secure federated learning for Alzheimer's disease detection. Front Aging Neurosci. 2024;16:1324032.

144. Hall DA, Voigt RM, Cantu-Jungles TM, Hamaker B, Engen PA, Shaikh M, et al. An open label, non-randomized study assessing a prebiotic fiber intervention in a small cohort of Parkinson's disease participants. Nat Commun. 2023;14(1):926.
145. Scheperjans F, Levo R, Bosch B, Lääperi M, Pereira PAB, Smolander OP, et al. Fecal microbiota transplantation for treatment of Parkinson disease: a randomized clinical trial. JAMA Neurol. 2024;81(9):925–38.
146. Sulthana A, Paneyala S, H.S. S, Narayan MK, Ahmed Hediyal T, Sharma S, et al. Gut microbiome-based dietary intervention in Parkinson disease subject: a case report. Int J Nutr Pharmacol Neurol Dis. 2023;13(3):158–65.
147. National Library of Medicine (US). Brain small chain fatty acid metabolism in Parkinson disease: tributyrin supplementation. 2022. [cited 2025 Jul 14]. Available from: https://ichgcp.net/clinical-trials-registry/NCT05446168.
148. Rodriguez J, Cordaillat-Simmons M, Pot B, Druart C. The regulatory framework for microbiome-based therapies: insights into European regulatory developments. NPJ Biofilms Microbiomes. 2025;11(1):53.
149. Microbiome Therapeutics Innovation Group, Barberio D. Navigating regulatory and analytical challenges in live biotherapeutic product development and manufacturing. Front Microbiomes [Internet]. 2024;3-2024. Available from: https://www.frontiersin.org/journals/microbiomes/articles/10.3389/frmbi.2024.1441290.
150. Thanush D, Basavaraj HC, Gowrav MP. Current regulation and initial considerations for successful development and commercialization of microbiome therapies. Adv Gut Microbiome Res. 2023;2023(1):6657515.

Gut Microbiome Changing Foods and Nutrition

16

Raja Shekhar Nunavath, Krishnaveni Nagappan, and Bhadram Kalyan Chekraverthy

16.1 Introduction

The gut microbiome (GMB) is a multi-trillion population of microorganisms, such as bacteria, archaea, fungi, viruses and protozoa, which live in the human gastrointestinal tract and create a complicated and dynamic community. Due to its extensive roles, it is called a virtual organ encoding over three million genes, vastly outnumbering the human genome's approximately 23,000 genes. The predominant phyla are *Firmicutes*, *Bacteroidetes*, *Actinobacteria*, *Proteobacteria* and *Verrucomicrobia* with the key genera being *Bifidobacterium*, *Lactobacillus*, *Faecalibacterium* and *Akkermansia* that maintain health [1–3]. The GMB performs critical functions, including the fermentation of non-digestible carbohydrates into short-chain fatty acids (SCFAs) such as butyrate, acetate and propionate, which regulate energy homeostasis, reduce inflammation and maintain gut barrier integrity. In addition, it synthesizes essential vitamins and modulates immune responses by training the mucosal immune system and preventing pathogen overgrowth through competitive exclusion. Recent advances in metagenomics and metabolomics have elucidated the GMB's role in host physiology, revealing its influence on metabolic, immune and

R. S. Nunavath
Department of Pharmaceutical Quality Assurance, SVKM NMIMS Global University, School of Pharmacy and Technology Management, Dhule, Maharashtra, India
e-mail: raja.nunavath@svkm.ac.in

K. Nagappan
Department of Pharmaceutical Analysis, JSS College of Pharmacy, JSS Academy of Higher Education & Research, Ooty, The Nilgiris, Tamil Nadu, India
e-mail: krisath@jssuni.edu.in

B. K. Chekraverthy (✉)
Department of Phytopharmaceuticals, School of Agriculture and Bioengineering, Centurion University of Technology and Management, R. Sitapur, Odisha, India
e-mail: bkalyan.chekraverthy@cutm.ac.in

M. H. Rahman, Muhasina K M (eds.), *The Microbiome's Impact on the Brain*,
https://doi.org/10.1007/978-981-95-7494-0_16

neurological pathways via the gut-brain axis. According to the study of these interactions, personalized nutrition strategies and microbiome-based therapies comprise essential elements in the health and disease risk reduction [4–8].

Moreover, nutrition is one of the main factors that define the GMB structure and its functionality and dramatically affects microbe diversity, metabolic production and human health. The dietary constituents such as macronutrients, micronutrients and the bioactive compounds affect the relative abundance of the microbial taxa and their metabolic processes directly. On the other hand, the weight loss diets with the characteristics of Western food that is rich in fat and low in fibre stimulate the growth of intestinal bile-tolerant bacteria, which in most cases predispose dysbiosis and enhanced intestinal permeability in addition to increased body system inflammation. Protein food also has a mixed effect on the microbiome, with vegetable protein helping the diversity of the microbiome and too much animal protein potentially leading to the production of potentially dangerous substances such as trimethylamine-N-oxide (TMAO). Microbial communities are additionally modulated through micronutrients, e.g. vitamins A and D, and plant-based foods, like polyphenols, which promote the growth of beneficial taxa and block the growth of pathogens. Diets in specific patterns are related to higher levels of microbial richness and production of SCFA, with the Mediterranean diet being one of them, while ultra-processed food containing sugars and emulsifiers breaks the microbial balance and decreases the levels of microbiome diversity, as well as its health of the gut. The dietary effects stress the opportunity provided by calorie-restricted diets to exert corrective pressure on the microbiome, making nutrition central to influence the development of the gut microbiome and subsequent impact upon host physiology and prevention of disease [9, 10].

The current book chapter investigates the complex relationship between food, nutrition and the gut microbiome and highlights their collective contributions to the determination of the microbial composition and functionality as well as host health outcomes. It discusses the effects of dietary pattern, macronutrients, micronutrients and the bioactive compounds on the diversity and metabolic functions of the GMB with an emphasis on short-chain fatty acid production, modulation of immune response and intestinal barrier integrity, interindividual differences and potential methodological shortcomings, as well as discusses the plans in the fields of precision nutrition and treatment with microbiome-specific approaches.

16.2 Gut Microbiome Composition and Functions

16.2.1 Diversity of Microbiome

The gut microbiome is notable for its microbial diversity, comprising bacteria, archaea, fungi, viruses and protozoa, with bacteria being the most abundant organisms in the microbiome. *Firmicutes*, *Bacteroidetes*, *Actinobacteria*, *Proteobacteria* and *Verrucomicrobia* are the most common bacterial phyla found in the human gut, and, together, these five phyla make up the largest moiety of the collection of microbes (around 90%). The genus of *Firmicutes* (such as *Clostridium*, *Faecalibacterium*, *Ruminococcus* and *Lactobacillus*) is mainly gram-positive, and their major role is to ferment carbohydrates and produce SCFAs including butyrate that maintain energy metabolism of colonocytes, increase the integrity of gut barriers and exert anti-inflammatory properties. *Bacteroidetes*, including genera such as *Bacteroides* and *Prevotella*, are gram-negative and show a high level of complex polysaccharide degradation, leading to production of SCFA (e.g. acetate and propionate) and modulating immunological activities via host receptor interactions (e.g. Toll-like receptors (TLRs)). *Actinobacteria*, which include *Bifidobacteria*, are vital in the fermentation of dietary fibres and the production of lactate that generates a positive gut habitat for other advantageous microbes. *Proteobacteria*, including *Escherichia* and *Enterobacter*, and *Verrucomicrobia*, especially *Akkermansia muciniphila*, assist with metabolic homeostasis and preserving a mucus layer, respectively, where their overgrowth indicates dysbiosis. This balance of microbial groups and especially the *Firmicutes/Bacteroidetes* (F/B) ratio are essential to health, and changes in the F/B ratio are associated with diseases such as obesity (increased F/B) and inflammatory bowel disease (IBD) (decreased F/B). These microbial communities have a dynamically interrelated relationship with dietary substrates, which affect host metabolism, immunity and disease predisposition, supporting the key role that they play in gut health [11–17].

16.2.2 Metabolic Functions

GMB is considered a crucial participant in host metabolism owing to its ability to aid in food-based nutrient dismantles, rendering bioactive metabolites and influencing energy tabs. Saccharolytic bacteria (e.g. *Bacteroides*, *Roseburia*, and *Faecalibacterium*) utilize complex carbohydrates, such as dietary fibres and resistant starches, which cannot be digested in the upper gastrointestinal tract and thus are fermented in the colon to SCFAs propionate, butyrate and acetate in molar ratios of 3:1:1–10:2:1. These SCFAs play pivotal roles in energy provision as the main source of energy in colonocyte through 2-oxidation, providing about 30% of its energy demands, whereas the acetate and propionate are distributed to extra-mucosal tissues including the liver in which they provide the process of gluconeogenesis and production of lipids [18–20]. The microbiome also metabolizes dietary proteins, producing branched-chain fatty acids (BCFAs), amines and ammonia, though these are less efficient energy sources compared to carbohydrates due to complex interconversion requirements. In addition, SCFAs regulate energy expenditure by augmenting thermogenesis in the brown adipose and skeletal tissue, via adenosine monophosphate (AMP)-activated protein kinase (AMPK) and peroxisome proliferator-activated receptor gamma coactivator 1-alpha pathways. The adverse effect of reduced SCFA-producing bacteria, i.e. dysbiosis, is related to the development of metabolic diseases, including obesity and type 2 diabetes, and emphasizes the importance of the microbiome in energy homeostasis and its therapeutic potential via nutrition [21–23].

16.3 The Gut Microbiome and Dietary Influences

16.3.1 Macronutrients and Microbial Response

The type and quantity of nutrients that people consume, known as dietary macronutrients, carbohydrates, proteins and fats, have significant effects on the structure and activity of the gut microbial communities, triggering unique microbial adaptations that have implications for the health of the person. Carbohydrates, especially indigestible fibres saccharolytic bacteria in the colon such as *Bifidobacterium*, *Prevotella* and *Faecalibacterium*, also ferment to produce SCFAs of butyrate, acetate and propionate [24, 25]. These SCFAs protect the integrity of the gut barrier, decrease inflammation and alter energy metabolism, and a higher fibre diet (e.g. the Mediterranean diet) is considered to enhance the diversity and abundance of organisms producing these SCFAs, as can be observed with an increased abundance of *Roseburia* and *Ruminococcus* when subjected to dietary fibre loads. Conversely, low-carbohydrate diets (<30% energy) are associated with reduced *Bifidobacterium* and SCFA levels, potentially compromising gut health [26]. Proteins, depending on their source and digestibility, elicit varied microbial responses; plant-based proteins promote SCFA-producing bacteria like *Lactobacillus* and *Bifidobacterium*, while animal-based proteins, particularly red meat, increase bile-tolerant genera such as *Bacteroides*, *Alistipes* and *Bilophila*, alongside potentially harmful metabolites like TMAO and BCFAs. High-protein, low-carbohydrate diets reduce butyrate-producing *Roseburia* and *Eubacterium rectale*, correlating with decreased SCFA production and potential detriments to colonic health. Dietary fats, particularly saturated fats, reduce microbial diversity and increase the *Firmicutes*/*Bacteroidetes* ratio, promoting bile-tolerant bacteria like *Bilophila wadsworthia*, which may exacerbate inflammation and metabolic disorders. In contrast, unsaturated fats, such as those in the Mediterranean diet, enrich *Bifidobacterium* and butyrate producers, supporting cardiometabolic health [27]. The interplay of macronutrient ratios, their chemical structures and host baseline microbiota underscores the complexity of dietary impacts on microbial ecosystems, necessitating personalized nutrition strategies to optimize microbial composition and host metabolic outcomes [28].

16.3.2 Carbohydrates and Fermentation

Dietary carbohydrates, particularly non-digestible fibres and prebiotics, are critical substrates for gut microbial fermentation, significantly influencing microbiome composition and host health. Dietary fibres, including non-starch polysaccharides (e.g. cellulose, hemicellulose, pectin) and resistant starches, escape digestion in the small intestine and reach the colon, where they are fermented by saccharolytic bacteria such as *Bifidobacterium*, *Lactobacillus* and *Faecalibacterium prausnitzii* [29, 30]. This fermentation process yields SCFAs, primarily acetate, propionate and butyrate, which serve as energy sources for colonocytes, modulate gut pH and regulate metabolic and immune functions.

Prebiotics, defined as selectively fermented substrates that promote the growth of beneficial bacteria, include fructo-oligosaccharides (FOS), galacto-oligosaccharides (GOS) and inulin, which specifically enhance *Bifidobacterium* and *Lactobacillus* populations, as evidenced by randomized controlled trials demonstrating increased SCFA production and improved glycaemic control in type 2 diabetes patients. The domain of fermentability parameters of fibres is dependent on physicochemical characteristics, which are solubility and chain length, whereby soluble fibres such as FOS and inulin are promptly fermentable in the proximal colon, and insoluble fibres such as cellulose are slower fermentable in the distal colon. Complicated fibre degradation is triggered by keystone species, like *Ruminococcus bromii* and resistant starch, that aid cross-feeding among microbial taxons and maximize the production of SCFA [31–33]. The low-fibre intake decreases microbial diversity and SCFA concentrations, diverting metabolism to less preferred substrates such as proteins, which generate potentially harmful metabolites that include ammonia and hydrogen sulphide. Therefore, dietary fibre and prebiotics have a central role in promoting the healthy microbiome balance with a rational fermentation process, and thus it has a therapeutic potential in the treatment of metabolic and gastrointestinal diseases [34, 35].

16.3.3 Proteins and Microbial Metabolites

Animal-based proteins and plant-based proteins exert different effects on gut microbial metabolism, thus leading to altered metabolite patterns that have serious effects on the health of the host. Protein sources derived from animal sources, e.g. meat, dairy and eggs, contain a high content of essential amino acids and are lowly digestible in the colon, probably because they are highly bioavailable in the small intestine and they provide fewer substrates to fermentation by intestinal microbes. However, their fermentation by gut microbes, particularly *Bacteroides*, *Clostridium* and *Bilophila*, produces metabolites such as BCFAs, ammonia, hydrogen sulphide and trimethylamine (TMA), which is oxidized in the liver to TMAO [36, 37]. Elevated TMAO levels are associated with increased cardiovascular risk, while ammonia and hydrogen sulphide may contribute to colonic inflammation and colorectal cancer risk, particularly with high red meat intake, as evidenced by studies showing dose-dependent increases in faecal ammonia concentrations. In contrast, plant-based proteins, derived from legumes, nuts and cereals, often contain anti-nutritional factors and lower levels of essential amino acids like lysine and methionine, leading to greater undigested protein reaching the colon (Table 16.1). This facilitates fermentation by saccharolytic bacteria such as *Bifidobacterium* and *Lactobacillus* to yield SCFAs, including butyrate and propionate, to maintain the gut barrier integrity and suppress inflammatory response [38, 39]. Nevertheless, the relatively lower Digestible Indispensable Amino Acid Score (<100%) of plant proteins than animal proteins indicates lower ileal digestibility that may decrease amino acid absorption, in favour of beneficial microbial, SCFA production. The disparities in microbial metabolism of plant-based and animal-based proteins indicate the

Table 16.1 Essential vitamins and minerals involved in energy metabolic pathways

Nutrient	Role/functions
Thiamine (B1)	Serves as a vital coenzyme in carbohydrate metabolism, supporting energy production. It is required for proper muscle and heart function and participates in oxidative carboxylation reactions, which also depend on manganese
Riboflavin (B2)	Acts as a cofactor within the mitochondrial respiratory chain, enabling energy release from nutrients. It forms part of the coenzymes flavin adenine dinucleotide and flavin mononucleotide
Niacin (vitamin B3/nicotinic acid)	Converts into nicotinamide adenine dinucleotide and nicotinamide adenine dinucleotide phosphate, which are central to redox reactions in cells. It aids in energy liberation through mitochondrial pathways
Pyridoxine (B6)	Functions in more than 100 enzymatic reactions, primarily associated with amino acid and protein metabolism, while also assisting in energy release from food
Cobalamin (B12)	Crucial for carbohydrate and fat metabolism, as well as protein synthesis. It works closely with folate in metabolic processes
Biotin (B7)	Acts as a coenzyme in fatty acid and amino acid metabolism and supports the utilization of other B vitamins
Ascorbic acid (vitamin C)	Required for the biosynthesis of carnitine and catecholamines. It enhances non-heme iron absorption, contributes to folate metabolism aids in cortisol formation and serves as a strong antioxidant
Folic acid (B9)	Transfers one-carbon units for the production of nucleotides such as thymidylate and purines and for methionine synthesis
Calcium	Essential for nerve signalling and muscle contraction. Activates fatty acid oxidation, supports mitochondrial adenosine triphosphate (ATP) transport and promotes insulin release in response to glucose
Phosphorus	A structural part of ATP and creatine phosphate, contributing to energy transfer and activation of biomolecules
Magnesium	Important for nerve and muscle activity. Participates in over 300 enzymatic reactions, including those linked to ATP metabolism and energy production
Copper	Serves as a cofactor for cytochrome c oxidase in the mitochondrial respiratory chain and plays a role in iron utilization
Chromium (III)	Enhances insulin's effectiveness, thereby facilitating glucose uptake by cells. Loss of chromium increases during intense physical activity
Iron	It's a key component of haemoglobin and myoglobin which enables oxygen transport. It participates in electron transfer within the respiratory chain, which is vital for ATP generation and red blood cell function
Manganese	Supports enzymes involved in carbohydrate metabolism and gluconeogenesis
Zinc	Integral to over 100 enzymes, including those regulating energy metabolism and various cellular functions

different health effects linked to them, where the digestion of animal protein in the gut can raise pro-inflammatory and pro-carcinogenic compounds, the digestion of plant protein can provide a microbial environment likely to produce anti-inflammatory compounds and a balanced approach of diet is important to maximize the microbiome-driven metabolic outcomes valuably [40–42].

16.3.4 Fats and Microbial Dysbiosis

Fats in foods like saturated fatty acids (SFAs) and unsaturated fatty acids (UFAs) have different influences on the composition of the gut microbiome, which corresponds greatly to the microbial dysbiosis and host health. High SFAs are animal foods like butter, lard and red meat that are associated with decreased microbial diversity and richness, which is usually characterized by an elevated F/B ratio, which is a characteristic of dysbiosis with both obesity and metabolic syndrome [44]. SFAs (including palmitic and stearic acids) are known to favour the proliferation of bile-tolerant, gram-negative organisms such as *Bilophila wadsworthia* and *Proteobacteria* which display the pro-inflammatory metabolic products such as lipopolysaccharides (LPS) resulting in enhanced intestinal permeability, endotoxemia and systemic inflammation as examined in both animal models and humans. These effects are mediated by SFA-induced activation of Toll-like receptor 4 (TLR4) signalling and reduced SCFA production, particularly butyrate, due to decreased abundance of SCFA-producing taxa like *Faecalibacterium* and *Roseburia* [45, 46]. Conversely, unsaturated fats, including monounsaturated fatty acids (MUFAs) from olive oil and polyunsaturated fatty acids (PUFAs) from fish oil, enhance microbial diversity and support beneficial bacteria such as *Bifidobacterium*, *Lactobacillus* and *Akkermansia muciniphila*. PUFAs, particularly omega-3 fatty acids (e.g. eicosapentaenoic and docosahexaenoic acids), exhibit antimicrobial properties that suppress pathogenic bacteria and promote SCFA production, improving gut barrier function and reducing inflammation. For instance, studies demonstrate that diets high in omega-3 PUFAs increase *Bacteroidaceae* and reduce hepatic steatosis, while MUFAs in Mediterranean diets correlate with improved cardiometabolic outcomes. However, high doses of omega-6 PUFAs, such as linoleic acid, may promote inflammation in certain contexts, highlighting the importance of fatty acid balance. These divergent microbial responses underscore the need for dietary strategies prioritizing unsaturated fats to mitigate dysbiosis and associated metabolic disorders [47].

16.3.5 Dietary Patterns

Either diet has a large impact on gut microbiome composition and activity, and a Western, Mediterranean or plant-based diet reveals different responses to microbiomes that influence the health of the host in different ways. Western diet (high consumption of saturated fat, refined sugar and ultra-processed foods, low dietary fibre intakes) is attributed to low microbial abundance and dysbiosis, the increased abundance of the *Firmicutes*/*Bacteroidetes* ratio and pro-inflammatory taxa *Bilophila wadsworthia* and *Collinsella* [48, 49]. This diet stimulates the formation of TMAO based on animal choline and carnitine, which increases the risk of cardiovascular diseases, and low production of SCFA results in inflammation and metabolic diseases, such as obesity and type 2 diabetes. Plant-based diets, including vegan and vegetarian patterns, similarly promote microbial diversity by providing fermentable fibres and polyphenols, supporting *Bifidobacterium* and *Lactobacillus* while reducing TMAO production due to lower animal protein intake. However, some studies report lower *Bacteroides* counts in vegans compared to omnivores, possibly due to methodological differences or dietary variability. While both Mediterranean and plant-based diets enhance microbiome health and protect against non-communicable diseases, the Western diet's detrimental effects underscore the need for dietary shifts towards fibre-rich, plant-centric patterns to optimize microbial ecosystems and host metabolic outcomes [50].

16.4 Prebiotics, Probiotics and Synbiotics

16.4.1 Prebiotics

Prebiotics are non-digestible dietary substrates that selectively stimulate the growth and activity of beneficial gut microorganisms, conferring health benefits to the host, as defined by the International Scientific Association for Probiotics and Prebiotics (ISAPP) (Table 16.2). These compounds, primarily oligosaccharides and polysaccharides, resist enzymatic hydrolysis in the upper gastrointestinal tract and are fermented by colonic microbiota, producing SCFAs that support gut health and systemic metabolism [51, 52]. The most well-established prebiotics include inulin-type fructans, such as inulin and FOS and GOS. Inulin, a linear β-(2→1) fructan with a degree of polymerization (DP) ranging from 2 to 60, is naturally abundant in chicory root, Jerusalem artichokes, garlic, onions, leeks, asparagus and wheat, with chicory serving as the primary industrial source due to its high inulin content (over 70% dry matter). FOS, or oligofructose, consists of shorter-chain fructans (DP 2–10), either extracted from inulin-rich plants or enzymatically synthesized from sucrose, and is found in foods like bananas and blue agave. GOS, composed of

Table 16.2 Overview of microbial-based strategies for gut and overall health [65]

Category	Examples	Key health roles
Probiotics	*Lactobacillus*, *Bifidobacterium*, *Saccharomyces boulardii*	Helps in maintaining intestinal balance, strengthens immunity, reduces the incidence of diarrhoea, eases symptoms of IBS and lowers gut inflammation
Prebiotics	Inulin, FOS, GOS, resistant starch	Stimulate the proliferation of beneficial microbes, improve digestive efficiency, enhance absorption of minerals and reinforce immune defences
Synbiotics	*Lactobacillus rhamnosus* + inulin; *Bifidobacterium* + GOS	Combine the effects of probiotics and prebiotics to optimize gut flora, regulate bowel activity, strengthen immune responses and reduce susceptibility to infections
Postbiotics	Short-chain fatty acids, microbial cell wall fragments, bioactive metabolites	Provide anti-inflammatory action, enhance the integrity of the intestinal barrier, fine-tune immune activity and contribute to better metabolic health

galactose polymers with a terminal glucose, is derived from lactose and is less prevalent in natural foods but used in fortified products. Emerging prebiotic candidates, such as xylooligosaccharides (XOS), isomalto-oligosaccharides, and resistant starches, show promise but require further clinical validation. These prebiotics promote the proliferation of *Bifidobacterium* and *Lactobacillus* species, enhancing SCFA production and microbial diversity, which are linked to improved gut barrier function, immune modulation and reduced risk of metabolic disorders, though their efficacy varies by chain length, dose (typically 3–10 g/day) and individual microbiota composition [53, 54].

16.4.2 Probiotics

Probiotics, defined as live microorganisms that confer health benefits when administered in adequate amounts, are predominantly represented by strains of *Lactobacillus* and *Bifidobacterium*, which modulate gut microbiota, enhance immune function and alleviate gastrointestinal disorders [55, 56]. *Lactobacillus* species, including *L. acidophilus*, *L. rhamnosus* GG, *L. casei*, *L. plantarum* and *L. reuteri*, are gram-positive, lactic acid-producing bacteria that colonize the gastrointestinal tract, producing antimicrobial peptides and competing with pathogens to reduce infections such as antibiotic-associated diarrhoea and *Helicobacter pylori*-related gastritis. *Bifidobacterium* species, such as *B. longum*, *B. bifidum*, *B. breve*, and *B. animalis* subsp. *lactis*, are gram-positive, anaerobic actinobacteria prevalent in the infant gut, known for fermenting oligosaccharides into SCFAs like acetate and lactate, which support gut barrier integrity and reduce inflammation in conditions like irritable bowel syndrome (IBS) and IBD [57]. These probiotics are commonly found in fermented dairy products, with yogurt containing *L. delbrueckii* subsp. *bulgaricus* and *Streptococcus thermophilus* (often combined with *L. acidophilus* or *B. lactis*), and kefir harbouring diverse strains including *L. kefiranofaciens*, *L. acidophilus* and *B. bifidum*. Non-dairy sources include fermented vegetables like sauerkraut and kimchi, which contain *Lactobacillus* species, and miso, a fermented soybean paste rich in *Lactobacillus* and occasionally *Bifidobacterium*. Recent studies emphasize the strain-specific efficacy of probiotics, with *L. rhamnosus* GG and *B. lactis* BB-12 demonstrating significant benefits in clinical trials for diarrhoea and IBS symptom relief, though their survival through gastric conditions often requires encapsulation techniques to ensure viability in food matrices [58–60].

16.4.3 Synbiotics

Synbiotics, defined as mixtures of live microorganisms (probiotics) and substrates selectively utilized by host microorganisms (prebiotics) that confer health benefits, leverage synergistic interactions to enhance gut microbiota functionality beyond the effects of their components [61]. The ISAPP distinguishes two types: complementary synbiotics, where probiotics (e.g.

Lactobacillus or *Bifidobacterium*) and prebiotics (e.g. inulin, FOS) act independently to target both exogenous and autochthonous microbes, and synergistic synbiotics, where the prebiotic is selectively metabolized by the co-administered probiotic to enhance its survival or activity. Synergistic synbiotics, such as *Lactobacillus rhamnosus* GG with FOS, improve probiotic viability in the gastrointestinal tract by providing fermentable substrates that enhance colonization and SCFA production, which supports gut barrier integrity and immune modulation [62, 63]. Emerging research highlights their efficacy in clinical contexts: for instance, synbiotics containing *Bifidobacterium longum* and inulin/FOS blends have improved histological outcomes in ulcerative colitis by reducing inflammatory markers like TNF-α and IL-1α, while *Lactiplantibacillus plantarum* with arabinoxylan ameliorates colitis in murine models by inhibiting NF-κB signalling. Recent studies also explore synbiotics in metabolic syndrome, showing modest effects on anthropometric parameters in adolescents, and in immunomodulation, with formulations including Chinese medicinal herbs increasing thymic indices and NK cell activity in immunosuppressed models. Challenges remain in demonstrating true synergy due to complex microbial interactions and variable host responses, necessitating advanced metagenomic approaches and rigorous clinical trials to optimize strain-substrate combinations and elucidate mechanisms underlying their anti-inflammatory, antioxidant and metabolic benefits [55, 64].

16.5 Food Processing and Its Impact

The degree of food processing significantly influences GMB composition and functionality, with minimally processed whole foods and ultra-processed foods eliciting divergent effects on microbial diversity and health outcomes. Such minimally processed whole foods like fruits, vegetables, whole grains and legumes are high in dietary fibres, micronutrients and bioactive compounds acting as prebiotic substrates to the beneficial bacteria, *Bifidobacterium*, *Faecalibacterium prausnitzii* and *Akkermansia muciniphila*. These are foods that have been known to encourage microbial fermentation to produce SCFA like butyrate, which leads to greater integrity of the gut barriers, decreases inflammation and promotes metabolic health due to the observed microbial diversity with diets rich in whole foods like the Mediterranean diet. Conversely, the energy-dense, nutrient-poor industrial formulations of the NOVA classification, UPFs with their additives (emulsifiers, sweeteners and preservatives), are linked to low microbial diversity and dysbiosis. Research, such as randomized controlled trials, presupposes that a consumption of UPFs reduces the abundance and diversity of *Faecalibacterium* and *Akkermansia* and increases taxa associated with a pro-inflammatory state, including *Bilophila wadsworthia* and *Proteobacteria*, which are facilitated by low-fibre content and additives, including carboxymethylcellulose (CMC) and polysorbate-80 (P80), which dismantles the thickness of the mucus layer and enhances intestinal permeability [66–69].

16.5.1 Food Additives and Preservatives

Food additives and preservatives, including emulsifiers, artificial sweeteners and preservatives, significantly alter gut microbiome composition and function, often contributing to dysbiosis and associated health risks. Emulsifiers such as CMC and P80, commonly found in ultra-processed foods like ice creams and sauces, disrupt microbial ecosystems by reducing microbial diversity and increasing pro-inflammatory taxa like *Bilophila wadsworthia* and *Proteobacteria*, as demonstrated in both murine models and ex vivo human microbiota studies. These compounds impair the intestinal mucus layer, increasing permeability and promoting LPS-induced endotoxemia, which is linked to IBD and metabolic syndrome [70, 71]. Artificial sweeteners, including sucralose, aspartame and saccharin, used in sugar-free beverages and desserts, induce glucose intolerance by altering microbial composition, increasing *Enterobacteriaceae* and reducing SCFA-producing bacteria like *Faecalibacterium*, as shown in mouse studies and limited human trials. These microbial shifts correlate with metabolic dysregulation, though human evidence remains sparse due to variability in gut microbiota responses. Preservatives like sodium sulphite and sodium benzoate, prevalent in processed meats and beverages, inhibit beneficial bacteria such as *Bifidobacterium* and *Clostridium coccoides*, reducing SCFA production and promoting an acidic gut environment that favours pathogenic taxa like *Escherichia/Shigella*. Conversely, some additives like acacia gum exhibit prebiotic-like effects, supporting microbial diversity. The cumulative impact of these additives, particularly in Western diets high in ultra-processed foods, underscores their role in microbiome-mediated chronic diseases, necessitating further human studies to establish causality and inform regulatory guidelines [72–74].

16.6 Nutritional Interventions for Microbiome Health

16.6.1 Personalized Nutrition

The implementation of such an approach to health-related interindividual microbial diversity, as personalized nutrition about the individual profile of the gut microbiome, will be considered a revolutionary one. Due to advances in metagenomics and metabolomics, it is clear that an individual variance in gut microbiota composition, which is determined by genetics, diet and environmental factors, has serious ramifications in the context of response to diet intakes [75–77].

16.6.2 Dietary Strategies for Disease Prevention

Dietary interventions targeting the gut microbiome offer promising strategies for preventing and managing chronic conditions like IBS, IBD, and metabolic syndrome, with efficacy dependent on modulating microbial composition and

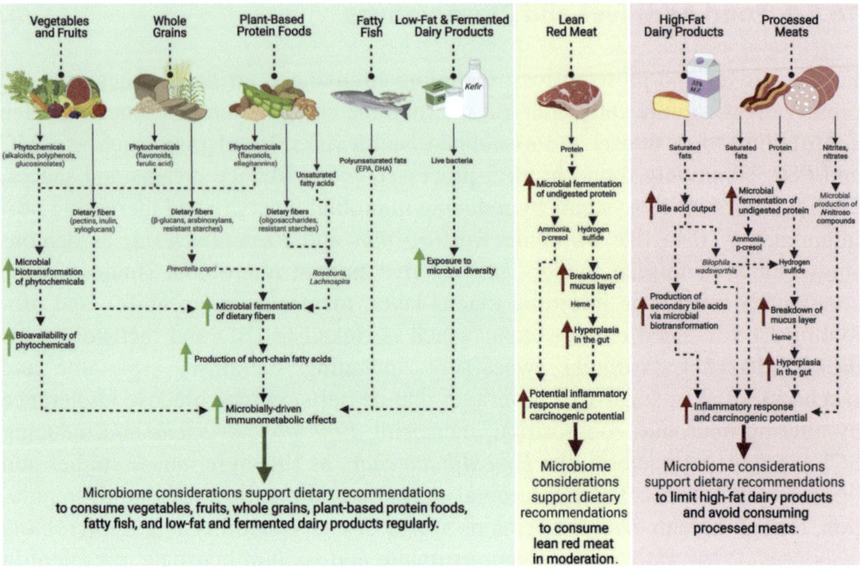

Fig. 16.1 Role of diet in shaping gut microbiota and its impact on human health. (Copyright (2022), reprinted with permission from Elsevier [86])

metabolite production [78–80]. For IBS, low-FODMAP diets, which restrict fermentable oligosaccharides, disaccharides, monosaccharides and polyols, reduce bloating and abdominal pain by limiting fermentation by *Ruminococcus* and *Clostridium* species, as shown in randomized controlled trials where symptom relief was significant in 50–70% of patients. In IBD, exclusive enteral nutrition is a first-line therapy for paediatric Crohn's disease, restoring microbial diversity by increasing *Faecalibacterium prausnitzii* and reducing pro-inflammatory *Escherichia coli*, achieving remission rates comparable to corticosteroids without side effects (Fig. 16.1) [81].

16.6.3 Emerging Trends in Functional Foods

Emerging trends in functional foods underscore the development of novel products enriched with bioactive compounds, probiotics, prebiotics and postbiotics to enhance gut microbiome health and promote systemic well-being [82]. Advances in microbiome science and omics technologies have facilitated the creation of precision-engineered foods, such as synbiotic-enriched beverages and snacks, which combine probiotics (e.g. *Lactobacillus rhamnosus* GG, *Bifidobacterium longum*) with prebiotics like inulin and XOS to synergistically boost *Bifidobacterium* and *Faecalibacterium* populations, enhancing SCFA production and reducing inflammation, as demonstrated in clinical trials for metabolic syndrome and IBS. Novel foods, including root and tuber-based prebiotic formulations (e.g. resistant starch from bioengineered cassava), leverage biotechnological approaches like CRISPR-based breeding to increase prebiotic content while reducing antinutrients, promoting microbial diversity and gut barrier integrity. Fermented foods, such as kimchi, kefir and kombucha, are being reformulated with specific microbial strains to target conditions like atopic dermatitis and metabolic disorders, with studies showing increased microbial diversity and reduced inflammatory markers after consumption [83–85].

16.7 Microbiome Research Challenges and Directions

Microbiome research faces significant challenges due to interindividual variability and methodological inconsistencies that hinder reproducibility and generalizability. Variability in gut microbiota composition, driven by factors such as genetics, diet, age and environmental exposures, complicates the identification of consistent microbial signatures across populations, as highlighted by studies showing divergent microbial responses to identical dietary interventions. Methodological challenges further exacerbate these issues, including differences in sample collection, DNA extraction protocols (e.g. kitome contamination from reagents) and sequencing techniques (e.g. 16S rRNA vs. shotgun metagenomics), which introduce biases in taxonomic and functional profiling. For instance, low-biomass samples are prone to contamination, and the lack of standardized bioinformatic pipelines leads to variability in data interpretation, with studies reporting up to 30% differences in microbial abundance estimates due to primer selection or software choice [87]. Advancements in multi-omics technologies, particularly metagenomics, metatranscriptomics, metaproteomics and metabolomics, have revolutionized microbiome analysis, offering unprecedented insights into microbial composition, function and host interactions. Metatranscriptomics and metaproteomics quantify gene expression and protein activity, respectively, providing functional insights into microbial responses to environmental cues, while metabolomics, using mass spectrometry and nuclear magnetic resonance, identifies microbiome-derived metabolites like SCFAs and bile acids that mediate host health. Artificial intelligence (AI) and machine learning enhance these analyses by integrating multi-omics data to predict microbial-metabolite interactions and host phenotypes, with tools like MelonnPan achieving high accuracy in metabolomic predictions from metagenomic data. However, challenges such as variable software pipelines and limited metadata standardization necessitate improved computational frameworks and curated databases to fully harness these technologies [88, 89].

16.7.1 Future of Nutrition and Microbiome Therapies

The future of nutrition and microbiome therapies lies in precision nutrition and microbiome-targeted interventions, leveraging multi-omics and AI to tailor dietary and therapeutic strategies to individual microbiome profiles. Precision nutrition, informed by metagenomic and metabolomic profiling, enables personalized dietary recommendations that account for microbial variability, as demonstrated by studies predicting glycaemic responses based on *Prevotella* and *Bacteroides* abundances. For example, tailored high-fibre diets have increased SCFA-producing bacteria in type 2 diabetes patients, improving glucose homeostasis. Microbiome-targeted therapies, including probiotics, prebiotics, synbiotics and postbiotics, show promise in managing conditions like IBS, IBD and metabolic syndrome, with clinical trials reporting reduced inflammation and enhanced gut barrier function using *Lactobacillus* and inulin combinations. Emerging interventions, such as faecal microbiota transplantation (FMT) and engineered microbial consortia, aim to restore eubiosis, though their efficacy remains variable due to donor-recipient microbiome mismatches [90–92].

16.8 Conclusion

The relationship between diet and gut microbiome in the health of humans is very complicated speaking to the importance of nutrition to influence microbial ecosystems and their implications to the physiology of the host. As discussed in this chapter, diet education and specifically nutrients, such as macronutrients, prebiotics, probiotics and fermented foods, play a significant role in shaping the composition, diversity and metabolic functions of the gut microbiome. The growth of beneficial SCFA-producing bacteria, including *Bifidobacterium* and *Faecalibacterium*, is stimulated by high-fibre diets, such as the Mediterranean diet, and the bacteria protect gut barrier integrity, suppress inflammation and protect against the development of chronic disease, including obesity, type 2 diabetes and inflammatory bowel disease. Conversely, Western diets that comprise of high proportion of saturated fats and ultra-processed food catalyse dysbiosis, which is characterized by impairment in microbial diversity and proliferation of pro-inflammatory taxa that lead towards metabolic and cardiovascular diseases. Synbiotics and fermented foods are new approaches that offer a more promising method to manipulate microbial communities in the provision of viable probiotics and bioactive components to support microbial communities by producing more SCFA and promoting immune regulation.

These developments are accompanied by limitations of microbiome studies such as differences between individuals and methodological discrepancies, which warrant common standards and studying longitudinally to improve reproducibility and generalizability. Future studies should focus on the use of multi-omics in combination with AI in the simplification of personalized nutrition or in the formulation of focused interventions, such as FMT and the use of fabricated microbial consortia. To translate these findings into the clinic, it will be important to establish strong regulatory frameworks and cost-effective sequencing technologies and ensure the widespread availability of microbiome-based interventions. By harnessing the dynamic interplay between diet and the gut microbiome, researchers, clinicians and policymakers can advance evidence-based dietary recommendations and therapeutic strategies to promote optimal health and prevent chronic diseases, ushering in a new era of precision nutrition.

Declarations

Competing Interests The authors have no competing interests to declare that are relevant to the content of this manuscript.

Ethics Approval The authors declare that this manuscript does not include primary research with human participants.

The authors declare that this manuscript does not include primary studies involving animals.

References

1. Hills RD, Pontefract BA, Mishcon HR, Black CA, Sutton SC, Theberge CR. Gut microbiome: profound implications for diet and disease. Nutrients. 2019. https://doi.org/10.3390/nu11071613.
2. Shreiner AB, Kao JY, Young VB. The gut microbiome in health and in disease. Curr Opin Gastroenterol. 2015. https://doi.org/10.1097/MOG.0000000000000139.
3. Afzaal M, Saeed F, Shah YA, et al. Human gut microbiota in health and disease: unveiling the relationship. Front Microbiol. 2022. https://doi.org/10.3389/fmicb.2022.999001.
4. Cani PD. Human gut microbiome: hopes, threats and promises. Gut. 2018. https://doi.org/10.1136/gutjnl-2018-316723.
5. Singh RK, Chang HW, Yan D, et al. Influence of diet on the gut microbiome and implications for human health. J Transl Med. 2017. https://doi.org/10.1186/s12967-017-1175-y.
6. Huang Z, Liu K, Ma W, Li D, Mo T, Liu Q. The gut microbiome in human health and disease—where are we and where are we going? A bibliometric analysis. Front Microbiol. 2022. https://doi.org/10.3389/fmicb.2022.1018594.
7. Hou K, Wu ZX, Chen XY, et al. Microbiota in health and diseases. Signal Transduct Target Ther. 2022. https://doi.org/10.1038/s41392-022-00974-4.
8. De Vos WM, Tilg H, Van Hul M, Cani PD. Gut microbiome and health: mechanistic insights. Gut. 2022. https://doi.org/10.1136/gutjnl-2021-326789.
9. Durack J, Lynch SV. The gut microbiome: relationships with disease and opportunities for therapy. J Exp Med. 2019. https://doi.org/10.1084/jem.20180448.
10. Salvadori M, Rosso G. Update on the gut microbiome in health and diseases. World J Methodol. 2024. https://doi.org/10.5662/wjm.v14.i1.89196.
11. Zhu Q, Jiang S, Du G. Effects of exercise frequency on the gut microbiota in elderly individuals. Microbiology. 2020. https://doi.org/10.1002/mbo3.1053.
12. Conlon MA, Bird AR. The impact of diet and lifestyle on gut microbiota and human health. Nutrients. 2015. https://doi.org/10.3390/nu7010017.
13. Johnson KVA. Gut microbiome composition and diversity are related to human personality traits. Hum Microb J. 2020. https://doi.org/10.1016/j.humic.2019.100069.
14. Milani C, Duranti S, Bottacini F, et al. The first microbial colonizers of the human gut: composition, activities, and health implications of the infant gut microbiota. Microbiol Mol Biol Rev. 2017. https://doi.org/10.1128/mmbr.00036-17.
15. Yamaguchi T, Nomura A, Matsubara A, Hisada T, Tamada Y, Mikami T, Ishida M. Effect of gut microbial composition and diversity on major inhaled allergen sensitization and onset of allergic rhinitis. Allergol Int. 2023. https://doi.org/10.1016/j.alit.2022.06.005.
16. Maher SE, O'Brien EC, Moore RL, Byrne DF, Geraghty AA, Saldova R, Murphy EF, Van Sinderen D, Cotter PD, McAuliffe FM. The association between the maternal diet and

the maternal and infant gut microbiome: a systematic review. Br J Nutr. 2023. https://doi.org/10.1017/S0007114520000847.
17. Nguyen TT, Zhang X, Wu TC, Liu J, Le C, Tu XM, Knight R, Jeste DV. Association of loneliness and wisdom with gut microbial diversity and composition: an exploratory study. Front Psych. 2021. https://doi.org/10.3389/fpsyt.2021.648475.
18. Blaak EE, Canfora EE, Theis S, et al. Short chain fatty acids in human gut and metabolic health. Benefic Microbes. 2020. https://doi.org/10.3920/BM2020.0057.
19. Rosario D, Bidkhori G, Lee S, et al. Systematic analysis of gut microbiome reveals the role of bacterial folate and homocysteine metabolism in Parkinson's disease. Cell Rep. 2021. https://doi.org/10.1016/j.celrep.2021.108807.
20. Blakeley-Ruiz JA, Erickson AR, Cantarel BL, Xiong W, Adams R, Jansson JK, Fraser CM, Hettich RL. Metaproteomics reveals persistent and phylum-redundant metabolic functional stability in adult human gut microbiomes of Crohn's remission patients despite temporal variations in microbial taxa, genomes, and proteomes. Microbiome. 2019. https://doi.org/10.1186/s40168-019-0631-8.
21. Portincasa P, Bonfrate L, Vacca M, De Angelis M, Farella I, Lanza E, Khalil M, Wang DQH, Sperandio M, Di Ciaula A. Gut microbiota and short chain fatty acids: implications in glucose homeostasis. Int J Mol Sci. 2022. https://doi.org/10.3390/ijms23031105.
22. Li HY, Zhou DD, Gan RY, Huang SY, Zhao CN, Shang A, Xu XY, Li HB. Effects and mechanisms of probiotics, prebiotics, synbiotics, and postbiotics on metabolic diseases targeting gut microbiota: a narrative review. Nutrients. 2021. https://doi.org/10.3390/nu13093211.
23. Rowland I, Gibson G, Heinken A, Scott K, Swann J, Thiele I, Tuohy K. Gut microbiota functions: metabolism of nutrients and other food components. Eur J Nutr. 2018. https://doi.org/10.1007/s00394-017-1445-8.
24. Jardon KM, Canfora EE, Goossens GH, Blaak EE. Dietary macronutrients and the gut microbiome: a precision nutrition approach to improve cardiometabolic health. Gut. 2022. https://doi.org/10.1136/gutjnl-2020-323715.
25. Zaher S. Nutrition and the gut microbiome during critical illness: a new insight of nutritional therapy. Saudi J Gastroenterol. 2020. https://doi.org/10.4103/sjg.SJG_352_20.
26. Nastasi C, Candela M, Bonefeld CM, et al. The effect of short-chain fatty acids on human monocyte-derived dendritic cells. Sci Rep. 2015. https://doi.org/10.1038/srep16148.
27. Zhu C, Sawrey-Kubicek L, Beals E, Rhodes CH, Houts HE, Sacchi R, Zivkovic AM. Human gut microbiome composition and tryptophan metabolites were changed differently by fast food and Mediterranean diet in 4 days: a pilot study. Nutr Res. 2020. https://doi.org/10.1016/j.nutres.2020.03.005.
28. Tily H, Patridge E, Cai Y, et al. Gut microbiome activity contributes to prediction of individual variation in glycemic response in adults. Diabetes Ther. 2022. https://doi.org/10.1007/s13300-021-01174-z.
29. David LA, Maurice CF, Carmody RN, et al. Diet rapidly and reproducibly alters the human gut microbiome. Nature. 2014. https://doi.org/10.1038/nature12820.
30. Cait A, Cardenas E, Dimitriu PA, et al. Reduced genetic potential for butyrate fermentation in the gut microbiome of infants who develop allergic sensitization. J Allergy Clin Immunol. 2019. https://doi.org/10.1016/j.jaci.2019.06.029.
31. Canfora EE, Meex RCR, Venema K, Blaak EE. Gut microbial metabolites in obesity, NAFLD and T2DM. Nat Rev Endocrinol. 2019. https://doi.org/10.1038/s41574-019-0156-z.
32. Pantoja-Feliciano IG, Karl JP, Perisin M, et al. In vitro gut microbiome response to carbohydrate supplementation is acutely affected by a sudden change in diet. BMC Microbiol. 2023. https://doi.org/10.1186/s12866-023-02776-2.
33. Oliphant K, Allen-Vercoe E. Macronutrient metabolism by the human gut microbiome: major fermentation by-products and their impact on host health. Microbiome. 2019. https://doi.org/10.1186/s40168-019-0704-8.
34. Mckenney EA. Gut microbial ecology. In: Microbial fermentations in nature and as designed processes; 2023. https://doi.org/10.1002/9781119850007.ch3.

35. Zeng X, Xing X, Gupta M, et al. Gut bacterial nutrient preferences quantified in vivo. Cell. 2022. https://doi.org/10.1016/j.cell.2022.07.020.
36. Sanders JG, Beichman AC, Roman J, Scott JJ, Emerson D, McCarthy JJ, Girguis PR. Baleen whales host a unique gut microbiome with similarities to both carnivores and herbivores. Nat Commun. 2015. https://doi.org/10.1038/ncomms9285.
37. Mills S, Stanton C, Lane JA, Smith GJ, Ross RP. Precision nutrition and the microbiome, part I: current state of the science. Nutrients. 2019. https://doi.org/10.3390/nu11040923.
38. Bouchard J, Malalgoda M, Storsley J, Malunga L, Netticadan T, Thandapilly SJ. Health benefits of cereal grain-and pulse-derived proteins. Molecules. 2022. https://doi.org/10.3390/molecules27123746.
39. Richter CK, Skulas-Ray AC, Champagne CM, Kris-Etherton PM. Plant protein and animal proteins: do they differentially affect cardiovascular disease risk? Adv Nutr. 2015. https://doi.org/10.3945/an.115.009654.
40. Mantrana IG, Royo MS, Baena CA, Amores MC. Relevance of diet and specific dietary components on human gut microbiota. Ann Nutr Metab. 2018.72:63.
41. Besser AC, Manlick PJ, Blevins CM, Takacs-Vesbach CD, Newsome SD. Variation in gut microbial contribution of essential amino acids to host protein metabolism in a wild small mammal community. Ecol Lett. 2023. https://doi.org/10.1111/ele.14246.
42. Garcia-Mantrana I, Selma-Royo M, Alcantara C, Collado MC. Shifts on gut microbiota associated to mediterranean diet adherence and specific dietary intakes on general adult population. Front Microbiol. 2018. https://doi.org/10.3389/fmicb.2018.00890.
43. Huskisson E, Maggini S, Ruf M. The role of vitamins and minerals in energy metabolism and well-being. J Int Med Res. 2007. https://doi.org/10.1177/147323000703500301.
44. Perler BK, Friedman ES, Wu GD. The role of the gut microbiota in the relationship between diet and human health. Annu Rev Physiol. 2023. https://doi.org/10.1146/annurev-physiol-031522-092054.
45. Figueiredo PS, Inada AC, Marcelino G, Cardozo CML, de Cássia Freitas K, de Cássia Avellaneda Guimarães R, de Castro AP, do Nascimento VA, Hiane PA. Fatty acids consumption: the role metabolic aspects involved in obesity and its associated disorders. Nutrients. 2017. https://doi.org/10.3390/nu9101158.
46. Yap SY, Voon PT, Cheah YK, Lee VKM, Selvaduray KR. Association of dietary fats with gut microbiota profile: how does palm oil fit IN? J Oil Palm Res. 2022. https://doi.org/10.21894/jopr.2021.0004.
47. Kimura I, Ichimura A, Ohue-Kitano R, Igarashi M. Free fatty acid receptors in health and disease. Physiol Rev. 2020. https://doi.org/10.1152/physrev.00041.2018.
48. Sakkas H, Bozidis P, Touzios C, Kolios D, Athanasiou G, Athanasopoulou E, Gerou I, Gartzonika C. Nutritional status and the influence of the vegan diet on the gut microbiota and human health. Medicina (Lithuania). 2020. https://doi.org/10.3390/medicina56020088.
49. Miao Z, Du W, Xiao C, et al. Gut microbiota signatures of long-term and short-term plant-based dietary pattern and cardiometabolic health: a prospective cohort study. BMC Med. 2022. https://doi.org/10.1186/s12916-022-02402-4.
50. Peters BA, Xing J, Chen GC, et al. Healthy dietary patterns are associated with the gut microbiome in the Hispanic Community Health Study/Study of Latinos. Am J Clin Nutr. 2023. https://doi.org/10.1016/j.ajcnut.2022.11.020.
51. Karakan T, Tuohy KM, Janssen-van Solingen G. Low-dose lactulose as a prebiotic for improved gut health and enhanced mineral absorption. Front Nutr. 2021. https://doi.org/10.3389/fnut.2021.672925.
52. Tuohy KM, Probert HM, Smejkal CW, Gibson GR. Using probiotics and prebiotics to improve gut health. Drug Discov Today. 2003. https://doi.org/10.1016/S1359-6446(03)02746-6.
53. Gibson PR, Halmos EP, Muir JG. Review article: FODMAPS, prebiotics and gut health-the FODMAP hypothesis revisited. Aliment Pharmacol Ther. 2020. https://doi.org/10.1111/apt.15818.
54. Wichienchot S, Youravong W, Prueksasri S, Ngampanya B. Recent researches on prebiotics for gut health in Thailand. Funct Foods Health Dis. 2015. https://doi.org/10.31989/ffhd.v5i11.214.

55. Cunningham M, Azcarate-Peril MA, Barnard A, et al. Shaping the future of probiotics and prebiotics. Trends Microbiol. 2021. https://doi.org/10.1016/j.tim.2021.01.003.
56. Lu K, Dong S, Wu X, Jin R, Chen H. Probiotics in cancer. Front Oncol. 2021. https://doi.org/10.3389/fonc.2021.638148.
57. Bermudez-Brito M, Plaza-Díaz J, Muñoz-Quezada S, Gómez-Llorente C, Gil A. Probiotic mechanisms of action. Ann Nutr Metab. 2012. https://doi.org/10.1159/000342079.
58. Merenstein D, Pot B, Leyer G, et al. Emerging issues in probiotic safety: 2023 perspectives. Gut Microbes. 2023. https://doi.org/10.1080/19490976.2023.2185034.
59. Maldonado Galdeano C, Cazorla SI, Lemme Dumit JM, Vélez E, Perdigón G. Beneficial effects of probiotic consumption on the immune system. Ann Nutr Metab. 2019. https://doi.org/10.1159/000496426.
60. Wieërs G, Belkhir L, Enaud R, Leclercq S, Philippart de Foy JM, Dequenne I, de Timary P, Cani PD. How probiotics affect the microbiota. Front Cell Infect Microbiol. 2020. https://doi.org/10.3389/fcimb.2019.00454.
61. Markowiak P, Ślizewska K. The role of probiotics, prebiotics and synbiotics in animal nutrition. Gut Pathog. 2018. https://doi.org/10.1186/s13099-018-0250-0.
62. Markowiak P, Ślizewska K. Effects of probiotics, prebiotics, and synbiotics on human health. Nutrients. 2017. https://doi.org/10.3390/nu9091021.
63. Madabushi JS, Khurana P, Gupta N, Gupta M. Gut biome and mental health: do probiotics work? Cureus. 2023. https://doi.org/10.7759/cureus.40293.
64. Shimizu K, Yamada T, Ogura H, et al. Synbiotics modulate gut microbiota and reduce enteritis and ventilator-associated pneumonia in patients with sepsis: a randomized controlled trial. Crit Care. 2018. https://doi.org/10.1186/s13054-018-2167-x.
65. Al-Habsi N, Al-Khalili M, Haque SA, Elias M, Al Olqi N, Al Uraimi T. Health benefits of prebiotics, probiotics, synbiotics, and postbiotics. Nutrients. 2024.16:3955.
66. Smith C, Van Haute MJ, Rose DJ. Processing has differential effects on microbiota-accessible carbohydrates in whole grains during in vitro fermentation. Appl Environ Microbiol. 2020. https://doi.org/10.1128/AEM.01705-20.
67. Ostrowski MP, La Rosa SL, Kunath BJ, et al. Mechanistic insights into consumption of the food additive xanthan gum by the human gut microbiota. Nat Microbiol. 2022. https://doi.org/10.1038/s41564-022-01093-0.
68. La Rosa SL, Ostrowski MP, Vera-Ponce de León A, McKee LS, Larsbrink J, Eijsink VG, Lowe EC, Martens EC, Pope PB. Glycan processing in gut microbiomes. Curr Opin Microbiol. 2022. https://doi.org/10.1016/j.mib.2022.102143.
69. Schmidt V, Amaral-Zettler L, Davidson J, Summerfelt S, Good C. Influence of fishmeal-free diets on microbial communities in Atlantic salmon (Salmo salar) recirculation aquaculture systems. Appl Environ Microbiol. 2016. https://doi.org/10.1128/AEM.00902-16.
70. Rinninella E, Cintoni M, Raoul P, Gasbarrini A, Mele MC. Food additives, gut microbiota, and irritable bowel syndrome: a hidden track. Int J Environ Res Public Health. 2020. https://doi.org/10.3390/ijerph17238816.
71. Hrncirova L, Machova V, Trckova E, Krejsek J, Hrncir T. Food preservatives induce proteobacteria dysbiosis in human-microbiota associated Nod2-deficient mice. Microorganisms. 2019. https://doi.org/10.3390/microorganisms7100383.
72. Zhou X, Qiao K, Wu H, Zhang Y. The impact of food additives on the abundance and composition of gut microbiota. Molecules. 2023. https://doi.org/10.3390/molecules28020631.
73. Liu C, Zhan S, Tian Z, Li N, Li T, Wu D, Zeng Z, Zhuang X. Food additives associated with gut microbiota alterations in inflammatory bowel disease: friends or enemies? Nutrients. 2022. https://doi.org/10.3390/nu14153049.
74. Cao Y, Liu H, Qin N, Ren X, Zhu B, Xia X. Impact of food additives on the composition and function of gut microbiota: a review. Trends Food Sci Technol. 2020. https://doi.org/10.1016/j.tifs.2020.03.006.
75. Di Renzo L, Gualtieri P, Romano L, Marrone G, Noce A, Pujia A, Perrone MA, Aiello V, Colica C, De Lorenzo A. Role of personalized nutrition in chronic-degenerative diseases. Nutrients. 2019. https://doi.org/10.3390/nu11081707.

76. Liu Z, de Vries B, Gerritsen J, Smidt H, Zoetendal EG. Microbiome-based stratification to guide dietary interventions to improve human health. Nutr Res. 2020. https://doi.org/10.1016/j.nutres.2020.07.004.
77. Berding K, Vlckova K, Marx W, Schellekens H, Stanton C, Clarke G, Jacka F, Dinan TG, Cryan JF. Diet and the microbiota-gut-brain axis: sowing the seeds of good mental health. Adv Nutr. 2021. https://doi.org/10.1093/advances/nmaa181.
78. Link VM, Hall K, Belkaid Y. Highly controlled nutritional interventions uncover defined immune and microbiome responses in human. J Immunol. 2023. https://doi.org/10.4049/jimmunol.210.supp.66.01.
79. Abdul Rahim MBH, Chilloux J, Martinez-Gili L, Neves AL, Myridakis A, Gooderham N, Dumas ME. Diet-induced metabolic changes of the human gut microbiome: importance of short-chain fatty acids, methylamines and indoles. Acta Diabetol. 2019. https://doi.org/10.1007/s00592-019-01312-x.
80. Catassi G, Aloi M, Giorgio V, Gasbarrini A, Cammarota G, Ianiro G. The role of diet and nutritional interventions for the infant gut microbiome. Nutrients. 2024. https://doi.org/10.3390/nu16030400.
81. Mehta S, Huey SL, McDonald D, Knight R, Finkelstein JL. Nutritional interventions and the gut microbiome in children. Annu Rev Nutr. 2021. https://doi.org/10.1146/annurev-nutr-021020-025755.
82. Gupta A, Sanwal N, Bareen MA, Barua S, Sharma N, Joshua Olatunji O, Prakash Nirmal N, Sahu JK. Trends in functional beverages: functional ingredients, processing technologies, stability, health benefits, and consumer perspective. Food Res Int. 2023. https://doi.org/10.1016/j.foodres.2023.113046.
83. Yemenicioğlu A, Farris S, Turkyilmaz M, Gulec S. A review of current and future food applications of natural hydrocolloids. Int J Food Sci Technol. 2020. https://doi.org/10.1111/ijfs.14363.
84. Basak S, Singhal RS. The potential of coextraction as an emerging trend in the development of new functional food ingredients. eFood. 2024. https://doi.org/10.1002/efd2.140.
85. Corbo MR, Bevilacqua A, Petruzzi L, Casanova FP, Sinigaglia M. Functional beverages: the emerging side of functional foods: commercial trends, research, and health implications. Compr Rev Food Sci Food Saf. 2014. https://doi.org/10.1111/1541-4337.12109.
86. Armet AM, Deehan EC, O'Sullivan AF, Mota JF, Field CJ, Prado CM, Lucey AJ, Walter J. Rethinking healthy eating in light of the gut microbiome. Cell Host Microbe. 2022. https://doi.org/10.1016/j.chom.2022.04.016.
87. Saxena R, Sharma V, Saxena AR, Patel A. Harnessing AI and gut microbiome research for precision health. J Artif Intell Gen Sci (JAIGS) ISSN:3006-4023. 2024. https://doi.org/10.60087/jaigs.v3i1.68.
88. Wen C, Wang Q, Gu S, Jin J, Yang N. Emerging perspectives in the gut–muscle axis: the gut microbiota and its metabolites as important modulators of meat quality. Microb Biotechnol. 2024. https://doi.org/10.1111/1751-7915.14361.
89. Puig-Castellví F, Pacheco-Tapia R, Deslande M, Jia M, Andrikopoulos P, Chechi K, Bonnefond A, Froguel P, Dumas ME. Advances in the integration of metabolomics and metagenomics for human gut microbiome and their clinical applications. TrAC Trends Anal Chem. 2023. https://doi.org/10.1016/j.trac.2023.117248.
90. Xia Y, Sun J. Hypothesis testing and statistical analysis of microbiome. Genes Dis. 2017. https://doi.org/10.1016/j.gendis.2017.06.001.
91. Cuna A, Morowitz MJ, Ahmed I, Umar S, Sampath V. Dynamics of the preterm gut microbiome in health and disease. Am J Physiol Gastrointest Liver Physiol. 2021. https://doi.org/10.1152/AJPGI.00399.2020.
92. Anderson BD, Bisanz JE. Challenges and opportunities of strain diversity in gut microbiome research. Front Microbiol. 2023. https://doi.org/10.3389/fmicb.2023.1117122.

MIX
Papier aus verantwortungsvollen Quellen
Paper from responsible sources
FSC® C105338

If you have any concerns about our products,
you can contact us on
ProductSafety@springernature.com

In case Publisher is established outside the EU,
the EU authorized representative is:
Springer Nature Customer Service Center GmbH
Europaplatz 3, 69115 Heidelberg, Germany

Printed by Libri Plureos GmbH
in Hamburg, Germany